INERT GASES

HALOGENS

NONMETALS, METALLOIDS

	IB	IIB	IIIA	IVA	VA	VIA	VIIA	0
								2 He Helium 4.003
			5 B Boron 10.811	6 C Carbon 12.011	7 N Nitrogen 14.007	8 O Oxygen 15.999	9 F Fluorine 18.998	10 Ne Neon 20.183
			13 Al Aluminum 26.982	14 Si Silicon 28.086	15 P Phosphorus 30.974	16 S Sulfur 32.064	17 Cl Chlorine 35.453	18 Ar Argon 39.948
28 Ni Nickel 58.71	29 Cu Copper 63.546	30 Zn Zinc 65.37	31 Ga Gallium 69.72	32 Ge Germanium 72.59	33 As Arsenic 74.922	34 Se Selenium 78.96	35 Br Bromine 79.904	36 Kr Krypton 83.80
46 Pd Palladium 105.4	47 Ag Silver 107.868	48 Cd Cadmium 112.40	49 In Indium 114.82	50 Sn Tin 118.69	51 Sb Antimony 121.75	52 Te Tellurium 127.60	53 I Iodine 126.904	54 Xe Xenon 131.30
78 Pt Platinum 195.09	79 Au Gold 196.967	80 Hg Mercury 200.59	81 Tl Thallium 204.37	82 Pb Lead 207.19	83 Bi Bismuth 208.98	84 Po Polonium 210	85 At Astatine 210	86 Rn Radon 222

RARE EARTH METALS

63 Eu Europium 151.96	64 Gd Gadolinium 157.25	65 Tb Terbium 158.92	66 Dy Dysprosium 162.5	67 Ho Holmium 164.93	68 Er Erbium 167.26	69 Tm Thulium 168.93	70 Yb Ytterbium 173.04	71 Lu Lutetium 174.97
95 Am Americium 243	96 Cm Curium 245	97 Bk Berkelium 245	98 Cf Californium 248	99 Es Einsteinium 253	100 Fm Fermium 254	101 Md Men- delevium 256	102 No Nobelium 253	103 Lw Lawrencium 257

CHEMICAL PRINCIPLES FOR THE LIFE SCIENCES

CHEMICAL PRINCIPLES FOR THE LIFE SCIENCES

second edition

Ralph J. Fessenden
Joan S. Fessenden

University of Montana

ALLYN AND BACON, INC. Boston London Sydney Toronto

Production Editor: Renee Burke
Interior Designer: Libby Griffiths
Manufacturing Buyer: Karen Mason

Printed in the United States of America.

Library of Congress Cataloging in Publication Data

Fessenden, Ralph J.
Chemical principles for the life sciences.
Includes index.
1. Chemistry. I. Fessenden, Joan S., joint author. II. Title.
QD31.2.F49 1979 540 78-21279

ISBN 0-205-06506-6

CONTENTS

PREFACE

In revising this text, our primary goal was to include more practical biochemistry. Rather than simply lengthening the book, we have shortened the general chemistry and organic chemistry sections. We have moved atomic radii and electronegativities from Chapter 4 to Chapter 3. We have deleted some historical discussions in Chapter 4. We have combined the chemistry of the metals and the nonmetals into one chapter (Chapter 9).

In the organic chapters, we have deleted some topics of interest only to organic chemists (for example, Grignard reactions). Instead, we have slanted the subject material toward the biochemical topics that are covered in the later chapters.

The biochemistry chapters have been expanded to include more biochemical diagrams, more on bioenergetics, and a new chapter on "Blood and Urine," including both chemistry and clinical tests.

Throughout this new edition, we have added more problems and have aimed discussions toward biological applications. For example, phenylketonuria (PKU) is included to exemplify aromatic substitution.

We are grateful to our friend Ronald Newman and to our son Doug for their many suggestions for improvement. We must also thank the reviewers; our typist, Carrie Avshalomov; and our editor, John Shine, for their help.

Ralph J. Fessenden
Joan S. Fessenden

PREFACE TO THE FIRST EDITION

Our present era might well be called the Age of Chemistry. From the plastic wrap in the kitchen to nuclear power plants, we are surrounded by the results of chemical research and development. The student pursuing almost any area of study should have at least a rudimentary grasp of this vast field of study.

This text is designed for college or university students who will start and complete their study of chemistry in one academic year. Of primary importance is our desire to provide an elementary, readable text for students pursuing degrees in Health, Nursing, and other life sciences. The topics include those traditional to general chemistry, organic chemistry, and biochemistry. Because these areas are all together under one cover, we have tried to integrate them by way of examples. The approach is descriptive and, we hope, pertinent to today's needs—How does the Cottrell precipitator help control industrial air pollution? What is a nuclear breeder reactor? How do birth control pills work? Why has the use of DDT been banned?

Students who find themselves in a one-year chemistry course may not have extensive mathematical backgrounds. However, they need a knowledge of simple algebra to handle gas-law problems and problems involving dilution and concentration. An appendix giving a brief review in algebra is included as a convenience for the student.

We must thank the many readers who helped us in this endeavor. A special thank you goes to Professor James Cox, University of Montana, for the time he spent on the manuscript and for his many excellent suggestions. Thanks must also go to our many typists and to our very patient editor, William Roberts.

Ralph J. Fessenden
Joan S. Fessenden

CHEMICAL PRINCIPLES FOR THE LIFE SCIENCES

CHAPTER 1 OF SCIENCE AND MEASUREMENT

OBJECTIVES

What Is Chemistry? *Define chemistry. * Describe the major areas of study within chemistry.*

Approaching a Scientific Problem *List the steps in the scientific method for approaching a problem. * Differentiate between a hypothesis, a theory, and a scientific law.*

Matter *Define and list some examples of matter. * Name the different states of matter.*

Energy *Name and describe the different forms of energy. * State the Law of Conservation of Matter and Energy.*

The Need for Uniform Measurements *Compare briefly the English system of measures with the metric system.*

International System of Units *Name the metric units used for measuring length, mass, volume, temperature, and heat. * Do simple conversions between the units of measurement within the metric system and between the metric system and the English system. * Given the weight and the volume, calculate density and specific gravity.*

Science may be loosely defined as the study of things and what makes them tick. Our innate curiosity and desire to improve the environment has led to our present degree of technological advancement.

Scientific research may be of the *pure,* or theoretical, type—research that has no immediate practical value. Studies concerning the composition of the stars or how protein fibers are synthesized within an organism fall into this category. Of course, pure research can quickly become practical. Who could have guessed that the work of Marie and Pierre Curie on radioactive materials in the early 1900's would eventually lead to nuclear weapons and power plants?

Applied research is directed toward an immediate, practical goal. Trying to devise a detergent that is not only efficient but also non-polluting and biodegradable is an example of applied research.

Both types of scientific endeavor together have supplied us with the technology to feed and shelter the world's burgeoning population. Without tractors, chemical fertilizers, and insecticides, a worldwide famine would be likely. Without dams and nuclear power plants, there would not be enough electrical energy to operate machinery. Certainly, problems have arisen. Insecticides may indirectly kill birds, fertilizer may be washed into streams and cause rampant algae growth, and dams may destroy nesting areas of waterbirds.

Considering the increasing population of the world and the ambitious nature of the human species, there is no way to undo any good or bad that we have already done. To solve our problems, we must look forward—to the development of insecticides that control harmful insects without damaging other life forms, to new sources of energy, to cures for some of our dreaded diseases, and to the development of ideas and solutions to problems that have not yet been conceived.

WHAT IS CHEMISTRY?

Over the years, science has been compartmentalized into different areas of study. Physics and chemistry are two such areas. Zoology, botany, geology, and human physiology are others. Each of these different areas of study depends heavily upon the others for tools and knowledge. The dividing line between these areas of science is often fuzzy at best. In general, chemistry is considered to be the study of the composition of things and of changes in their composition.

Chemistry itself has been subdivided into different areas. *Inorganic chemistry* is the study of substances that have never been alive: minerals, metals, and so forth.

Historically, *organic chemistry* was the study of substances that came from living things—substances such as petroleum and extracts from plants. However, organic chemistry today embraces a host of drugs, plastics, and other substances that have never been part of a living organism, but are synthesized in the laboratory. A better definition of organic chemistry might be the "study of substances that contain the element *carbon* in combination with itself and other elements." (Thus, *inorganic chemistry* might be defined as "the study of substances that do *not* contain carbon.") We will find that these definitions also have limitations in that limestone and some other inorganic materials also contain carbon.

Biochemistry, as its name implies, is the study of substances found in living things and how these substances interact within the living organism. Studies of the composition of viruses or of the chemistry of DNA (which carries our genetic code) are examples of biochemical research.

Physical chemistry is the application of the principles of physics to chemistry. What happens to substances at extremely low temperatures? What are the effects of expansion on a gas? How much energy can be obtained from a gram of uranium in a nuclear reactor? These are some topics of physical chemistry.

The boundaries between the subdivisions of chemistry are indeed nebulous. In which area would we classify a study of the effect of organic mercury compounds on fish? Mercury is an *inorganic* moiety in combination with *organic* materials. The effect on fish is a *biochemical* (or even *zoological*) study. How the mercury compound behaves in the water is a *physical chemical* study. Much of contemporary scientific endeavor is equally hard to pinpoint as part of any historical area of study. Instead, our most exciting research embraces a number of fields at once.

In beginning a study of any field, we must have a starting point from which to develop the fundamentals of the field. While it would be intriguing to start our study of chemistry by considering the effects of organomercury compounds on aquatic life, we would soon be confused by the vocabulary and would miss the salient features of the discussion. Before we apply chemistry, we must learn fundamentals.

In this text, we will first approach *general chemistry,* in which we will discuss primarily the principles of physical and inorganic chemistry. The second main topic will be *organic chemistry,* a necessary base for the third main topic, *biochemistry.* However, even within these subdivisions, we will find ourselves drawing upon the others for examples and applications.

APPROACHING A SCIENTIFIC PROBLEM

In any field, research is simply the study of something new. Doing an effective job of designing experiments and interpreting their results requires a certain amount of logical thought and mental discipline. Thus, certain rules have been developed for approaching scientific problems. These rules, called *the scientific method,* list the steps one should logically take in an investigation:

1. State the problem. (For example, "the door squeaks.")
2. Gather and organize data pertaining to the problem. ("Swing the door back and forth, listen to each hinge, and note which one squeaks.")
3. Formulate a hypothesis, that is, a tentative idea of your conclusion. ("I think it is the top hinge that squeaks.")
4. Test the hypothesis. ("I will oil the top hinge and see if the door stops squeaking.")
5. Draw a conclusion. ("I was right!" or "I was wrong.")

As with most sets of rules, there are exceptions. The antibiotic activity of penicillin, for example, was discovered quite by accident in 1928. The Englishman Alexander Fleming was studying variants of staphylococcus when he observed that a contaminant in his cultures (a variety of *Penicillium* mold) inhibited the growth of his cultures. Unfortunately, it was another twelve years before this observation was put to practical, life-saving use. Similarly, vulcanized rubber was the result of accidental discovery and not of logical, preplanned research. Albert Einstein formulated his equations of mass, energy, space, and time relationships by means of brilliant intuition and not by the routine gathering of data. (However, later observations are still being used to test Einstein's ideas.) Therefore, imagination, observation, and intuition might well be added to the need for disciplined thinking in our listing of qualities needed to approach a scientific problem.

Research leads to ideas, or *hypotheses.* (*Hypothesis* is a Greek word from *hypo-,* under, and *tithenai,* to place.) A hypothesis is a supposition, or the groundwork, used to explain certain facts and to guide further investigation. When a hypothesis has been tested repeatedly and judged to be sound, it may graduate to being called a *theory.* (We will shortly encounter the atomic theory.) Whereas a hypothesis is an idea, a theory is an explanation accepted by most scientists because it is supported by considerable evidence. In any scientific text we also find statements called scientific laws. A *law* is a generalization that describes the actual behavior of some aspect of the physical universe. (Newton's *Law of Gravity* is an example.)

MATTER

Chemistry encompasses the study of matter and energy and their relationships. Matter is something that has mass and occupies space. We can see and feel our shoe; it is matter. So is the coffee in our cup. And so is the air around us, even though we cannot see it.

Matter is composed of extremely tiny particles called *atoms* and *molecules*. All of chemistry, from the manufacture of a good house paint to the workings of the human body, is based upon the behavior of these minute particles. Most of the topics covered in this text will be approached from the basis of atomic and molecular theory.

By virtue of physical interactions between its atoms and molecules, matter can exist in different *states*. The three states of matter are (1) *gas*, (2) *liquid*, and (3) *solid*. When matter goes from one state to another, we refer to the change as a *change in the state of matter*. When water freezes, its state has changed from liquid to solid.

Matter may be changed from one state to another, but only in unusual circumstances (as in the sun or in a nuclear reactor) is mass actually lost.

ENERGY

Matter such as a shoe is easy to perceive. Although energy is not quite as tangible, it is real nonetheless. Energy is defined as the *capacity to do work*. Electricity, heat, and light are all forms of energy. Some important general forms of energy are *potential energy* (such as energy possessed by water behind a dam), *kinetic energy* (the energy of motion), and *radiant energy* (light).

In chemistry, energy is considered as important as matter. The atoms and molecules in matter are in motion; they have kinetic energy. If the temperature is increased, the atoms or molecules will move faster. Heat is a form of kinetic energy. The molecules in natural gas contain *chemical energy*, a type of potential energy. When ignited in the presence of oxygen, natural gas burns and its potential energy is converted to heat and light.

Energy can be changed from one form to another. The burning of natural gas is one example. The opening of the flood gates of a dam is another. The water behind the dam has potential energy. When the water goes through the turbine, the potential energy is changed to kinetic energy. Some of this kinetic energy is used to turn the turbine and generate electricity. The electricity is transported to our homes, where it can be changed to heat and light. In all these processes, no energy is gained or lost. The energy has merely been changed in form. Energy is rarely gained or lost. However, in the sun and nuclear reactors, matter is converted to energy. In these cases, mass is lost and energy is gained.

Although energy is rarely actually lost, it is often "lost" in a practical sense because of the inefficiency of energy conversions. For example, the purpose of a light bulb is to convert electrical energy to light. However, not all the electrical energy ends up as light. Some of the energy is "lost" as heat.

We can generalize what we know about the fixed amount of matter and energy in the following law.

LAW OF CONSERVATION OF MATTER AND ENERGY: *The total quantity of matter and energy in the universe is fixed, even though there may be some interconversions.*

THE NEED FOR UNIFORM MEASUREMENTS

Put yourself in the place of a king in days of old. You are sitting on your throne listening to two of your barons squabble over a piece of land. Interwoven through such germane remarks as "My knight can beat your knight," you hear evidence that the disputed land was measured by feet and you realize that the two barons have different sized feet. Being kingly and wise, you would probably settle the problem by adopting a standard foot—yours.

We need standard, uniform units of measurement. In the United States, the system of measurement currently used is the *English system of measurement*. The units of measurement are the foot, mile, pound, quart, and so forth. We are used to thinking in terms of these units. If you hear that a kitten weighs 3 pounds and is one foot long, you conjure up an immediate picture. This picture is different from that of a cat that is 150 pounds and 6 feet long.

The English system of measurement has one distinct disadvantage. Consider the following question. How many inches are there in $5\frac{1}{4}$ miles? You can make the conversion, but probably not in your head. Arithmetic is difficult to carry out in the English system. Our money system is different. It is based upon units of ten. There are ten pennies in a dime and ten dimes in a dollar. How many pennies are there in $5\frac{1}{4}$ dollars? You can do this calculation in your head.

The system of measurement used in science is the *metric system*. Like the dollar system, it is based upon units of ten. An advantage of the metric system is the same as that of the dollar system. Many calculations can easily be performed in our heads.

THE INTERNATIONAL SYSTEM OF UNITS

The *International System of Units* (SI), which is based on the metric system, was devised by the French Academy of Science at the close of the French Revolution. Over the years, this system has slowly been replacing the unwieldy English system of measurement. At the present time, the metric system is used universally by all scientists; it has been adopted by the U.S. Armed Forces, and in the 1960's, by Great Britain. Of the great industrial nations, only the United States still uses the English system of measurement for official weights and measurements. However, a change to the metric system is now being accomplished in the United States.

We will be using units and measures throughout this text. Therefore, we will briefly outline the units commonly encountered for *length, mass, volume, temperature,* and *heat*. We will also mention *density* and *specific gravity,* although they are actually combinations of mass and volume.

LENGTH. The basic unit of length in the metric system and SI is the *meter,* abbreviated m, which is just a little longer than the yard. One of the track events in the Olympic games is the 100-meter dash. American athletes use the 110-yard dash to train for this event.

One-hundredth of a meter is a *centimeter,* abbreviated cm. There are 2.54 centimeters to an inch. One tenth of a centimeter is called a *millimeter* (one thousandth of a meter), which is abbreviated mm. If we need to use a unit larger than the meter, we use the *kilometer* (km), which is 1000 meters.

Table 1-1. Some Common Units of Length

Unit	Metric or SI Equivalent	English System Equivalent
kilometer (km)	1000 m	0.6214 mile
meter (m)	1.0 m	39.37 inches
centimeter (cm)	0.01 m	0.3937 inch
millimeter (mm)	0.001 m	0.03937 inch
micrometer (μm) or micron (μ)	10^{-6} m	—
nanometer (nm)	10^{-9} m	—
Angstrom (Å)	10^{-10} m	—

Because chemists often deal with very small units of length, such as the wavelengths of light, we also need some designation for these. Thus, we may encounter the *micron* (μ, or μm), 0.000001 m, or 10^{-6} m; the *Angstrom* (Å), 10^{-10} m; and the *nanometer* (nm), 10^{-9} m. For quick reference to these units, see Table 1-1. For an explanation of the prefixes, see Table 1-2.

Table 1-2. Prefixes in SI

Prefix	Meaning
centi-	$\frac{1}{100}$
milli-	$\frac{1}{1000}$
kilo-	1000

MASS. In everyday usage, the terms *weight* and *mass* are used interchangeably; however, there is a difference between the two. The *mass* of an object is defined as the amount of matter in that object. (Theoretically, this mass is a measure of the inertia of the object—its resistance to being moved, or its resistance to a change in its motion once it is moving.) *Weight* is our measure of an object's mass under the influence of the earth's gravity. The mass of an astronaut who goes to the moon does not change. However, the astronaut's weight is only one-fifth of what it would be on earth.

The chemist, who deals with measured quantities of matter, is therefore interested in the mass of that matter rather than its weight, which is a changeable measure. However, most of our chemistry is done on earth. If a chemist says *weight* while actually meaning *mass,* it is a forgiveable error.

The basic unit of mass in the metric system is the *gram,* abbreviated g. One thousand grams is a *kilogram* (kg). There are 454 grams in one pound; therefore, one kilogram contains 2.2 pounds.

If chemists are dealing in small quantities of a substance, they may report the mass in *milligrams* (mg), where one milligram is one thousandth of a gram. While most drugs are dispensed by pharmacists or doctors in milligram quantities, some are dispensed in *grains*. Grain is an English system unit; therefore, it is not used by most chemists. There certainly is at least a possibility of confusion between gram (g) and grain (gr) by those involved in the medical sciences. Table 1-3 lists some common units of mass in the metric and English systems.

Table 1-3. Some Common Units of Mass and Volume

Unit	Metric or SI Equivalent	English System Equivalent
Mass:		
gram (g)	—	0.03527 ounce
milligram (mg)	0.001 g	0.01543 grain
kilogram (kg)	1000 g	2.205 pounds
Volume:		
liter (l or L)	1000 cc, or 1000 ml	1.057 quarts
milliliter (ml or mL)	1.0 cc, or 0.001 L	0.0338 fluid ounce

VOLUME. Volume is not truly a fundamental measure because it is based upon length. The volume of a cube that has sides 1.0 cm long is 1.0 cubic centimeter. The cubic centimeter, abbreviated cc or cm^3, is the basic unit of volume in the metric system. The cubic centimeter is also referred to as a milliliter (ml), and 1000 cc is called a *liter* (L). One liter is just slightly larger than a quart. (The equal relationship between a cubic centimeter and a milliliter is true only for water at 4°C; however, the difference between the two at other temperatures is quite small. In practice, cc and ml are used interchangeably.)

TEMPERATURE. The *temperature* of an object is a measure of how hot or cold that object is. We use a thermometer to measure this. Most objects expand when they are heated. Alcohol and mercury, the two common liquids used in standard glass thermometers, expand to a greater extent than glass when heated. Thus, the liquid within the thermometer may be seen to rise in the thermometer as heat is applied.

In the metric system, there are two temperature scales, the *Kelvin scale* (abbreviated °K) and the *Celsius,* or *centigrade scale* (abbreviated °C). The *Fahrenheit scale* (°F) is used in the English system.

The Celsius scale takes as its two reference points the boiling point and the freezing point of water under rigidly defined conditions. The freezing point is set at 0°C and the boiling point at 100°C. Therefore, one degree on the Celsius scale is $\frac{1}{100}$ of the temperature change between these two reference points.

The Fahrenheit scale was designed by the German physicist G. D. Fahrenheit around 1700. At that time, there were no accurate measuring devices at his disposal. The two reference points he chose were body temperature, which he set at 100°F, and the temperature of an ice-salt-water bath, which he set at 0°F. Unfortunately, he missed the true temperatures slightly at both ends. The Fahrenheit scale has 180 divisions between the freezing point of water (32°F) and the boiling point of water (212°F). (See Table 1-4 and Figure 1-1.)

Table 1-4. Some Comparisons between the Fahrenheit and Celsius Scales

	°F	°C
Absolute zero	−459.67	−273.15
Freezing water	32	0
Room temperature	72	22
Body temperature	98	37
Boiling water	212	100

When conversions between the Celsius and Fahrenheit scales are necessary, the following formulas are useful.

$$°F = \tfrac{9}{5}°C + 32° \qquad °C = \tfrac{5}{9}(°F - 32°)$$

In the first formula, the fraction $\frac{9}{5}$ relates the size of a °C to that of a °F. (The °C is $\frac{180}{100}$, or $\frac{9}{5}$, larger.) The term $+32°$ is the difference that the two scales are offset from each other. The second formula is derived similarly.

The following examples show how the two formulas may be used to convert a temperature from one scale to the other.

Example: If we wish to convert 30°C to °F, we substitute in the first formula:

$$\begin{aligned} °F &= \tfrac{9}{5}(30°) + 32° \\ &= 54° + 32° \\ &= 86° \end{aligned}$$

Figure 1-1. Comparison of the Celsius, Kelvin, and Fahrenheit temperature scales.

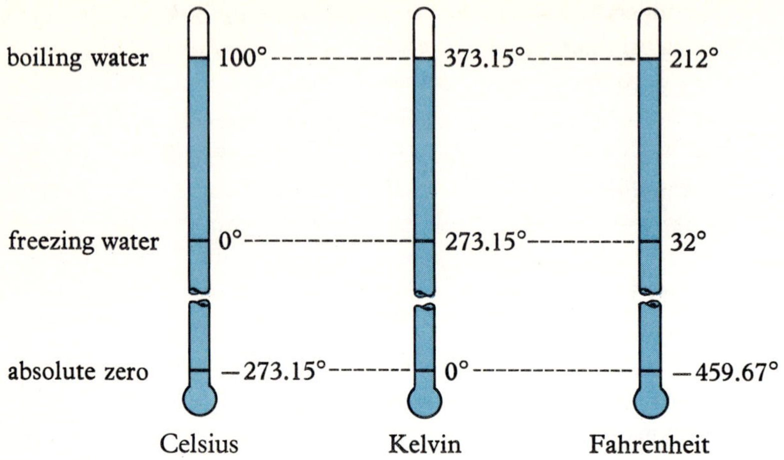

Example: If we wish to convert 65°F to °C, we substitute in the second formula:

$$
\begin{aligned}
{}^\circ C &= \tfrac{5}{9}(65^\circ - 32^\circ) \\
&= \tfrac{5}{9}(33^\circ) \\
&= 18^\circ
\end{aligned}
$$

Another temperature scale that we will use in Chapter 2 is called the *Kelvin,* or *absolute,* scale. This scale is closely related to the Celsius scale. (See Figure 1-1.) The only difference between the two scales is that zero is set at a different place on the scale. In the Kelvin scale, 0°K is placed at the lowest temperature possible, called *absolute zero,* which is approximately −273°C. Therefore, to convert °C to °K, we simply add 273°.

$$ {}^\circ K = {}^\circ C + 273^\circ $$

Example: We wish to convert 30°C to °K.

Substitute:

$$
\begin{aligned}
{}^\circ K &= 30^\circ + 273^\circ \\
&= 303^\circ
\end{aligned}
$$

Example: We wish to convert 50°K to °C.

Substitute:

$$ 50^\circ K = {}^\circ C + 273^\circ $$

Rearrange:

$$
\begin{aligned}
{}^\circ C &= 50^\circ - 273^\circ \\
&= -223^\circ
\end{aligned}
$$

HEAT. We need a measure not only for temperature but also for heat. The unit we use is the *calorie* (cal). A calorie is defined as the amount of heat that is required to raise the temperature of one gram of water from 14.5°C to 15.5°C. (For practical purposes, one calorie raises the temperature of water one degree at any temperature.)

Probably everyone has heard of Calories in connection with diet. For example, one piece of cheese pizza contains about 185 Calories. But these calories are "Calories" with a capital C; they are truly kilocalories, or 1000 true calories. In this book, we use only the true calorie (lower case c). If we need a larger unit, we will use kilocalories.

Sample Problem: How many calories would be required to raise the temperature of 350 g water from 14°C to 15°C? (The use of proportions is reviewed in the Appendix.)

Solution: *Set up a proportion:* $\frac{1.0\text{ cal}}{1.0\text{ g}} = \frac{x\text{ cal}}{350\text{ g}}$

Rearrange and solve: $x\text{ cal} = \frac{(1.0\text{ cal})(350\text{ g})}{1.0\text{ g}}$

$= 350\text{ cal}$

DENSITY AND SPECIFIC GRAVITY. Another value for which we need a unit of measurement is the *density* of a substance. A cubic centimeter of lead weighs more than a cubic centimeter of goose down. Therefore, we say that lead has a greater density. Density is defined as the *mass per volume*. One cubic centimeter of lead weighs 11.3 grams. The density of lead is thus 11.3 g/cc. The densities of some solid materials are found in Table 1-5.

Table 1-5. Densities of Some Common Solids

Substance	Density in g/cc at 20°C[a]
Aluminum	2.70
Copper	8.92
Cork	0.22-0.26
Iron	7.86
Lead	11.3
Silver	10.5
Sodium chloride[b]	2.16
Sucrose[c]	1.58

[a] Values taken from *Handbook of Chemistry and Physics*, 52nd edition, The Chemical Rubber Co., Cleveland, Ohio, 1971.
[b] Common table salt.
[c] Table sugar.

$$\text{Density} = \frac{\text{Mass in g}}{\text{Volume in cc}}$$

Closely related to density is *specific gravity.* The specific gravity of a substance is *the ratio of its mass to the mass of an equal volume of water.* To find a specific gravity, a given volume of the substance is weighed and then the same volume of water is weighed. To arrive at the value for the specific gravity, the weight of the substance is divided by the weight of the water. Specific gravity is useful for comparing the relative weights of liquids, while density is best for solids. See Table 1-6 for some representative specific gravities.

$$\text{Specific gravity} = \frac{\text{Mass of a substance in g}}{\text{Mass of the same volume of water in g}}$$

Table 1-6. Specific Gravities of Some Common Liquids

Substance	Specific Gravity at 20° C[a]
Acetone	0.79
Benzene	0.90
Ethyl alcohol	0.79
Gasoline	0.66–0.69
Mercury	13.6
Olive oil	0.91
Pure water	1.00
Sea water	1.025

[a] Values taken from *Handbook of Chemistry and Physics,* 52nd edition, The Chemical Rubber Co., Cleveland, Ohio, 1971.

Both density and specific gravity vary with temperature. Although the masses do not change, the volumes of liquids and solids change somewhat with temperature. Therefore, comparisons should be made at the same temperature. For practical purposes, the weight of 1.0 cc of water at room temperature is 1.0 g. Therefore, a liquid with a specific gravity of 0.79 has an approximate density of 0.79 g/cc.

Specific gravities are conveniently measured with a *hydrometer,* of which a variety of types are available. For example, a wine hydrometer is specially calibrated to measure the specific gravity of the "must" (the fermenting mixture). Water containing sugar is more dense than pure water. The hydrometer can tell the winemaker if the concentration of sugar is sufficient to form the desired alcohol content. As sugar in the must is converted to ethyl alcohol, the specific gravity of the must decreases, allowing the winemaker to follow the progress of the fermentation.

Service station mechanics use another type of hydrometer to see if battery acid is of proper strength, and another type to check the antifreeze in the radiator.

Figure 1-2. A hydrometer.

Hydrometers are also used to check the specific gravities of urine specimens. A higher specific gravity than normal may be an indication of sugar in the urine, a condition that can arise from *diabetes mellitus*. (See Chapter 29.)

A hydrometer may be compared to a boat. A boat in water sinks only far enough to displace a quantity of water equal in mass to its own mass. A hydrometer also sinks just enough to displace liquid to equal its mass. In a more dense liquid, the hydrometer will not sink as far as it will in a less dense liquid. For example, syrup is more dense than water. Therefore, a hydrometer will not sink as far in syrup as in water. (See Figure 1-2.)

STUDY QUESTIONS

1-1. Identify each of the following statements as a hypothesis, a theory, or a law:

(a) The amount of matter and energy in the universe is fixed, although there may be some interconversions.

(b) In all probability, there are other planets besides Earth in the universe that are capable of supporting life.

1-2. Identify the change in state involved in the formation of each of the following substances from liquid water:

(a) rain (b) frost (c) snow (d) steam

1-3. Identify the type of energy contained in or exhibited by the following:

(a) a boulder resting at the top of a cliff

(b) an avalanche

1-4. When a campfire burns, the __________ energy of the wood is being converted to __________ and __________.

1-5. Carry out the following conversions:

(a) 5.0 miles to kilometers

(b) 1.0 ft to centimeters

(c) 1.0 m to microns

(d) 1.94 Å to microns

(e) 1.94 Å to inches
(f) 1.0 in. to Angstroms

1-6. Carry out the following conversions:
(a) 5.0 lb to kilograms
(b) 1.0 g to pounds
(c) 1500 g to kilograms

1-7. Carry out the following conversions:
(a) 1.0 qt to liters (b) 1.0 gal to liters
(c) 22.4 liters to gallons (d) 5.0 ml to liters

1-8. Listed below are the boiling points of a few common solvents. Express these boiling points in °F, °C, and °K.
(a) water, 100°C
(b) ethyl ether, 95°F
(c) ethyl alcohol, 174°F
(d) gasoline, 39°-204°C

1-9. Normal body temperature is 98.6°F. Convert this value to °C.

1-10. How many calories would be required to raise the temperature of 200 g of water from 14°C to 15°C?

1-11. How many calories would be needed to heat 100 g of water from 0°C to room temperature (see Table 1-4)? How many kilocalories?

1-12. How much more dense is:
(a) silver than copper?
(b) sodium chloride than sucrose?
(c) lead than cork?

1-13. Using Table 1-6, determine the approximate mass of 2.10 liters of:
(a) sea water (b) acetone
(c) pure water (d) mercury

1-14. What is the approximate mass of one gallon of:
(a) water? (b) gasoline? (c) lead?

1-15. A metal reputed to be pure silver displaces 3.00 ml of water and weighs 33.9 g. Is the metal pure silver?

1-16. A home brewer of beer finds that two tubs of malt-sugar-water mixture have different specific gravities of 1.03 and 1.07. Which of the two tubs will likely yield the beer of higher alcohol content?

CHAPTER 2 THE THREE STATES OF MATTER

OBJECTIVES

Composition of Matter *Name the particles that comprise matter. * Differentiate between an atom and a molecule. Give an example of each. * Differentiate between an element, a pure compound, and a mixture. List some examples. * Define physical change.*

Changes in State *Draw a heating curve and explain its relation to the heating of a solid and its subsequent changes in state. * Define heat of fusion and heat of vaporization, and explain where the heat goes when ice is heated, melted, and then vaporized.*

The Solid State *Contrast a crystalline solid to an amorphous solid. List some examples of each. * Define heat capacity, specific heat, and melting point.*

The Liquid State *Contrast the motion of the molecules of a liquid to the motion of those of a solid. * Define viscosity, surface tension, and diffusion. Describe how these properties vary in different liquids and how they are affected by temperature changes. * Describe the process of evaporation. * Define vapor pressure and equilibrium vapor pressure. Define equilibrium as a general term. * Describe the process of boiling.*

The Gaseous State *Contrast the motion of the molecules of a gas to the motion of the molecules of a liquid. * List the assumptions in the kinetic molecular theory of gases. * Define standard pressure. Solve problems using Boyle's Law, Charles' Law, the combined gas laws, and the Law of Partial Pressures.*

COMPOSITION OF MATTER

We have already defined matter as something that has mass and occupies space. But what is matter made of? Water is matter. In your mind's eye, picture a drop of water. Now enlarge the drop until it is the size of the earth. Look very closely at the earth-sized drop. You will see that it is made up entirely of identical particles, each one about the size of a hamburger. These particles are the *molecules* of water. Although a molecule is tiny beyond comprehension, a molecule has mass and occupies space. It is matter. If we add up the masses of all the water molecules in the water drop, the value would equal the mass of the drop.

If we examine a single molecule of water, we will find that it is composed of three smaller particles, which are called *atoms*. Two of these atoms are the same and are called hydrogen (H) atoms, while the other atom is called an oxygen (O) atom. The three atoms are *bonded* together to form the molecule of water. We represent the water molecule as H—O—H, or H_2O.

According to modern atomic theory, all matter is composed of particles that are in motion. These particles may be individual atoms or they may be groups of atoms bonded together as molecules. The differences we observe between two types of matter, such as table salt and water, are due to the fact that they are composed of different types of molecules. That is, table salt and water differ in the kinds of atoms they contain and how these atoms are bonded together. We will discuss atoms in more detail in Chapter 3, and how atoms are bonded together in Chapter 4.

ELEMENTS AND COMPOUNDS. A substance composed of just one kind of atom is called an *element*. In one type of element, the smallest individual particles are simply the atoms themselves. Neon gas (Ne) is composed of only neon atoms and is said to be *monatomic* (*mono-*, meaning one). The other type of element is composed of molecules, but the molecules contain only one kind of atom. An example of this type of element is illustrated by oxygen (O_2), a molecule that is composed of two oxygen atoms. The molecules of oxygen are *diatomic* (*di-*, meaning two). Other elements may

Figure 2-1. Elements and compounds.

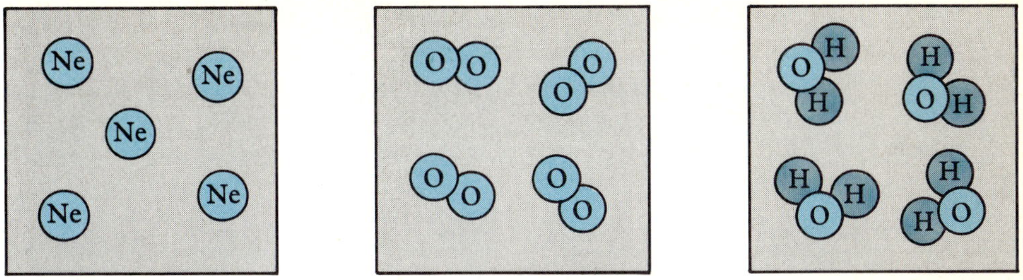

Neon (Ne) is an element composed of neon atoms.

Oxygen (O_2) is an element composed of oxygen molecules.

Water (H_2O) is a compound composed of water molecules.

bond together into molecules that contain more than two atoms. For example, a phosphorus molecule contains four atoms (P_4), a sulfur molecule contains eight atoms (S_8), and a diamond, which is a giant molecule of carbon (C), contains an uncountable number of atoms.

A *compound*, such as water, is a substance made up of molecules that contain more than one kind of atom. The molecule is the smallest particle of a compound that still has the properties of the compound. The properties that we attribute to a compound, such as density and color, are due to the collective effect of all the identical molecules. (See Figure 2-1.)

PURE COMPOUNDS AND MIXTURES. Pure water is a *pure compound*. If we enlarge a water drop, we find only one kind of molecule, H_2O molecules. The sugar you put into your coffee is also a pure compound. If we enlarged a sugar crystal sufficiently, we would find that all the molecules are composed of carbon, hydrogen, and oxygen atoms and that all the molecules are identical.

A pure compound cannot be changed to another compound or broken into its elements by ordinary physical means. We cannot make hydrogen and oxygen from water by cleaving it with an ax or by pulling it apart with tweezers or even by boiling it. (We will find later that a compound can be changed by *chemical* means, however.)

If we mix pure iron filings and pure sugar, we will no longer have a pure substance, but a *mixture* of two substances. The two components keep their identity, however, and can be separated by physical means. We could use a pair of tweezers or a magnet to separate the sugar-iron mixture.

We can also mix sugar and water. In this case, the sugar dissolves in the water. Yet, if we could enlarge a drop of the water with the dissolved sugar, we would find two types of molecules—sugar molecules and water molecules. Each kind of molecule has retained its identity. Again, we have a mixture and could separate the two by physical means. For example, the water could be boiled off, which is a physical process. (See Figure 2-2.)

We can formalize the definitions of pure compounds and mixtures:

A *pure compound* cannot be broken down into its elements by physical means.

A *mixture* of components can be separated into its pure components by physical means.

Figure 2-2. Sugar dissolved in water is a mixture of two types of molecules.

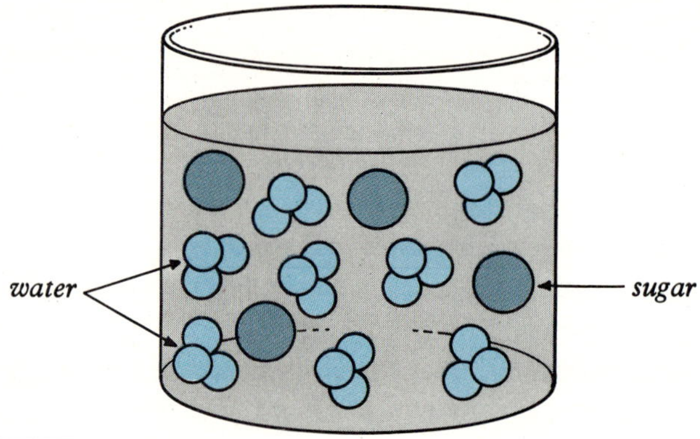

PHYSICAL CHANGE VERSUS CHEMICAL CHANGE. The liquid compound water can freeze and become ice. Ice, however, is still a form of water. If we could enlarge ice, we would find that it is composed of the same water molecules of which liquid water is composed. Water can evaporate and exist as a gas that we call water vapor. The gaseous form is still water; the molecules have not changed. A change in the state of matter is called a *physical change.*

If water (H_2O) were changed to its elements, hydrogen gas (H_2) and oxygen gas (O_2), we would have a *chemical change.* A chemical change involves a change in the structure of a molecule and, therefore, changes in the properties of the substance or substances.

We will devote most of this chapter to the three states of matter and their interconversions. Chemical changes will be covered in later chapters.

CHANGES IN STATE

When a substance is heated, its atoms or molecules move faster, or gain kinetic energy; thus, its energy content is increased. When a substance is cooled, its energy or heat content is decreased. Temperature is a measure of the heat content of a substance. As we heat water, its heat content and its temperature rise up to the *boiling point,* the temperature at which it starts to bubble vigorously. Once the boiling point is reached, as we apply more heat, the temperature of the liquid does not continue to rise, but remains constant. In other words, the temperature of boiling water does not increase if the burner is turned up; the water merely boils more vigorously. During the process of boiling, the energy absorbed is used to convert liquid water to water vapor rather than to increase the energy content of the water. Water vapor at 100°C has a higher energy content than liquid water at 100°C. Once all the water has boiled away, additional heat will cause a rise in temperature of the water vapor.

The amount of heat required to change a substance from a liquid to a gas at its boiling point is called the *heat of vaporization* (ΔH_v) and is the difference in energy content between the liquid and the gas at this temperature.

$$\Delta H_v = H_{\text{gas}} - H_{\text{liquid}}$$

A similar phenomenon is observed as a solid melts. A solid may be heated up to its *melting point,* which is defined as the temperature at which it becomes a liquid. However, at the melting point, energy is required to change the solid to liquid. The temperature of the substance will not increase again until the solid has entirely melted.

The amount of heat needed for the transition from solid to liquid is called the *heat of fusion* (ΔH_f) and is the difference in energy content between the liquid and the solid at the melting point.

$$\Delta H_f = H_{\text{liquid}} - H_{\text{solid}}$$

We can draw a graph (Figure 2-3) that describes the relationship between the temperature of a substance and the application of heat. The graph, called a *heating curve,* shows the increase in temperature as a substance is heated and goes through changes in state.

We will now consider some properties of matter in the solid, liquid, and gaseous states. Although we will discuss these topics one at a time, we will find that they are actually interrelated.

THE SOLID STATE

A solid is a substance that holds its shape and volume even when it is not in a container. The reason a solid holds its shape is that the molecules are close together and do not move about. A molecule of a solid just sits in its prescribed spot and wiggles, or oscillates.

A solid may be heated and the energy of the molecules increased. The molecules

Figure 2-3. Heating curve for a pure substance.

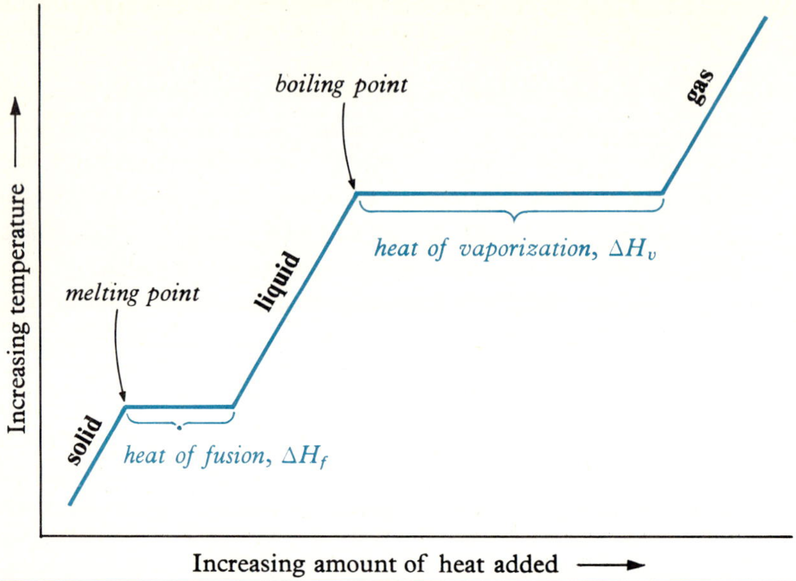

will then oscillate to a greater extent. If the temperature reaches the point that the molecules break out of their positions and move about, the solid melts and becomes a liquid.

Some solid substances have a tendency to go directly from the solid state to the gaseous state without ever becoming a liquid. This process is called *sublimation.* Dry ice (solid carbon dioxide) and old-fashioned mothballs (naphthalene) are examples of solids that sublime.

TYPES OF SOLID. There are two principal types of solid. A *crystalline* solid is one in which the molecules or atoms are in an ordered arrangement, called a *crystal lattice.* An *amorphous* solid is one in which the molecules or atoms are clumped together in a disordered fashion. (See Figure 2-4.)

We can often tell by sight if a solid is crystalline or not because the overall ordered arrangement of a single crystal may be large enough to see. Crystals have regular shapes and often appear shiny and translucent.

An amorphous solid presents no ordered arrangement to the eye. A lump of flour in gravy is an amorphous aggregation of starch molecules. (See Figure 2-5.) In some respects, an amorphous solid may be regarded as a supercooled liquid and may exhibit some characteristics of a liquid, such as flow under pressure. Amorphous solids are often called *glasses.* Common window glass is an example of an amorphous solid.

Figure 2–4. Molecular arrangements in a crystalline solid and in an amorphous solid.

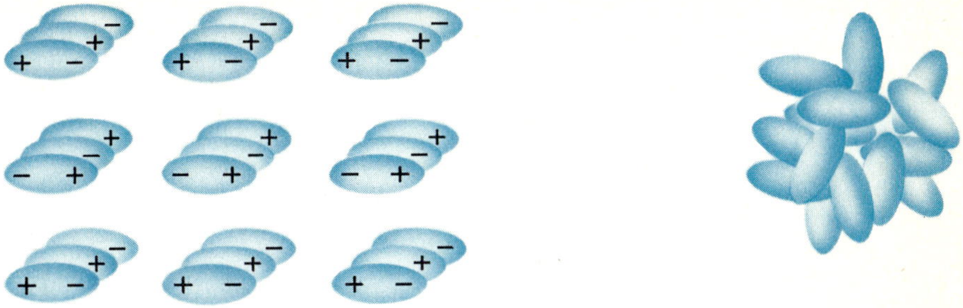

ordered arrangement, as in a crystal

disordered arrangement, as in an amorphous solid

Figure 2–5. Crystals often display their ordered structure to the eye.

quartz crystals

snowflake (a water crystal)

an amorphous lump of clay

HEAT CAPACITY AND SPECIFIC HEAT. The *heat capacity* of a substance is defined as the *quantity of heat required to raise the temperature of the substance 1* °C. The greater the mass present, the greater the heat capacity. A lake has a greater heat capacity than a pond. To compare heat capacities of substances, we must specify a given mass; for example, 1.00 g. The quantity of heat required to raise the temperature of 1.00 g of a substance 1 °C is referred to as the *specific heat* of that substance. The specific heat of 1.00 g of water is 1.0 calorie/gram-degree.

Different substances have different specific heat values. (See Table 2-1.) Water has one of the highest specific heats, while silver and copper are at the low end of the scale. We can use specific heat values to calculate the number of calories that would be required to heat a substance from one temperature to another or to calculate the temperature rise when a known number of calories is absorbed by a substance.

Table 2-1. Some Specific Heats

Substance	Specific Heat (cal/g-deg) at 25°C
Water (at 14°C)	1.00
Iron metal (Fe)	0.106
Silver metal (Ag)	0.057
Copper metal (Cu)	0.092
Helium gas (He)	1.24

Sample Problem: How many calories are required to heat 454 g (1 lb) of iron from room temperature (25°C) to its melting point, 1535°C?

Solution: Heat required = Mass × Specific heat × Temperature change

$$= 454\text{ g} \times 0.106\frac{\text{cal}}{\text{g-deg}} \times (1535° - 25°)$$

$$= 72{,}700\text{ cal}$$

MELTING POINT AND FREEZING POINT. As we observed in the heating curve, as a solid is heated the temperature increases up to the melting point. Different solids have different melting points. In organic chemistry, melting points are one of the physical properties used to identify compounds. Some representative melting points are found in Table 2-2.

When a liquid is cooled, a temperature is eventually reached at which it will solidify. This temperature is the *freezing point* of that substance. The melting point and the freezing point of a substance are the same. The difference between them is one of viewpoint. Many liquids have the ability to become supercooled, that is, to exist as a liquid below their freezing points. The reason for this behavior lies in the fact that crystals have an ordered arrangement of molecules. The transition from the random liquid state to the ordered crystalline state is difficult. However, a seed crystal may be added to a supercooled liquid to provide a template for the growth of the crystals. Solidification will then occur rapidly.

Table 2-2. Some Melting Points and Heats of Fusion

Substance	Mp (°C)	ΔH_f (cal/g)
Water (H_2O)	0	79.7
Iron (Fe)	1535	63.7
Silver (Ag)	961	25.0
Copper (Cu)	1083	49.0
Sodium chloride (NaCl)	801	123.5
Tristearin (common fat)	55	18.6

HEAT OF FUSION. Heat of fusion was defined earlier in this chapter as the energy required to effect the change from solid to liquid at the melting point. Some typical heats of fusion are given in Table 2-2. Note, as in the case of specific heat, that water is at the high end of the scale. Also note that the values are much higher than the values for specific heat. Changes in state of a substance require more energy than small increases in its temperature.

Heat of fusion values may be used to calculate the energy needed for the transition of a substance from solid to liquid.

Sample Problem: How much heat is needed to melt 1.00 pound of iron at its melting point?

Solution:

$$\begin{aligned} 1.00\ \text{lb} &= 454\ \text{g} \\ \text{Heat required} &= \text{Mass} \times \text{Heat of fusion} \\ &= 454\ \text{g} \times 63.7\ \text{cal/g} \\ &= 28{,}900\ \text{cal} \end{aligned}$$

We have previously calculated that 72,700 calories are needed to raise the temperature of one pound of iron from room temperature to its relatively high melting point of 1535°C. Now we find that it takes 28,900 calories to convert one pound of iron from solid to liquid with the temperature remaining 1535°C.

THE LIQUID STATE

A liquid, such as water, is fluid and therefore assumes the shape of its container. The reason for this fluidity is that the molecules in a liquid are in constant motion and do not have the fixed, ordered arrangement that is characteristic of a solid. The molecules of a liquid are close together; therefore, a liquid resists being compressed.

The molecules in a liquid are held together by attractions between the molecules. These attractions between molecules are a type of *intermolecular force.* The reason for these forces of attraction is that molecules are not homogeneous blobs. Instead, they have within themselves portions that are positively and negatively charged. The charges are due to the types of atoms in the molecules, the shapes of the molecules, and so forth—topics we will discuss in more detail in later chapters.

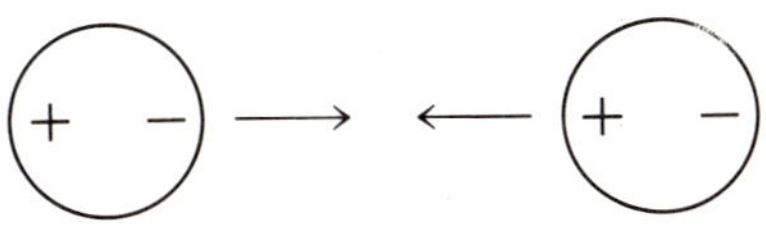

Attraction between + and −.

Figure 2–6. A liquid with a greater viscosity has a greater resistance to flow.

VISCOSITY. *Viscosity* is the resistance to flow, or the syruplike qualities, of a liquid. (See Figure 2–6.) Because of differences in the attractions between molecules in liquids, we observe differences in viscosity among liquids. Gasoline flows easily; it is not very viscous. Water is more viscous than gasoline because the attractions between its molecules are greater. Salad oil is more viscous than water—again, greater attractions.

Viscosity is inversely proportional to temperature. If a liquid is heated, the molecules move more rapidly and the liquid becomes less viscous. If a liquid is cooled, its viscosity increases.

Automotive lubricating oils are marketed in terms of *viscosity numbers.* No. 10 weight motor oil is fairly fluid, while No. 50 weight oil is quite viscous. The all-weather, multigrade motor oils, such as 10–30 weight, are constant-viscosity oils. These oils contain large coiled molecules that become uncoiled and intertwined with an increase in the temperature. In this way, the normal effect of temperature on viscosity is counteracted, and the oil can retain a relatively constant viscosity over a temperature range.

SURFACE TENSION. Related to viscosity is *surface tension,* the resistance to expansion of the surface of a liquid. Surface tension is the phenomenon that causes a drop of a liquid to hold its shape rather than flatten out on a surface. Surface tension is also caused by the attractive forces among molecules. The molecules on the surface of a drop are attracted to molecules inside the drop; there are no similar attractions outside to pull the surface molecules away from each other.

A similar effect may be observed if we fill a clean container with water; if we are careful, we can fill it "fuller than full." If we continue adding water to the container,

Figure 2–7. The surface tension of a liquid allows us to fill a glass "fuller than full."

the force of gravity eventually overcomes the surface tension, and the excess water flows out. (See Figure 2–7.)

Surface tension varies with the liquid. Water has a fairly large surface tension, while gasoline has a comparatively small one. Surface tension also varies inversely with temperature. If a liquid is heated, its surface tension decreases. Surface tension may also be decreased by the addition of a *wetting agent,* such as a detergent. A wetting agent is attracted to a surface (glass, cloth, or other surfaces) and also attracts water molecules. These attractions overcome the intermolecular attractions at the surface of the water droplets and allow the water to spread out and wet the solid surface. The action of a detergent depends on this wetting action and also on its ability to emulsify oils, a phenomenon that will be described in Chapter 6.

DIFFUSION. Because molecules of a liquid are able to move about more or less at random, liquids that are soluble in one another (that is, *miscible*) can undergo *diffusion,* or spreading out, through one another. Pour a bit of milk into a cup of coffee and you can watch the diffusion of the two liquids. If a liquid is heated, the molecules move faster, and diffusion becomes more rapid.

EVAPORATION AND VAPOR PRESSURE. Some molecules within a liquid may have enough energy that, when they strike the surface of the liquid, they will escape the attractive forces of the other liquid molecules. Such a molecule escapes from the liquid and becomes a molecule of gas. This process is called *evaporation.* (See Figure 2–8.)

The rate of evaporation depends upon the attractive forces within the liquid and upon the temperature of the liquid. At the same temperature, water evaporates more slowly than gasoline because of greater attractive forces among water molecules. When heated, the energy of the molecules in a liquid is increased, more molecules gain

Figure 2-8. Evaporation is the process of molecules escaping from a liquid and becoming a gas.

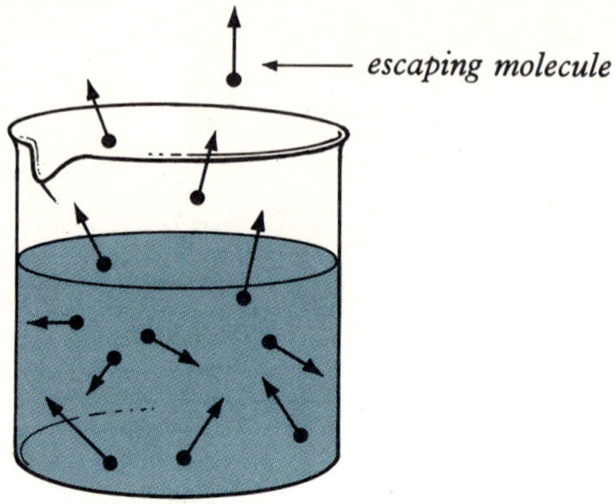

enough energy to escape, and the rate of evaporation increases. The escaping molecules take their energy with them. As a consequence, the evaporating liquid loses energy and becomes cooler. We observe this effect when we put rubbing alcohol on our skin. As the rubbing alcohol evaporates, we can feel the remaining liquid becoming cooler and cooling the skin.

If a container with some liquid in it is closed, some of the liquid evaporates. When a large number of molecules have become gaseous, some of them will strike the surface of the liquid and become liquid again. The process of a gas becoming a liquid is called *condensation*. After a period of time, in a closed container, the number of molecules condensing from the gaseous state equals the number of molecules evaporating from the liquid. At this point, the total number of molecules in the gaseous state remains constant, even though molecules are switching back and forth between the liquid and gaseous states. This is called an *equilibrium* between the gas and the liquid. (See Figure 2-9.)

We will encounter other equilibria in this text. Just what is an equilibrium? A dictionary defines an equilibrium as a *state of balance*. In chemistry, an equilibrium is a situation in which the number of molecules going one way (in this case, from liquid to gas) is equal to the number of molecules going the reverse way (from gas to liquid).

Because the molecules in the gaseous state collide with the surface of the liquid, they exert a pressure on the surface. This pressure is called the *vapor pressure*. In a closed container, at the equilibrium point, we have an *equilibrium vapor pressure*. At the equilibrium, the pressure exerted by escaping molecules is the same as the pressure exerted by returning molecules. Strictly speaking, the term "vapor pressure" means the pressure of the gaseous molecules over the liquid. However, it is convenient to speak also of the vapor pressure of the liquid, that is, the pressure exerted on the gaseous molecules above the liquid by molecules escaping from the liquid.

Figure 2–9. In a closed container, some of the escaped molecules will return to the liquid state. At equilibrium, molecules escape and return at the same rate.

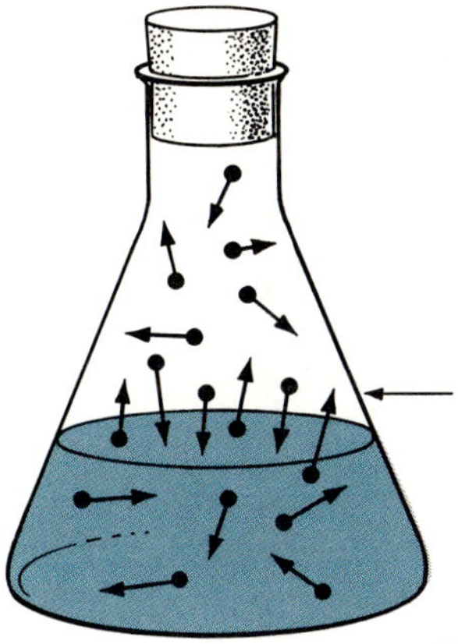

BOILING POINT. As water is heated, a point is reached at which the water boils and becomes water vapor. While evaporation is a surface phenomenon, boiling is an internal one. The molecules within the liquid have enough kinetic energy to become gaseous. The boiling point of a liquid is reached when the vapor pressure of that liquid equals the atmospheric pressure. That is, the molecules leaving the liquid exert as much pressure on the atmosphere as the atmosphere exerts on the liquid. (See Figure 2–10.) If the atmospheric pressure is reduced, such as on a mountain top, the boiling point is reached at a lower temperature. It takes slightly longer to boil an egg or cook stew in Missoula, Montana, than it does in San Francisco because Missoula is about 1000 meters higher in altitude.

Different liquids have different boiling points. Gasoline, with its fewer attractive forces, boils at a lower temperature than water. Salad oil boils at a higher temperature than water. The boiling point is one of the physical properties commonly used to help identify an unknown liquid.

The amount of energy required to vaporize a liquid at its boiling point is, as we have said, the heat of vaporization. Some heats of vaporization are in Table 2–3. Again, note that water has a relatively high heat of vaporization. This is another indication of the attraction that water molecules have for each other.

Sample Problem: How much heat is required to change 100 g of water to water vapor at its boiling point?

Solution: Heat required = Mass × Heat of vaporization

$$= 100\text{ g} \times 539.6\text{ cal/g}$$
$$= 54{,}000\text{ cal}$$

Figure 2–10. The boiling point is reached when the pressure of the molecules escaping from the liquid is equal to the atmospheric pressure.

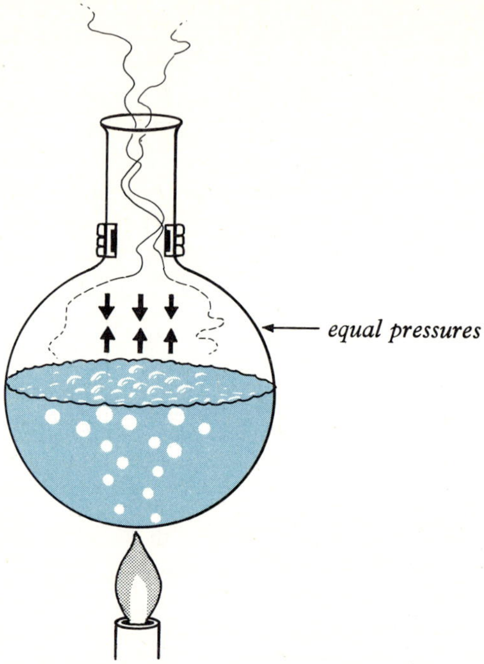

Table 2–3. Some Boiling Points and Heats of Vaporization at the Normal Boiling Point

Substance	Bp (°C)	ΔH_v (cal/g)
Water (H_2O)	100.0	540
Ethyl alcohol (CH_3CH_2OH)	78.5	204
Hydrogen sulfide (H_2S)	−60.7	132
Acetic acid (CH_3CO_2H)	117.9	97
Mercury (Hg)	356.6	71

THE GASEOUS STATE

A gas is the least dense form of matter. While one cubic centimeter of liquid water weighs 1.00 gram, one cubic centimeter of oxygen gas at sea level weighs 0.0014 gram. The same amount of nitrogen gas (N_2) weighs 0.00125 gram. Helium (He) is even less dense and weighs only 0.00018 gram per cubic centimeter. The extremely low density of helium explains why a balloon filled with this gas rises in the air, which is composed mainly of nitrogen and oxygen.

The molecules or atoms in a gas are far apart and can move in any direction.

Thus, the molecules of a gas can diffuse readily and assume the shape of any container.

Most gases, such as nitrogen or oxygen, are invisible to us. Even though we may not be able to see a gas, a gas has physical properties. For example, a gas can exert pressure. It is the gas in a champagne bottle exerting pressure against the cork that can cause the cork to fly across the room if the bottle is opened carelessly. A gas may be heated and cooled. It may *expand,* or occupy a larger volume. Unlike a liquid or a solid, it may be *compressed,* or forced to occupy a smaller volume.

THE KINETIC MOLECULAR THEORY. The physical properties of a gas may be explained by the *kinetic molecular theory of gases*. The following assumptions are made in this theory:

1. Gases are composed of atoms or molecules.
2. The atoms or molecules are in continuous rapid motion and move in straight lines and in random directions.
3. The moving atoms or molecules can collide with each other and with their container.
4. The average kinetic energy of molecules of all gases is the same at a given temperature and will increase with an increase in temperature.
5. There is no attraction or repulsion among the molecules of a gas.

Let us look at these assumptions in more detail. The molecules of a gas are very tiny and are separated by comparatively large distances. The molecules are moving very rapidly in space, almost 1000 miles per hour at room temperature, and in straight lines.

Occasionally, the molecules will collide with each other or with the walls of the container. A collision results in a change of direction of the movement of a molecule; it then moves along in another, different, straight line until it collides with something else. Then it rebounds in a different direction. This process is not too different from that of billiard balls bounding and rebounding around a billiard table. Because the molecules collide with their container, they exert pressure, or force, on the container walls.

At a given temperature, the average kinetic energy of molecules of all gases is the same. (Note the term "average" kinetic energy. When molecules collide, some may rebound faster and some slower. However, the *average* kinetic energy at a constant temperature is a constant value.) If a gas is heated, the average kinetic energy increases. This means that the molecules move more rapidly, hitting each other and the walls of the container more frequently and with greater impact. The greater force of the impact and the increased number of impacts means an increase in the pressure of a gas. If you heat a closed container containing a gas, such as an aerosol can, the increase in pressure may cause the container to explode.

The kinetic theory includes the assumption that gas molecules neither attract nor repel each other. That is, a molecule moves in a straight line until it hits something, and does not swerve to the right or left or up or down to seek another molecule or to avoid it.

UNITS OF PRESSURE. At sea level and at 0 °C, the atmosphere exerts a pressure that will support a column of mercury that is, on the average, 76.0 cm high. This pressure is defined as the standard atmospheric pressure and is called *one atmosphere of pressure*, or *standard pressure*. In everyday terms, the amount of pressure exerted by the atmosphere at sea level is 14.7 pounds per square inch.

A barometer is a device used to measure atmospheric pressure. One common type of barometer is simply an inverted tube of mercury. The tube is usually calibrated so that the pressure may be read directly at the top of the column of mercury.

Other liquids could be used, but most liquids are too light in weight to be practical. As we have mentioned, the density of mercury is 13.5 g/cc, while the density of water is 1.0 g/cc. Standard atmospheric pressure can support a column of water 34 feet high. Since most scientists care not to use a ladder to read a barometer, mercury is the preferred liquid.

PRESSURE–VOLUME RELATIONSHIP (BOYLE'S LAW). When a gas is compressed to a smaller volume, the molecules have less volume in which to move, and they strike the surface of the container more frequently. Therefore, a decrease in the volume gives rise to an increase in pressure. In 1662, the English physicist and chemist Robert Boyle quantified this relationship into what we now call:

BOYLE'S LAW: *At a constant temperature, the pressure of a given mass of a gas is inversely proportional to the volume.*

If we let P represent the pressure of a given mass of gas, and V represent its volume, then we can state Boyle's law:

$$P = \frac{C}{V} \quad \text{or} \quad PV = C \text{ at constant } T$$

where C is a constant that is characteristic of the system at hand.

Sample Problem: If we have 5.0 liters of a gas at 1.0 atmosphere of pressure, what will be the pressure if the gas is compressed to 2.0 liters?

Solution: *Step 1:* Calculate the constant.

$$PV = C$$
$$(1.0 \text{ atm})(5.0 \text{ liters}) = 5.0 \text{ liter–atm}$$

Step 2: Using the constant, calculate the unknown pressure at the new volume.

$$PV = C$$

$$(P)(2.0 \text{ liters}) = 5.0 \text{ liter–atm}$$

$$P = \frac{5.0 \text{ liter–atm}}{2.0 \text{ liters}}$$

$$= 2.5 \text{ atm}$$

TEMPERATURE–VOLUME RELATIONSHIP (CHARLES' LAW). For the sake of explanation, let us assume we have a gas in a container that can expand or contract in such a way that the gas is always at a constant pressure. When we heat the container and its contents, we observe that the container gets larger—the volume of the gas has increased. When we cool the container, it contracts—the volume of the gas has decreased.

Consider the following mental experiment. Since the volume of a gas decreases with decreasing temperature, let us continue to decrease the temperature until the volume is zero. Now if we decrease the temperature a little more, the volume will be less than zero; that is, the gas will have a negative volume. A negative volume or even a volume of zero is, of course, absurd. Matter must occupy space and thus must have volume. For this reason, there must be a minimum temperature to which a gas can be cooled. This is the temperature at which a gas would have zero volume and at which there is no molecular motion. This temperature is defined as absolute zero. Although a temperature of absolute zero is not really attainable, we use this temperature as a reference point in gas calculations. The temperature scale that uses absolute zero as its reference point is the Kelvin scale, where $°K = °C + 273$.

In 1787, J. A. C. Charles, a French physicist, formulated what is now known as Charles' Law, which relates the volume of a gas to the temperature.

CHARLES' LAW: *For a given mass of gas at constant pressure, the volume is directly proportional to the absolute temperature (temperature in °K).*

If we let V be the volume of a gas and T be its absolute temperature in degrees Kelvin (°K), then Charles' Law can be stated:

$$V = CT \qquad \text{or} \qquad \frac{V}{T} = C \text{ at constant } P$$

where C is a constant for a given system.

Sample Problem: If we have 100 liters of gas at 200°K, what will its volume be at 300°K?

Solution: *Step 1:* Find the constant, C:

$$\frac{V}{T} = C$$

$$\frac{100\text{ liters}}{200^\circ\text{K}} = C$$

$$C = 0.500\frac{\text{liters}}{^\circ\text{K}}$$

Step 2: Using C, find the new volume:

$$\frac{V}{T} = C$$

$$\frac{V}{300^\circ\text{K}} = 0.500\frac{\text{liters}}{^\circ\text{K}}$$

$$V = 0.500\frac{\text{liters}}{^\circ\text{K}} \times 300^\circ\text{K}$$

$$= 150\text{ liters}$$

THE COMBINED GAS LAWS. Boyle's Law and Charles' Law can be put together into a combined gas law. Once again letting P = pressure, V = volume, and T = absolute temperature, then

$$\frac{PV}{T} = \text{a constant value for a given system}$$

A given mass of a gas thus can be compressed or expanded, heated or cooled, or put under more or less pressure. For that system, however, the relationship PV/T will remain at the constant value, even though the values for P, V, and T will change.

Start with a given mass of gas under conditions we will call P_1, V_1, and T_1:

$$\frac{P_1V_1}{T_1} = \text{a constant}$$

Change any one of the values (P, V, or T) and the other two will be altered so that the relationship is constant. Call the new values P_2, V_2, and T_2:

$$\frac{P_2V_2}{T_2} = \text{the same constant}$$

Therefore, for a given system

$$\frac{P_1V_1}{T_1} = \frac{P_2V_2}{T_2}$$

This form of the combined gas law allows us to calculate a new pressure, volume, or temperature when all of the conditions have been changed.

Sample Problem: Given 1.0 liter of a gas at 1.0 atm and 0.0°C, calculate the volume when the pressure has been increased to 3.0 atm and the temperature has been increased to 100°C.

Solution: *Step 1:* Change °C to °K.

$$T_1 = 0.0°\text{C} + 273°\text{C} = 273°\text{K}$$
$$T_2 = 100°\text{C} + 273°\text{C} = 373°\text{K}$$

Step 2: Define the two sets of conditions.

$$P_1 = 1.0 \text{ atm} \qquad P_2 = 3.0 \text{ atm}$$
$$V_1 = 1.0 \text{ liter} \qquad V_2 = ?$$
$$T_1 = 273°\text{K} \qquad T_2 = 373°\text{K}$$

Step 3: Substitute and solve.

$$\frac{P_1V_1}{T_1} = \frac{P_2V_2}{T_2}$$

$$\frac{(1.0 \text{ atm})(1.0 \text{ liter})}{(273°\text{K})} = \frac{(3.0 \text{ atm})(V_2)}{373°\text{K}}$$

$$V_2 = \frac{(1.0 \text{ atm})(1.0 \text{ liter})(373°\text{K})}{(273°\text{K})(3.0 \text{ atm})}$$
$$= 0.46 \text{ liter}$$

THE LAW OF PARTIAL PRESSURES. Molecules of a gas behave independently of each other. If two gases are mixed, each behaves as if the other were not there. Suppose we take a one-liter jug of oxygen at one atmosphere of pressure and then pump in two liters of nitrogen, also at one atmosphere. Now the combined pressure of the gases in the jug will be *three* atmospheres (1 atm from O_2 and 2 atm from N_2).

This is an example of the law observed by the Englishman John Dalton in 1801.

LAW OF PARTIAL PRESSURES: *The total pressure of a mixture of gases is equal to the sum of the partial pressures of the individual gases in the mixture.*

P_{total} = the sum of the individual pressures of all of the gases in the mixture

Sample Problem: A pressurized cylinder contains a mixture of nitrogen and oxygen in the ratio of 3.00 volumes of nitrogen to 1.00 volume of oxygen. The total pressure of the gases is 5.00 atmospheres. What is the pressure of the oxygen in the cylinder?

Solution: *Step 1:* Write the equation:

$$P_{O_2} + P_{N_2} = P_{\text{total}}$$

Step 2: Determine the relationship between P_{O_2} and P_{N_2}. There is 3.00 times as much N_2 as O_2, therefore:

$$P_{N_2} = 3\,P_{O_2}$$

Step 3: Substitute and solve:

$$P_{O_2} + 3\,P_{O_2} = P_{total}$$
$$4\,P_{O_2} = 5.00 \text{ atm}$$
$$P_{O_2} = \frac{5.00 \text{ atm}}{4}, \text{ or } 1.25 \text{ atm}$$

STUDY QUESTIONS

2-1. Identify each of the following substances as a monatomic element, a diatomic element, or a compound:
(a) N_2 (b) CO_2 (c) $C_6H_{12}O_6$ (d) H_2

2-2. Which of the following common substances are pure substances and which are mixtures?
(a) gasoline (b) river water
(c) mercury (d) ethyl alcohol (C_2H_5OH)

2-3. Match each item on the left with its description on the right.
(a) $N_2 + O_2$ (1) a pure compound
(b) $C_6H_{12}O_6$ (2) a mixture of compounds
(c) a solution of ethyl alcohol in H_2O (3) an element
(d) a chocolate chip cookie (4) a mixture of elements

2-4. Which of the following processes involves a chemical change?
(a) the melting of iron metal
(b) the spoilage of food
(c) the flaming of brandy on plum pudding
(d) melting ice by means of rock salt

2-5.
(a) A solid is heated to its melting point and, as the solid is melting, the temperature (does/does not) increase.
(b) After the solid becomes a liquid, added heat (does/does not) cause an increase in temperature.
(c) When the liquid reaches its boiling point, added heat (does/does not) raise the temperature as the liquid boils.
(d) As heat is added to a gas, the temperature (does/does not) increase.

2-6. Which of the following statements could be used to describe a solid?
(a) The molecules move in random directions.
(b) Intermolecular forces hold the molecules in their places.
(c) Adding heat does not increase the temperature because the molecules cannot move.
(d) The density is usually higher than that of a liquid or gas.

2-7. How many calories are required to increase the temperature of 100 g of each of the following substances from 22°C to 100°C?
(a) water (b) silver (c) helium gas

2-8. If 25 g of a substance requires 750 cal for its temperature to be increased by 10°C, what is its heat capacity in that temperature range?

2-9. How many calories must be removed from one liter of water to cool it from its boiling point to its freezing point?

2-10.
(a) How many calories are required to raise

1.0 kg of copper from 22°C to its melting point?

(b) How many calories are required to melt 1.0 kg of solid copper at its melting point?

(c) How many calories are required to melt 1.0 kg of copper starting from room temperature?

2-11. The body contains approximately 5.7 liters of blood. Assuming blood and water have the same heat capacities, how many kcal are required to raise the temperature of this amount of blood 1.0°C?

2-12. Which of the following statements is true?

(a) A liquid is easily compressed.

(b) The molecules of a liquid do not attract each other.

(c) A "runny" liquid is said to be less viscous than a syrupy liquid.

(d) Gasoline has a greater surface tension than molasses.

(e) If we put 1.0 calorie of heat into any two liquids, the temperatures of both the liquids will rise by the same amount.

2-13. Some water is placed in a flask, and the flask is corked.

(a) Some of the water molecules become gaseous in a process called ______.

(b) Some of the gaseous water molecules go back to liquid in a process called ______.

(c) When equal numbers of molecules are shifting back and forth, the situation is called ______.

2-14. Rank the following mixtures in order of ease of separation by distillation (most difficult first):

(a) a mixture of sodium chloride and sugar

(b) a mixture of two liquids that have boiling points of 96.5°C and 100°C, respectively

(c) a mixture of two liquids that have boiling points of 96.5°C and 182°C, respectively

2-15. Which of the following phrases describes a gas?

(a) may be compressed to a smaller volume

(b) is a type of matter

(c) can hold its shape

(d) is very dense

(e) has atoms or molecules that move in random directions

(f) can undergo a change in temperature

(g) consists of particles that can undergo an increase in kinetic energy when they are heated

2-16. Carbon dioxide (25.0 liters at 1.00 atm) has been compressed into a 5.00-liter cylinder. At the same temperature, what is the pressure of the carbon dioxide in the cylinder?

2-17. A balloon contains 3.00 liters of helium at 30.0°C. What would be the volume of the helium after chilling to 15.0°C without altering its pressure?

2-18. If 5.0 liters of a gas is allowed to expand to 10 liters at constant pressure, what will the temperature be if the initial temperature was 23°C? Explain your answer.

2-19. If 1.0 liter of O_2 and 1.0 liter of N_2, both at 760 mm pressure, are compressed together into a 1.0-liter tank at constant temperature, what will be the pressure of *each* gas in the tank?

2-20. If 1.00 liter of oxygen at 5.00 atm pressure and 10.0°C is allowed to expand to 3.00 liters and warm to 27.0°C, what will be its pressure?

2-21. A cylinder containing 12.0 liters of compressed air at 3.0 atm and 20°C is opened. The air escapes into the room (1.0 atm) and warms to 27°C. What is the final volume of the air from the cylinder?

2-22. The air is composed of approximately 80 percent N_2. At 1.0 atm of total air pressure, what is the pressure exerted by the nitrogen?

2-23. If a gas contained in a 10.0-liter tank at a pressure of 2,000 pounds per square inch is allowed to come to atmospheric pressure, what will be its volume if the temperature remains constant?

CHAPTER 3 A CLOSER LOOK AT THE ATOM

OBJECTIVES

Atomic Symbols *Define atomic symbol and formula. * Name some common elements and give their symbols.*

The Periodic Table—A Preview *Give the reason for the development of the Periodic Table of the Elements.*

Parts of the Atom *Diagram, name, and describe the principal parts of an atom.*

Ions *Define ion and write equations for the ways in which ions can be formed. * Define cation and anion, and list some examples.*

Atomic Number *Define atomic number.*

Isotopes *Define isotope.*

Atomic Mass *Define atomic mass.*

Molecular Weight *Define molecular weight and calculate a molecular weight from the formula.*

Gram-molecular Weight and Moles *Define and describe the relationships between gram-molecular weight and mole. * Calculate the weight of one mole of a substance from its chemical formula.*

Electronic Structure of the Atom *Describe the spatial relationship between the nucleus and the shells of the electrons in an atom. * Show how the* *electron shells fill with electrons as the atomic number increases.* ∗ *Define atomic radius and electronegativity, and show how they vary.* ∗ *Describe and diagram the* 1s, 2s, *and* 2p *orbitals.* ∗ *Explain why an orbital can contain a maximum of two electrons.* ∗ *Diagram the filling of the orbitals in the first two shells.*

The Periodic Table Revisited *In terms of electronic structure and the Periodic Table, compare the reactivities of hydrogen; the alkali metals; the alkaline earth metals; the boron, carbon, and nitrogen groups; the oxygen group; the halogens; and the inert gases.* (*Use equations.*)

The Greek philosopher Democritus (about 460–320 B.C.) is credited with the concept that matter is composed of atoms, a concept that lay dormant through the Middle Ages. The idea of atoms was brought to life again about 1800 by John Dalton, the son of a poor Quaker weaver. From Dalton's time to the present, the atomic theory has undergone many modifications as new facts have emerged and improved our understanding of nature.

We have already used some concepts of atomic theory in discussing physical changes in matter. We will find that understanding the atom is even more important in studying chemical change. Indeed, without the atomic theory, chemistry would be not much further along than it was in Medieval times.

ATOMIC SYMBOLS

Each of the 106 elements known today has a name as well as an alphabetic symbol, called the *atomic symbol,* which can be used to represent either the element or one of its atoms. When elements are combined in compounds, the shorthand representation, or *formula,* for the compound is the combination of the atomic symbols for the elements found in that compound.

H — *atomic symbol for hydrogen*

O — *atomic symbol for oxygen*

H—O—H or H_2O — *formula for water*

subscript 2 means two hydrogen atoms

Some elements are represented by just one capital letter of the alphabet, and others are represented by two alphabetic characters. In the latter cases, the first letter is always capitalized, and the second letter is lower case. For example, Co is the atomic symbol for cobalt, but CO is the formula for the compound carbon monoxide.

Table 3-1 lists some common elements and their atomic symbols. The majority of the symbols (such as Al and C) are abbreviations of the English name for the element. Several of the atomic symbols are derived from their Latin names. For example, the symbol for iron, Fe, is taken from the Latin *ferrum*.

Table 3–1. Some Common Elements and Their Atomic Symbols

Name	Symbol	Name	Symbol
Aluminum	Al	Lithium	Li
Bromine	Br	Magnesium	Mg
Calcium	Ca	Mercury	Hg
Carbon	C	Neon	Ne
Chlorine	Cl	Nitrogen	N
Chromium	Cr	Oxygen	O
Copper	Cu	Phosphorus	P
Fluorine	F	Potassium	K
Gold	Au	Silicon	Si
Helium	He	Silver	Ag
Hydrogen	H	Sodium	Na
Iodine	I	Sulfur	S
Iron	Fe	Zinc	Zn

THE PERIODIC TABLE—A PREVIEW

Early chemists were struck by the similarities in chemical behavior among certain elements. There were many attempts to group the different elements in some pictorial way to emphasize these similarities. These efforts culminated in what we now call *The Periodic Table of the Elements* (Figure 3-1). Credit for the Periodic Table is given to both Dimitri Mendeleev, a Russian, and Lothar Meyer, a German. Mendeleev and Meyer jointly received the Royal Society's Davy medal in England in 1882 for their work on the Periodic Table. (It is an interesting historical sidelight that the Periodic Table is the epitaph on Mendeleev's tomb.)

The Periodic Table contains eight main *groups* arranged in columns. The elements in each group generally behave in a similar way chemically. For example, the elements in the group on the far right in the Periodic Table (Group 0) are collectively referred to as the *inert gases*. These gases are all chemically inert, or unreactive.

The elements in the first group, Group IA (Li, Na, K, Rb, Cs, and Fr), are called

Figure 3-1. The Periodic Table of the Elements.

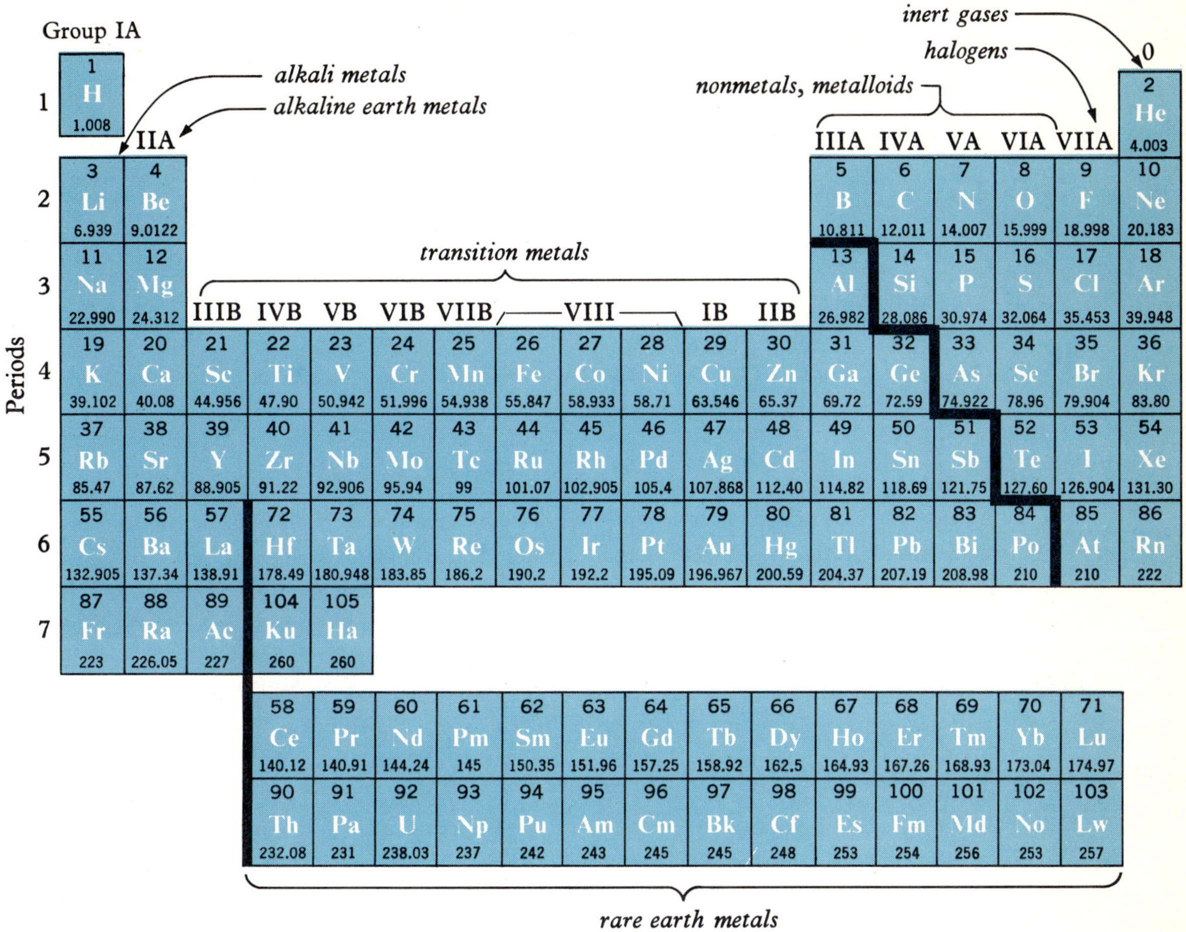

the *alkali metals*. The elements in Group IIA are referred to as the *alkaline earth metals*. The next group is referred to as the *transition metals* and is further divided into ten subgroups (Groups IIIB through IIB). Following these are four groups that contain other metals, nonmetals, and *metalloids* (substances intermediate between metals and nonmetals in their properties). The next to the last group (Group VIIA) is called the *halogens*. The chemical similarities within each of these groups will be discussed later in this chapter.

The horizontal rows in the Periodic Table are called *periods*, or simply *rows*. The elements in a row do not show the striking chemical similarities as the elements in one group.

Figure 3-2. The principal parts of an atom.

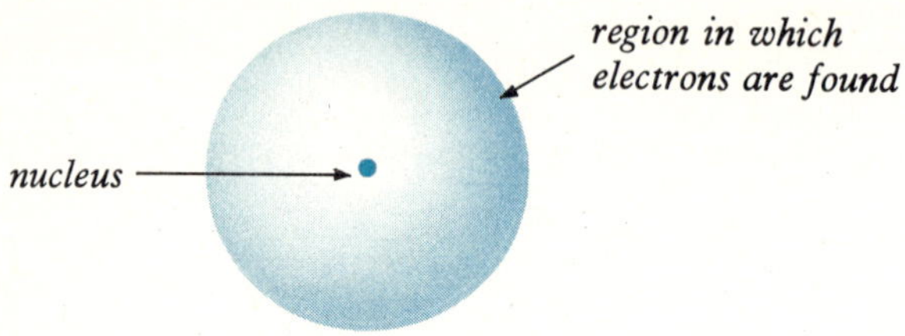

PARTS OF THE ATOM

In spite of the fact that the atom is minute, it is composed of even smaller particles. The atom is composed of two main parts—an inner heavy kernel called the *nucleus* and, surrounding the nucleus, the *electrons*. (See Figure 3-2.)

Changes in the composition of the nucleus of an atom are the subject of a field known as *nuclear chemistry*. Atomic energy is a result of nuclear change. We will deal with this subject in Chapter 10.

The number of electrons and how they are arranged around the nucleus of an atom determine the chemical properties of that atom. In this book, we will spend most of our time discussing the ramifications of the electronic structure of atoms and molecules.

THE NUCLEUS. The nucleus of an atom is very tiny and very heavy compared to the atom as a whole. The nucleus is not homogeneous; rather, it is composed of yet smaller particles. There are a large number of nuclear particles known, and all are important in nuclear chemistry; however, there are only two nuclear particles that we need to consider. These are the *neutron* and the *proton*.

A neutron and a proton have essentially equal masses. Their masses are so small that it would be impractical to discuss these masses in terms of grams. (For example, the mass of a hydrogen atom is 1.6×10^{-24} grams.) Instead, we say that each proton and each neutron has a mass of one *atomic mass unit* (amu). What is the difference between a proton and a neutron? A proton (p^+) carries a positive electrical charge, while a neutron (n^0) is electrically neutral.

Nuclei of all atoms are made up of varying numbers of protons and neutrons. Therefore, nuclei vary in mass and in the amount of positive charge they carry. For example, a helium nucleus contains two protons and two neutrons and carries an electrostatic charge of $+2$, while a carbon nucleus contains six protons and six neutrons and carries a charge of $+6$.

ELECTRONS. An electron (e^-) carries a negative charge and has a mass of only 1/1836 amu. The charges on a single proton and a single electron are equal, but opposite. If an atom has the same number of electrons as protons, the atom has no net electrical charge. A neutral atom of helium (He) has two protons in the nucleus and two electrons outside the nucleus. A neutral atom of carbon (C), with six protons in the nucleus, must have six electrons around the nucleus.

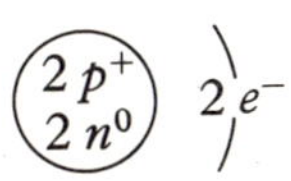

A neutral helium atom has two electrons outside the nucleus.

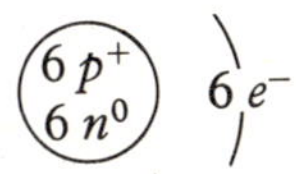

A neutral carbon atom has six electrons outside the nucleus.

IONS

When an atom has an equal number of protons and electrons, it is electrically neutral. Outside of a nuclear bomb or reactor, human beings cannot change the nucleus of an element; however, the number of electrons outside the nucleus can be easily changed. One major facet of chemistry involves the study of *electron transfer,* or the transferring of electrons from one atom to another. When electrons are transferred, a chemical reaction takes place.

There are two possible ways for an atom to behave in the process of electron transfer. (1) An atom may give up, or *lose,* one or more electrons, which is called *oxidation.* (2) An atom may *gain* one or more electrons, which is called *reduction.* (An electron can never be truly lost. If one atom loses electrons, then some other atom must gain the electrons.)

Let us look at the result of the oxidation of a hydrogen atom. A neutral atom of hydrogen has one proton in the nucleus and one electron outside the nucleus. When the neutral hydrogen atom is oxidized, it loses its electron and becomes a hydrogen atom that is no longer neutral, but has a positive charge, or H^+. An atom that has an electrical charge is called an *ion.* (The hydrogen ion is often called a proton because, in this case, that is what is left after the electron is removed.)

A neutral hydrogen atom (H):

$$1\,p^+ + 1\,e^- = 0 \text{ electrical charge}$$

A positively charged hydrogen ion (H^+):

$$1\,p^+ = +1 \text{ electrical charge}$$

We can represent a hydrogen atom losing an electron and becoming a hydrogen ion by a *chemical equation.*

$$\underset{\text{hydrogen atom}}{H} \longrightarrow \underset{\text{hydrogen ion}}{H^+} + e^-$$

The equation is read, "A hydrogen atom yields a hydrogen ion and an electron."

An atom can lose more than one electron and become an ion with more than one positive charge. Calcium is an atom that loses two electrons; it becomes a calcium ion (Ca^{2+}) with a charge of $+2$ because it now contains two more protons in its nucleus than electrons around the nucleus.

$$\underset{\text{calcium atom}}{Ca} \longrightarrow \underset{\text{calcium ion}}{Ca^{2+}} + 2\,e^-$$

Now let us look at the *reduction* of the hydrogen atom. When a neutral hydrogen atom gains an electron, it contains one proton and two electrons and now has a negative charge.

$$\underset{\text{hydrogen atom}}{H} + e^- \longrightarrow \underset{\text{hydride ion}}{H^-}$$

Ions that are positively charged are called *cations,* while ions that are negatively charged are called *anions.*

Some typical cations: H^+, Na^+, Li^+, Ca^{2+}

Some typical anions: Cl^-, F^-, Br^-, S^{2-}

ATOMIC NUMBER

Atoms of different elements differ from one another in the number of electrons, protons, and neutrons they contain. The simplest atom is that of the element hydrogen (H), with one proton in the nucleus and one electron outside the nucleus. The next lightest element is helium (He). The third lightest element is lithium (Li). A lithium atom contains three protons and four neutrons in its nucleus and three electrons outside the nucleus. As we go to the next element, we add another proton, another electron, and some more neutrons. We can continue this process right up to the heaviest element.

It is the number of protons in the nucleus that determines to which element the atom belongs. An atom with 79 protons in the nucleus is gold (Au), regardless of what else the nucleus might contain. If an atom has 78 protons, it is platinum (Pt). If it has 80 protons, then it is mercury (Hg). The number of protons in one atom of an element is called the *atomic number* of that element. In the Periodic Table (Figure 3–1), you will find the atomic number above the atomic symbol.

5	6	7	8	9
B	C	N	O	F
10.811	12.011	14.007	15.999	18.998

atomic number
(the number of protons in the nucleus)

ISOTOPES

All atoms that have one proton are said to be hydrogen. In nature we find that the majority of hydrogen atoms have only one proton in their nuclei. However, there are a few hydrogen atoms that have one proton and one neutron in their nuclei. These are still hydrogen atoms because they contain only one proton, but they are twice as heavy as most hydrogen atoms. We call these two types of hydrogen *isotopes* of each other. *Isotopes are forms of an element in which the atoms have the same number of protons but different numbers of neutrons.* The common isotope of hydrogen is called hydrogen-1, or simply *hydrogen.* The isotope that has an additional neutron is called hydrogen-2, or *deuterium.* The number following the name (the "2" in hydrogen-2) refers to the total number of atomic mass units in the nucleus, in this case a proton plus a neutron. Hydrogen has a third isotope, called hydrogen-3 or *tritium,* that contains one proton and two neutrons. Tritium is unstable and undergoes a slow and spontaneous disintegration in which the tritium nuclei break up and give off particles and energy called *radiation.* Tritium is said to be *radioactive.*

ATOMIC MASS

The *atomic mass* of an atom is the sum of the masses of the protons and neutrons in the nucleus. The atomic mass of helium is usually said to be four atomic mass units (the sum of the masses of two protons plus two neutrons). The atomic mass of carbon is 12 atomic mass units (the sum of the masses of six protons plus six neutrons). In the Periodic Table (Figure 3–1), the atomic mass of each element is beneath its atomic symbol.

5 B 10.811	6 C 12.011	7 N 14.007	8 O 15.999	9 F 18.998

atomic mass
(the sum of the masses of
the protons and the neutrons)

You may have noticed that the atomic masses as given in the Periodic Table are not usually whole numbers. Carbon, for example, has an atomic mass of 12.011. The reason for this is that natural carbon is a mixture of two isotopes. Most carbon atoms are carbon-12, with a mass of 12.000. But some of the carbon atoms in nature are carbon-13, which contains seven neutrons (plus six protons) in its nucleus and has an atomic mass of 13.000. The atomic mass of carbon (12.011) as given in the Periodic Table is an *average* value for the naturally occurring isotopes.

MOLECULAR WEIGHT

The mass of a molecule is the sum of the atomic masses of the atoms within that molecule and is usually referred to as its *formula weight,* or its *molecular weight.* Here the term "mass" would be more appropriate than the term "weight"; however, the term "molecular weight" is in such common use that we will use it throughout this book. The molecular weight of a hydrogen molecule (H_2) is the sum of the atomic masses of two hydrogen atoms. The molecular weight of water (H_2O) is the sum of the atomic masses of two hydrogen atoms and one oxygen atom.

$$\begin{aligned}\text{Molecular weight of } H_2 &= 2\ (\text{atomic mass of H}) \\ &= 2\ (1.008\ \text{amu}) \\ &= 2.016\ \text{amu}\end{aligned}$$

$$\begin{aligned}\text{Molecular weight of } H_2O &= 2\ (\text{atomic mass of H}) + (\text{atomic mass of O}) \\ &= 2\ (1.008\ \text{amu}) + (15.999\ \text{amu}) \\ &= 18.015\ \text{amu}\end{aligned}$$

GRAM-MOLECULAR WEIGHT AND MOLES

The *gram-molecular weight* of a substance is the molecular weight expressed as grams. For example, water (H_2O) has a molecular weight of 18.0 amu; its gram-molecular weight is 18.0 grams. The molecular weight of hydrogen (H_2) is 2.0 amu; its gram-molecular weight is 2.0 grams.

In order to find the gram-molecular weight of any substance, first calculate the molecular weight in atomic mass units, then convert to grams. A gram-molecular weight of a substance (for example, 18.0 grams of water) is also referred to as the weight of one *mole* of that substance.

Molecular weight in grams = Gram-molecular weight = Weight of 1.0 mole

Sample Problem: Find the gram-molecular weight (the weight of 1.0 mole) of H_2SO_4.

Solution: *Step 1:* Find the molecular weight, using the atomic masses (to one decimal place) from the Periodic Table.

$$\begin{aligned} 2\,\text{H} &= 2 \times 1.0 = 2.0 \text{ amu} \\ 1\,\text{S} &= 1 \times 32.1 = 32.1 \text{ amu} \\ 4\,\text{O} &= 4 \times 16.0 = \underline{64.0} \text{ amu} \\ \text{Molecular weight } H_2SO_4 &= 98.1 \text{ amu} \end{aligned}$$

Step 2: Express the molecular weight in grams.

The gram-molecular weight of H_2SO_4 = 98.1 g

Sample Problem: How many grams are in (a) 1.0 mole of uranium (U)? (b) 1.0 mole of oxygen gas (O_2)?

Solution: (a) 238.0 (the same as the atomic mass, expressed in g) (b) 32.0 (twice the atomic mass, expressed in g)

The most important ramification of the concept of moles is that *1.0 mole of any substance contains the same number of particles (molecules or atoms) as 1.0 mole of any other substance.* For example, a bottle that contains 18.0 grams (1.0 mole) of water and another bottle that contains 98.1 grams (1.0 mole) of sulfuric acid (H_2SO_4), each contains 6×10^{23} molecules. The number 6×10^{23} is called *Avogadro's number* in honor of the Italian Amadeo Avogadro, an early worker in the study of atomic theory.

1.0 mole of a substance contains 6×10^{23} molecules

The fact that a mole of any substance contains the same number of particles is of extreme importance in chemistry. We cannot work with individual atoms or molecules because they are too small; however, we can work with moles. The concept of a mole provides the bridge between our representations of molecules on paper and what we can actually do in the laboratory. For example, we know that two molecules of hydrogen (H_2) undergo reaction with one molecule of oxygen (O_2) to give two molecules of water (H_2O). But, if that is the case, two *moles* (12×10^{23} molecules) of hydrogen undergo reaction with one *mole* (6×10^{23} molecules) of oxygen to give two *moles* (12×10^{23} molecules) of water. (In other words, the number of molecules undergoing reaction has been multiplied up to the point at which a chemist can see them.)

Since we know that two moles of hydrogen weigh 4.0 grams and that one mole of oxygen weighs 32.0 grams, we can conclude that 4.0 grams of hydrogen will react completely with 32.0 grams of oxygen. Furthermore, we can predict that 36.0 grams of water will be produced. Indeed, if we carry out the reaction in the laboratory, 36.0 grams is the weight of the water that is formed.

$$2\,H_2 + O_2 \longrightarrow 2\,H_2O$$

$$2 \text{ moles } H_2 + 1 \text{ mole } O_2 \longrightarrow 2 \text{ moles } H_2O$$

$$4.0 \text{ g } H_2 + 32.0 \text{ g } O_2 \longrightarrow 36.0 \text{ g } H_2O$$

ELECTRONIC STRUCTURE OF THE ATOM

When we talk about mass and mass relationships among the atoms, we are talking about the nuclei of the atoms since the mass of an atom is due primarily to the mass of the nucleus. However, an entire atom is a nucleus surrounded by *electrons*. Even though the electrons contribute very little to the mass of the atom, it is these electrons that control the chemical properties of the elements. Consequently, we will consider the electrons in some detail.

SIZES AND DISTANCES WITHIN AN ATOM. Although the nucleus is heavy, it is very small in comparison to the overall volume of space occupied by the atom. It has been calculated that the nucleus is only 1/10,000 to 1/100,000 the size of the atom as a whole, which is only one to two Angstroms in diameter. To give you an idea of this size relationship, consider the nucleus of an atom enlarged to the size of a bee hive. We will consider the electrons as the bees. In our enlarged example, the electron bees range two to twenty miles away from the nuclear hive. Both in our enlarged atom and in a real atom as well, most of the volume occupied by the atom is empty space.

ELECTRON SHELLS. The electrons of an atom are not scattered at random around the nucleus, nor are the electrons free to wander as they please. Rather, the electrons

Figure 3–3. Two-dimensional diagram of the first three shells in which electrons are found around a nucleus. In three dimensions, the shells might be compared to balloons within balloons.

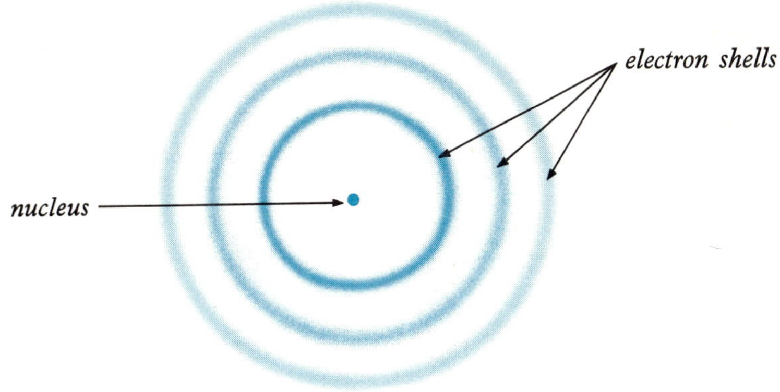

are restricted to certain specified regions of space, called *shells,* around the nucleus. The different shells in which the electrons are found are referred to as the first shell (which is closest to the nucleus), the second shell, (which is slightly farther away), up to the seventh shell (which is the farthest from the nucleus). (See Figure 3–3.)

Attractions between oppositely charged particles, like gravitational attractions, are stronger when the particles are closer together and decrease with increasing distance. Therefore, electrons in the first shell are more strongly attracted by the positively charged nucleus than electrons that are farther away. The electrons in the first shell are more difficult to remove from the atom. We say that the electrons in the first shell have *lower energy* and are at the *first energy level.*

The second shell is farther away from the nucleus than the first. The electrons in the second shell are easier to remove from the atom than electrons in the first shell. The electrons in the second shell are said to be at the *second energy level* and are of slightly higher energy than electrons in the first shell. The third shell is farther out than the second and is of higher energy than the second. Each succeeding shell through the seventh is of successively higher energy.

Each of the different shells can accommodate a limited number of electrons. The first shell can hold no more than two electrons. When there are three electrons around a nucleus, two of the electrons are in the first shell and the third is in the second shell. The second shell can have no more than eight electrons; the third, 18; and the fourth, 32. (See Table 3–2.) The reasons for these limits will become apparent shortly.

FILLING THE SHELLS. Because the positively charged nucleus attracts the negatively charged electrons as closely as possible, electrons seek the closest shell and the

Table 3–2. Maximum Number of Electrons in the First Four Shells

Shell	Number of Electrons
1	2
2	8
3	18
4	32

lowest energy level possible. As the number of protons is increased in the nucleus, the closest shells fill with electrons first, then the farther shells fill in succession. Table 3–3 shows the electron structures of some of the elements. You can see how the shells are filled when the nucleus contains an increasing number of protons.

Table 3–3. Electronic Structures of Some Elements

		Number of Electrons in Each Shell		
Element	Atomic Number	1st Shell	2nd Shell	3rd Shell
Hydrogen (H)	1	1	—	—
Helium (He)	2	2	—	—
Lithium (Li)	3	2	1	—
Beryllium (Be)	4	2	2	—
Boron (B)	5	2	3	—
Carbon (C)	6	2	4	—
Nitrogen (N)	7	2	5	—
Oxygen (O)	8	2	6	—
Fluorine (F)	9	2	7	—
Neon (Ne)	10	2	8	—
Sodium (Na)	11	2	8	1
Magnesium (Mg)	12	2	8	2

ATOMIC RADII. Atoms of different elements vary in size. The term we use to compare sizes of atoms is the *atomic radius,* which is defined as the distance from the center of the nucleus to the outermost electrons.

The radius of an atom is determined, in part, by the number of shells containing electrons. The greater the number of shells, the greater is the atomic radius. As we go down any group in the Periodic Table, we are in effect adding a shell, and we observe that the atomic radii increase. For example, the atomic radius of lithium is 1.225 Angstroms, while that of sodium is 1.572 Angstroms. (See Table 3–4.)

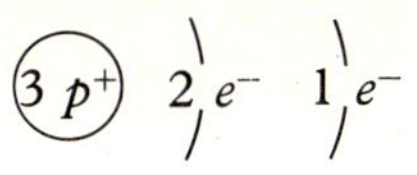

lithium (Li)

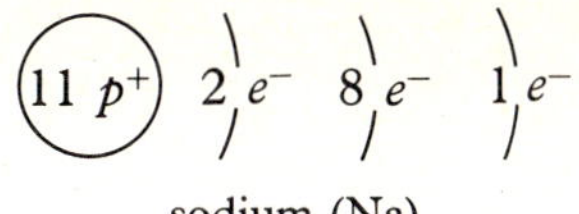

sodium (Na)

Table 3-4. Atomic Radii for Some Elements[a]

Atomic Radii (Å) of Group IA Elements		Atomic Radii (Å) of Group VIB Elements		Atomic Radii (Å) of Group VIIB Elements	
Li	1.225	O	0.74	F	0.72
Na	1.572	S	1.04	Cl	0.994
K	2.025	Se	1.17	Br	1.142
Rb	2.16	Te	1.37	I	1.334
Cs	2.35	Po	1.53		

[a] Values taken from Therald Moeller, *Inorganic Chemistry,* John Wiley & Sons, Inc., New York, 1952.

Besides the number of shells present, the atomic radius is affected by the attraction of the protons in the nucleus for the electrons around the nucleus. If the attraction is *greater,* the nucleus pulls more strongly on its electrons and holds them closer; the atomic radius is smaller.

A nucleus with a greater positive charge has a greater attraction for electrons than a nucleus with a lesser charge. Therefore, the number of protons in the nucleus is a major factor in the amount of attraction the nucleus has for its electrons. The greater the number of protons, the greater is the pull on all the electrons, including the outermost electrons. As we go from left to right in any period of the Periodic Table, we do not add any more shells, but we do add one proton to the nucleus (and one more electron to the outer shell). We observe a *decrease* in the atomic radii. Thus, oxygen with eight protons in its nucleus (and eight electrons) has an atomic radius of 0.74 Angstroms, while fluorine, with nine protons (and nine electrons) has an atomic radius of 0.72 Angstroms.

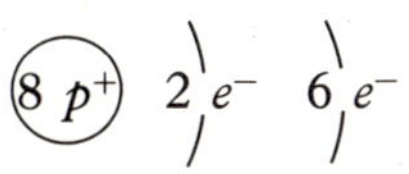

oxygen (O)

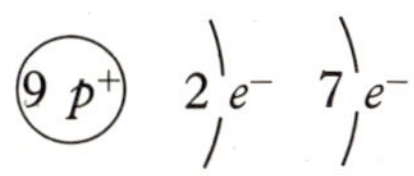

fluorine (F)

In summary, we can say:

Atomic radii *increase* as we go down any one group in the Periodic Table because of added shells.

Atomic radii *decrease* as we go across any one row in the Periodic Table because of the increasing number of protons.

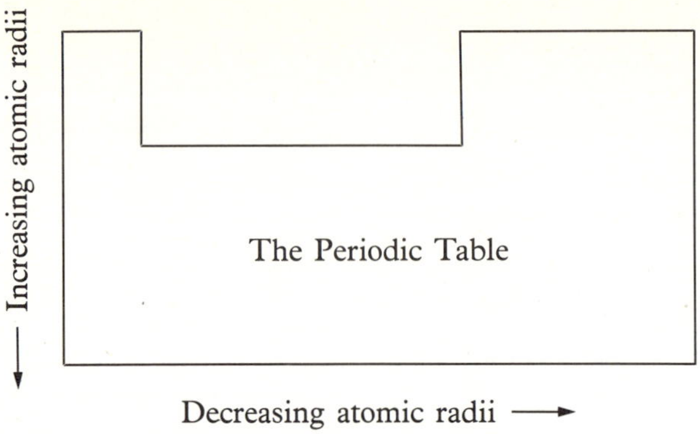

ELECTRONEGATIVITY. Electrons are held by their respective nuclei because of the electrical attraction between the protons and the electrons. However, any electrical attraction *decreases* with increasing distance between the positive and negative charges. Thus, atoms with smaller atomic radii have a greater hold on the outer electrons. Atoms with larger atomic radii have a looser hold on the outer electrons.

In order to discuss the bonds between atoms with different attractions for their electrons, we need a *measure* of the ability of an atom to attract electrons. This measure is called the *electronegativity* of the element. An element with atoms that have

Figure 3–4. Electronegativities of the elements. The elements that are in the more highly shaded areas are the more electronegative. Adapted from Garth L. Lee, Harry O. Van Orden, and Ronald O. Ragsdale, *General and Organic Chemistry,* W. B. Saunders Co., Philadelphia, 1971, p. 71.

1 H 2.1																		2 He
3 Li 1.0	4 Be 1.5												5 B 2.0	6 C 2.5	7 N 3.0	8 O 3.5	9 F 4.0	10 Ne
11 Na 0.9	12 Mg 1.2												13 Al 1.5	14 Si 1.8	15 P 2.1	16 S 2.5	17 Cl 3.0	18 Ar
19 K 0.8	20 Ca 1.0	21 Sc 1.3		22 Ti 1.5	23 V 1.6	24 Cr 1.6	25 Mn 1.5	26 Fe 1.8	27 Co 1.8	28 Ni 1.8	29 Cu 1.9	30 Zn 1.6	31 Ga 1.6	32 Ge 1.8	33 As 2.0	34 Se 2.4	35 Br 2.8	36 Kr
37 Rb 0.8	38 Sr 1.0	39 Y 1.3		40 Zr 1.4	41 Nb 1.6	42 Mo 1.8	43 Tc 1.9	44 Ru 2.2	45 Rh 2.2	46 Pd 2.2	47 Ag 1.9	48 Cd 1.7	49 In 1.7	50 Sn 1.8	51 Sb 1.9	52 Te 2.1	53 I 2.5	54 Xe
55 Cs 0.7	56 Ba 0.9	57 La 1.1	58 71 Ce→Lu	72 Hf 1.3	73 Ta 1.5	74 W 1.7	75 Re 1.9	76 Os 2.2	77 Ir 2.2	78 Pt 2.2	79 Au 2.4	80 Hg 1.9	81 Tl 1.8	82 Pb 1.8	83 Bi 1.9	84 Po 2.0	85 At 2.2	86 Rn
87 Fr 0.7	88 Ra 0.9	89 Ac 1.1	90 103 Th→Lr	104 Ku	105 Ha													

a greater ability to attract their outermost electrons is said to have a greater electronegativity. The smaller fluorine atom has a greater electronegativity than the larger chlorine atom.

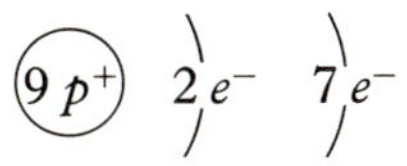

fluorine (F)

Smaller atomic radius; greater electronegativity.

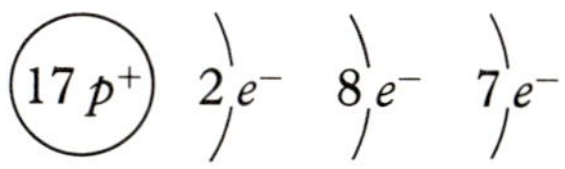

chlorine (Cl)

Larger atomic radius; lesser electronegativity.

The elements have been rated according to their electronegativities. Figure 3–4 shows the elements in the Periodic Table with numerical values (underneath the atomic symbols) given for their relative electronegativities. These electronegativity values are based upon a scale that gives lithium (Li) a value of 1.0 (low electronegativity) and fluorine (F) a value of 4.0 (high electronegativity). The other elements are rated in comparison with these two.

ATOMIC ORBITALS. Each of the shells around a nucleus is composed of *orbitals*. An orbital is defined as the *region in space around the nucleus in which a particular electron is likely to be found.*

The first shell contains only one orbital, called the 1*s* orbital. The number 1 refers to the first shell and the letter *s* refers to a spherical orbital with the nucleus in the center.

The hydrogen atom has only one electron, which is found somewhere in the 1*s* orbital. Usually, we think of the one electron, not as a discrete particle, but as a fuzzy cloud of electronic charge, as shown in Figure 3–5.

Figure 3–5. The 1*s* atomic orbital of a hydrogen atom.

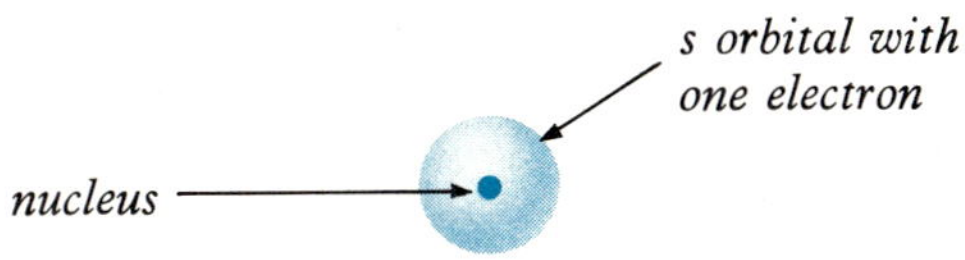

The second shell also contains one *s* orbital, called the 2*s* orbital. Like the 1*s* orbital, the 2*s* orbital is a fuzzy sphere. Because the 2*s* orbital is farther from the nucleus than the 1*s* orbital, it surrounds the 1*s* orbital. (See Figure 3–6.)

The second shell also contains three *p* orbitals, called 2*p* orbitals, that are of slightly higher energy than the 2*s* orbital. Thus, the second shell contains a total of *four* orbitals: 2*s*, 2*p*, 2*p*, and 2*p*.

Figure 3-6. The spherical 2*s* orbital.

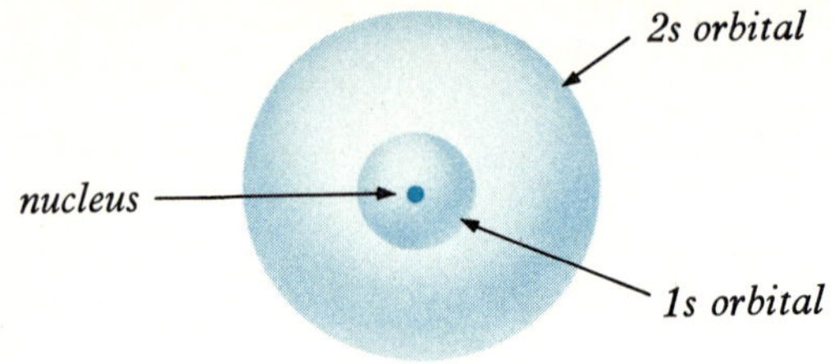

A *p* orbital (Figure 3-7) is not spherical. Rather, it is shaped like a dumbbell. A *p* orbital has two lobes with the nucleus in the center. The most likely place to find a 2*p* electron is in this dumbbell-shaped region in space.

Figure 3-7. The 2*p* orbitals.

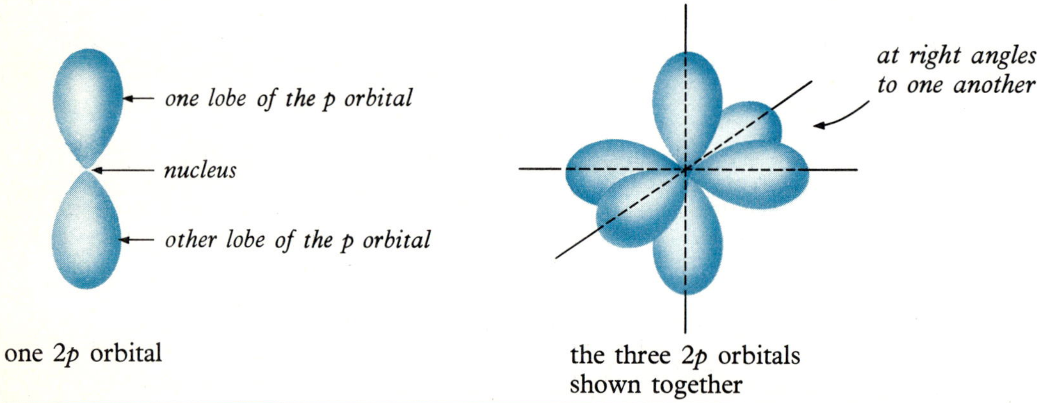

The third shell has one 3*s* orbital, three 3*p* orbitals, and five 3*d* orbitals, for a total of nine orbitals. The fourth shell has sixteen orbitals: one 4*s*, three 4*p*, five 4*d*, and seven 4*f* orbitals. Larger shells contain even greater numbers of orbitals. However, most of the chemistry that we will be discussing is concerned with the first two shells only; therefore, we will have very little to say about orbitals other than the *s* and *p* orbitals. Table 3-5 lists the orbitals in the first three shells.

Electrons spin around their axes, much like a child's spinning top. An electron can spin in a clockwise or counterclockwise direction. We say that these electrons have *opposite spin*. Each orbital can contain a maximum of two electrons, and these two electrons must be of opposite spin.

Table 3-5. Atomic Orbitals in the First Three Shells

Shell	Orbitals
1	1*s*
2	2*s* 2*p*, 2*p*, 2*p*
3	3*s* 3*p*, 3*p*, 3*p* 3*d*, 3*d*, 3*d*, 3*d*, 3*d*

two electrons of opposite spin

Because of the pairing of electrons in orbitals, the first shell, with only a 1*s* orbital, can hold up to two electrons. The second shell, with one 2*s* orbital and three 2*p* orbitals, can hold up to eight electrons. (These maximum numbers of electrons are the same as those listed earlier in Table 3-2.)

First shell (one orbital): maximum of 2 e^-

Second shell (four orbitals): maximum of 8 e^-

Third shell (nine orbitals): maximum of 18 e^-

Electron spin is often represented by arrows. An up-pointing arrow (↑) shows a spin that is opposite from that shown by a down-pointing arrow (↓). If we represent an orbital by a box, we can then represent (1) an orbital that contains no electrons (no arrows in the box), (2) an orbital that contains one electron (one arrow in the box), and (3) an orbital that is filled (an up-pointing arrow and a down-pointing arrow in the box).

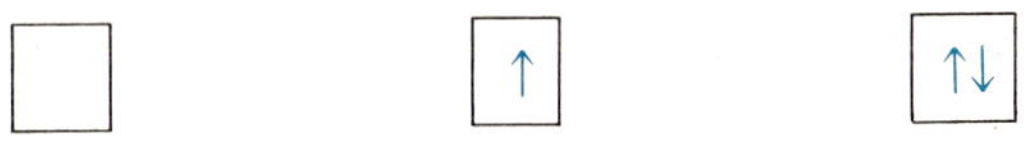

An empty orbital—an orbital with no electrons in it. *An orbital with one electron.* *A filled orbital—an orbital with two electrons of opposite spin.*

FILLING THE ORBITALS. We have mentioned the energy differences between shells and between the orbitals within a shell. Now we will consider how the orbitals

Figure 3–8. Orbitals of the first four shells.

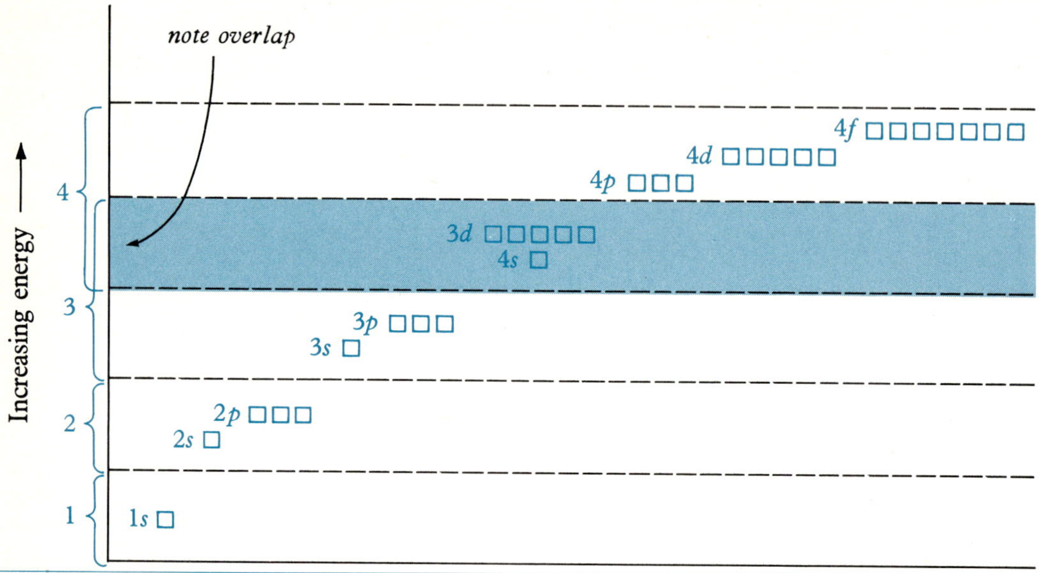

fill with electrons in relation to these energy differences. To do this, we must picture successive filling of the orbitals with electrons as the number of protons in the nucleus increases. We must remember that electrons tend to go to the lowest energy level possible, which is the orbital closest to the nucleus. We must also keep in mind that one orbital can hold a maximum of two electrons.

Figure 3–8 shows the relative energies of the orbitals. The diagram shows that the first shell has one orbital, the 1*s* orbital. This orbital is the lowest energy orbital available. The next lowest energy orbital is in the second shell; this is the 2*s* orbital. Of slightly higher energy than the 2*s* orbitals are the next available orbitals, the three 2*p* orbitals. These three orbitals are of *equal energy*; that is, electrons that occupy these 2*p* orbitals all have the same energy.

What happens when we fill these orbitals with electrons? A neutral atom of hydrogen has one proton in the nucleus and, therefore, one electron outside the nucleus. This electron goes into the lowest energy orbital, the 1*s* orbital. A helium atom (He) has two electrons; both go into the 1*s* orbital. Now the 1*s* orbital is filled. Lithium (Li) has three electrons; two fill the 1*s* orbital, the third occupies the 2*s* orbital, which is the next lowest orbital, and so on until we get to neon (Ne). Neon has ten electrons: two in the 1*s* orbital, two in the 2*s* orbital, and two in each of the 2*p* orbitals. In neon, the first and second shells are completely filled; any more electrons must go into the third shell.

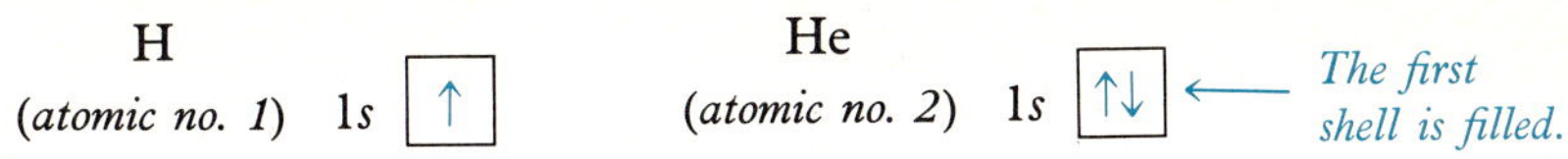

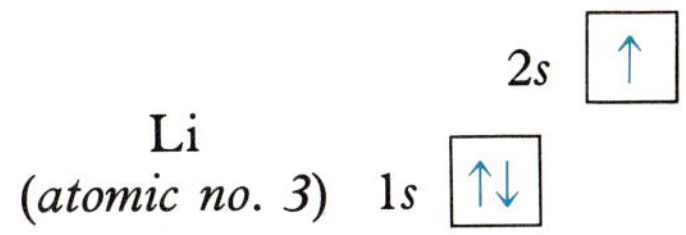

Sample Problem: Fill the orbital boxes for neon, as we have just done for H, He, and Li.

Solution:

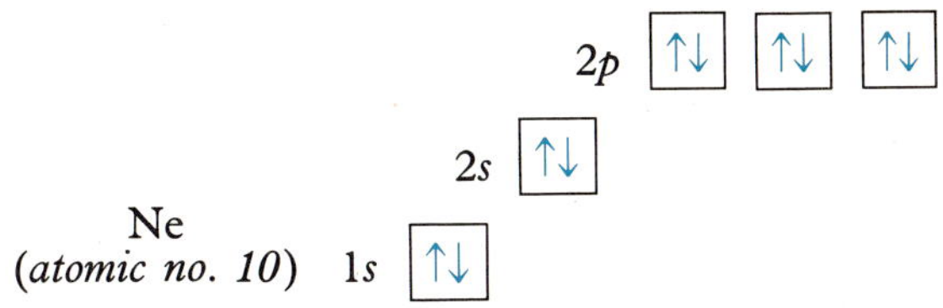

Consider the orbitals of the third and fourth shells in Figure 3-8. The third shell has three types of orbital: 3*s*, 3*p*, and 3*d*. The third shell can hold up to 18 electrons since it has a total of nine orbitals. The fourth shell has a total of 16 orbitals and can accomodate up to 32 electrons.

As the shells get farther from the nucleus, their energies start to *overlap* each other. Note that the 4*s* orbital has approximately the same energy as the 3*d* orbitals. The 4*s* orbital usually fills with electrons before the 3*d* orbitals. The electron structure of potassium (K) shows this.

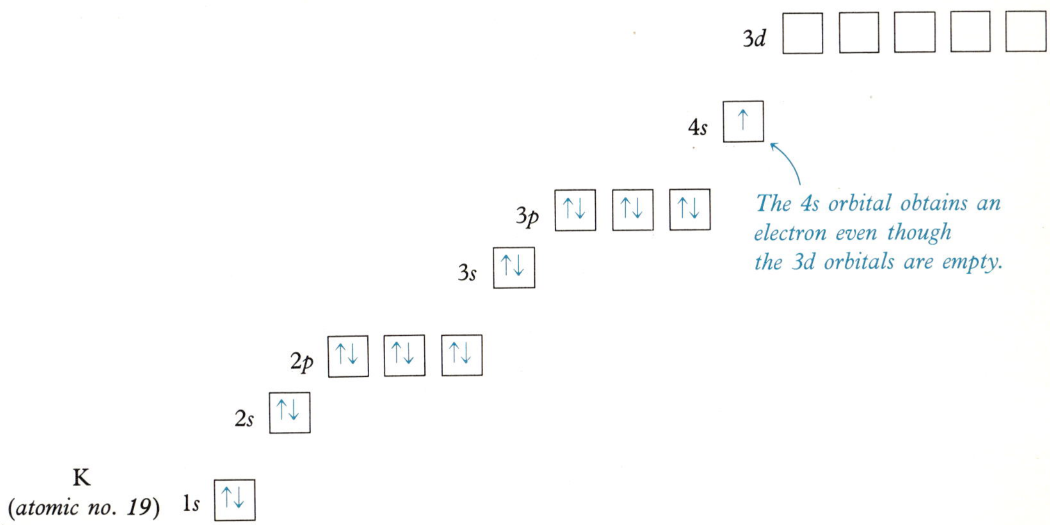

THE PERIODIC TABLE REVISITED

The outermost electrons of an atom govern its chemical behavior. A configuration of filled shells is stable compared to a configuration in which the outermost shell is unfilled. As a result, there is a driving force to have each orbital in a shell filled with paired electrons. Atoms lose, gain, or share electrons in order to attain this stable arrangement of filled shells. Atoms which already have only filled shells are unusually stable, and are therefore chemically unreactive.

A quick pass through the elements in the Periodic Table shows us why the elements in each group exhibit similar chemical behavior. Starting with hydrogen, we find one proton in the nucleus and one electron, which is in the 1*s* orbital. Because of the instability of a lone electron, a hydrogen atom tends to lose this electron and become a hydrogen ion.

$$\underset{\text{hydrogen atom}}{H} \longrightarrow \underset{\text{hydrogen ion}}{H^+} + e^-$$

The next element after hydrogen is helium (He), which has two protons and two electrons. Both of these electrons are in the 1*s* orbital, and the first shell is filled. Helium, then, is chemically inert and is the first of the group called *inert gases*.

The next element in line has an atomic number of 3. This is lithium (Li), the first member of the group called the *alkali metals*. Two of the three electrons of lithium fill the first shell, and the third electron occupies the 2*s* orbital. If a lithium atom could lose this third electron, then it would have a filled outermost shell (the first shell). In chemical reactions, we observe that the main characteristic of lithium is to lose an electron and become a cation.

Group IA:

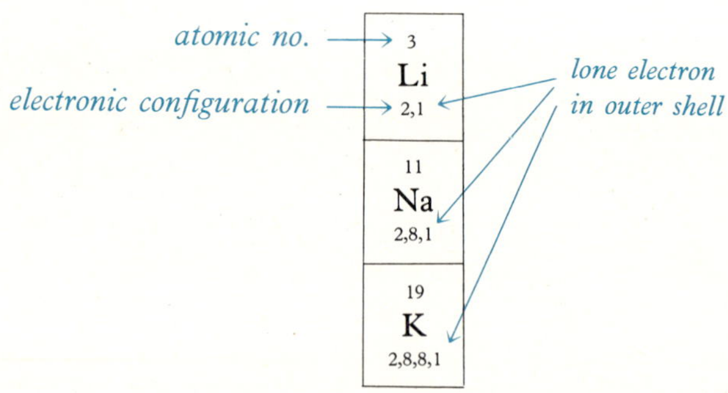

In the preceding diagram, we can see that each of the alkali metals has one lone electron in the outer shell. Also, recall that these elements have low electronegativities—that is, they have small attractions for their outermost electrons. Like lithium, sodium (Na) tends to lose its lone electron in the outer shell. Potassium (K) reacts in a similar fashion. The characteristic chemistry for each of the elements in this group is the loss of one electron and the formation of a cation.

$$\underset{\text{lithium}}{Li} \longrightarrow \underset{\text{lithium ion}}{Li^+} + e^-$$

$$\underset{\text{sodium}}{Na} \longrightarrow \underset{\text{sodium ion}}{Na^+} + e^-$$

$$\underset{\text{potassium}}{K} \longrightarrow \underset{\text{potassium ion}}{K^+} + e^-$$

Beryllium (Be), the element after lithium, has the atomic number 4, and is the first member of the *alkaline earth metals.* It has four electrons, two of which fill the first shell and two more that are in the 2*s* orbital. The other elements in the alkaline earth group have the same type of electronic configuration. Magnesium (Mg) has two lone electrons in the 3*s* orbital; calcium (Ca) has two in the 4*s*; and strontium (Sr) has two in the 5*s*.

Group IIA:

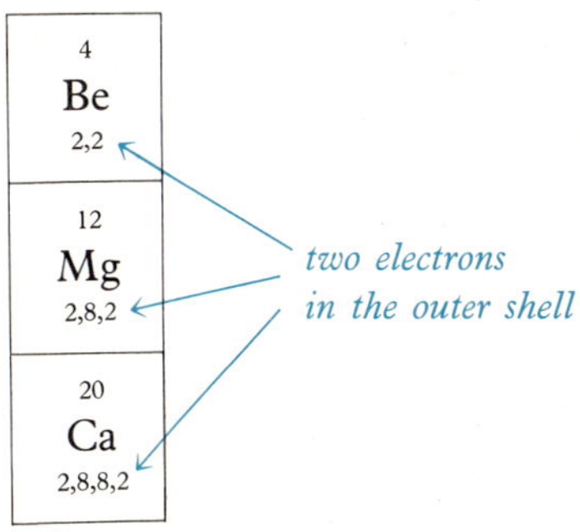

All members of the alkaline earth family behave in the same general way. That is, they tend to lose the two electrons in the outermost unfilled shell.

$$\underset{\text{magnesium}}{Mg} \longrightarrow \underset{\text{magnesium ion}}{Mg^{2+}} + 2\,e^-$$

$$\underset{\text{calcium}}{Ca} \longrightarrow \underset{\text{calcium ion}}{Ca^{2+}} + 2\,e^-$$

$$\underset{\text{strontium}}{Sr} \longrightarrow \underset{\text{strontium ion}}{Sr^{2+}} + 2\,e^-$$

Leaving beryllium and its relatives, we add more protons and electrons to move across the Periodic Table to the right. Boron (B) has three electrons in its outermost shell, carbon (C) has four electrons in its outer shell, nitrogen (N) has five electrons in its outer shell, and oxygen (O) has six electrons in its outer shell.

IA	IIA	IIIA	IVA	VA	VIA	VIIA	0
3 Li 2,1	4 Be 2,2	5 B 2,3	6 C 2,4	7 N 2,5	8 O 2,6	9 F 2,7	10 Ne 2,8

The energy required for an atom to lose three or four electrons (instead of just one or two) to gain a filled outer shell is high. We will discuss the reasons for this in the next chapter. Boron, carbon, and nitrogen do not tend to lose electrons. Rather, they tend to bond to other atoms by *sharing* electrons to attain a filled outer shell.

Next to nitrogen in the Periodic Table is oxygen, which has six electrons in the outer shell. Oxygen also tends to share electrons, but it has another choice. Oxygen has six electrons in an outer shell that would be filled if it had eight electrons. Oxygen also has a high electronegativity; it has a great attraction for electrons. Oxygen can *pick up* two electrons from another atom to fill this outer shell. In so doing, an oxygen atom becomes an *anion.*

$$\underset{\text{oxygen}}{O} + 2e^- \longrightarrow \underset{\text{oxide ion}}{O^{2-}}$$

Proceeding on to fluorine (F), we come to an atom that has nine electrons: two in the first shell and seven in the second. Fluorine needs only one electron to fill out the second shell with a full complement of eight electrons.

$$\underset{\text{fluorine}}{F} + e^- \longrightarrow \underset{\text{fluoride ion}}{F^-}$$

Group VIIA:

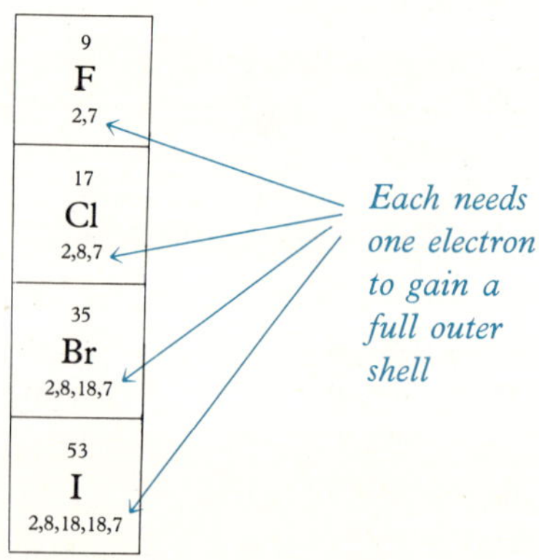

Fluorine is the first member of the family of the halogens. The other halogens, chlorine (Cl), bromine (Br), and iodine (I) have similar types of electronic configurations. Each of these halogens needs one electron to fill the outer shell, and each can form an anion. We will find that the halogens can also fill their outer shells by sharing electrons.

$$\underset{\text{chlorine}}{Cl} + e^- \longrightarrow \underset{\text{chloride ion}}{Cl^-}$$

$$\underset{\text{bromine}}{Br} + e^- \longrightarrow \underset{\text{bromide ion}}{Br^-}$$

$$\underset{\text{iodine}}{I} + e^- \longrightarrow \underset{\text{iodide ion}}{I^-}$$

The end of the second period in the Periodic Table is neon (Ne). Neon has ten electrons, a sufficient number to fill the first and second shells completely. Like helium, neon is chemically inert.

The third row in the Periodic Table begins another cycle with sodium (Na) with 11 electrons and only one in its outer shell, just like lithium. Now we begin to see the periodicity of the elements—*why* certain elements tend to behave alike; neon like helium, sodium like lithium, and chlorine like fluorine. The periodicity depends on the number of electrons in the outer shells.

STUDY QUESTIONS

3-1. Identify each of the following as an atomic symbol or as a formula:
(a) C_6H_6 (b) H_2 (c) Ne (d) CO

3-2. Write the name for each of the following elements:
(a) Na (b) K (c) Fe (d) P

3-3. Label each of the following phrases as applicable to a neutron, a proton, or neither of these:
(a) is found in the nucleus of an atom
(b) has a negative electrical charge
(c) has no electrical charge
(d) has a mass of 1 amu

3-4. Refer to the Periodic Table and give (1) the number of protons in the nucleus, (2) the atomic number, and (3) the atomic mass of each of the following elements:
(a) nitrogen (b) aluminum
(c) carbon (d) copper

3-5. A pair of isotopes of an element have the same:
(a) atomic weight
(b) atomic number
(c) number of electrons
(d) number of protons

3-6. Refer to the Periodic Table, and tell how many electrons surround the nucleus of each of the following ions:
(a) Fe^{3+} (b) O^{2-} (c) K^+ (d) Cl^-

3-7. Identify each of the following reactions as oxidation or reduction:
(a) $Cu^{2+} + 2\,e^- \longrightarrow Cu$
(b) $Cl_2 + 2\,e^- \longrightarrow 2\,Cl^-$
(c) $Fe \longrightarrow Fe^{2+} + 2\,e^-$
(d) $Cr \longrightarrow Cr^{3+} + 3\,e^-$

3-8. Identify each of the following particles as an anion, a cation, or a neutral atom.
(a) H^- (b) K^+ (c) Na (d) S^{2-}

3-9. The atomic mass of nitrogen as shown in the Periodic Table is 14.007. From this number we might deduce that:
(a) most naturally occurring nitrogen atoms have an atomic mass of 14.000 amu
(b) not all naturally occurring nitrogen atoms have an atomic mass of 14.000 amu
(c) the mass of the electrons in a nitrogen atom is 0.007 amu
(d) nitrogen has a naturally occurring isotope that is heavier than 14.000

3-10. The molecular weight of methane (CH_4) is ______.

3-11. One mole of iron (Fe) is:
(a) 1.0 g of iron
(b) 6×10^{23} atoms of iron
(c) 55.8 amu
(d) 55.8 g of iron

3-12. What is the weight of two moles of sodium hydroxide (NaOH)?

3-13. How many shells in each of the following atoms or ions contain electrons?
(a) H^+ (b) Na (c) Na^+ (d) C

3-14. Which has the smaller atomic radius?
(a) Ca or Co (b) Ni or Pt

3-15. Refer to Figure 3-4 (page 50) and rank each of the following sets of atoms in order of increasing electronegativity. (Start with the least electronegative and proceed to the more electronegative.)
(a) O, S, Se, Te
(b) Na, Mg, Al, P
(c) F, Cl, Br, I
(d) Cu, Fe, Ca, K

3-16. The first electron shell can hold a maximum of ______ electrons.

3-17. The second shell can hold a maximum of ______ electrons.

3-18. The second shell contains ______ orbitals. (There may be more than one correct answer.)
(a) two (b) three (c) four (d) *s* and *p*

3-19. A 2*s* atomic orbital:
(a) is a region in space in which electrons may be found
(b) is in the first energy level
(c) can hold a maximum of eight electrons
(d) may, in some cases, contain only one electron

3-20. Using orbital diagrams similar to those on page 55, draw electronic configurations of:
(a) carbon (b) nitrogen (c) oxygen

3-21. Which of the following elements would be least likely to form a cation? Li, Mg, Al, Cl

3-22. Which of the following elements would be least likely to form an anion? Mg, O, S, Br

CHAPTER 4 HOW ATOMS ARE JOINED TOGETHER

OBJECTIVES

Types of Bond *List and define the three principal types of chemical bond.*

Ionic Bonds *Compare the groups in the Periodic Table in terms of the abilities of their members to form cations or anions. * Write a chemical equation showing the formation of an ionic compound from the elements, and describe the fate of the outermost electrons in the atoms involved. * Define an ionic compound and describe its structure in the solid state and in water solution. * Define oxidation number and demonstrate how it can be determined for the cation in an ionic compound such as* $CuCl_2$ *or* Fe_2O_3.

Covalent Bonds *Describe how two atoms such as fluorine atoms can attain filled outer shells by forming a covalent bond.*

Orbital Theory of Covalent Bonds *Define and exemplify a sigma bond. * Describe the three ways that a carbon atom can hybridize its orbitals. Describe the orientation of the orbitals in each type of hybridization, and give examples of compounds that illustrate each type. * Contrast a structural formula, a condensed structural formula, and a dimensional formula. (Use examples.) * Define and exemplify a pi bond. * Describe the aromatic pi cloud of benzene. * Draw the resonance struc-*

tures of benzene. * *Describe the sp³ hybridization of a nitrogen atom and an oxygen atom, and show the orbital structures of ammonia, water, and the hydroxide ion.*

Polar Covalent Bonds *Differentiate between a polar covalent bond and a nonpolar covalent bond. Give examples of each.*

A compound is composed of molecules, which are atoms joined together by *chemical bonds*. There are an almost infinite number of compounds that can be formed from the elements. The variety of ways in which atoms can be bonded together gives rise to the variety of compounds. Before we can gain an understanding of how molecules can change their bonds and undergo chemical reactions, we must first have an understanding of the chemical bonds within the molecules. In other words, we must answer the question, "How are atoms joined together within a molecule?"

TYPES OF BOND

Since the inner shells fill with electrons first and these inner electrons are tightly held, bonding involves only electrons in the outermost shell of an atom. An atom can achieve a full shell by gaining one or more electrons to fill partially filled orbitals ($Cl + e^- \longrightarrow Cl^-$), by losing one or more electrons to empty partially filled orbitals ($Na \longrightarrow Na^+ + e^-$), or by sharing electrons ($H\cdot + H\cdot \longrightarrow H:H$).

There are three principal types of chemical bond: (1) ionic bonds, (2) covalent bonds, and (3) metallic bonds. *Ionic bonds* are formed by attractions between oppositely charged ions. *Covalent bonds* are formed by the sharing of electrons by atoms. Most of the chemistry in this text will be involved with these two types of bond. The *metallic bond*, which plays an important role in the properties of metallic substances, is formed by attractions between free electrons and metal cations. Since the metallic bond is limited to metals, we will discuss it only briefly.

Ionic bond:	*Covalent bond:*	*Metallic bond:*
$Na^+ Cl^-$	$H:H$	$e^- Na^+ e^- Na^+ e^- Na^+$
electrostatic attraction	*sharing of two* e^-	*electrostatic attractions*

IONIC BONDS

CATION FORMATION. To remove an electron from the outer shell of an atom requires a certain amount of energy—enough energy to break the attraction between the nucleus and the electron. An atom with a low electronegativity (Na, for example) holds its outer electron or electrons more loosely than does an atom with a high electronegativity (F, for example). Thus, less energy is required to remove an electron from a sodium atom. It is easy for Na to lose an electron and become Na^+. It is not likely that F would lose seven electrons and become F^{7+}. Magnesium (Mg), the element next to sodium in the Periodic Table, is slightly more electronegative than sodium (see Figure 3-4). Magnesium can lose its two outer electrons to become the magnesium ion (Mg^{2+}), but not quite as easily as Na.

$$Na \xrightarrow{\text{easy}} Na^+ + e^- \qquad Mg \xrightarrow{\text{less easy}} Mg^{2+} + 2\,e^-$$

$$F \not\rightarrow F^{7+} + 7\,e^- \quad \text{(F does not form a } F^{7+} \text{ ion)}$$

In chemical reactions, we observe that the elements in Group IA (Li, Na, K) with their low electronegativities lose their outer electrons and form cations with ease. The alkaline earth metals (Be, Mg, Ca) in Group IIA also form cations, but not quite as readily as the Group IA elements. The halogens (F, Cl, Br, I) with their high electronegativities do not form cations. (See Figure 4-1 for a representation of the elements in the Periodic Table that undergo cation formation.)

Figure 4-1. Elements that will form cations.

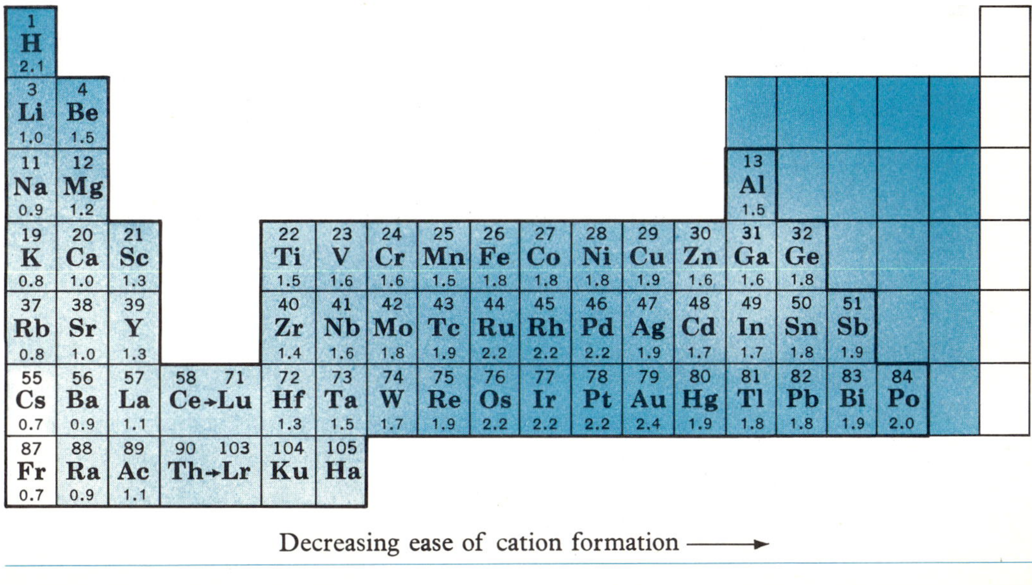

Decreasing ease of cation formation ⟶

With the exception of hydrogen, the elements that form cations are all *metals* in their elemental states. Metals are distinguished from nonmetals by their luster, their hardness, and their ability to conduct electricity. These properties are due to the way in which metal atoms are bonded together in the elemental state. The low electronegativities of metal atoms allow one of these atoms to give up its bonding electrons to a general pool of electrons. From an atomic viewpoint, a piece of metal is a closely packed group of metal cations awash with electrons. The bonding electrons in a metal are not tied to any one nucleus but can wander around from one atom to another. The strong electrical attractions between the positively charged cations and the negatively charged electrons hold the metal together and constitute the metallic bond.

The wandering electrons of the metallic bond enable metals to conduct electricity. A metal wire is electrically neutral. Within the wire, the positive charges of the metal cations and the negative charges of the bonding electrons are equal to each other. If electrons are added to one end of a wire by an electric current, the wire does not become negatively charged. Instead, an equal number of electrons is pushed off the other end of the wire, and a flow of electrons occurs through the wire. This flow of electrons is an electrical current. (See Figure 4-2.)

Figure 4-2. An electric current passing through a silver (Ag) wire.

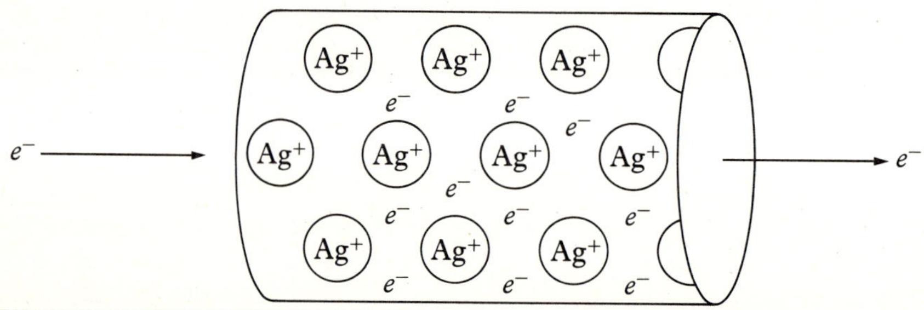

ANION FORMATION. The atoms of elements with high electronegativities have strong attractions for their outer electrons. The attraction of some kinds of atoms for electrons is strong enough that they tend to *gain* electrons and become anions. We say these elements have an *electron affinity*. The halogens, in Group VIIA, have high electronegativities and high electron affinities. The tendency in chemical reactions is for a halogen atom to gain one electron and form an anion. The elements in Group VIA (O and S) can gain two electrons and also become anions, though not as readily as the halogens. (See Figure 4-3.)

$$\underset{\text{chlorine}}{Cl} + e^- \xrightarrow{\text{easy}} \underset{\text{chloride ion}}{Cl^-}$$

$$\underset{\text{sulfur}}{S} + 2\,e^- \xrightarrow{\text{less easy}} \underset{\text{sulfide ion}}{S^{2-}}$$

$$\underset{\text{aluminum}}{Al} + 3\,e^- \not\longrightarrow Al^{3-} \quad \text{(Al does not form an anion.)}$$

Figure 4-3. Elements that will form anions.

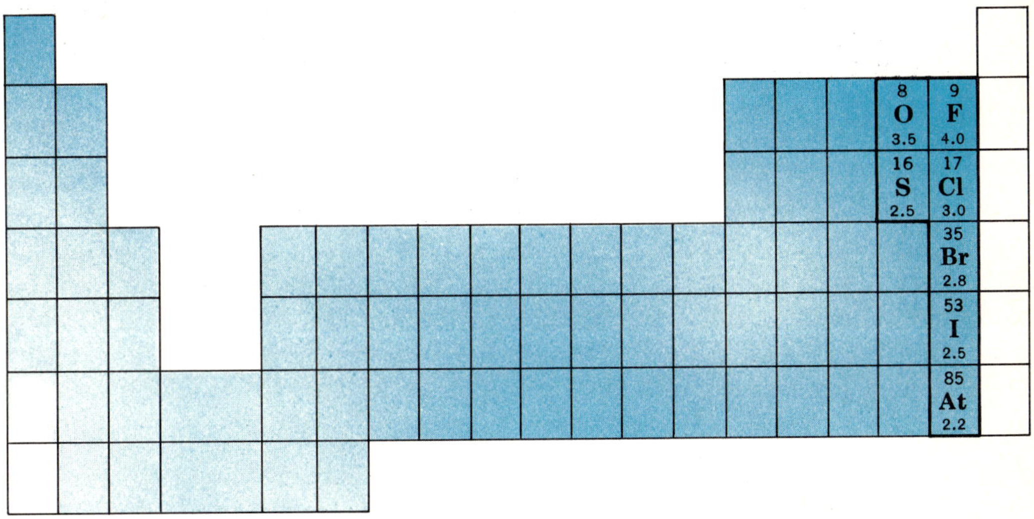

IONIC COMPOUNDS. An ionic bond is formed when there is a complete transfer of an electron from one atom to another. A classic example of a compound that has an ionic bond is sodium chloride (NaCl, or Na^+Cl^-), or common table salt. There are several ways we can represent the transfer of an electron from a neutral sodium atom to a neutral chlorine atom to give sodium chloride:

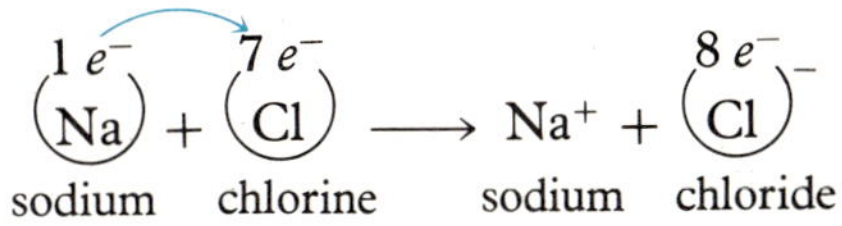

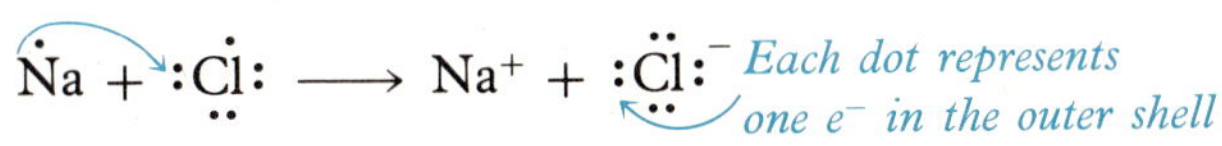

$$Na + Cl \longrightarrow Na^+ + Cl^-$$

All three preceding equations mean the same thing. An atom of sodium gives up an electron to an atom of chlorine to yield a positively charged sodium cation (Na^+) and a negatively charged chloride anion (Cl^-). The positively charged sodium ion is attracted to the negatively charged chloride ion by the attraction between opposite electrical charges. The attraction between a positive ion and a negative ion is called an *ionic bond.*

Most ionic compounds are crystalline solids at room temperature. A crystal of sodium chloride large enough to see, of course, must contain more than one pair of ions. In fact, there must be trillions upon trillions of ions present for us to see a crystal. In the crystal, the total number of sodium ions and the total number of chloride ions must be equal so that the positive and negative charges are balanced. There is no way to separate sodium chloride crystals into one pile of sodium ions and another pile of chloride ions. The amount of energy required to separate the opposite charges and then to bring together like charges would be enormous.

How are the huge number of ions arranged in a crystal? An ionic crystal is formed from positive and negative ions alternating in a three-dimensional crystal lattice (see Figure 4-4). In the crystal lattice of sodium chloride, each sodium ion is surrounded by chloride ions; and each chloride ion is surrounded by sodium ions. Each charged ion exerts an attraction toward all the oppositely charged ions surrounding it.

Figure 4-4. An ionic compound in the solid state.

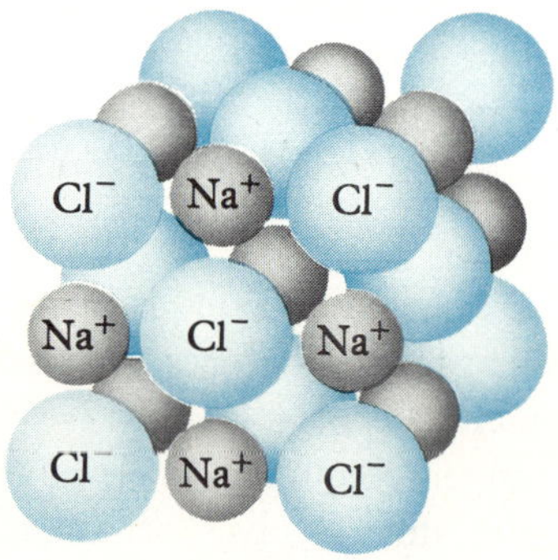

Because of this crystal lattice, ionic compounds tend to have high melting points when compared to covalent compounds of the same molecular weight. For example, sodium chloride melts at 804°C, while butane (C_4H_{10}), a covalently bonded compound, melts at −135°C.

When an ionic compound is dissolved in water, the cations (Na^+) and the anions (Cl^-) become separated from each other and become surrounded by water molecules.

Figure 4-5. An ionic compound dissolved in water.

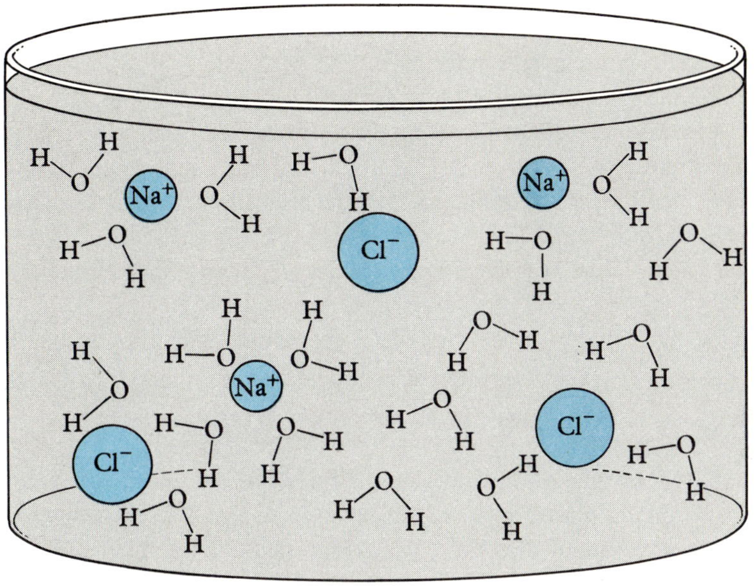

(See Figure 4-5.) There still must be equal numbers of cations and anions present in a solution.

Although ions are only attracted to each other and not truly attached, we still represent ionic compounds by formulas showing the atomic symbols in combination and in the proper proportions. In sodium chloride, the numbers of sodium ions and chloride ions are equal; therefore, we write its formula as NaCl, or Na^+ Cl^-. Calcium chloride must contain *two* chloride ions for each calcium ion in order for the compound to be electrically neutral. Thus, the formula for calcium chloride is $CaCl_2$, or $Ca^{2+} + 2\,Cl^-$.

NAMING IONIC COMPOUNDS. In the name of an ionic compound, the cation's name is given first, followed by the anion's name as a separate word. Generally, the cation is a hydrogen ion or a metal ion, and it goes by its atomic name. The name of the anion is derived from the atomic name, but the ending is changed to tell us that it is an anion. For example, chlorine forms a chlor*ide* ion (Cl^-); sulfur forms a sulf*ide* ion (S^{2-}); oxygen forms an ox*ide* ion (O^{2-}). The ending *-ide* is generally used when the anion contains only one element. (An exception is the hydroxide ion, OH^-.)

LiBr	CaO
lithium bromide	calcium oxide

OXIDATION NUMBERS. The amount of positive or negative charge a single atom is carrying is called the *oxidation number*. Sometimes the terms *oxidation state, oxidation level,* or *valence* are used; these terms mean the same as oxidation number.

A neutral atom has no positive or negative charge; therefore, a neutral atom has an oxidation number of zero. An ion with a single positive charge, such as Na^+, has an oxidation number of $+1$. An ion carrying two positive charges (Ca^{2+}) has an oxidation number of $+2$. An ion carrying one negative charge (Cl^-) has an oxidation number of -1.

The oxidation numbers of the atoms in a formula for a neutral compound must add up to zero. This is another way of saying that there must be equal numbers of positive and negative electrical charges in a neutral compound.

$$\begin{aligned} CaCl_2 \text{ contains } Ca^{2+} &+ 2\,Cl^- \\ \text{The sum of the oxidation numbers} &= +2 + 2(-1) \\ &= 0 \end{aligned}$$

Certain metals, notably the transition metals, can form ions that have different oxidation numbers. Iron, for example, is found in some compounds as Fe^{2+} and in other compounds as Fe^{3+}. When naming a compound of iron, we need to differentiate between the two possibilities. Iron chloride is an incorrect name because we cannot tell if $FeCl_2$ or $FeCl_3$ is the formula for the compound. There are two common ways of differentiating the different oxidation numbers in a name. One is by using a Roman numeral. Using this method, Fe^{2+} is called the iron(II) ion, and Fe^{3+} is called the iron(III) ion. The other method is the use of different suffixes in the names. For example, the name ferr*ic* is used for the Fe^{3+} ion, and ferr*ous* is used for the Fe^{2+} ion.

$FeCl_2$	$FeCl_3$
iron(II) chloride, or ferrous chloride	iron(III) chloride, or ferric chloride
(Iron has an oxidation number of +2.)	*(Iron has an oxidation number of +3.)*

Mercury can form compounds by losing either one electron or two electrons. In compounds in which mercury has lost one electron, mercury has an oxidation number of $+1$. In its compounds, two Hg^+ ions are joined together so that the cation is $Hg_2{}^{2+}$. The $Hg_2{}^{2+}$ ion can be called either the mercury(I) ion or the mercurous ion. In compounds in which mercury has lost two electrons, its oxidation number is $+2$. The Hg^{2+} ion can be called the mercury(II) ion or the mercuric ion.

Note that the *ic* ending is used for the ion of higher oxidation state and the *ous* ending is used for the lower oxidation state. (In the following two oxides of mercury, oxygen has an oxidation number of -2.)

Hg_2O	HgO
mercury(I) oxide, or mercurous oxide	mercury(II) oxide, or mercuric oxide

We can often determine the likely oxidation number for an element in compounds from its position in the Periodic Table. The halogens need one electron to attain a full outer shell and have an oxidation number of -1 in their simpler compounds. The alkali metals have one electron that is easily lost and have a usual oxidation number of $+1$ in compounds. For convenience, we have tabulated the oxidation numbers for some of the more common ions in Table 4-1.

Table 4-1. Oxidation Numbers for Some Elements Commonly Found in Ionic Compounds

Element	Usual Oxidation Number	Example of Compound	Name
Group IA:			
H	$+1$	HCl	hydrogen chloride
Na	$+1$	NaCl	sodium chloride
K	$+1$	KCl	potassium chloride
Group IIA:			
Mg	$+2$	$MgCl_2$	magnesium chloride
Ca	$+2$	$CaCl_2$	calcium chloride
Transition Metals:			
Fe	$+2$, $+3$	$FeCl_2$, $FeCl_3$	iron(II) chloride and iron(III) chloride
Ag	$+1$	AgCl	silver chloride
Group IIIA:			
Al	$+3$	$AlCl_3$	aluminum chloride
Group VIA:			
O	-2	Ag_2O	silver oxide
S	-2	Na_2S	sodium sulfide
Group VIIA:			
F	-1	NaF	sodium fluoride
Cl	-1	NaCl	sodium chloride
Br	-1	NaBr	sodium bromide
I	-1	NaI	sodium iodide

Sample Problem: What is the formula for copper(II) chloride?

Solution: Copper(II) is Cu^{2+}; its oxidation number is $+2$. Chloride is Cl^-; its oxidation number is -1. Two chloride ions are needed for each Cu^{2+} ion. The formula is $CuCl_2$.

Sample Problem: Name the compound with the formula Fe_2O_3.

Solution: The oxidation number of oxygen is -2. Three O^{2-} ions have a total charge of -6. Two iron ions must, therefore, have a charge of +6. Each iron ion has a charge of +3. Thus, the compound is named iron(III) oxide, or ferric oxide.

Sample Problem: (a) From the Periodic Table, predict the oxidation number of lithium (Li) in compounds. (b) Write the formula for lithium iodide.

Solution: (a) +1; (b) $Li^+ I^-$ or LiI

COVALENT BONDS

Carbon (C), nitrogen (N), and phosphorus (P) are three important elements that almost always form covalent bonds instead of ionic bonds. Oxygen, hydrogen, and the halogens also form covalent bonds readily.

A covalent bond is the sharing of two electrons by two atoms. We use a long dash (—) between two atomic symbols to represent the two electrons being shared by the two atoms (for example, H—H). Atoms joined by ionic bonds are free to leave each other, or dissociate, when they are in solution. With only a few exceptions, atoms joined by covalent bonds do not dissociate in water because the atoms are firmly attached to each other.

$$\underset{\text{hydrogen}}{\text{H—H}} \xrightarrow{H_2O} \text{no ion formation}$$

The distance between the nuclei of covalently bonded atoms is called the *bond distance.* When there are more than two atoms sharing electrons within a single molecule, the angle between two covalent bonds is called the *bond angle.* The bond distances and bond angles are more or less fixed within a given molecule. Depending upon the compound, bond angles vary from about 60° to 180°, and bond distances vary from about 0.9 Å to 1.6 Å.

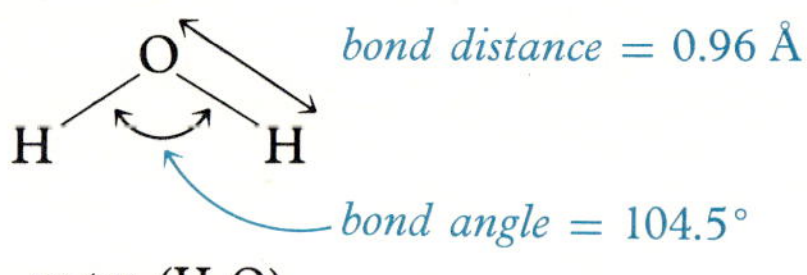

water (H_2O)

WHY ATOMS SHARE ELECTRONS. An atom with a completely filled (or completely empty) outer shell is stable. Many atoms can attain a filled outer shell by losing or gaining electrons. Other atoms must rely on sharing electrons.

Hydrogen, the simplest of atoms, forms covalent bonds by sharing its single electron with another atom. The hydrogen molecule (H_2) consists of two hydrogen atoms sharing their combined two electrons. By sharing two electrons, each hydrogen atom has gained a filled first shell. We can represent the sharing of electrons between two H atoms in H_2 in several ways.

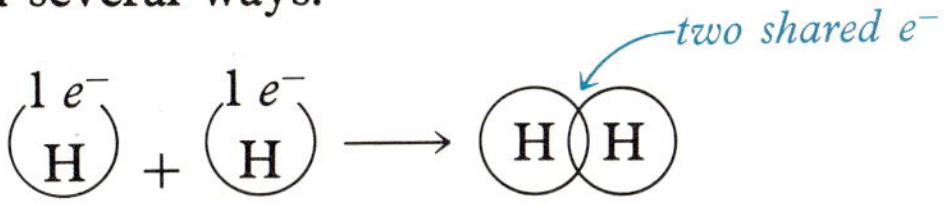

hydrogen atoms — hydrogen molecule

$$\mathrm{H\cdot} + \mathrm{H\cdot} \longrightarrow \mathrm{H{:}H} \quad \textit{(two shared } e^-\textit{)}$$

$$\mathrm{H} + \mathrm{H} \longrightarrow \mathrm{H{-}H} \quad \textit{(two shared } e^-\textit{)}$$

Fluorine is also capable of forming covalent compounds. A fluorine atom has seven electrons in the second shell, but eight are needed to fill the shell. A molecule of fluorine (F_2) is formed from two atoms of fluorine sharing one pair of electrons.

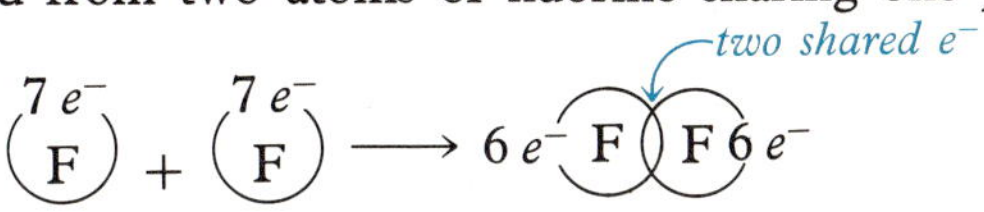

fluorine atoms — fluorine molecule

$$:\overset{\cdot\cdot}{\underset{\cdot\cdot}{\mathrm{F}}}\cdot + :\overset{\cdot\cdot}{\underset{\cdot\cdot}{\mathrm{F}}}\cdot \longrightarrow :\overset{\cdot\cdot}{\underset{\cdot\cdot}{\mathrm{F}}}:\overset{\cdot\cdot}{\underset{\cdot\cdot}{\mathrm{F}}}: \quad \textit{(two shared } e^-\textit{)}$$

$$\mathrm{F} + \mathrm{F} \longrightarrow \mathrm{F{-}F} \quad \textit{(two shared } e^-\textit{)}$$

In the fluorine molecule, each atom has given up one of its seven electrons to form the covalent bond. Each fluorine atom still has six unshared electrons in the outer shell, but also now has two shared electrons for a total of eight electrons.

Sample Problem: Using dots to represent outer electrons, draw the formulas for the following covalent compounds: (a) HCl and (b) H_2O

Solution: (a) $\mathrm{H\cdot} + \cdot\overset{\cdot\cdot}{\underset{\cdot\cdot}{\mathrm{Cl}}}: \longrightarrow \mathrm{H}:\overset{\cdot\cdot}{\underset{\cdot\cdot}{\mathrm{Cl}}}:$

(b) $\mathrm{H\cdot} + \cdot\overset{\cdot\cdot}{\underset{\cdot\cdot}{\mathrm{O}}}\cdot + \cdot\mathrm{H} \longrightarrow \mathrm{H}:\overset{\cdot\cdot}{\underset{\cdot\cdot}{\mathrm{O}}}:\mathrm{H}$

ORBITAL THEORY OF COVALENT BONDS

Let us now consider the *molecular orbital theory of covalent bonding.* Electrons around the nucleus of an atom are found in orbitals. When atoms share two electrons, they overlap their orbitals; the shared electrons become paired and occupy the same region in space.

We call an orbital of a nonbonded atom an *atomic orbital.* When two atoms overlap their atomic orbitals, these atomic orbitals merge together to form a new orbital that is shared jointly by both atoms. This new orbital is called a *molecular orbital* and is the covalent bond. Like the atomic orbital, the molecular orbital can hold up to two electrons of opposite spin.

MOLECULAR ORBITAL OF H_2. An atom of hydrogen contains a single electron in a 1*s* orbital. When one hydrogen atom bonds with another hydrogen atom, the 1*s* orbital of each atom merges with that of the other to form a molecular orbital. The molecular orbital in H_2 is a fuzzy cloud of electronic charge that may be likened to a slightly elongated balloon, or possibly to a peanut shell, as illustrated in Figure 4-6. This molecular orbital contains two electrons of opposite spin.

Figure 4-6. The bonding between the 1*s* orbitals of two hydrogen atoms in the hydrogen molecule (H_2).

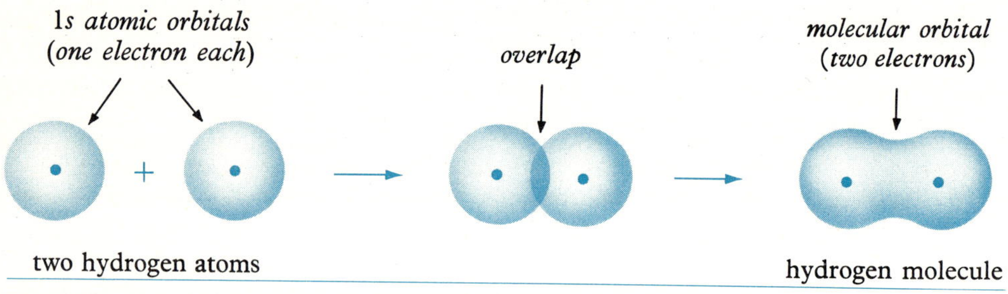

The covalent bond in the hydrogen molecule is called a *sigma bond,* which is one type of covalent bond. A sigma bond is defined as a covalent bond that is *symmetrical around an imaginary axis joining the two nuclei.* The symmetry is pictured in Figure 4-7.

Figure 4-7. A sigma bond is one that is symmetrical around an imaginary axis joining the two nuclei.

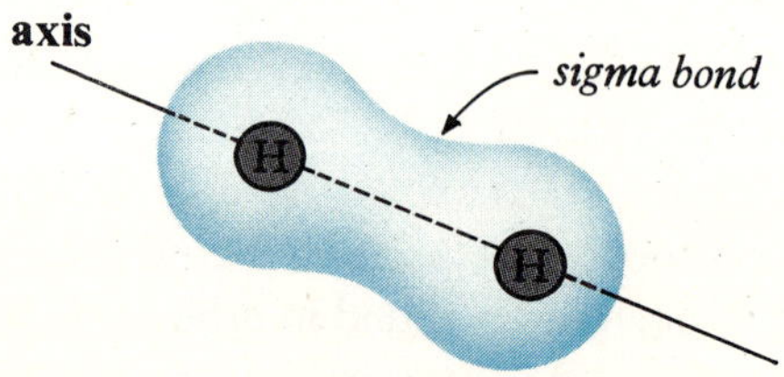

For examples of other objects symmetrical around an imaginary axis, picture a frankfurter or other uncurved sausage, a ball, or a child's top (Figure 4–8). If any of these objects were whirled around its axis, its appearance would not change because each is symmetrical around its axis.

Figure 4-8. Examples of objects symmetrical around axes.

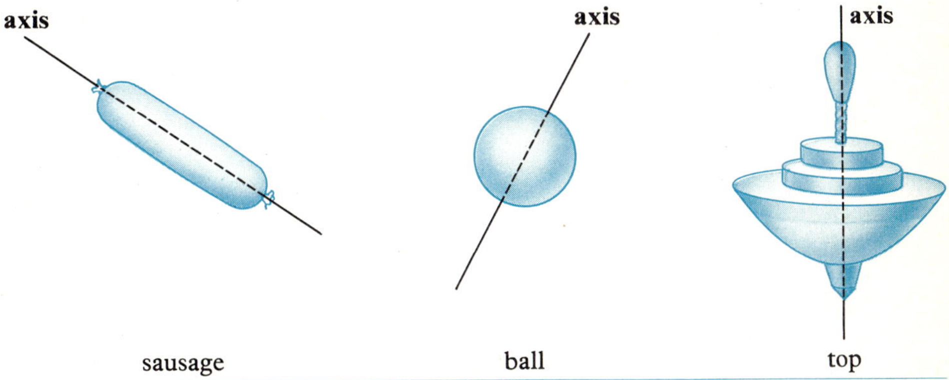

HYBRID ORBITALS OF CARBON. Hydrogen uses its 1*s* atomic orbital for bonding. Since all the other elements have the 1*s* orbital filled, they must use electrons at the second or higher energy level (whichever is the outermost unfilled shell) for bonding.

In the second shell, there are both *s* and *p* orbitals, but atoms with their outer electrons in the second shell do not always use these *s* and *p* atomic orbitals as such for bonding. Instead, these atoms frequently blend the *s* and *p* orbitals in a process called *hybridization*. Hybrid corn is a cross between two different types of corn. A hybrid orbital is a cross between two or more different types of atomic orbitals. The atom uses these hybrid orbitals to form covalent bonds with other atoms.

Figure 4–9. Hybridization of atomic orbitals.

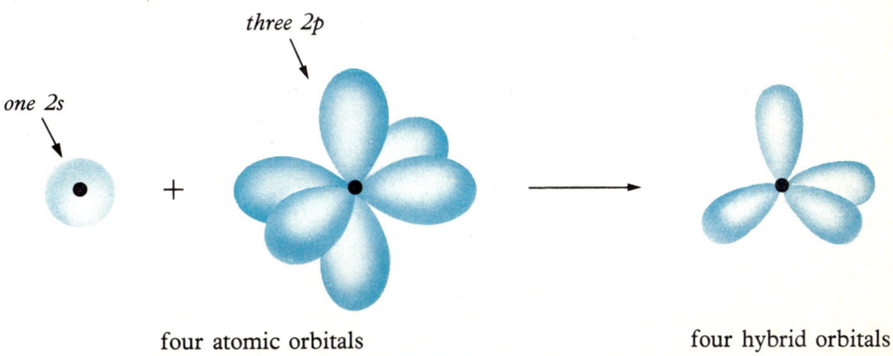

Carbon is the best example of an element that uses hybrid orbitals for bond formation. A carbon atom has a total of six electrons, two in the filled first shell and four bonding electrons (the electrons in the unfilled second shell). To attain a filled second shell, a carbon atom must share four electrons from other atoms. In other words, *a carbon atom forms four covalent bonds.* When a carbon atom forms these four covalent bonds, its four atomic orbitals are hybridized in one of three different ways.

Carbon hybridizes its orbitals in one fashion when it is bonded to *four other atoms,* as in methane.

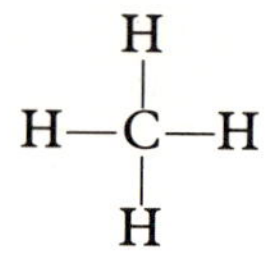

methane

(*The carbon atom has four single bonds.*)

Carbon hybridizes its orbitals in two different other ways if it is bonded to only *three* or to only *two* other atoms. Yet, in each case, each carbon atom is always bonded by a total of four covalent bonds.

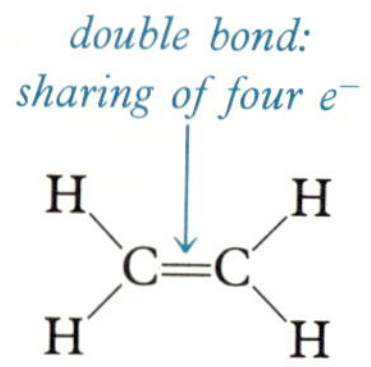

ethylene

(*Each carbon has two single bonds and a double bond—a total of four covalent bonds.*)

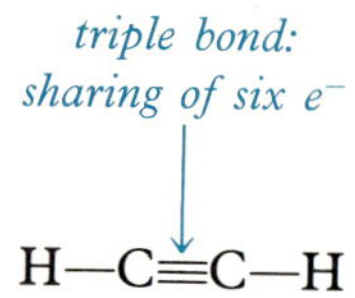

acetylene

(*Each carbon has one single bond and one triple bond—a total of four covalent bonds.*)

sp^3 HYBRIDIZATION. The four electrons in the second shell of a carbon atom would be expected to be in the four second-level orbitals: the 2*s*, 2*p*, 2*p*, and 2*p* orbitals. However, carbon can hybridize these orbitals to form four new orbitals that are equivalent in shape and energy. These four equivalent orbitals form four single bonds to four other atoms, as in methane (CH_4). Carbon starts with four atomic orbitals and ends up with four hybrid orbitals, each containing one electron and each with the ability to share one more electron from another atom to fill the outer shell. These four equivalent orbitals are called sp^3 (ess-pee-three) orbitals. The term sp^3 means that the hybrid orbitals are formed with *one s* atomic orbital and *three p* atomic orbitals. Figure

Figure 4–10. The sp^3 hybridization of the orbitals of carbon.

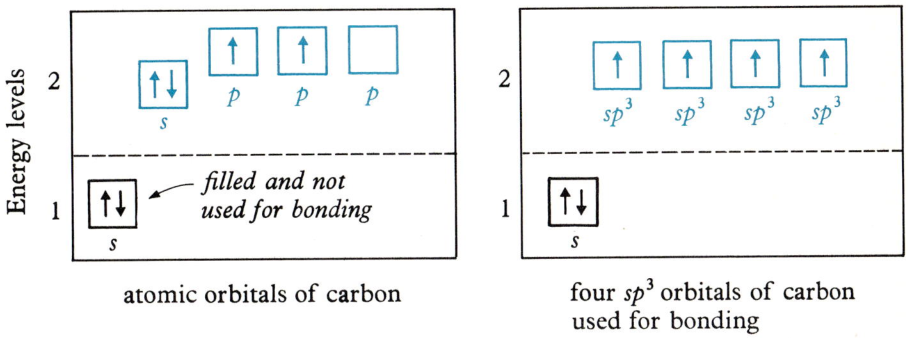

4–10 shows the relative energies of the sp^3 orbitals compared to the *s* and *p* atomic orbitals.

Each of the four sp^3 orbitals is shaped like a bowling pin with the nucleus at the neck (see Figure 4–11). An electron in an sp^3 orbital is most likely to be found within this shaped volume in space. It is the larger end of the bowling pin shape that is used for bonding.

Figure 4–11. The shapes of the sp^3 orbitals of carbon.

one sp^3 orbital
(The other three are not shown.)

the four sp^3 orbitals
(The small lobes are not shown.)

Electrons that are not paired repel each other. Because of the repulsions between electrons in separate orbitals, orbitals around a nucleus lie as far apart as possible while still fanning out from the nucleus. Thus, the four sp^3 orbitals point toward the corners of a regular tetrahedron (Greek *tetra-*, four, and *hedra,* side or face). If you blow up four balloons and tie their ends together, you will find that they too point toward the corners of a tetrahedron. (See Figure 4–12 for a representation of the tetrahedral sp^3 orbitals.)

Figure 4–12. The four sp^3 orbitals of carbon point toward the corners of a regular tetrahedron.

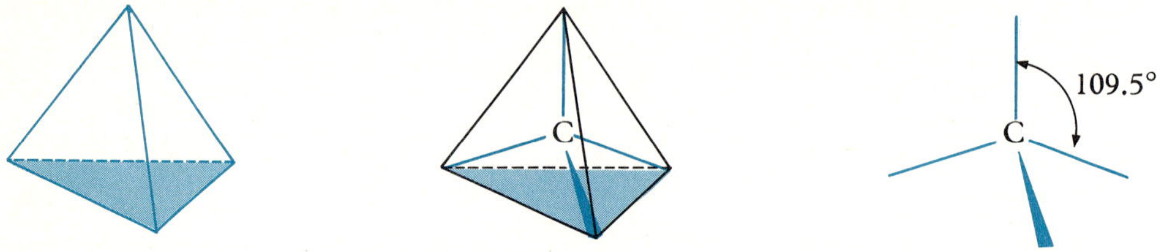

a regular tetrahedron

carbon in the center with the sp^3 orbitals, represented by lines, pointing toward the corners

the angle between any two of the sp^3 orbitals is 109.5°

If you draw lines from the center of a tetrahedron to each of its four corners, then measure the angles between these lines, you will find each of the four angles to be 109.5°. We also find the bond angles between any two sp^3 orbitals to be 109.5° ideally.

In methane, each of the sp^3 orbitals of carbon, each containing one electron, overlaps with a 1*s* orbital of hydrogen also containing one electron. Thus, each new molecular orbital contains two electrons. Each hydrogen atom has access to two electrons, and its outer shell is filled. The carbon atom has access to a total of eight electrons, and its outer shell is also filled. Figure 4–13 shows pictorially how the four molecular orbitals between carbon and hydrogen in methane are formed.

Figure 4–13. The four sp^3 orbitals of carbon overlap with four *s* orbitals of hydrogen to form methane (CH_4).

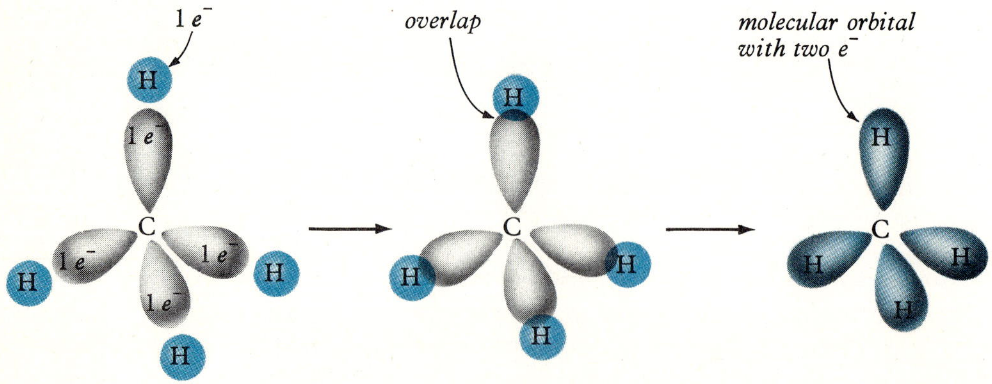

This type of covalent bond between carbon and hydrogen is a sigma bond because it is symmetrical around an imaginary axis between the two nuclei just as the H—H bond is. (See Figure 4-14.)

Figure 4-14. Each carbon-hydrogen bond in methane is a sigma bond, symmetrical around an axis joining the two nuclei.

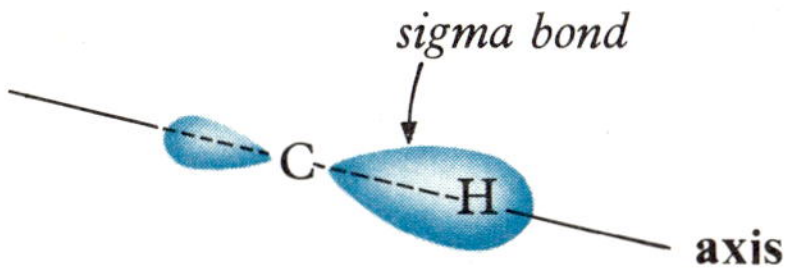

As we have mentioned, we usually represent a covalent bond by a line. Methane is a sufficiently complex molecule that we can use several types of formula to show us its structure. A formula with lines for bonds shows the order of attachment of the atoms and is called a *structural formula.* A formula may be abbreviated by omitting the lines (as in the formula CH_4); this type of formula is called a *condensed structural formula.* Although structural formulas show the order of attachment, they do not show the bond angles. A *dimensional formula* may be used when the geometry of a structure is to be stressed. In such a formula, a wedge represents a bond pointing in the direction of the viewer, and a dashed line represents a bond pointing away from the viewer. The other two bonds in the dimensional formula are in the plane of the paper.

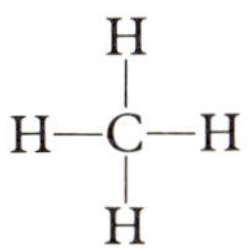

structural formula

CH_4

condensed structural formula

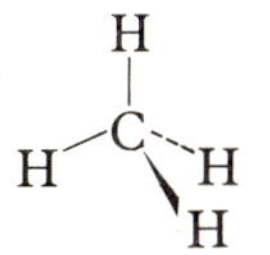

dimensional formula

***sp*² HYBRIDIZATION.** When carbon forms two bonds to one other atom, as to another carbon in ethylene, the orbitals are hybridized differently from those in methane. Ethylene contains two carbon atoms that share four electrons between them in a double bond. (Each line in the double bond represents two electrons.) We will first look at just one carbon atom at a time, and then put the two of them together shortly.

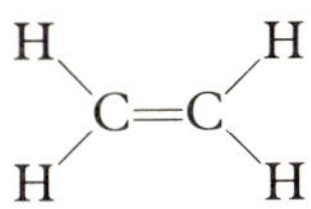

ethylene

To form a double bond, a carbon atom uses only the 2*s* orbital, and *two* of the 2*p* orbitals, for its hybrid orbitals. One *p* orbital is left over. The carbon atom still has four orbitals, three sp^2-hybrid orbitals and the "leftover" *p* orbital. (See Figure 4-15.)

Figure 4-15. The sp^2 hybridization of the orbitals of carbon.

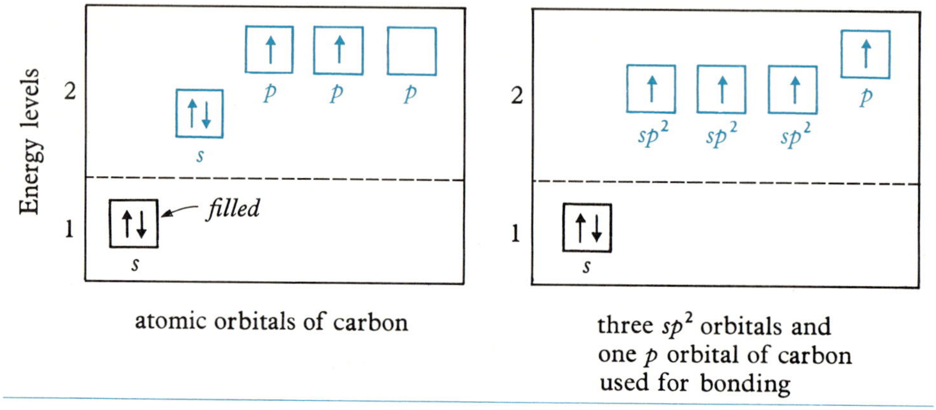

The shape of an individual sp^2 orbital is similar to that of an individual sp^3 orbital. The principal difference between the two types is in the orientation of the hybrid orbitals in reference to each other. The three sp^2 orbitals lie in a *plane* with angles of 120° between one another (see Figure 4-16). If you blow up three balloons and tie them together, you will see that they will be as far apart as possible and will have the same planar geometry as the three sp^2 orbitals.

Figure 4-16. The shape and orientation of the sp^2 orbitals of carbon.

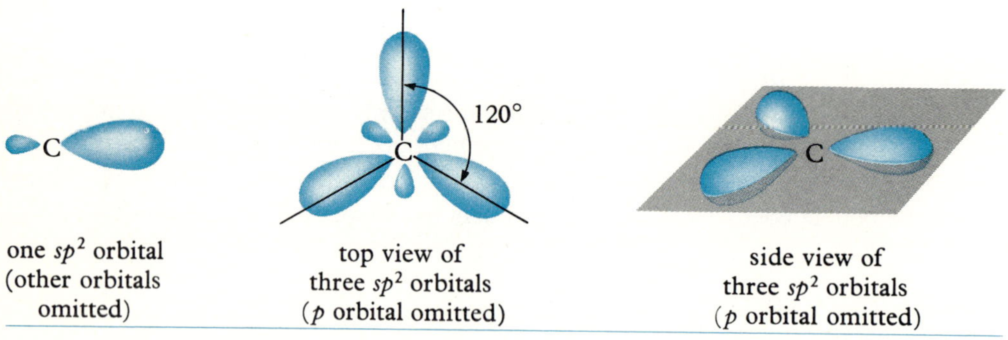

Each sp^2 orbital has one electron and can overlap with an orbital of another atom to share a pair of electrons. An sp^2 orbital of carbon can overlap with the 1*s* orbital of hydrogen to form a sigma bond; an sp^2 orbital can also overlap with an sp^2 orbital from another carbon atom to form a sigma bond. (See Figure 4-17.)

Figure 4-17. The sp^2 orbitals of a carbon atom can overlap with orbitals of other atoms to form sigma bonds. (The other orbitals of the carbon atoms have been omitted.)

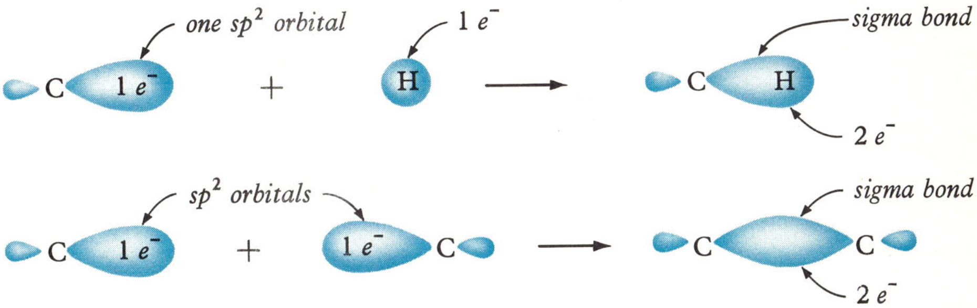

In ethylene, each carbon atom has three sp^2 orbitals forming covalent bonds—two to hydrogen atoms, one to the other carbon atom. These sp^2 bonds all lie in a plane. Each carbon atom also has one *p* orbital, each with one electron. These two *p* orbitals overlap to form another bond. This type of overlap is *side to side*, as opposed to the others we have discussed, which are *end to end*. (See Figure 4-18.)

Figure 4-18. The bonding in ethylene ($CH_2{=}CH_2$).

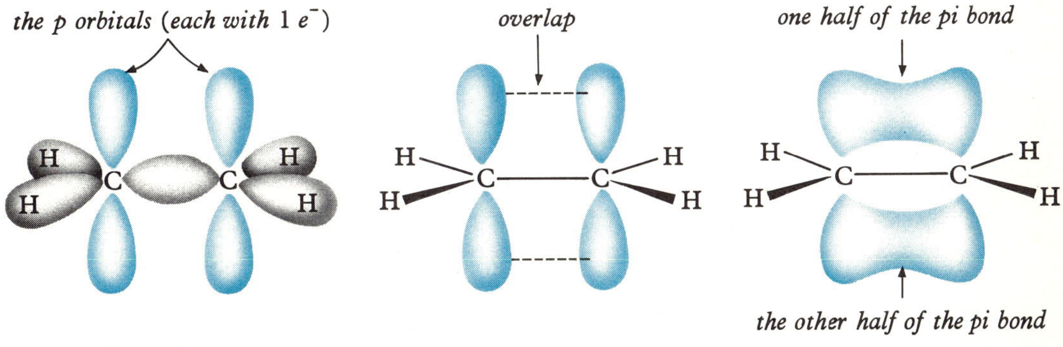

Each carbon atom has three sigma bonds, two to hydrogen atoms and one to the other carbon atom. Each carbon atom also has one *p* orbital.

The *p* orbitals overlap with each other. (The sigma bonds are represented by lines.)

The pi bond, which contains two electrons, between the carbon atoms.

The bond formed from the side-to-side overlap of two *p* orbitals is called a *pi bond*. Unlike a sigma bond, the pi bond is *not* symmetrical around the axis joining the two carbon atoms. Instead, it is a fuzzy cloud in two parts, half above the plane of the sp^2 sigma bonds and half below. The pi bond contains two electrons, one from each of the overlapping *p* orbitals. Thus, the two carbon atoms in ethylene are joined by both a sigma bond and a pi bond and share four electrons between them—two in the sigma bond and two in the pi bond. Each carbon atom in ethylene also shares four electrons with two hydrogen atoms; therefore, each carbon has access to eight electrons and has a full outer shell.

The usual formula for the structure of ethylene uses lines for all the bonds. In the double bond between the two carbon atoms, each line represents two electrons being shared. The symbolism for a double bond is indeed unfortunate because it implies that both bonds of the double bond are equivalent. They are not. We will find that the pi bond is highly reactive, while the sigma bond is quite inert.

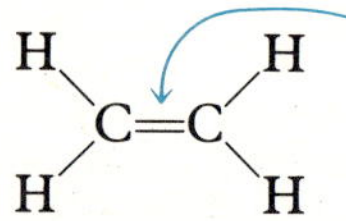

structural formula

In the double bond, one line represents the reactive pi bond; the other line represents the stable sigma bond.

$$CH_2{=}CH_2$$

condensed structural formula

***sp* HYBRIDIZATION.** Carbon can hybridize its four atomic orbitals in yet a third way in order to form a *triple bond.*

$$H{-}C{\equiv}C{-}H$$

triple bond, six shared e^-

acetylene

Carbon forms a triple bond by mixing the 2*s* orbital and only *one* of the 2*p* orbitals. The hybrid orbitals formed from this process are called *sp* orbitals (one part *s*, one part *p*). Since two atomic orbitals were used, two *sp*-hybrid orbitals result. When carbon forms *sp* orbitals, two *p* orbitals remain. (See Figure 4-19.)

The shape of an *sp* orbital is similar to that of the sp^3 or sp^2 orbital. The two *sp* orbitals lie as far apart from each other as possible; thus, they lie in a line, 180° apart. (See Figure 4-20.)

What of the *p* orbitals? One *p* orbital lies above and below the line of the two *sp* orbitals. The other is in the front and the back. Each *p* orbital forms 90° angles with the line of the *sp* orbitals. Remember that *p* orbitals are also at 90° angles to each other. The *p* orbitals of a carbon atom in the *sp*-hybrid state are shown in Figure 4-21.

Each *sp* orbital can form a sigma bond with another atom. In acetylene, each carbon atom has two *sp* orbitals, and each carbon bonds to the other carbon atom and to one hydrogen atom. The *p* orbitals undergo side-to-side overlap. Since each carbon

Figure 4–19. The *sp* hybridization of the orbitals of carbon.

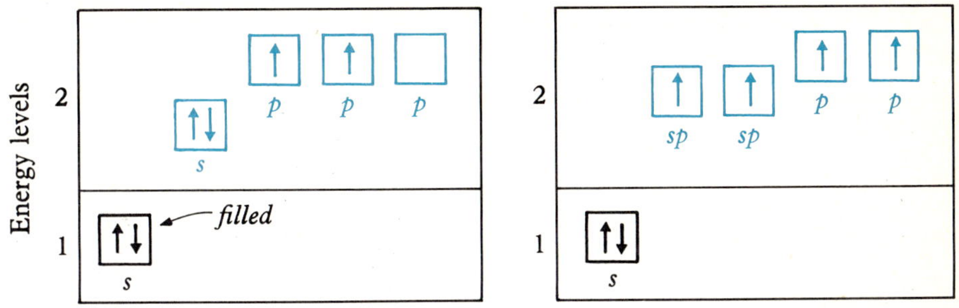

atomic orbitals of carbon

two *sp* orbitals and two *p* orbitals of carbon used for bonding

Figure 4–20. The shape and orientation of the *sp* orbitals of carbon. (Do not confuse the two *sp* orbitals and one *p* orbital just because they are drawn similarly.)

one *sp* orbital (other orbitals omitted)

two *sp* orbitals (*p* orbitals omitted)

Figure 4–21. The orientation of the *p* orbitals when carbon is in the *sp*-hybrid state.

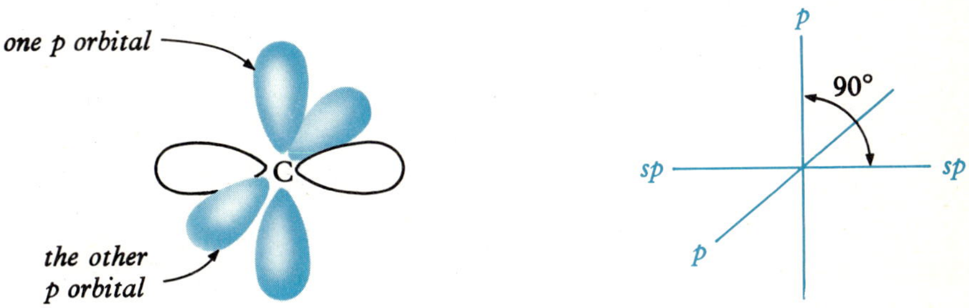

The carbon atom has two *sp* orbitals (white) and two *p* orbitals (color).

The *p* orbitals are at right angles to each other and to the *sp* orbitals.

atom has *two p* orbitals, *two* pi bonds are formed. Each of the pi bonds has two lobes. One pi bond is on the front and the back of the line of the sigma bonds; the other pi bond is on the top and bottom. (See Figure 4–22.)

Figure 4–22. The bonding in acetylene (CH≡CH).

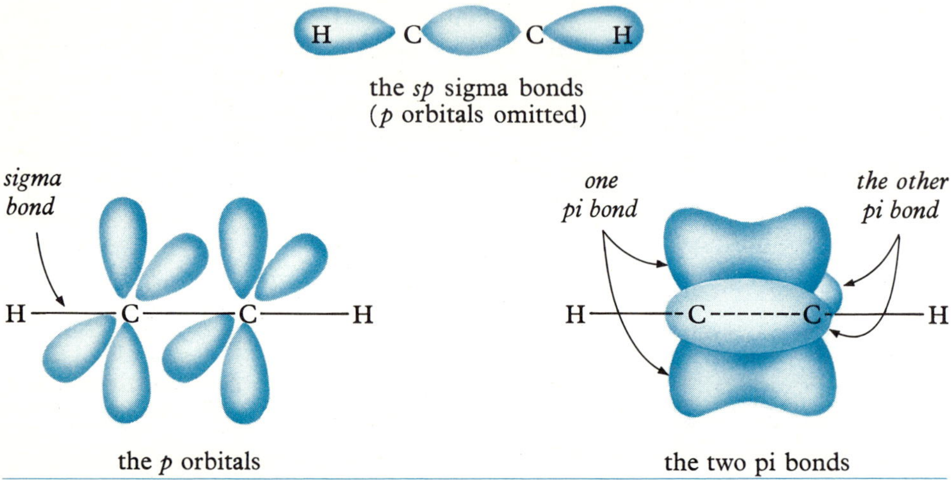

Each pi bond contains two electrons, and each sigma bond also contains two electrons. In acetylene, the carbon atoms share six electrons between them and each carbon atom shares two electrons with a hydrogen atom. The carbon and hydrogen atoms have all gained complete outer shells.

The usual representation for a triple bond is three lines joining the two atoms. As in the case of the double bond, each line represents two electrons being shared. Keep in mind that the triple bond contains two types of bonds—one sigma bond and two pi bonds.

one sigma bond and two pi bonds

H—C≡C—H	CH≡CH
structural formula	*condensed structural formula*

Let us summarize the hybrid orbitals of carbon.

1. sp^3 hybridization, used when C forms four single bonds:

tetrahedral

2. sp^2 hybridization, used when C forms a double bond:

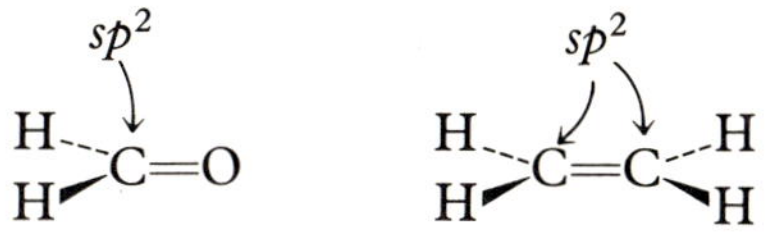

planar

3. sp hybridization, used when C forms a triple bond:

linear

Sample Problem: What is the hybridization of each carbon atom in the following structure?

$$\overset{(a)}{CH_2}=\overset{(b)}{CH}-\overset{(c)}{CH}=\overset{(d)}{CH}-\overset{(e)}{CH_2}-\overset{(f)}{C}\equiv\overset{(g)}{CH}$$

Solution: (a), (b), (c), and (d) are sp^2; (e) is sp^3; (f) and (g) are sp.

THE AROMATIC PI CLOUD. Benzene (C_6H_6) is an important compound of carbon and hydrogen that contains six carbon atoms joined in a ring. Each of these carbon atoms is in the sp^2 state and has sigma bonds to its carbon neighbors and to a hydrogen atom. All the carbon and hydrogen atoms lie in a plane.

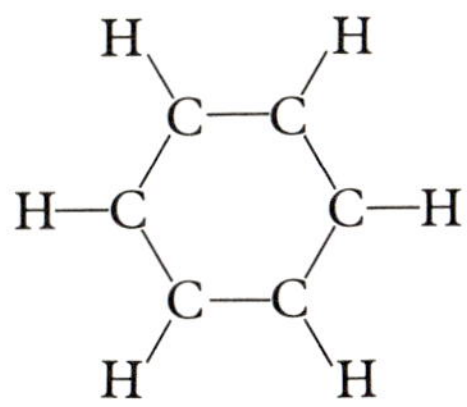

The sigma bonds in benzene.

Because the carbon atoms are in the sp^2 state, each one has a p orbital above and below the plane of the ring. Each of these p orbitals contains one electron. These p orbitals do not pair up to form ordinary pi bonds. Instead, each p orbital overlaps equally with each of its two neighbors. Because of this equal overlap, the six p electrons (one from each carbon) are completely shared in a two-lobed cloud over and under the ring (see p. 323). This cloud of electrons is called an *aromatic pi cloud.*

The term *aromatic* is used because compounds containing the aromatic pi cloud often have distinctive odors. *Pi* is used because of the similarity to a pi bond, and it is called a *cloud* because this describes how chemists think this type of orbital must look.

RESONANCE STRUCTURES. Benzene is an example of a compound that we cannot properly represent by conventional line formulas. Let us try to write a conventional formula for benzene and then consider why it is inadequate. Each carbon atom in the ring has a *p* orbital with one electron. The total number of *p* electrons in a benzene molecule is six. A line bond represents two electrons. Therefore, we should add three double-bond lines (and only three) to show the six *p* electrons.

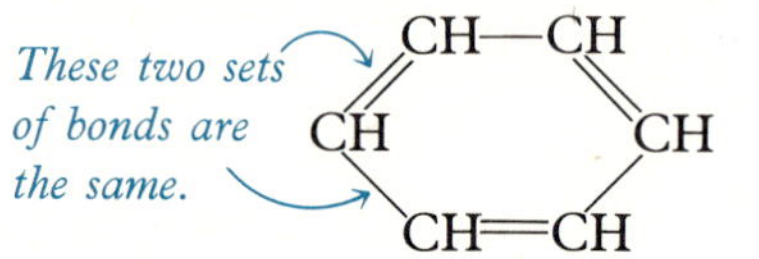

The formula that results is not a correct representation of benzene. The orbital picture shows us that the bonding between each pair of carbon atoms in the benzene ring is identical. The above structure erroneously shows two types of carbon–carbon bonding in the ring instead of only one type. Early chemists devised a way to make the bonding between the carbon atoms look the same on paper. They drew *two* structures and said "The real structure of benzene is like *both* of these structures." These early chemists then said, "We can think of the molecule of benzene as shifting back and forth between these two structures."

The structure of a molecule of benzene does not really shift back and forth at all—the true structure of benzene is something in between the two resonance structures. The situation is similar to that of trying to describe a mule using only pictures of a horse and a jackass. A mule does not switch back and forth from being a horse to being a jackass. Although a mule has some of the characteristics of both, a mule is still a mule.

When two or more structures are necessary to describe the true bonding in a molecule, we say that the structures are in *resonance*. A single, double-headed arrow is used to show the two structures in resonance.

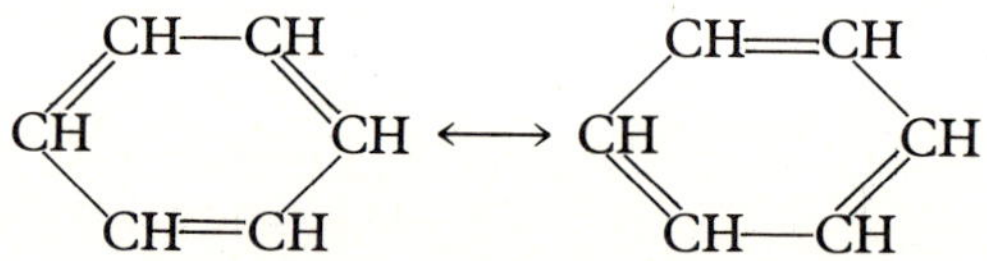

Resonance structures of benzene.

When we discuss benzene chemistry in Chapter 19, we will use a convention different from resonance structures to represent the structure of benzene. To show that

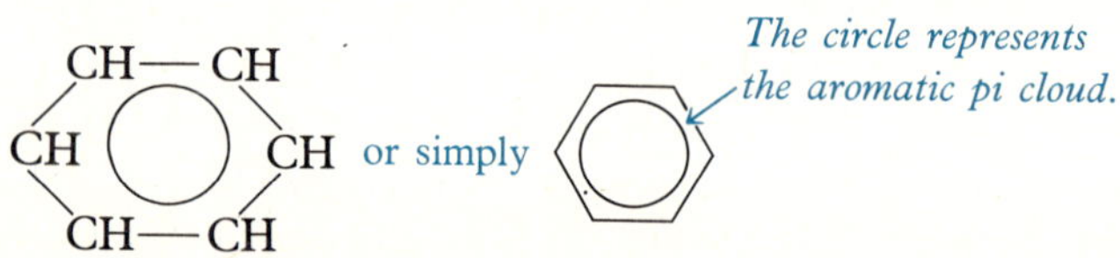

benzene

the pi cloud bonds all carbons equally, we will use a circle in the ring to represent the six electrons in the aromatic pi cloud.

HYBRID ORBITALS OF NITROGEN. The other elements besides carbon in the second period of the Periodic Table also use electrons in the second shell for bonding. To the right of carbon in the Periodic Table, we find nitrogen. In many respects, the bonding in nitrogen compounds is similar to that in carbon compounds. However, a neutral nitrogen atom has seven electrons: two in the filled first shell and five in the second shell (the shell used for bond formation). Similar to the orbitals of carbon, the $2s$ orbital and the three $2p$ orbitals of nitrogen may hybridize to form four sp^3 orbitals. However, since there are *five* electrons in these four orbitals, one of these hybrid orbitals is *filled* with two electrons. Nitrogen needs to share electrons with only *three* other atoms to attain a full second shell. (See Figure 4-23.)

Figure 4-23. The sp^3 hybridization of the orbitals of nitrogen.

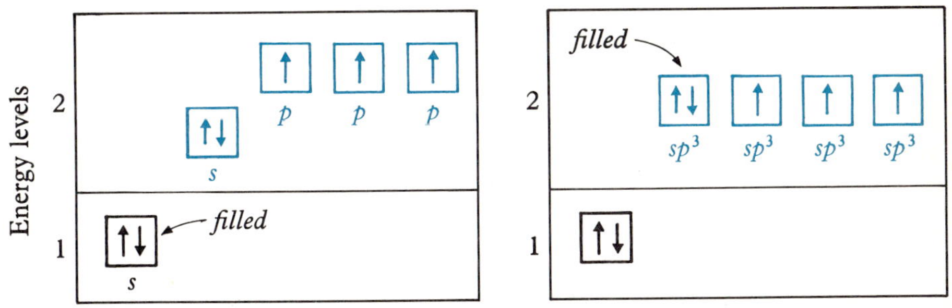

atomic orbitals of nitrogen   four sp^3 orbitals of nitrogen

One of the most common compounds of nitrogen is ammonia (NH_3), a compound used in household cleaning solutions and as a fertilizer. The nitrogen in ammonia may be thought of as being in a sp^3-hybrid state. The three sp^3 orbitals that contain one electron each overlap with three hydrogen $1s$ orbitals to form three sigma bonds. The remaining sp^3 orbital with two electrons remains as a filled orbital. (See Figure 4-24.)

The following are other ways of representing ammonia.

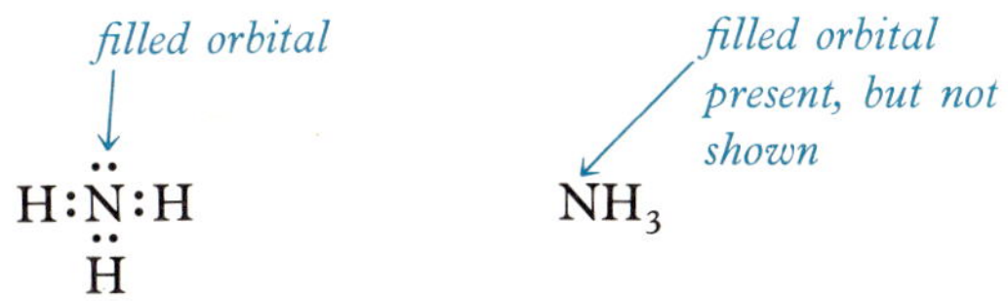

Figure 4–24. The overlap of three sp^3 orbitals of nitrogen with the 1*s* orbitals of hydrogen to form ammonia (NH_3).

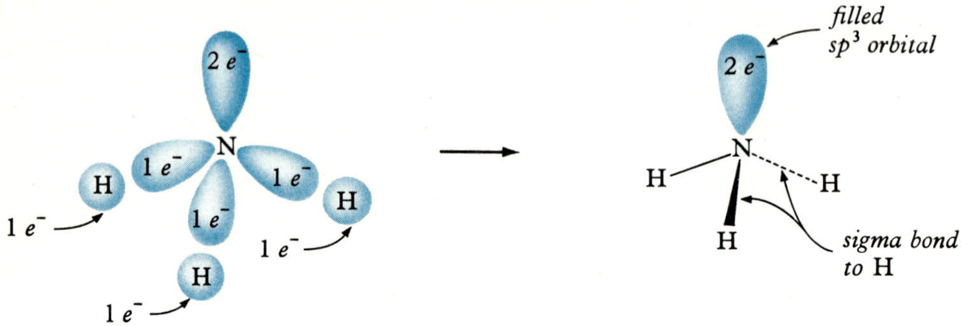

HYBRID ORBITALS OF OXYGEN. The bonding of oxygen is similar to that of carbon and nitrogen. At the second energy level, oxygen has six electrons, one more than nitrogen.

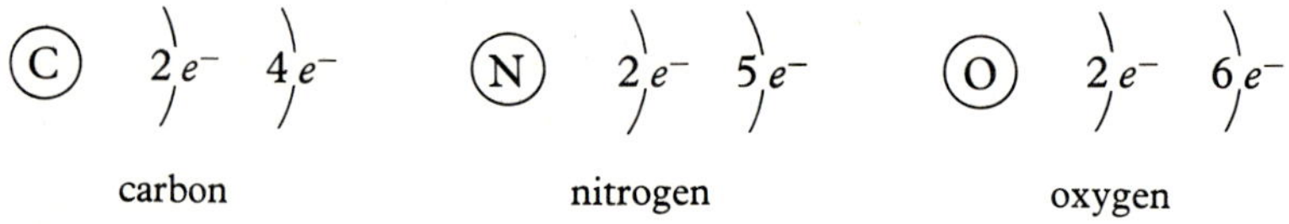

Oxygen can form bonds with sp^3 orbitals; however, *two* of the four sp^3 orbitals are filled (see Figure 4–25). Therefore, oxygen can form compounds by bonding to *two* other atoms.

Figure 4–25. The sp^3 hybridization of the orbitals of oxygen.

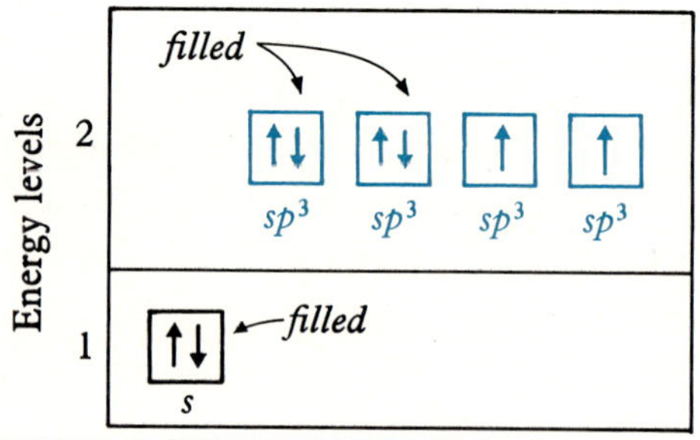

Water is an example of a typical covalently bonded oxygen compound. An orbital representation of water is shown in Figure 4–26.

We may represent the structure of water in other ways.

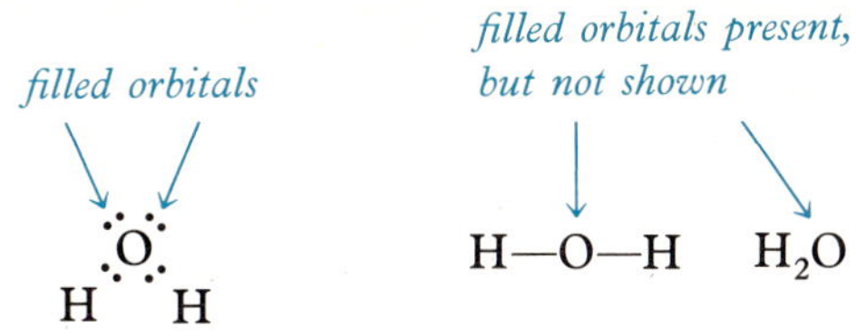

Figure 4–26. The overlap of two sp^3 orbitals of oxygen with 1*s* orbitals of hydrogen to form water (H_2O).

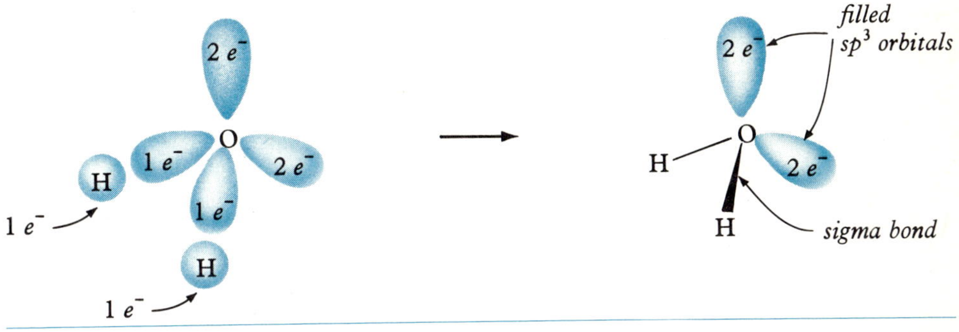

The hydroxide ion plays an important role in chemistry. Examples of compounds containing this ion are sodium hydroxide, NaOH (or $Na^+\ OH^-$), and potassium hydroxide, KOH (or $K^+\ OH^-$). In the hydroxide ion, we find that oxygen has one covalent sigma bond to hydrogen and also carries a negative ionic charge. We may think of the hydroxide ion as a molecule of water that has lost a hydrogen ion. (See Figure 4–27.)

$$\mathrm{H:\ddot{\underset{..}{O}}:H} \longrightarrow \mathrm{H^+} + \mathrm{^-:\ddot{\underset{..}{O}}:H}$$

Figure 4–27. Orbital picture of the hydroxide ion ($^-\!:\ddot{\underset{..}{O}}\!:\!H$, or OH^-).

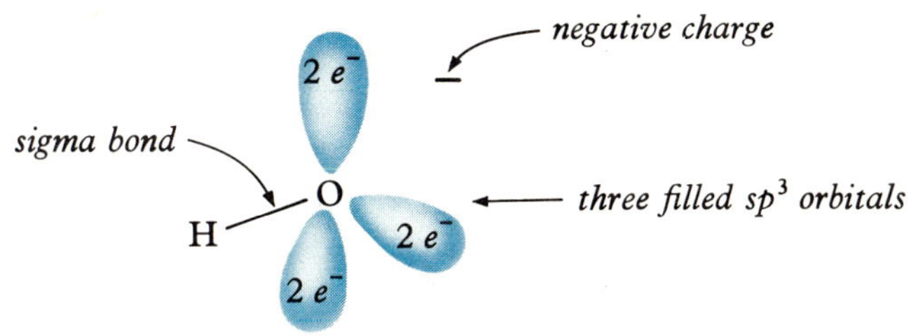

The hydroxide ion is an example of a *complex ion,* an ion composed of a group of atoms held together by covalent bonds. Since the group has an electrical charge, the entire group is an ion.

Table 4–2. Some Common Complex Ions

Name of Ion	Formula	Electrical Charge	Example of Compound
Hydroxide	OH^-	−1	NaOH
Sulfate	SO_4^{2-}	−2	Na_2SO_4
Nitrate	NO_3^-	−1	$NaNO_3$
Phosphate	PO_4^{3-}	−3	Na_3PO_4
Permanganate	MnO_4^-	−1	$KMnO_4$
Ammonium	NH_4^+	+1	NH_4Cl

Oxygen is especially important in complex ion formation. As may be seen in the list of some common complex ions in Table 4–2, the names of most anions containing oxygen end in *-ate,* as opposed to name for monatomic (one atom) ions, which usually end in *-ide.* As we have mentioned, the hydroxide ion is an exception.

$KMnO_4$ — potassium permanganate

Na_2SO_4 — sodium sulfate

The ending -ate usually signifies oxygen in the anionic portion of a compound.

POLAR COVALENT BONDS

We have seen that an element with a great electron affinity can remove an electron from an element with a low electron affinity, the result being the formation of a pair of ions.

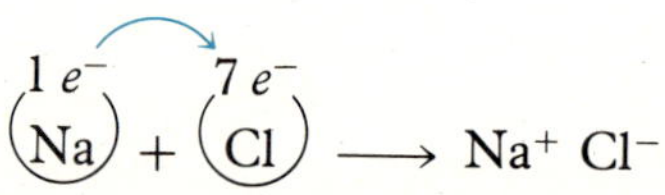

But what of covalent bonds formed by elements with differing electronegativities? In its gaseous state, hydrogen chloride (HCl) contains a covalent bond. (The fact that HCl is a gas, and not a crystalline solid, is one of the bits of evidence that this compound is indeed covalent.) Chlorine is more electronegative than hydrogen and has a greater pull on the shared electrons in hydrogen chloride. A bond in which the atoms do not pull equally on the bonding electrons is called a *polar covalent bond,* a type of bond that is exceedingly important in the chemistry of covalently bonded compounds.

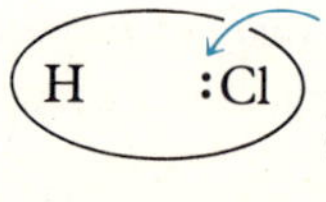

The shared electrons spend more of their time closer to the Cl; thus, the Cl is more negative than the H.

hydrogen chloride (HCl)

In the case of HCl, the bonding electrons lie closer to the Cl than to the H. Thus, hydrogen may be said to have a *partial positive charge* and chlorine, a *partial negative charge*. We symbolize a partial positive charge by the Greek letter delta followed by a plus sign: $\delta+$. A partial negative charge is shown as $\delta-$.

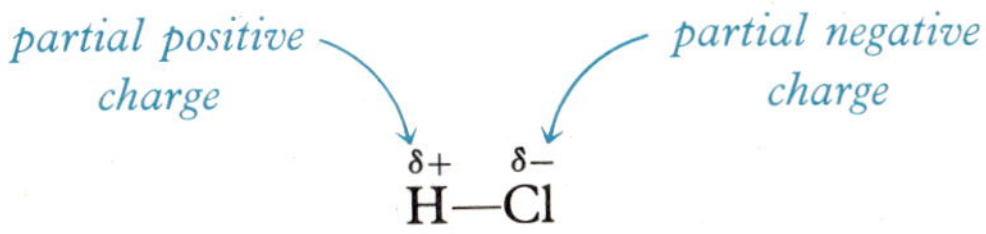

Some other terms are used to describe a molecule with an unequal distribution of electronic charge. Any object, including a molecule, with a charge separation may be called a *dipole*. Such a molecule may also be called a *polar molecule*. A polar covalent bond, as in HCl, may be said to have some *ionic character*.

Water is another example of a compound with polar molecules. Because oxygen is more electronegative than hydrogen, the oxygen atom draws the bonding electrons toward itself and away from the hydrogens. The polarity of water molecules makes water a good solvent for ionic compounds and for other polar covalent compounds. These are topics that will be discussed at greater length in later chapters.

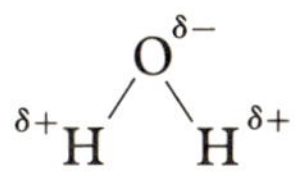

Water is a polar molecule.

When two carbon atoms are bonded together, they both have an equal pull on shared electrons. Since both atoms have the same electronegativity, there is an equal distribution of electrons between the two atoms. We call such a bond a *nonpolar bond*.

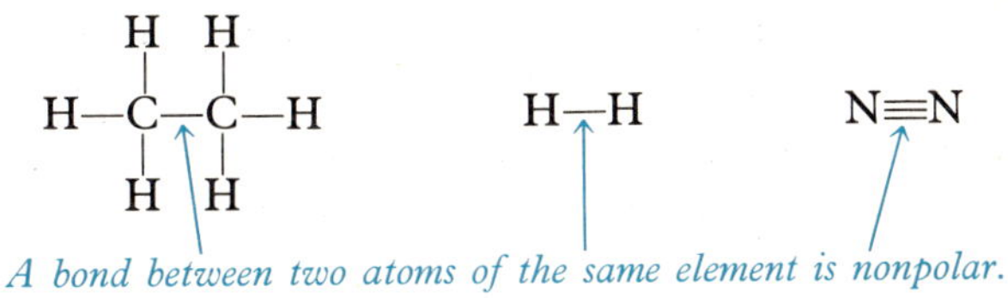

SUMMARY

We can summarize our discussion of ionic bonds and covalent bonds in the following statements.

1. Atoms undergo ionic or covalent compound formation in order to pair electrons in unfilled orbitals and to fill (or empty) the outermost shell.
2. In ion formation, one or more electrons are transferred from one atom to another. Then the cations and the anions are held together by a mutual electrostatic attraction.

3. In covalent bond formation between atoms, pairs of electrons are shared in mutually held molecular orbitals, formed by the overlap of either atomic orbitals or hybrid orbitals. A covalent bond between atoms of differing electronegativities is polar.
4. Hybrid orbitals in the second energy level are formed by some combination of the 2*s* and three 2*p* orbitals to give (a) four sp^3 orbitals, (b) three sp^2 orbitals plus one *p* orbital, or (c) two *sp* orbitals plus two *p* orbitals.

STUDY QUESTIONS

4-1. Identify whether each of the following substances has ionic bonds, covalent bonds, or metallic bonds.
(a) iron (Fe)
(b) potassium iodide (KI)
(c) methylamine (CH_3NH_2)
(d) sodium hydroxide (NaOH)

4-2. Which of the following elements would form cations fairly readily?
(a) magnesium (Mg)
(b) nickel (Ni)
(c) oxygen (O)
(d) iodine (I)

4-3. Name the following ionic compounds:
(a) CaO (b) CuF_2 (c) Fe_2S_3

4-4. Write formulas for the following compounds:
(a) magnesium iodide
(b) mercury(II) iodide
(c) aluminum oxide

4-5. Which of the following elements has the greatest electron affinity? Ca, Mn, C, O

4-6. Write the equation for the formation of lithium bromide (LiBr) from lithium atoms and bromine atoms. In your equation, indicate the electrons in the outer shell of each species with dots.

4-7. Calculate the molecular weights of (a) LiBr and (b) Na_2S.

4-8. Give the oxidation number of each of the following ions: Sr^{2+}, Cr^{3+}, S^{2-}, I^-.

4-9. Determine the oxidation number for each of the ions in the following compounds: $ZnBr_2$, FeS, Al_2O_3, Hg_2Cl_2.

4-10. Write a chemical equation for the formation of a chlorine molecule from two chlorine atoms. Indicate the number of electrons in the outer shell of each chlorine atom before and after their combination.

4-11. Give the numbers of electrons where indicated.

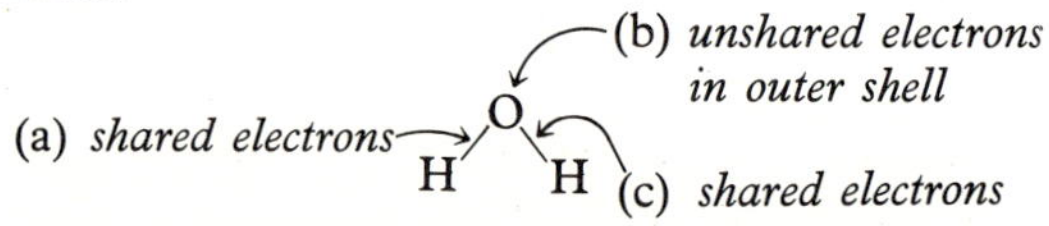

4-12. Use a formula with dots to represent electrons to show the electronic structure of the covalently bonded compound hydrogen fluoride (HF).

4-13. Give the number of sigma bonds in one molecule of each of the following:
(a) H_2 (b) CH_3CH_3 (c) CH_3OH (d) $CH{\equiv}CH$

4-14. How many bonding electrons does a carbon atom have?

4-15. How many sp^3 orbitals does the carbon atom in methane have?

4-16. Describe the geometry of the sp^3 orbitals of carbon in methane.

4-17. In methane, each of the molecular orbitals has arisen from overlap of an ______ orbital of carbon and an ______ orbital of hydrogen.

4-18. In methane, how many electrons does each of the molecular orbitals contain?

4-19. Each carbon atom in ethylene has four bonding orbitals. Identify each orbital as to type.

4-20. Describe the geometry of the sp^2 orbitals of the carbon atoms in ethylene.

4-21. Describe the geometry of the pi bond in ethylene in relation to the sigma bonds.

4-22. Describe the geometry of the sp orbitals of the carbon atoms in acetylene.

4-23. Name the bonds (sigma or pi):

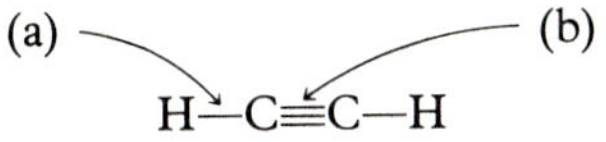

4-24. Indicate the hybridization of each atom (except *H*) in the following formula:

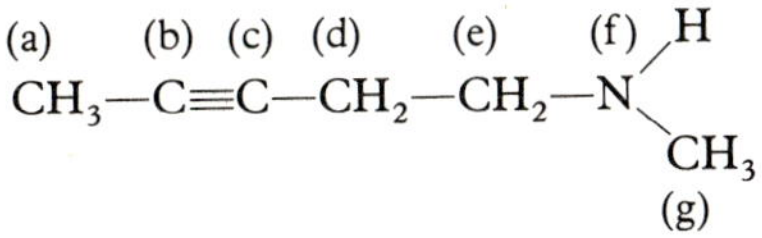

4-25. Describe the geometry of the sigma bonds in benzene.

4-26. What is the total number of electrons in the p orbitals of benzene?

4-27. The aromatic pi cloud in benzene contains a total of ______ electrons.

4-28. Give another resonance structure for benzene.

4-29. Tell how many electrons are indicated:

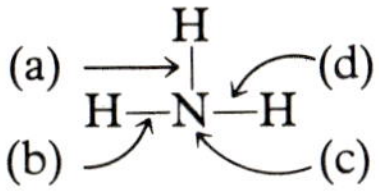

4-30. Draw the orbital picture of the hydroxide ion and indicate the number of electrons in each orbital.

4-31. Place partial charges ($\delta+$ and $\delta-$) where appropriate in the following structures:

(a) H—Br (b) Cl—Cl

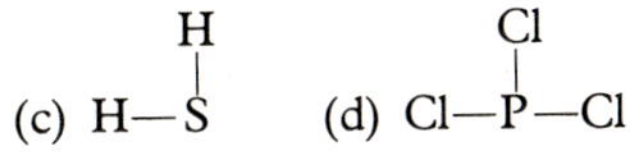

4-32. Rewrite each of the following formulas, showing each bond.

Example:

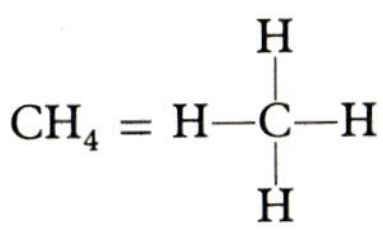

(a) CH_3CH_3 (b) $CH_3CH_2CH_3$

(c)

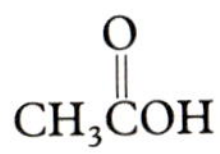

CHAPTER 5 CHEMICAL REACTIONS

OBJECTIVES

How Does a Reaction Occur? *Define reaction mechanism and tell upon what two premises a reaction mechanism is based. * Describe how the average kinetic and internal energies of molecules affect the course of a chemical reaction. * Construct an energy diagram and identify its features. * Describe the effects of a catalyst, a temperature change, and a change in concentration of reactants upon a reaction. * Define and write equations to exemplify side reactions and reversible reactions.*

How We Represent Chemical Reactions *Identify the different types of chemical formulas, and give examples of each type. * Calculate the percent by weight of an element in a compound from the formula. * Balance a chemical equation by inspection. * Use a balanced equation to calculate proportions of reactants and products in a reaction.*

When iron rusts or when the grass grows, *chemical reactions* are taking place. Atoms and molecules are getting together, breaking bonds, and changing their partners. New substances are being formed from old. Atoms are not being lost; they are just being rearranged.

A chemical reaction is represented on paper by a *chemical equation,* a shorthand way of showing what goes on in a test tube, in our bodies, or in a blast furnace. We have already encountered some chemical equations; for example, $2\,H_2 + O_2 \longrightarrow 2\,H_2O$. In this reaction, we start with two gases, hydrogen and oxygen. We call these the *reactants,* and they appear on the left side of the equation. On the other side of the arrow we find the *products,* in this case, water. The arrow means *yield* or *yields* and, in a chemical equation, is used to separate the reactants from the products. We read the equation *hydrogen and oxygen yield water,* or *hydrogen undergoes reaction with oxygen to yield water.*

A chemical equation: $$\underbrace{2\,H_2 + O_2}_{\textit{reactants}} \longrightarrow \underbrace{2\,H_2O}_{\textit{product}}$$

HOW DOES A REACTION OCCUR?

A chemical equation is a convenient device for representing a chemical reaction. However, the chemical equation does not tell us the whole story about a reaction. The equation does not tell us if two chemicals will explode when they are mixed or if they will just sit there quietly until they are heated. The equation does not tell us if *other* reactions might occur between the reactants. Furthermore, the equation does not tell us *how* a particular reaction occurs, that is, how reactant molecules get together and form products. The explanation of *how* a reaction occurs is called a *reaction mechanism* and is based on the following premises:

1. Molecules must collide with one another in order for a reaction to occur.
2. Colliding molecules need a certain minimum amount of energy if the collision is going to result in reaction.

COLLISION BETWEEN MOLECULES. The first of the preceding two premises is fairly obvious. If two molecules are five miles apart, they are not going to undergo reaction with each other. Molecules must collide with each other in order to react. The exception would be a molecule that can react all by itself with an internal rearrangement of its atoms. The explosion of ammonium nitrate, frequently used in dynamite, is a reaction in which collisions between different molecules are not needed for reaction to occur. (An explosive is a compound that can make a quick conversion from a solid or liquid, which has a small volume, to gases, which have large volumes.)

$$\underset{\substack{\text{ammonium}\\ \text{nitrate}\\ \textit{solid}}}{2\,NH_4NO_3} + \text{Pressure} \xrightarrow{\text{fast}} \underbrace{\underset{\text{nitrogen}}{2\,N_2} + \underset{\text{oxygen}}{O_2} + \underset{\text{water vapor}}{4\,H_2O}}_{\textit{gases}}$$

Molecules in the gaseous state are in a continuous state of movement and collide frequently with each other. Molecules in a liquid also move around and collide. Molecules in a solid are restricted in their motion and are less likely to collide with each other. We find that reactions occur more readily in the gaseous and liquid states than in the solid state because the molecules are in motion. Reactions of a solid generally occur only on the surface of the solid. For example, an iron bar rusts on its surface, and the rusting works its way in. The rusting does not begin at the center of the bar.

In the gaseous and liquid states, collisions between molecules are commonplace. Yet, not all these collisions result in a chemical reaction. Most collisions between molecules result only in a change in direction of the movement of the molecules. This brings us to our second premise—for a reaction to occur, colliding molecules require a certain minimum amount of *energy*.

ENERGY OF COLLIDING MOLECULES. Molecules have two general types of energy, kinetic energy (energy of motion) and internal energy. The kinetic energy of molecules may be increased by adding heat. The heat energy is converted to kinetic energy and the molecules move faster. Not all the molecules in a substance have the same kinetic energy. Rather, the kinetic energy is distributed unevenly throughout the group of molecules. As a result, we speak of the average kinetic energy of molecules of a substance. When we heat the substance, we increase the average kinetic energy.

Internal energy is the energy within a molecule. We can best picture internal energy as vibrations of the bonds in a molecule. A bond is not a completely rigid entity; a bond can stretch, shrink, and bend, even though it continues to bond two atoms.

Think of two atoms connected by a covalent bond as if they were two balls connected by a rubber band. The balls can jiggle and dance around and still be connected by the rubber band. However, if the balls are jiggled hard enough, the rubber band may break. The same is true of two atoms connected by a covalent bond. When the molecule has sufficient internal energy, the atoms may vibrate vigorously enough that the bond can break. (See Figure 5-1.)

The reason that all colliding molecules do not react is that they do not have enough internal energy for bonds to break at the time of collision. For many reactions,

Figure 5-1. The bonds in a molecule are capable of vibrating and breaking.

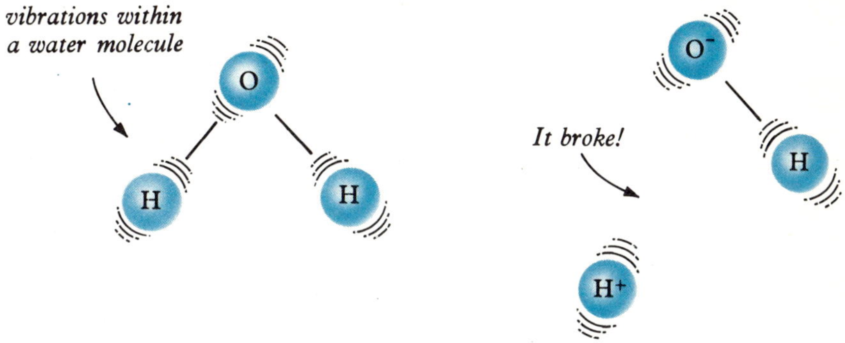

we must supply outside energy to the molecules for a reaction to occur. We do this usually (but not always) by heating. The heat increases the average kinetic energy, and this increased energy causes more-frequent collisions and also collisions of greater impact. Some of this kinetic energy is transferred to internal energy. When the internal energy of colliding molecules is great enough, a reaction can take place. Since the individual energy of the molecules in a group is not all the same, the entire group of molecules will not all have sufficient internal energy to undergo reaction at the same time. Rather, some collisions will result in reaction; other collisions will not.

An example of a chemical reaction is the reaction of methane and oxygen:

$$\underset{\text{methane}}{CH_4} + \underset{\text{oxygen}}{2\,O_2} \longrightarrow \underset{\text{carbon dioxide}}{CO_2} + \underset{\text{water}}{2\,H_2O} + \text{Heat}$$

Methane is the principal ingredient of natural gas. We know that it will burn readily in the presence of oxygen and form carbon dioxide and water. However, if we just mix methane and oxygen in a bottle, the methane does not burn; none of the molecules in the bottle has enough internal energy to break its bonds and form a new molecule. The molecules just move around and collide with each other at random, and no reaction takes place.

Now let us add a little energy, a spark, to the methane-oxygen mixture. The energy of the spark is transferred to a few of the molecules. These few molecules now have enough internal energy so that when one of these molecules of methane collides with one of these molecules of oxygen, they can undergo reaction. We say they have enough energy to cross the *energy barrier* and go to products.

As methane and oxygen cross the barrier and go to carbon dioxide and water, energy is released. Some of this energy is transferred to other molecules of methane and oxygen, and, when they collide, they too can cross the energy barrier and give up still more energy. Finally, the entire mixture of methane and oxygen has undergone reaction. We can see and feel the excess energy released from the overall process as light and heat.

THE ENERGY DIAGRAM. Picture a bear or, better yet, a crowd of bears in a mountain valley. On the other side of the mountains is another valley, and the two valleys are connected by a pass. Any one bear needs energy to go over the pass. He may start up the pass, but he will not make it to the top if he does not have enough energy. The pass is the energy barrier.

Let us increase the average energy of the bears, perhaps with a forest fire behind them. Bears will begin climbing toward the pass more energetically than before. Eventually, some will make it to the top and come down into the valley on the other side. Since all bears do not have the same energy, it would be unlikely that all the bears would reach the top of the pass at the same time.

Back to our reacting molecules. They, too, begin at a low point of energy, which we can call the *average energy of the reactants*. If we increase the average energy of the reactants, eventually some of the molecules of the reactants will have the minimum amount of internal energy needed to cross the energy barrier and go to products. We can illustrate the relationship between energy and the progress of the reaction by an *energy diagram*. (See Figure 5-2.)

Looking at the energy diagram for the burning of methane (Figure 5-3) we see that methane and oxygen have a certain amount of energy at room temperature, but

Figure 5-2. Energy diagrams.

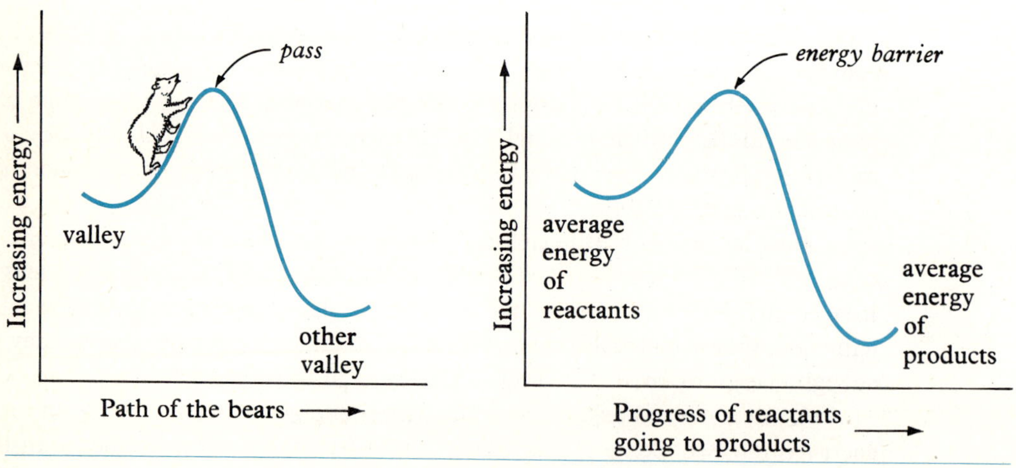

Figure 5–3. The energy diagram for the reaction of methane and oxygen.

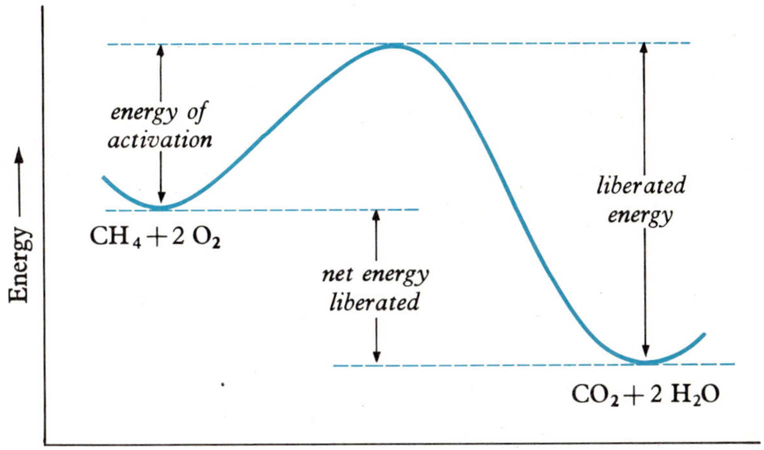

not enough to undergo reaction. We must supply a certain amount of energy to *activate* some of the molecules. The amount of energy needed to go from the energy of the reactants to the top of the energy barrier is called the *energy of activation.* Then, after passing the energy barrier, as the reactants are going down the other side and becoming products, energy is *liberated.*

Not all substances have the same amount of energy. From the energy diagram, we can see that carbon dioxide and water have less energy than methane and oxygen. When the products of a reaction have less energy than the reactants, there is a net release of energy during the course of the reaction. When methane burns, heat and light are given off. A reaction that gives off energy as it proceeds is called an *exothermic reaction.* (*Exo-* means *out,* as in the word "exit." *Therm* means *heat,* as in thermal underwear.)

If the products of a reaction contain more energy than the reactants, the reaction is called an *endothermic* reaction. (*Endo-* comes from the Greek word *endon,* meaning *within.*) We must supply heat continuously to an endothermic reaction to keep it going. The heat both helps the molecules over the energy barrier and makes up the energy difference between the reactants and the products. (See Figure 5–4.)

Not all reactions have energy barriers of the same height. Since sodium and water explode when they are put in contact with each other at room temperature, we know the energy of activation for the reaction of sodium and water is not very high. (See Figure 5–5.)

Figure 5-4. Energy diagrams showing the differences between an exothermic reaction and an endothermic reaction.

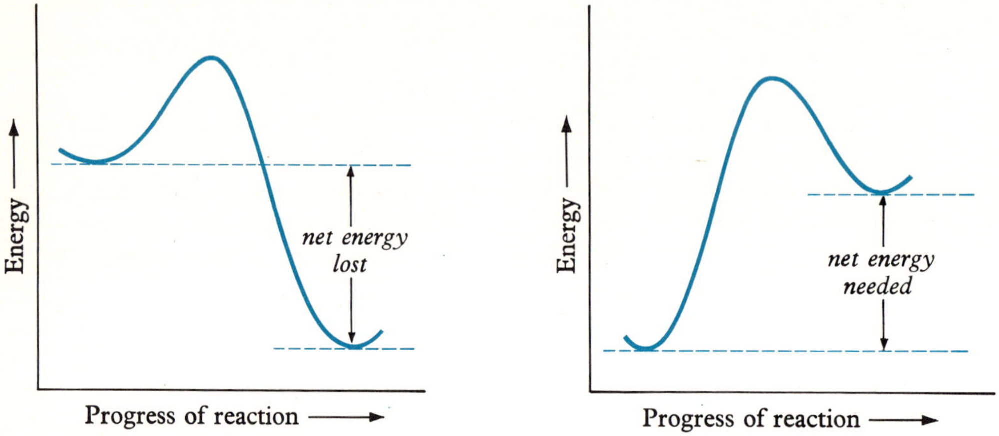

(an exothermic reaction) (an endothermic reaction)

Figure 5-5. Energy diagram for the reaction of sodium and water.

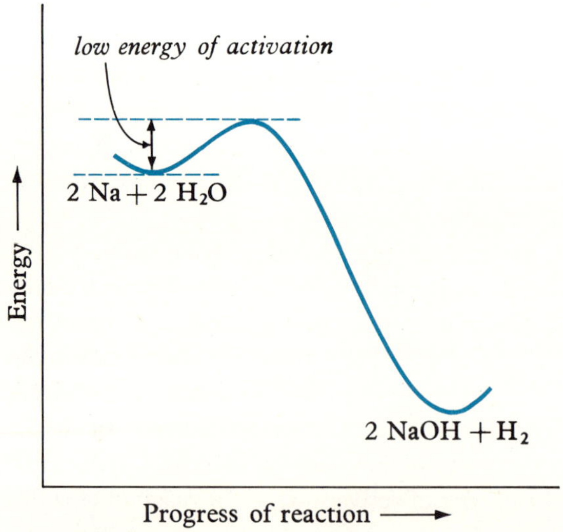

$$2\ Na + 2\ H_2O \longrightarrow 2\ NaOH + H_2 + \text{Energy}$$

CATALYSTS. Some reactions have energy barriers so high that we cannot make the reactions proceed; however, the same reactions might proceed easily when a *catalyst* is present. A catalyst is a substance that does not itself undergo reaction under the reaction conditions, but lowers the energy barrier for the reaction of other substances. Since it does not undergo reaction itself, only a small amount of a catalyst is needed to catalyze a reaction of comparatively large amounts of reactants.

Living organisms could not function without catalysts, which in living systems are called *enzymes*. An enzyme permits a reaction to take place quickly at body temperature under mild conditions, when the same reaction may require hours of heating in the laboratory. For example, hydrolysis (cleavage with water) of protein molecules (which is an essential part of the digestion of meats, eggs, etc.) requires boiling with a strong acid in the laboratory. In the human digestive tract, this same reaction occurs at 37°C (body temperature) and under only slightly acidic conditions because the reaction is catalyzed by enzymes (pepsin, trypsin, and others).

$$\text{A large protein molecule} + H_2O \xrightarrow{\text{enzymes}} \text{Many small molecules of amino acids}$$

Figure 5-6 shows two typical energy diagrams for a reaction with and without a catalyst. It may be seen that the energy of the reactants and that of the products are not changed by the addition of the catalyst; only the height of the energy barrier is changed.

Figure 5-6. A catalyst lowers the energy barrier for a reaction.

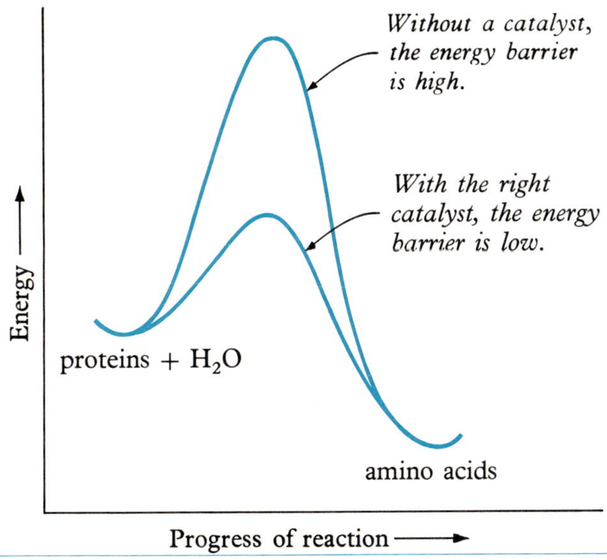

THE RATE OF A REACTION. Some reactions proceed quickly and some slowly. The speed at which a reaction proceeds is called the *rate of the reaction.* The rate of any reaction depends upon the number of molecules of reactants passing over the energy barrier in a given period of time.

Experimentally, we find that certain *types* of reactions proceed at different rates. Reactions of ionic compounds in solution usually proceed quickly because of the mobility of the ions and the low energy of activation for combination. In reactions of covalently bonded compounds, bonds must be broken and reformed. In general, reactions of covalent compounds have higher energies of activation than do reactions of ionic compounds, and these reactions generally proceed slowly in comparison to ionic reactions.

The rate of a reaction may be increased in a number of ways. An *increase in temperature* usually causes an increase in the rate of a reaction because the added heat increases the average kinetic and internal energies of the reactant molecules. The increased energy allows molecules to cross the energy barrier at a greater rate because of the increased number of collisions of molecules and because the collisions are more likely to result in reaction. Roughly, a temperature increase of 10°C doubles the rate of a reaction.

Conversely, a *decrease in temperature* usually causes a decrease in the rate of a reaction because there are fewer collisions and because the molecules have a lower internal energy. Thus, the likelihood of a collision yielding products is diminished.

The *concentration* of a substance is the amount of that substance in a given volume of a mixture. If we dissolve a pinch of table salt in a pan of water, there is a certain concentration of salt in the water. If we dissolve a cup of salt in the same amount of water, we have a greater concentration of salt. Concentration may be expressed in many ways—grams per liter, percent by weight, percent by volume, and in other ways that we will discuss in Chapter 6.

An *increase in the concentration* of a reactant causes an increase in the rate of the reaction. The increased concentration means simply that there are more molecules of a reactant in a given volume. Therefore, there are a greater number of collisions between reactant molecules and more molecules can cross the energy barrier in a given period of time.

A reaction rate can also be increased by the *addition of a catalyst.* As we have mentioned, a catalyst lowers the energy barrier and allows molecules to cross more easily.

SIDE REACTIONS. A side reaction is a reaction that can occur simultaneously with the main, or expected, reaction. When methane burns, we expect to get carbon dioxide and water, but carbon monoxide (CO) may also be formed.

Since the formation of carbon monoxide requires less oxygen per CH_4 molecule than the formation of carbon dioxide, this side reaction is especially prevalent when methane is burned in an inadequate oxygen supply. Because carbon monoxide is poisonous, most states have laws that require proper venting of gas furnaces so that any carbon monoxide formed will pass outside.

The expected reaction: $CH_4 + 2\,O_2 \longrightarrow CO_2 + 2\,H_2O$
(carbon dioxide)

A side reaction: $2\,CH_4 + 3\,O_2 \longrightarrow 2\,CO + 4\,H_2O$
(carbon monoxide)

REVERSIBLE REACTIONS. The tart taste of a carbonated beverage is due in part to the presence of carbonic acid (H_2CO_3). When a beverage is carbonated, carbon dioxide gas is forced under pressure into the solution. The dissolved carbon dioxide undergoes reaction with the water to form carbonic acid.

$$CO_2 + H_2O \longrightarrow H_2CO_3$$
(carbonic acid)

When a bottle of soda is opened, it fizzes. When the pressure is released, carbon dioxide is formed from the dissociation of carbonic acid in the soda.

$$H_2CO_3 \longrightarrow CO_2 + H_2O$$

These two reactions are just the reverse of each other. Carbon dioxide and water form carbonic acid. Carbonic acid forms carbon dioxide and water. The two reactions can proceed either backward or forward. Such a reaction is called a *reversible reaction.* We use two arrows ($\rightleftharpoons$), instead of just one, to show the reversibility of a reaction.

$$CO_2 + H_2O \rightleftharpoons H_2CO_3$$

The double arrow means a reversible reaction.

For a reaction to be reversible, the energies of activation must be fairly low for both the forward reaction and the reverse reaction. This means that the energy of the reactants and that of the products must be fairly close to each other. An energy diagram for the reversible reaction of carbon dioxide and water is shown in Figure 5-7.

Figure 5–7. Energy diagrams for a reversible reaction and an irreversible reaction.

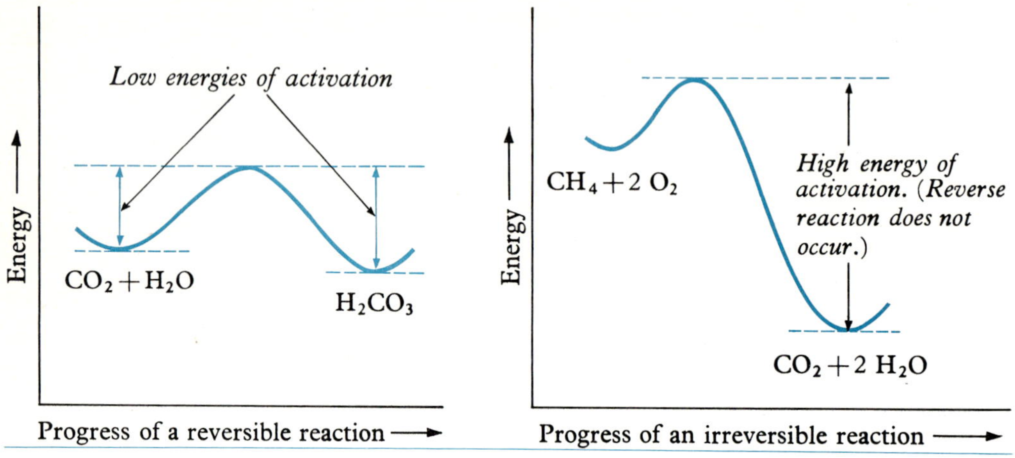

If the energy of the products of a reaction is much lower than the energy of the reactants, there is a high energy of activation for the reverse reaction. In this case, the reverse reaction does not proceed. For example, the burning of methane in oxygen produces carbon dioxide and water. The energy that would be required to get back to methane and oxygen from carbon dioxide and water is prohibitive, and no reverse reaction occurs. Therefore, we say that the reaction is *irreversible*.

HOW WE REPRESENT CHEMICAL REACTIONS

Let us now take a more detailed look at chemical equations, the formulas that go into them, and how we use them.

THE CHEMICAL FORMULA. An equation is made up from *chemical formulas,* which are symbols or groups of symbols representing molecules. The formula for neon is the same as its atomic symbol, Ne. This is because neon exists as single atoms. The formula for hydrogen is H_2, and that for water, H_2O. A formula may be simple or complex, and often we can represent the same compound in more than one way. As we have already mentioned, a formula that shows how all the atoms in a molecule are attached is called a *structural formula.*

If an atom appears more than once in a molecule, we may use a subscript number after the appropriate atomic symbol to shorten the formula. If the order of attachment of the atoms can still be deduced from the formula, we have a *condensed structural formula.*

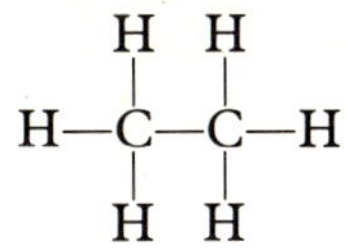

structural formula for ethane

CH_3CH_3

condensed structural formula for ethane

A formula that shows the composition of a molecule, but not the order of the attachment of atoms, is called a *molecular formula*. Another type of formula sometimes encountered is the *empirical formula,* a formula that shows the *relative* numbers (not necessarily the actual numbers) of atoms of each type in a molecule.

C_2H_6	CH_3
molecular formula for ethane	*empirical formula for ethane*

Table 5-1 lists some other examples of different types of formula for a few compounds.

Table 5-1. Different Types of Chemical Formula

Name of Compound	Structural Formula	Condensed Structural Formula	Molecular Formula	Empirical Formula
Water	H—O—H	H_2O	H_2O	H_2O
Carbonic acid	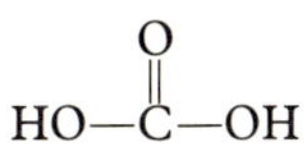	$HOCO_2H$	H_2CO_3	H_2CO_3
Ethylene	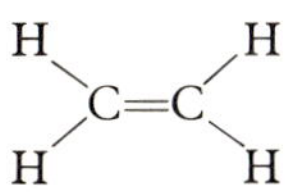	$CH_2{=}CH_2$	C_2H_4	CH_2

Sample Problem: Write the molecular and empirical formulas for (a) butane ($CH_3CH_2CH_2CH_3$) and (b) acetylene ($CH{\equiv}CH$).

Solution: (a) C_4H_{10} is the molecular formula and C_2H_5 is the empirical formula for butane.

(b) C_2H_2 is the molecular formula and CH is the empirical formula for acetylene.

Any of these formulas tells us the composition of a molecule, that is, each kind of atom in a molecule and its ratio to the other atoms. Using atomic masses, we also can calculate the *percent by weight* of any of the elements within a compound.

Sample Problem: What percentage of the weight of CO_2 is contributed by carbon?

Solution: Calculate the molecular weight of CO_2 by adding the atomic masses:

$$\begin{aligned} C &= 12.0 \\ O &= 16.0 \\ \underline{O} &\underline{= 16.0} \\ \text{Molecular weight} &= 44.0 \end{aligned}$$

Now the problem becomes: Carbon (atomic mass 12.0) is what percentage of the total mass of a CO_2 molecule (molecular weight 44.0)?

$$\frac{12.0}{44.0} \times 100 = 27.3\% \text{ C in } CO_2$$

READING AN EQUATION. In an equation, the formulas are arranged to show a chemical change. An equation is a statement.

$$H_2 + Cl_2 \longrightarrow 2\,HCl$$

Hydrogen undergoes reaction with chlorine to yield hydrogen chloride.

An equation tells us more than just this statement of fact. We *balance* an equation to show the proportions of reactants and products. The proportions can be read in a variety of ways, depending on what you want to know from the equation. You will recall that a *mole* is the molecular weight of a substance expressed in grams, and that 1.0 mole of any substance contains 6×10^{23} molecules. The number of moles of a substance is therefore proportional to the number of molecules. For this reason, an equation can be read as "molecules" or as "moles."

$$H_2 + Cl_2 \longrightarrow 2\,HCl$$

One molecule of H_2 undergoes reaction with one molecule of Cl_2 to yield two molecules of HCl.

or

One mole of H_2 undergoes reaction with one mole of Cl_2 to yield two moles of HCl.

BALANCING AN EQUATION. Mass is neither gained nor lost in a chemical reaction: "What goes in has to come out." Both sides of a chemical equation must have the same number of each type of atom. If there are four hydrogen atoms in the reactant side of an equation, there must be four hydrogen atoms on the product side.

$$2\,H_2 + O_2 \longrightarrow 2\,H_2O$$

reactants *product*

Each side of the equation shows four H's and two O's.

A subscript number in a formula cannot be changed to balance an equation, because changing a subscript amounts to changing a formula. Rather, we use a number *before* a formula to give the proper proportions in an equation.

Let us consider the reaction of sodium metal with chlorine gas. Because it has been investigated in the laboratory, we know that sodium, a very reactive metal that explodes on contact with water, and chlorine, a toxic gas used in very small quantities to kill bacteria in drinking water, undergo reaction to give us ordinary table salt, or sodium chloride.

Not balanced: $$\underset{\text{sodium}}{Na} + \underset{\text{chlorine}}{Cl_2} \longrightarrow \underset{\text{sodium chloride}}{NaCl}$$

The preceding equation is not balanced. We do not have the same number of chlorine atoms on each side. We can put a 2 in front of NaCl. This gives us two Cl atoms on each side.

Not balanced: $$\underset{\text{one Na atom}}{Na} + \underset{\text{two Cl atoms}}{Cl_2} \longrightarrow 2\,NaCl$$

(2 NaCl: two Cl atoms; two Na atoms)

Now the sodium is not balanced. We must put a 2 in front of Na on the left side of the equation. Now all the atoms are balanced.

Balanced: $$2\,Na + Cl_2 \longrightarrow 2\,NaCl$$

Most chemical equations may be balanced by just such a hit-or-miss method. We call this method of balancing equations *balancing by inspection.*

USING A BALANCED EQUATION. A balanced equation gives a variety of information—the structures of the reactants and of the products and, importantly, the proportions of the reactants and the products.

In the preceding reaction, the ratio of Na to Cl_2 is 2:1. We do not need to use exactly two atoms of sodium and one molecule of chlorine, or exactly two moles of sodium and one mole of chlorine for complete reaction to occur. We can vary the amounts of reactants we use as long as the proportions are the same as in the balanced equation. In this case, as long as the molar ratio (not the weight ratio) of sodium to chlorine is 2:1, we have the proper proportions of sodium and chlorine for complete reaction to occur.

$$2\,Na + Cl_2$$

2 moles + 1 mole

2000 moles + 1000 moles

0.2 mole + 0.1 mole

0.1 mole + 0.05 mole

0.05 mole + 0.025 mole

Given the number of moles of any reactant or product, we can calculate how many moles of the other substances will be needed or produced by using a balanced equation.

Sample Problem: Iron metal undergoes reaction with oxygen to form ferric oxide (Fe_2O_3), or ordinary rust. How many moles of oxygen would be used for the complete rusting of 1.0 mole of Fe?

Solution:

$$\underset{1.0\ \text{mole}}{4\,Fe} + \underset{x\ \text{moles}}{3\,O_2} \longrightarrow 2\,Fe_2O_3$$

The ratio of Fe to O_2 is 4:3. To keep this ratio of Fe to O_2, we set up a proportion: 4 is to 3 as 1.0 is to x.

$$\frac{4\ \text{moles Fe}}{3\ \text{moles}\ O_2} = \frac{1.0\ \text{mole Fe}}{x\ \text{moles}\ O_2}$$

$$x = (\tfrac{3}{4})(1.0)$$

$$= 0.75\ \text{mole}\ O_2$$

In setting up a proportion like this, we must be careful not to invert one side. For this reason, it is a good idea to carry along the units (for example, moles O_2). Then we can check to see that the units are the same on both sides of the equation.

$$\frac{m\ \text{moles of}\ A}{n\ \text{moles of}\ B} = \frac{x\ \text{moles of}\ A}{y\ \text{moles of}\ B}$$

CONVERTING MOLES TO GRAMS. If we have an analytical balance, we can weigh a certain number of grams of a substance, but we cannot directly weigh a mole. However, the number of moles of reactants and products can be converted to grams. A chemist can thus calculate exactly how many grams of each reactant are necessary for complete reaction,

$$\underset{\substack{1.0\ \text{mole}\\ \text{or}\ 2.0\ \text{g}}}{H_2} + \underset{\substack{1.0\ \text{mole}\\ \text{or}\ 71.0\ \text{g}}}{Cl_2} \longrightarrow \underset{\substack{2.0\ \text{moles}\\ \text{or}\ 73.0\ \text{g}}}{2\,HCl}$$

What if a chemist desires to use 0.1 mole or 100 moles instead of simply 1.0 mole of H_2? The chemist would need to know the weight of a fraction of a mole or of several moles. To convert a certain number of moles to grams, we can use the following proportion:

$$\frac{x\ \text{grams}}{y\ \text{moles}} = \frac{\text{Molecular weight in grams}}{1.0\ \text{mole}}$$

Sample Problem: How many grams of HCl are present in 0.33 mole?

Solution: *Step 1:* Find the molecular weight of HCl.

$$1.0 + 35.5 = 36.5$$

Step 2: Set up the equation:

$$\frac{x \text{ grams}}{y \text{ moles}} = \frac{\text{Molecular weight in grams}}{1.0 \text{ mole}}$$

Step 3: Substitute and solve.

$$\frac{x \text{ grams}}{0.33 \text{ mole}} = \frac{36.5 \text{ grams}}{1.0 \text{ mole}}$$

$$x = \frac{(36.5 \text{ g})(0.33 \text{ mole})}{(1.0 \text{ mole})}$$

$$= 12 \text{ g of HCl}$$

Sample Problem: How many grams of oxygen (O_2) are present in 0.75 mole?

Solution: *Step 1:* Find the molecular weight of O_2.

$$16.0 + 16.0 = 32.0$$

Step 2: Set up the equation.

$$\frac{x \text{ g}}{y \text{ mole}} = \frac{\text{Molecular weight in g}}{1.0 \text{ mole}}$$

Step 3: Substitute and solve.

$$\frac{x \text{ g}}{0.75 \text{ mole}} = \frac{32.0 \text{ g}}{1.0 \text{ mole}}$$

$$x = \frac{(32.0 \text{ g})(0.75 \text{ mole})}{(1.0 \text{ mole})}$$

$$= 24.0 \text{ g } O_2$$

CONVERTING GRAMS TO MOLES. To convert a certain number of grams of a substance to moles, we use the following formula. (This is the same as the formula in the preceding section; it is simply rearranged for convenience.)

$$\text{Number of moles} = \frac{\text{Weight of sample in grams}}{\text{Weight of 1.0 mole}}$$

Sample Problem: How many moles of NaOH are in 10.0 g?

Solution: *Step 1:* Find the molecular weight of NaOH.

$$23.0 + 16.0 + 1.0 = 40.0$$

Step 2: Weight of 1.0 mole = 40.0 g

Step 3: Number of moles in 10.0 g NaOH $= \dfrac{10.0 \text{ g}}{40.0 \text{ g}}$

$$= 0.25 \text{ mole}$$

USE OF MOLES AND GRAMS IN AN EQUATION. Now we can look at chemical reactions on a weight basis. Given the weight of any one substance in an equation, whether it is a reactant or a product, we can predict what the other weights should be. However, we still must use moles because only the moles give us the right proportions of reactants and products. This is the procedure that we follow:

1. Balance the equation.
2. Convert grams to moles.
3. Calculate the desired number of moles from the proportions in the balanced equation.
4. Finally, convert the desired number of moles to grams.

Sample Problem: Starches and sugars are converted to glucose in the digestive system. The glucose is then absorbed into the bloodstream and carried to the cells of the body, where some of it is oxidized to carbon dioxide and water. (The series of reactions in this conversion gives off energy that can be used by the body.) How many grams of carbon dioxide would be produced from 5.0 g of glucose?

$$\underset{\text{glucose}}{C_6H_{12}O_6} + \underset{\text{oxygen}}{6\,O_2} \longrightarrow \underset{\text{carbon dioxide}}{6\,CO_2} + \underset{\text{water}}{6\,H_2O} + 686\ \text{kcal/mole}$$

Solution: (1) The equation is balanced.

(2) Molecular weight of glucose $= 6(12.0) + 12(1.0) + 6(16.0) = 180$

The weight of 1.0 mole of glucose is therefore 180 g.

$$\text{No. of moles of glucose} = \frac{5.0\ \text{g}}{180\ \text{g per mole}} = 0.028$$

(3) No. of moles of CO_2 produced $= 6 \times$ (moles of glucose) $= 6(0.028) = 0.17$

(4) Molecular weight of $CO_2 = 12.0 + 2(16.0) = 44.0$. The weight of 1.0 mole of CO_2 is therefore 44.0 g. To find the number of grams (x) of CO_2 produced,

$$\frac{x}{0.17\ \text{mole}} = \frac{44.0\ \text{grams}}{1.0\ \text{mole}}$$

$$x = \frac{(44.0\ \text{g})(0.17\ \text{mole})}{(1.0\ \text{mole})} = 7.5\ \text{g}\ CO_2$$

STUDY QUESTIONS

5-1. Express the following equations in words:
(a) $2\,Na + 2\,H_2O \longrightarrow 2\,NaOH + H_2$

(b) $2\,NH_4NO_3 + \text{Pressure} \xrightarrow{\text{fast}} 2\,N_2 + O_2 + 4\,H_2O$

5-2. Label the indicated parts of the following energy diagram:

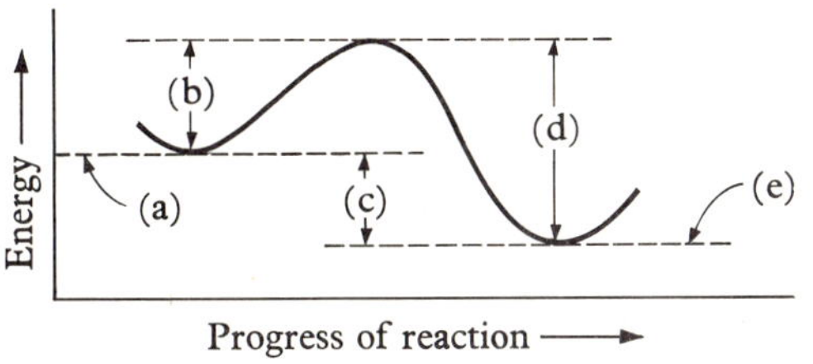

$2\,H_2 + O_2 \longrightarrow 2\,H_2O$

5-3. The reaction of hydrochloric acid and sodium hydroxide is very fast and exothermic. Draw an approximate energy diagram for the reaction.

5-4. The reaction of HCl and NaOH occurs almost instantly while the reaction of NaOH and $CH_3CO_2CH_2CH_3$ (ethyl acetate) requires several minutes depending upon the reaction conditions. Which reaction has the faster rate of reaction?

5-5. Which of the following reactions might be a side reaction in the burning of propane:
$CH_3CH_2CH_3 + 5\,O_2 \longrightarrow 3\,CO_2 + 4\,H_2O$?

(a) $CH_4 + 2\,O_2 \longrightarrow CO_2 + 2\,H_2O$
(b) $CH_4 + O_2 \longrightarrow CO + 2\,H_2O$
(c) $CH_3CH_2CH_3 + 2\,O_2 \longrightarrow 3\,C + 4\,H_2O$
(d) $2\,CH_3CH_2CH_3 + 7\,O_2 \longrightarrow 6\,CO + 8\,H_2O$

5-6. Explain briefly how a reversible reaction may be recognized by its energy diagram.

5-7. For each of the following formulas give (1) a condensed structural formula, (2) the molecular formula, and (3) the empirical formula:
(a) H—C≡C—H

(b) 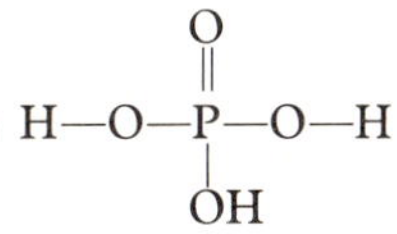

(c) 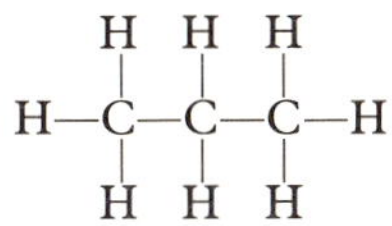

(d) 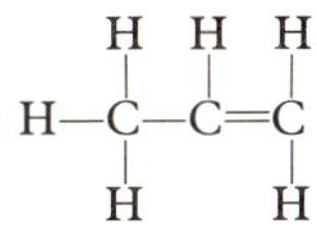

(e) 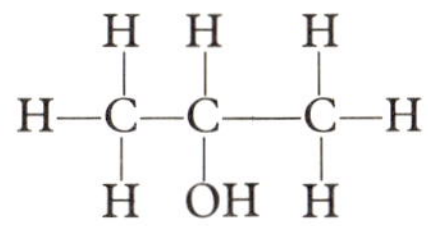

5-8. What percent by weight of H_2S is contributed by hydrogen?

5-9. What percent by weight of each of the following compounds is contributed by carbon?
(a) CH_4 (b) CH_3CH_3
(c) CH_3CH_2OH (d) $CH{\equiv}CH$
(e) $HO_2C{-}CO_2H$

5-10. What percent by weight of chloroform ($CHCl_3$) is contributed by chlorine?

5-11. Balance the following equations:
(a) $K + H_2O \longrightarrow KOH + H_2$
(b) $H_2 + Br_2 \longrightarrow HBr$
(c) $CH_3CH_3 + O_2 \longrightarrow CO_2 + H_2O$
(d) $Zn + HCl \longrightarrow ZnCl_2 + H_2$

5-12. Given the following equations, what would be the weight ratio of the reactants in each case?
(a) $HCl + NaOH \longrightarrow NaCl + H_2O$
(b) $H_2SO_4 + 2\,NaOH \longrightarrow Na_2SO_4 + 2\,H_2O$
(c) $H_3PO_4 + 3\,NaOH \longrightarrow Na_3PO_4 + 3\,H_2O$

5-13. How many grams are present in 0.125 mole of each of the following compounds?
(a) CH_4 (b) $Ag(NH_3)_2Cl$
(c) $Ca_3(PO_4)_2$ (d) $CHCl_3$

5-14. Fill in the blanks:

$$\underset{\text{(a) moles}}{HCl} + \underset{\text{(b) moles}}{NaOH} \longrightarrow \underset{\text{0.325 mole}}{NaCl} + \underset{\text{(c) moles}}{H_2O}$$

$$\underset{\text{10 moles}}{H_2SO_4} + \underset{\text{(d) moles}}{2\,NaOH} \longrightarrow \underset{\text{(e) moles}}{Na_2SO_4} + \underset{\text{(f) moles}}{2\,H_2O}$$

5-15. In the preceding problem, how many grams of HCl would be required for the complete reaction to 0.325 mole NaCl?

5-16.

(a) Balance the following equations:

$$AgNO_3 + NaCl \longrightarrow AgCl + NaNO_3$$
$$H_3PO_4 + NaOH \longrightarrow Na_3PO_4 + H_2O$$
$$(NH_4)_2SO_4 + KOH \longrightarrow NH_3 + K_2SO_4 + H_2O$$

(b) In the examples in (a), starting with 0.50 g of the reagent on the far left, calculate the number of grams of each product formed.

(c) Starting with 0.50 g of *both* reactants, calculate the number of grams of each product that would be formed.

5-17. Chlorine gas (Cl_2) can be produced from molten sodium chloride (NaCl) by an electric current. How many grams of sodium chloride would be required to produce 142 grams of chlorine gas?

5-18. In the presence of a catalyst, methane (CH_4) can react with chlorine gas to produce carbon tetrachloride (CCl_4) and hydrogen chloride gas (HCl). If 4.00 grams of methane were converted to CCl_4 by this procedure, how many grams of HCl would be produced?

CHAPTER 6 SOLUTIONS AND COLLOIDS

OBJECTIVES

Solutions *Give examples of different types of solution.*

Terms Used in Discussing Solutions *Define and exemplify solvent, solute, soluble, and insoluble.*

Liquid-in-Liquid Solutions *Define miscible and immiscible liquids, giving examples.*

Solid-in-Liquid Solutions *Define saturated solution, describe the equilibrium involved, and describe the effect of temperature changes upon the equilibrium.* * *Describe a supersaturated solution.*

Gas-in-Liquid Solutions *State Henry's Law.*

Concentrations of Solutions *Define the following: %(volume/volume), %(weight/weight), %(weight/volume), mg/liter, ppm,* M. * *Given the weight of solute and the volume of solution, calculate molarity.* * *Calculate equivalent weight and normality, given the weight of solute and the volume of solution.*

Dilution Problems *Calculate the volume of solvent needed to dilute a solution from one concentration to another.*

Properties of Solutions *Describe how a solute affects the freezing point and the boiling point of a solvent.* * *Describe the process of osmosis.*

Colloids *Define and give examples of a colloid.*

Particle Size of Colloids *Describe the differences between a colloid, a solution, and an ordinary mixture.*

Types of Colloid *Define and exemplify the different types of colloid.*

Emulsifying Agents *Describe and illustrate the function of an emulsifying agent.*

Properties of Colloids *Define the Tyndall effect and Brownian movement.*

Dialysis *Describe and exemplify the process of dialysis.*

SOLUTIONS

Solutions are encountered everywhere—in living organisms, in industry, in the laboratory, and in the kitchen. Drinking water is a solution of small amounts of minerals dissolved in water. Soda water is a solution of the gas carbon dioxide in water. Vinegar is a solution of the organic compound acetic acid in water. A *solution* is defined as a *homogeneous,* or *uniform, mixture of two or more kinds of atoms, molecules, or ions.* (See Figure 6-1.)

When a substance goes into solution, or dissolves, the individual molecules or ions separate from each other. These particles are too small to reflect light. Thus, with

Figure 6-1. A solution is a homogeneous mixture of two or more substances.

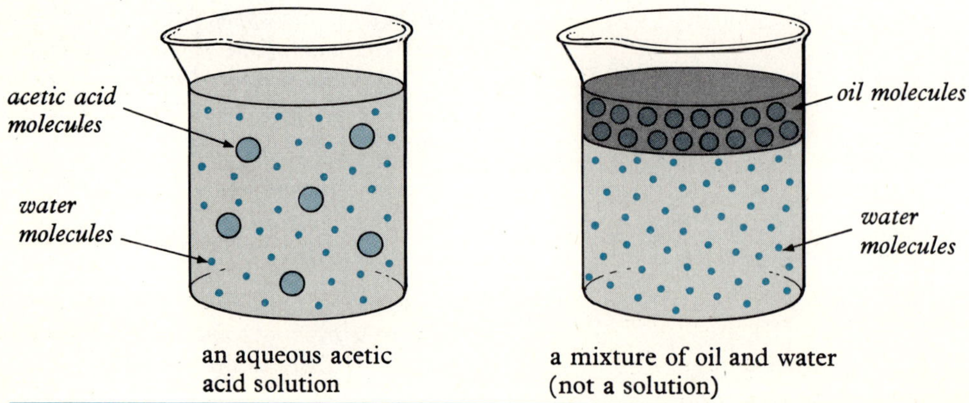

rare exceptions, solutions are transparent. We can see through a solution of salt in water as well as we can see through water. (Although transparent, a solution may be colored, as is red table wine.) If we put some flour in water and mix the two thoroughly, the mixture is cloudy; it is not a solution. The mixture is a colloid, a subject we will discuss later in this chapter.

Gravity cannot pull dissolved material from a solution. For example, dissolved sugar will not settle out of your morning coffee.

Our definition of the term "solution" is broad. Although we commonly think of a solution as a substance dissolved in a liquid, this is only one type of solution. The atmosphere is a gaseous solution; it is a homogeneous mixture of a number of different gases. In Table 6-1, we have listed several types of solution between the different states of matter. While there are many different types of solution, we will discuss primarily solutions of solids, liquids, or gases in liquids.

Table 6-1. Some Common Types of Solution

Type	Example
Liquid in liquid	Vinegar (acetic acid in water)
Solid in liquid	Brine and saline solution (sodium chloride in water); syrup (sugar in water)
Gas in liquid	Soda water (carbon dioxide in water)
Gas in gas	Air (oxygen and other gases in nitrogen)

TERMS USED IN DISCUSSING SOLUTIONS

When a particular substance can be dissolved in a liquid, we say that the substance is *soluble* in that liquid. We call the liquid the *solvent* and the material being dissolved the *solute*. When you add sugar to your coffee, sugar is the solute and water is the solvent.

Substances exhibit varying solubilities in solvents. When a large amount of a substance can dissolve in a given amount of solvent, we say that the substance is *very soluble* in that solvent. Sodium chloride is very soluble in water. When only a small quantity of a substance can dissolve in a given quantity of a solvent, we say that the substance is *slightly soluble* in that solvent. When a substance will not dissolve to any appreciable extent in a solvent, such as glass in water, we say that the substance is *insoluble* in that solvent.

LIQUID-IN-LIQUID SOLUTIONS

The definitions of solute and solvent become hazy when we speak of dissolving a liquid in a liquid. Ethyl alcohol (CH_3CH_2OH) is an organic compound that is a liquid and is *infinitely soluble* in water. That is, any amount of ethyl alcohol will dissolve in any amount of water, and any amount of water will dissolve in any amount of ethyl alcohol. Two liquids that are infinitely soluble in each other are said to be *completely miscible.* Since two completely miscible liquids can dissolve in each other to any extent, the distinction between the solvent and the solute in such a case is rather meaningless. Generally, we call the liquid that is present to the greater extent the solvent.

Liquids vary in their solubilities in each other. Because diethyl ether ($CH_3CH_2OCH_2CH_3$) is very slightly soluble in water, we say that diethyl ether and water are *partially miscible.* If more ether is added to a quantity of water than can dissolve in that amount of water, the excess ether forms a layer on top of the water. Gasoline and motor oil are insoluble in water. We say that these substances are *immiscible* with water.

SOLID-IN-LIQUID SOLUTIONS

SATURATION. When we dissolve a solid solute in a given amount of liquid solvent, we find that there is a limit to the amount of solute that will dissolve. Let us put some water in a pan, add some salt, and stir. The salt dissolves. We add more salt and stir. If we keep this up, we will arrive at a point at which, no matter how long or hard we stir, not all of the salt will dissolve. When a solvent has dissolved as much of a solute as it is able, we say that the solvent is saturated with solute, or that we have a *saturated solution* of the solute in the solvent.

A saturated solution is in equilibrium. In the example of a saturated salt solution, salt is dissolving in the water, but an equal amount of salt is coming out of the solution. In other words, as sodium ions and chloride ions break away from the salt crystals and dissolve, other sodium ions and chloride ions join the crystals as a

Figure 6–2. A saturated solution. The same number of ions or molecules are leaving the solution as are dissolving.

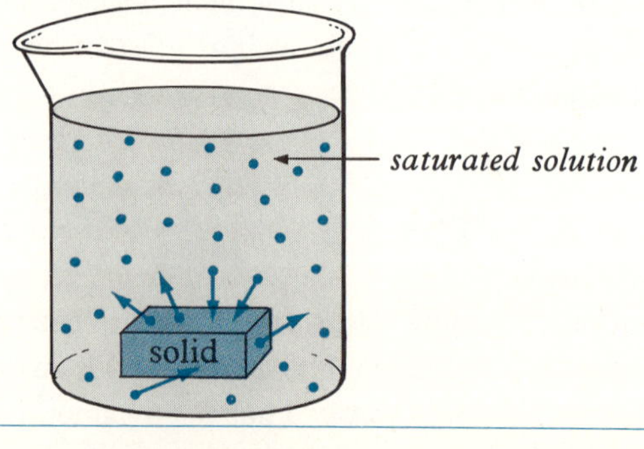

Table 6–2. Some Typical Solubilities in Water

Solute	Solubility in 100 g of Water at 20°C	at 100°C
Sodium chloride	36.0 g	39.8 g
Sodium hydroxide	109.0	347.0
Copper(II) sulfate	20.7	75.4
Ethyl alcohol	∞[a]	∞[a]

[a] Infinitely soluble, or miscible.

solid. (See Figure 6–2.) However, the *number* of sodium ions and chloride ions in the solution remains constant.

Now we will put our saturated salt solution on the stove and heat it. Part of the excess solid salt dissolves. If we cool the hot salt solution, some salt will come back out of, or *crystallize* from, the solution. The amount of salt that will dissolve in a given amount of water depends upon the temperature. Therefore, when we talk about the amount of solute needed to saturate a given amount of solvent, we must specify the temperature.

We find that 36.0 grams of sodium chloride will dissolve in 100 grams of H_2O at 20°C and that 39.8 grams will dissolve at 100°C. It is generally true that a greater amount of a solid or liquid solute will dissolve in a solvent as the temperature is increased. The maximum amount of a substance that will dissolve in a given amount of solvent at a particular temperature is called the *solubility* of that substance in that solvent at that temperature. Some typical solubilities of compounds in water are found in Table 6–2.

SUPERSATURATION. When a hot saturated solution of a substance is cooled, some of the solute may come out of solution. Sometimes, however, the solute does *not* come out of solution as we would expect. We call such a solution a *supersaturated solution.* The reason that a solution can be supersaturated is that the excess molecules or ions have no crystal lattice present that they can join. Once a crystal lattice is present, ions or molecules can easily come out of solution, but for molecules or ions to start a crystal lattice requires energy.

To start crystallization from a supersaturated solution, a crystal of the solute, called a *seed crystal,* may be added. The seed crystal provides a crystal lattice that can easily be joined by excess solute molecules that are in the solution. (See Figure 6–3.)

Seeding is a technique used in rainmaking. Pieces of dry ice or crystals of silver iodide are used as seeds and dropped into clouds to attract water molecules. Once a small droplet of water is formed on the seed, it supplies a base for other molecules of water to aggregate. When the droplet becomes heavy enough, it falls to the ground as rain.

Figure 6–3. Planting a seed crystal in a supersaturated solution allows the excess solute to crystallize.

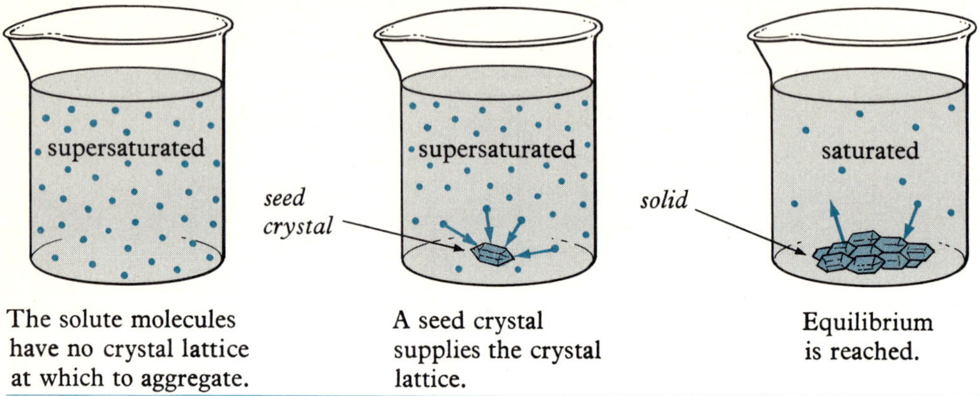

GAS-IN-LIQUID SOLUTIONS

When a scuba diver ascends too rapidly from deep water, the diver may develop a painful and sometimes fatal malady known as the "bends." To prevent or treat the bends, the diver is put under pressure in a decompression chamber so that the pressure on the body can be brought slowly back from the greater pressure of the water to atmospheric pressure. The bends are caused by the formation of bubbles of nitrogen gas in the blood. Where do the gas bubbles come from? Why do they appear only when the diver ascends in the water? Why do they not appear when the pressure on the diver is decreased slowly?

The quantity of gas that dissolves in a liquid is dependent upon the pressure upon the gas. The greater the pressure, the more gas that can dissolve; the lower the pressure, the less gas that can dissolve. (See Figure 6–4.) A more formal statement of this relationship is the following:

HENRY'S LAW: *The solubility of a gas in a liquid is directly proportional to the pressure of the gas above the liquid.*

When a scuba diver descends in the water, the pressure of the water upon the diver's body increases the pressure of the gases in the lungs. This increase in gas pressure, in turn, increases the solubility of air in the blood. When the diver ascends, the pressure of the water upon the diver's body decreases; the pressure of the air in the lungs decreases; and the solubility of air, primarily nitrogen, in the blood decreases. If the ascent is slow (or if a decompression chamber is used), an equilibrium can be maintained, and the excess air can be discharged back into the lungs and

Figure 6-4. The solubility of a gas is directly proportional to the pressure of the gas above the liquid.

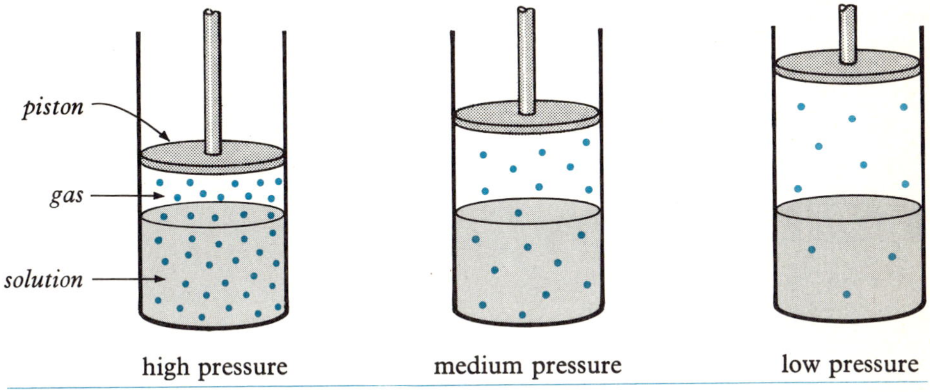

exhaled. On the other hand, if the ascent is rapid, not all the excess air can be discharged quickly enough by way of the lungs, and air bubbles appear in the capillaries. These bubbles cause the symptoms of the bends.

The fizzing of beer or a carbonated drink is another example of Henry's Law. These beverages each contain carbon dioxide maintained in solution by the pressure of the gas above the liquid. When a bottle is opened, the pressure is decreased. Some of the dissolved carbon dioxide bubbles out of solution.

CONCENTRATIONS OF SOLUTIONS

Most solutions with which we deal are not saturated solutions, but rather solutions that are less than saturated in varying degrees. Therefore, we need a way of expressing how much solute is present in the solution. Qualitatively, we may say that a solution is *dilute* (of low concentration) or that it is *concentrated* (of high concentration). For example, we can describe vinegar as a dilute solution of acetic acid, and shampoo as a concentrated solution of soap.

Since chemical reactions take place in definite proportions, we need a *quantitative,* rather than qualitative, method of expressing concentration. We find that a number of methods have been devised and are in common use.

PERCENT (VOLUME/VOLUME). One convenient way to mix a liquid-liquid solution is by measuring the volume of the solute, then diluting with solvent to the desired volume. The concentration of the solute then can be expressed as a percent of the solution by volume. If 5 ml of pure ethyl alcohol is diluted to 100 ml with water, the final solution is *5 percent of ethyl alcohol by volume, 5 percent* (*volume/volume*), or *5* *percent* (v/v). The concentration may be expressed by any of these terms, all of which mean the same thing.

$$\% \text{ by volume} = \% \text{ (volume/volume)} = \frac{\text{ml of solute}}{\text{Total ml of solution}} \times 100$$

Note that the volume of solute is divided by the *total volume of solution,* and *not* by the volume of solvent added. When we are mixing our 5 percent (volume/volume) solution of ethyl alcohol, we must dilute 5 ml *to* 100 ml with water. If we diluted 5 ml of ethyl alcohol *with* 100 ml of water, we would have slightly less than a 5 percent (volume/volume) solution.

Sample Problem: Find the percent by volume of ethyl alcohol in a solution prepared by diluting 30.0 ml of ethyl alcohol *to* 170 ml with water.

Solution:

$$\frac{30.0 \text{ ml solute}}{170 \text{ ml solution}} \times 100 = 17.6\% \text{ (volume/volume)}$$

Sample Problem: Find the percent by volume of ethyl alcohol after 30 ml of ethyl alcohol has been diluted *to* 200 ml (that is, approximately 170 ml of water has been added).

Solution:

$$\frac{30 \text{ ml solute}}{200 \text{ ml solution}} \times 100 = 15\% \text{ (volume/volume)}$$

PERCENT (WEIGHT/WEIGHT). Expressing a concentration in *percent by weight, percent* (*weight/weight*), or *percent* (w/w) is similar to expressing a percent by volume, except that grams are used instead of milliliters.

$$\% \text{ by weight} = \% \text{ (weight/weight)} = \frac{\text{g of solute}}{\text{Total g of solution}} \times 100$$

Once again, we must pay attention to whether the solute is diluted *to* or diluted *with* a given weight of solvent.

Sample Problem: What is the percent by weight of ammonium chloride (NH_4Cl) in a solution made from 20.0 g of ammonium chloride diluted *with* 300 g of water?

Solution:

$$\frac{20.0 \text{ g } NH_4Cl}{(20.0 \text{ g} + 300 \text{ g}) \text{ solution}} \times 100 = 6.25\% \text{ (weight/weight)}$$

Sample Problem: What is the percent by weight of NH_4Cl in a solution made by dissolving 20.0 g of NH_4Cl in water and diluting *to* 300 g?

Solution:

$$\frac{20.0 \text{ g } NH_4Cl}{300 \text{ g solution}} \times 100 = 6.67\% \text{ (weight/weight)}$$

We can use the percent (weight/weight) formula to calculate other values as well. If we know the weight of the solution and the percent by weight of solute, we can calculate the actual weight of the solute in grams.

Sample Problem: How many grams of sodium chloride are present in 60 g of a 5.0 percent (weight/weight) solution?

Solution:

$$\%\ (\text{weight/weight}) = \frac{\text{g of solute}}{\text{g of solution}} \times 100$$

Substitute:

$$5.0\% = \frac{\text{g NaCl}}{60\ \text{g solution}} \times 100$$

Rearrange:

$$\text{g NaCl} = \frac{5.0\% \times 60\ \text{g}}{100}$$

$$\textit{Solve} = 3.0$$

PERCENT (WEIGHT/VOLUME). A widely used method of expressing concentration is a combination of weight and volume. When working with a solid solute and a liquid solvent, it is convenient to weigh the solute in grams and to measure the volume of solution in milliliters. In this case, we speak of concentration in terms of *percent (weight/volume)*, or *percent* (w/v).

$$\%\ (\text{weight/volume}) = \frac{\text{g of solute}}{\text{Total ml of solution}} \times 100$$

Sample Problem: If 2.0 g of NaCl is diluted with water to 200 ml of solution, what is the percent (weight/volume) of NaCl in the solution?

Solution:

$$\frac{2.0\ \text{g NaCl}}{200\ \text{ml solution}} \times 100 = 1.0\%\ (\text{weight/volume})$$

Sample Problem: *Physiological saline solution,* a 0.9% (w/v) solution of sodium chloride in water, has the same concentration of NaCl as blood. How many grams of sodium chloride are present in 1000 ml of physiological saline solution?

Solution:

$$\%\ \text{w/v} = \frac{\text{g of solute}}{\text{ml of solution}} \times 100$$

Substitute:

$$0.9\% = \frac{\text{g NaCl}}{1000\ \text{ml}} \times 100$$

Rearrange and solve:

$$\text{g NaCl} = \frac{0.9 \times 1000}{100}$$

$$= 9$$

Expressions of concentration as percentages are exceedingly common in our society, particularly in industry and in the practicing medical sciences. Unfortunately, these expressions are often misused. For example, when we read that a carton of milk contains 2 percent butterfat, is that percent (volume/volume), percent (weight/weight), or percent (weight/volume)? We do not know, because the label does not tell us what method was used for the calculation. Each method would give a slightly different percentage of butterfat in a quart of milk. Therefore, unless we know the method that was used, we can only use the figure "2 percent" as a semiquantitative estimate of the butterfat in the milk.

CONCENTRATION EXPRESSIONS FOR VERY DILUTE SOLUTIONS. When we are dealing with very dilute solutions, we need some other ways of expressing concentration. After all, it is not very convenient to say that drinking water should contain no more than 0.000195 percent by weight of sodium fluoride (the additive used to prevent tooth decay).

Milligrams per liter is a self-explanatory expression that refers to the number of milligrams of solute in one liter of solution.

Parts per million (ppm) is frequently encountered in pollution-control work and refers to the number of parts of solute per million parts of solution. There are two ways of calculating and interpreting the ppm expression, depending upon whether the solvent is a liquid or a gas. In liquids, the ppm means the number of milligrams of solute per liter of solution. For example, 1.0 liter that contains 5 ppm of cadmium chloride ($CdCl_2$) contains 5 mg of $CdCl_2$.

Sample Problem: If the effluent from an industrial plant contains 1.0 ppm of mercury ions and if the plant discharges the effluent at a rate of 500 gallons per minute, how many pounds of mercury ions will be released in 24 hours?

Solution: This problem contains mixed metric and English units, and it is necessary to convert all units to one system or the other. We will choose the metric system and leave it to the reader to solve the same problem in English units.

Step 1: Convert gallons/minute to liters/minute.

1 gal = 3.78 liters (obtained from a conversion table)

$$500 \frac{\text{gal}}{\text{min}} \times \frac{3.78 \text{ liters}}{\text{gal}} = 1890 \text{ liters/min}$$

Step 2: Find the total volume of effluent in 24 hours.

$$\frac{1890 \text{ liters}}{\text{min}} \times \frac{60 \text{ min}}{\text{h}} \times 24 \text{ h} = 2{,}720{,}000 \text{ liters}$$

Step 3: Find the total mg of Hg ions in 24 hours.

$$2{,}720{,}000 \times 1.0 \text{ ppm} = 2{,}720{,}000 \text{ mg}$$

Step 4: Convert to grams.

$$2{,}720{,}000 \text{ mg} \times 0.001 \frac{\text{g}}{\text{mg}} = 2720 \text{ g}$$

Step 5: Convert to pounds.

$$\frac{1.00 \text{ lb}}{454 \text{ g}} \times 2720 \text{ g} = 5.99 \text{ lb}$$

In air-pollution control, the term ppm is based on volume measurements, rather than weight measurements, and may be expressed as microliters (μl) of solute (where 1.0 μl = 0.001 ml, or 1.0×10^{-6} liter) per liter of air. For example, if one says that there are 5 ppm of sulfur dioxide (SO_2) in the air, this means that there are 5 μl, or 5×10^{-6} liter, of SO_2 per liter of air.

Sample Problem: If 1.00 cubic mile of air (at 1.0 atm pressure) is contaminated with 5.0 ppm of SO_2, what is the total volume of SO_2 present?

Solution: *Step 1:* Find the number of liters of air in 1.0 cubic mile.

$$1.00 \text{ mi} = 5280 \text{ ft (conversion table)}$$
$$1.00 \text{ cu ft} = 28.3 \text{ liters (conversion table)}$$

$$\frac{5280^3 \text{ cu ft}}{\text{cu mi}} \times \frac{28.3 \text{ liters}}{\text{cu ft}} = 4.166 \times 10^{12} \text{ liters/cu mi}$$

Step 2: Multiply.

$$4.166 \times 10^{12} \text{ liters air} \times \frac{5.0 \times 10^{-6} \text{ liter } SO_2}{\text{liter of air}} = 2.1 \times 10^{7} \text{ liters } SO_2$$

MOLARITY. Chemical reactions occur in definite proportions. When HCl undergoes reaction with NaOH, one mole of HCl undergoes reaction with one mole of NaOH to yield one mole of NaCl and one mole of water.

$$HCl + NaOH \longrightarrow NaCl + H_2O$$

If less than a mole or more than a mole of one reactant is used, then the amounts of the other reactants and of the expected products must be adjusted accordingly to keep the molar ratio in the proper proportions. Otherwise, any excess reactant will not undergo reaction and will just be left over when the reaction is completed. (A one-mole reaction of NaOH and HCl would require 40.0 g of NaOH and 37.5 g of HCl and would yield 58.5 g NaCl and 18.0 g H_2O.)

Chemical reactions are often carried out in solutions. It would be inconvenient to treat pure sodium hydroxide, which we buy as pellets, with pure hydrogen chloride, which is a gas, to effect the above reaction. It is far easier to mix an aqueous solution of sodium hydroxide with an aqueous solution of hydrochloric acid.

When we carry out a chemical reaction using solutions, it is convenient to express the concentration of the solutions in terms of the number of moles of solute present in a given volume of solution. The expression we use is *molarity,* which is defined as *the number of moles of solute in 1.0 liter of solution.*

$$\text{Molarity } (M) = \frac{\text{Number of moles of solute}}{\text{Number of liters of solution}}$$

A solution that contains 1.0 mole of NaOH in 1.0 liter of solution would therefore be a 1.0 molar (1.0 M) solution of NaOH. A solution that contains 2.0 moles of NaOH in 1.0 liter of solution would be a 2.0 molar (2.0 M) solution of NaOH.

Sample Problem: What is the molarity of a solution prepared by dissolving 4.0 g NaOH in water and diluting to 1.0 liter?

Solution: Molecular weight of NaOH = 23.0 + 16.0 + 1.0 = 40.0

$$\text{No. of moles NaOH} = \frac{\text{Wt. of NaOH}}{\text{Wt. of 1.0 mole of NaOH}}$$

$$\text{No. of moles NaOH in 4.0 g} = \frac{4.0 \text{ g}}{40.0 \text{ g}} = 0.10$$

$$M = \frac{\text{No. of moles of solute}}{\text{No. of liters of solution}} = \frac{0.10 \text{ mole}}{1.0 \text{ liter}} = 0.10$$

Sample Problem: How many moles of NaCl are present in 25 ml of a 3.0 M solution?

Solution:

$$M = \frac{\text{No. of moles of solute}}{\text{No. of liters of solution}}$$

Substitute: $3.0\ M = \dfrac{x}{0.025 \text{ liter}}$

Solve: $x = 0.075$ moles

Sample Problem: How many ml of 3.0 M NaCl solution would be required to give 0.50 mole?

Solution:

$$M = \frac{\text{No. of moles of solute}}{\text{No. of liters of solution}}$$

Substitute: $$3.0\ M = \frac{0.50\ \text{mole}}{x}$$

Solve: $$x = 0.17\ \text{liter} = 170\ \text{ml}$$

Sample Problem: What is the molarity of 200 ml of an aqueous solution that was made up from 2.0 g NaOH?

Solution:

$$\text{Moles NaOH} = \frac{2.0\ \text{g}}{40.0\ \text{g}} = 0.050$$

$$M = \frac{0.050\ \text{mole}}{0.200\ \text{liter}} = 0.25$$

EQUIVALENT WEIGHT. We have seen that 1.0 mole (36.5 g) of HCl undergoes reaction with 1.0 mole (40.0 g) of NaOH. As long as we keep a molar ratio of 1:1 between the two reactants, we will have a complete reaction. Therefore, 1.0 liter of 1.0 M HCl will react completely with 1.0 liter of 1.0 M NaOH. Or, 25.0 ml of 1.0 M HCl will react completely with 25.0 ml of 1.0 M NaOH.

Some reactions, however, do not occur in a 1:1 molar ratio. Only one-half a mole of H_2SO_4 is needed to undergo complete reaction with one mole of NaOH.

$$\underset{0.5\ \text{mole}}{H_2SO_4} + \underset{1.0\ \text{mole}}{2\,NaOH} \longrightarrow \underset{0.5\ \text{mole}}{Na_2SO_4} + \underset{1.0\ \text{mole}}{2\,H_2O}$$

The reason we need only half as many moles of H_2SO_4 as of HCl to react with a given amount of NaOH is that a molecule of H_2SO_4 can supply *two* hydrogen ions while a molecule of HCl can supply only one. Thus, we can say that half a mole of H_2SO_4 is *equivalent* to one mole of HCl or to one mole of NaOH. Because of this equivalency, the terms *equivalent* and *equivalent weight* were coined.†

> In a reaction involving H^+ and OH^-, one *equivalent* of a compound is the number of moles of that compound required to supply one mole of H^+ or OH^-.
>
> The *equivalent weight* of a compound is the weight in grams of one equivalent.

† The concept of equivalents is also used in other types of reactions, such as oxidation and reduction reactions. The number of equivalents per mole of a compound may vary depending on the reaction in question. However, in this chapter, we will speak of equivalency only in terms of H^+ and OH^-.

One mole of HCl can supply one mole of hydrogen ions; one mole of HCl is the same as one equivalent of HCl. One half mole of H_2SO_4 can supply one mole of hydrogen ions; one half mole of H_2SO_4 is one equivalent of H_2SO_4.

Sample Problem: How many equivalents of H_3PO_4 are present in one mole?

Solution: One mole of H_3PO_4 can supply three moles of H^+. Therefore, one mole of H_3PO_4 contains three equivalents.

Sample Problem: What is the equivalent weight of H_2SO_4?

Solution:

$$\text{Molecular weight of } H_2SO_4 = 98$$

$$\text{Equivalent weight} = \frac{98.0\text{ g}}{2}$$

$$= 49.0\text{ g}$$

Sample Problem: One equivalent of $Ca(OH)_2$ is what fraction of a mole?

Solution: One half mole of $Ca(OH)_2$ is needed to supply one mole of OH^-. Therefore, one equivalent of $Ca(OH)_2$ is 0.5 mole.

NORMALITY. Now that we have defined equivalent weight, we can discuss another commonly used term for expressing concentration—*normality*. Normality is defined as *the number of equivalents of a solute in 1.0 liter of solution.*

$$\text{Normality } (N) = \frac{\text{Number of equivalents of solute}}{\text{Number of liters of solution}}$$

For compounds that yield only one H^+ or OH^- (such as HCl, NaOH, or KOH), the normality and the molarity are identical. Only in those cases where one mole of an acid or base can yield two or more moles of H^+ or OH^- is normality different from molarity. The most commonly encountered cases are H_2SO_4 and H_3PO_4. In these cases, the normality of the solution is always larger than its molarity. For example, a 0.5 M solution of H_2SO_4 (49 g in 1.0 liter) is 1.0 N (one equivalent in 1.0 liter).

Sample Problem: What is the normality of 1.0 liter of an aqueous solution of 98.0 g of H_2SO_4?

Solution:

$$\text{Equivalent weight of } H_2SO_4 = 49.0\text{ g}$$

$$\text{Number of equivalents of } H_2SO_4 = \frac{98.0\text{ g}}{49.0\text{ g}} = 2.0$$

$$N = \frac{2.0\text{ equivalents}}{1.0\text{ liter}}$$

$$= 2.0$$

Sample Problem: What is the normality of 100 ml of an aqueous solution that contains 24.5 g of pure H_2SO_4?

Solution:

$$\text{Number of equivalents in 24.5 g} = \frac{24.5\text{ g}}{49.0\text{ g}} = 0.50$$

$$N = \frac{0.50\text{ equivalents}}{0.100\text{ liter}}$$

$$= 5.0$$

DILUTION PROBLEMS

We are often faced with the task of taking a solution of one concentration and diluting it to another concentration. For example, how much water would need to be added to 100 ml of a 1.2 M solution of $FeCl_3$ to change it to a 1.0 M solution? This type of problem is called a *dilution problem.* Our tool for solving a dilution problem is the following equation.†

$$(\text{Concentration}_{\text{old}})(\text{Volume}_{\text{old}}) = (\text{Concentration}_{\text{new}})(\text{Volume}_{\text{new}})$$

In all dilution problems, we have four values—the volume and concentration of the solution with which we are starting (the old) and the final volume and concentration of the solution we want (the new). In using the equation, wc must be careful to keep the units the same on both sides of the equation. For example, if we use normality on one side, the other side of the equation is also in normality. If we use milliliters on one side, we must use milliliters on the other.

$$N_{\text{old}}V_{\text{old}} = N_{\text{new}}V_{\text{new}} \text{ (where } V \text{ is in ml)}$$

We are not restricted, however, to normality and milliliters. Any unit of concentration and any unit of volume are acceptable, as long as we are consistent. The following equations also hold true when we are diluting a solution.

$$M_{\text{old}}V_{\text{old}} = M_{\text{new}}V_{\text{new}} \text{ (where } V \text{ is in liters)}$$

$$[\%\ (\text{v/v})_{\text{old}}][V_{\text{old}}] = [\%\ (\text{v/v})_{\text{new}}][V_{\text{new}}] \text{ (where } V \text{ is in buckets)}$$

† This equation is derived from the equation

$$M = \frac{\text{No. of moles of solute}}{\text{No. of liters of solution}}$$

Solve for No. of moles:

$$\text{No. of moles} = M \times V\text{(in liters)}$$

Since $(\text{No. of moles})_{\text{old}} = (\text{No. of moles})_{\text{new}}$,

$$M_{\text{old}}V_{\text{old}} = M_{\text{new}}V_{\text{new}}$$

Similar derivations may be made for other units of concentration and for other units of volume.

Sample Problem: To what volume should we dilute 100 ml of a 1.2 *M* solution of $FeCl_3$ to make the concentration 1.0 *M*?

Solution:

$$M_{old}V_{old} = M_{new}V_{new}$$

Substitute: $(1.2\ M)(100\ \text{ml}) = (1.0\ M)(V_{new})$

Solve: $V_{new} = \dfrac{(1.2\ M)(100\ \text{ml})}{(1.0\ M)} = 120\ \text{ml}$

120 ml is the *total* new volume. The volume of water to be *added* is approximately 20 ml.

Note that the formula for dilution makes use of only volume and concentration values. The fact that the solution in our sample problem was of iron(III) chloride was superfluous to finding the answer to the problem. In fact, we can use the formula for calculating dilutions of solutions of any substance—even essence of skunk, if we care to.

Sample Problem: We have 150 ml of a solution of essence of skunk in ethyl alcohol. The concentration of the solution is 0.60 *M*. What is the final concentration of essence of skunk if the solution is diluted to 200 ml?

Solution:

$$M_{old}V_{old} = M_{new}V_{new}$$

Substitute: $(0.60\ M)(150\ \text{ml}) = M_{new}\ (200\ \text{ml})$

Solve: $M_{new} = \dfrac{(0.60\ M)(150\ \text{ml})}{(200\ \text{ml})} = 0.45\ M$

PROPERTIES OF SOLUTIONS

The presence of a solute in a liquid solvent changes the physical properties of the solvent from those of pure solvent without the solute. The properties that we will discuss are referred to as the *colligative properties* of solutions. Colligative properties are those that depend upon the *number of dissolved particles* in a given mass of solvent.

One mole of any substance has the same number of molecules (6×10^{23}) as one mole of any other substance. However, ionic compounds break up into ions when in solution. When one mole of the covalent compound sugar is dissolved in water, 6×10^{23} particles (the sugar molecules) are added to the water. However, when one mole of sodium chloride is dissolved in water, 6×10^{23} sodium ions plus 6×10^{23}

chloride ions, or a total of 12×10^{23} particles, are put in solution. A mole of an ionic compound affects the colligative properties of a solution to a greater extent than does a mole of a covalently bonded compound.

FREEZING POINT DEPRESSION. One of the colligative properties of a solution is its *freezing point depression.* A solution solidifies, or freezes, at a *lower temperature* than the freezing point of the pure solvent. The freezing point of pure water is 0°C. When 1.0 mole of a nonionic solute, such as sugar, is placed in 1000 g of H_2O, the freezing point of the solution is lowered to −1.86°C. If 2.0 moles of sugar are dissolved in 1000 g of water, the freezing point is lowered twice as much, to −3.72°C. For each mole of dissolved nonionic substance in 1000 g of water, the freezing point is depressed 1.86°C.

Antifreeze used in automobile radiators is an example of freezing point depression in action. Antifreezes are nonvolatile (high boiling) organic compounds added to the water in the radiator to prevent the water from freezing. Ethylene glycol ($HOCH_2CH_2OH$) is a commonly used antifreeze. Enough moles of this compound are added to the water in the radiator to depress the freezing point of the water to the desired level. One mole of sodium chloride depresses the freezing point twice as much (3.72°C/mole) as a mole of a covalent compound like ethylene glycol. Because salts encourage corrosion of metals, they are not useful as antifreezes.

BOILING POINT ELEVATION. The addition of a solute to a solvent *raises* the boiling point of the solution over that of pure solvent. One mole of a solute, like sugar, in 1000 g of water raises the boiling point by 0.52°C. We observe that the solution boils at 100.52°C rather than at 100°C.

OSMOSIS. Recall from Chapter 2 that liquids are capable of *diffusion*, or mixing, because the molecules of the liquid are in motion. Solutions also diffuse. If we pour pure water into an aqueous salt solution, the water will diffuse through the solution (even though we do not stir the two), and eventually the entire solution will be homogeneous.

A *semipermeable membrane* is a thin sheet of material that contains microscopic holes so tiny that relatively large molecules cannot pass through, but large enough that molecules of a certain size and smaller are able to pass through. In most cases, water can flow through such a membrane, but not solute molecules or larger particles. A semipermeable membrane is somewhat like a strainer that is used for the straining of molecules. (See Figure 6–5.)

Ions in solution are solvated—that is, the ions attract water molecules and become surrounded by them. When a concentrated solution of sodium chloride is placed on one side of a semipermeable membrane with proper-sized holes and a dilute solution is placed on the other side, water molecules can pass freely through the

Figure 6–5. A semipermeable membrane contains holes of such a size that water molecules are permitted to pass through, but not larger molecules or solvated ions.

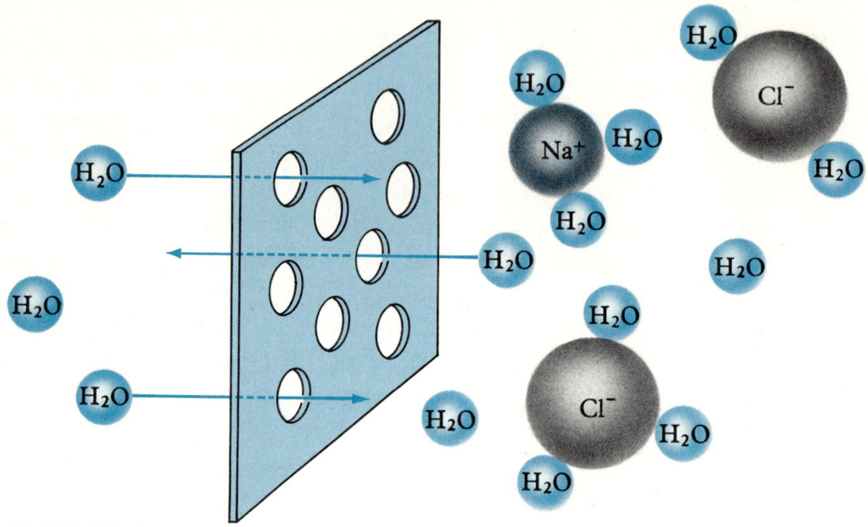

membrane. The ions with attracted solvent molecules cannot squeeze through the holes in the membrane.

Although the solvent molecules flow back and forth through the membrane, a greater number will flow in the direction of the more concentrated solution in an attempt to make the solutions of equal concentration. The net result is a flow of solvent from the dilute side to the concentrated side. The selective flow of a diffusible substance through a semipermeable membrane is called *osmosis,* and the pressure that causes the solvent to diffuse through the membrane is called *osmotic pressure.* (See Figure 6–6.)

As osmosis occurs, the concentrations of the two solutions on the two sides of the membrane change. The dilute solution loses solvent to the concentrated solution. Thus, the volume of the dilute solution decreases and it becomes more concentrated. At the same time, the more concentrated solution gains solvent; its volume increases and it is diluted. The added weight of the increased volume of the solution that gains solvent causes an increase in pressure on the sides and bottom of the container and on the membrane. This pressure can build up sufficiently to rupture the membrane, but if the membrane is strong enough an equilibrium will be reached. At the equilibrium, the pressure on the membrane caused by the increased volume on the one side equals the osmotic pressure on the other side, and the volumes of the solutions cease changing.

Figure 6–6. In the process of osmosis, solvent flows through a semipermeable membrane from the dilute solution toward the more concentrated solution. The flow continues until the pressure of the increased volume of the concentrated solution is equal to the pressure of the solvent flow.

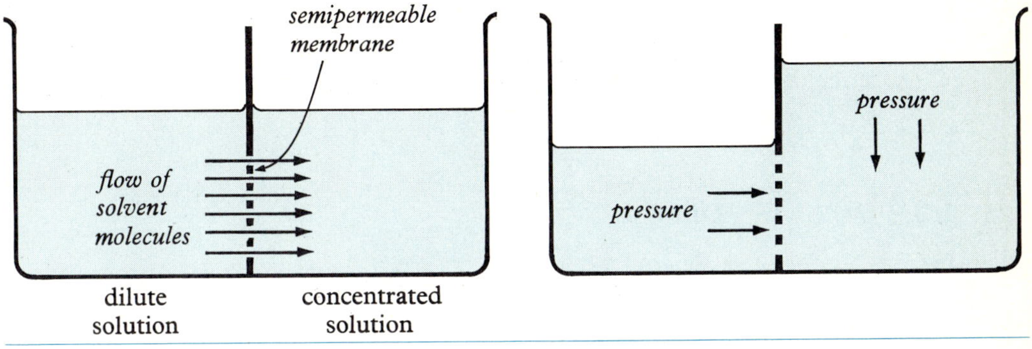

Osmosis and related phenomena are especially important in body processes. Our cell membranes are a type of semipermeable membrane. Our blood cells contain about 0.9 percent sodium chloride. If the blood cells are placed in pure water, water will pass into the cells by osmosis and eventually cause the cells to swell up and burst. This bursting of red blood cells is called *hemolysis*. On the other hand, if red blood cells are placed in a concentrated sodium chloride solution, water *leaves* the cells, and the cells become shrivelled. We say that the cells have undergone *crenation*.

Physiological saline solution is a sodium chloride solution of about 0.9 percent (w/v). Blood cells in this solution release as much water as they take in, so they neither swell nor shrink. Solutions with the same osmotic pressure (blood cells and physiological saline solution, for example) are said to be *isotonic solutions*.

COLLOIDS

The terms *colloid* and *colloidal dispersion* are synonymous and refer to a dispersion of small particles of one substance, the *dispersed phase,* in another substance, the *dispersion medium.* If we mix a bit of flour in water and shake up the mixture, the flour does not dissolve. However, not all the flour settles out either. The water has a cloudy appearance as evidence of suspended flour particles. In this example, flour is the dispersed phase and water is the dispersion medium.

The word *colloidal* comes from the Greek words *kolla,* meaning glue, and *eidos,* meaning "looks like." Colloidal means "looks like glue." Animal glue itself is a colloid formed from solid protein material dispersed in a solvent. Many other colloids resemble animal glue, but many do not. For example, smoke and marshmallows are colloids, but the physical resemblance of these to glue is remote.

The importance of colloids in our technology and in biochemical processes cannot be denied. A large part of air pollution, for example, is composed of solid particles, so-called particulate matter (dust, carbon, ash, and so forth), suspended in the atmosphere. Paint, aerosol sprays, and other commercial products are types of colloidal dispersion. Blood is also a colloidal dispersion. In this section, we will briefly describe the different types of colloid, their properties, and uses.

PARTICLE SIZE OF COLLOIDS

The principal difference between a colloid and a solution is the size of the particles of the dispersed phase. In a solution, the particles of solute are tiny, less than about 10 Angstroms in diameter. In a colloid, the particle size of the substance that forms the dispersed phase is larger, from about 10 to 1000 Angstroms in diameter. The particles are too small to see, too small to settle out, and even too small to be filtered out with filter paper. However, the particles are large enough to reflect and scatter light so that a colloid often appears cloudy. (See Figure 6-7.)

One way to destroy a colloidal dispersion is to allow the small colloidal particles to get together and form larger aggregates. The larger aggregates, too large to stay suspended in the dispersion medium, then will precipitate—that is, come out of solution.

Figure 6-7. The differences between a solution, a colloid, and an ordinary mixture.

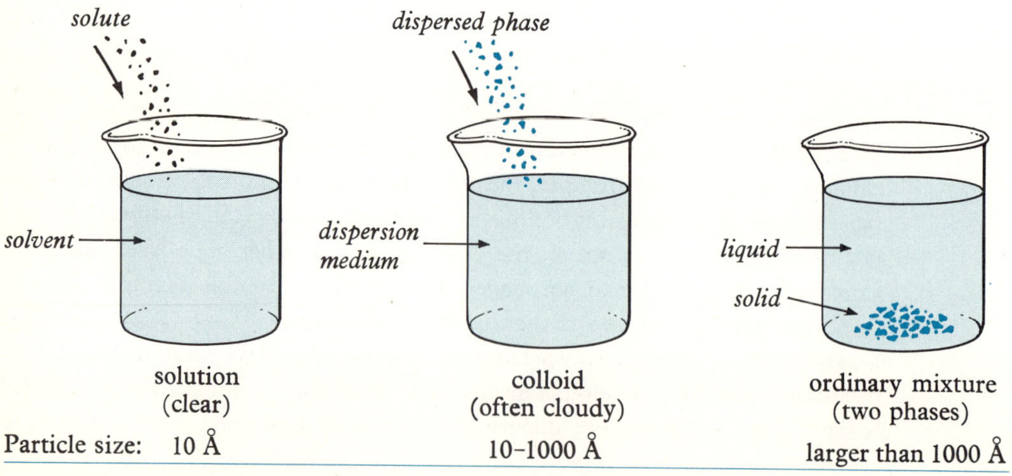

TYPES OF COLLOID

We have mentioned a variety of colloids that are quite different in type. Flour in water is an example of a colloidal dispersion of a *solid in a liquid;* this type of colloid is called a *sol.* Smoke is a dispersion of a *solid in a gas.* Some common types of colloid are listed in Table 6–3.

Table 6–3. Types of Colloid

Type	Name	Example
Gas in liquid	Foam	Whipped cream, shaving cream
Gas in solid		Marshmallows, foam rubber
Liquid in gas	Aerosol	Fog, aerosol sprays (insecticide, deodorant, and hair sprays)
Liquid in liquid	Emulsion	Milk, mayonnaise
Liquid in solid		Cheese
Solid in gas		Smoke
Solid in liquid	Sol	Paint, flour in water, ink, milk of magnesia, hot chocolate
Solid in solid		Colored glass

Gelatin desserts and jelly belong to a special class of sols called *gels,* which are semisolid, viscous materials. A gel is formed from a solid that, although insoluble, has a great attraction for the dispersion medium. In the case of jelly, the medium is water. The solid particles bind to water molecules, and the colloid "sets" by forming a three-dimensional network of particles and the water.

The ability to form a gel with water is not limited to any one type of compound. A large variety of solids have this ability. *Gelatin* is protein material derived from collagen (Chapter 23), and is obtained by boiling animal bones and skin in water. The water is then removed, and the residual gelatin is sold in the solid form. A gel is made by adding hot water to solid gelatin, then allowing the mixture to cool.

Pectin is a carbohydrate, a compound formed from large molecules that are made up of smaller sugar molecules. Pectin is found in unripe fruit such as apples and is indispensable in the making of jellies because of its gel-forming ability.

"Canned heat," a jelled cooking fuel sold in cans, is a gel formed from a mixture of a saturated aqueous solution of calcium acetate, $(CH_3CO_2)_2Ca$, and ethyl alcohol, CH_3CH_2OH. Many other metallic compounds, such as iron(III) hydroxide, $Fe(OH)_3$, and aluminum hydroxide, $Al(OH)_3$, form gels when they are precipitated from water.

EMULSIFYING AGENTS

An *emulsifying agent* is a substance that either keeps another substance emulsified (dispersed in a medium) or else causes another substance to form an emulsion. Soaps and detergents are emulsifying agents. A soap or detergent molecule is a molecule that has a long organic chain and an ionic end.

$$CH_3CH_2CH_2CH_2CH_2CH_2CH_2CH_2CH_2CH_2CH_2CH_2CH_2CH_2CH_2\overset{\overset{\displaystyle O}{\|}}{C}O^- \; Na^+$$

a soap

$$CH_3CH_2CH_2CH_2CH_2CH_2CH_2CH_2CH_2CH_2CH_2CH_2O\underset{\underset{\displaystyle O}{\|}}{\overset{\overset{\displaystyle O}{\|}}{S}}O^- \; Na^+$$

a detergent

Most large organic molecules are insoluble in water, but, because the ionic portion of a soap molecule is attracted to water, a soap readily forms a colloidal dispersion in water. Oils and greases are not soluble in water, but they are attracted to

Figure 6-8. Emulsification of an oil droplet by soap molecules.

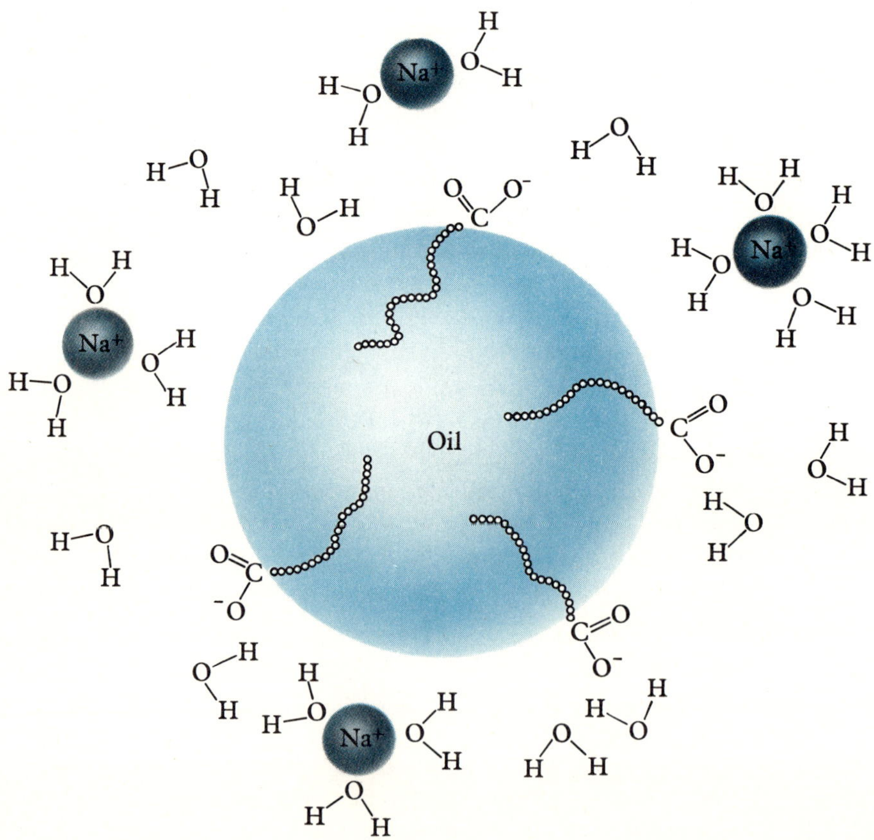

the organic end of a soap or detergent molecule. The result of these attractions is that the soap molecules pick up droplets of oil and grease and emulsify them. Once the oil droplets are dispersed in the water by the soap, they can be washed away. (See Figure 6-8.)

Egg yolks are rich in lecithins (Chapter 24), another type of compound that is formed from large organic molecules with ionic ends. The lecithins are also good emulsifying agents. Mayonnaise is an emulsion of water-insoluble vegetable oils in vinegar and is stabilized by the addition of egg yolk. The lecithins, like soap, dissolve partly in the droplets of oil. The ionic ends of the lecithin molecules are at the surface of the oil droplets where they are attracted to water molecules.

PROPERTIES OF COLLOIDS

TYNDALL EFFECT. Have you ever noticed a shaft of sunlight coming through the window of a dimly lit room? We cannot actually see a beam of light unless it is directed into our eyes. When we observe the shaft of sunlight, what we really are observing is the reflection and scattering of the light by minute dust particles floating in the air—dust particles that would also be invisible if it were not for the beam of light. The ability of a colloid to scatter light rays so that we can see a beam of light passing through it is known as the *Tyndall effect*.

If we shine a beam of light through a true solution in a jar, the light travels right through—we cannot see the track of the light. (See Figure 6-9.) However, if we shine a beam of light through a colloidal dispersion, even one that does not appear cloudy to us, we can see the path of the light just as we can see a sunbeam in a dusty room.

Figure 6-9. The Tyndall effect. A beam of light passes through a solution without being reflected. A beam of light passing through a colloidal dispersion is reflected to our eyes.

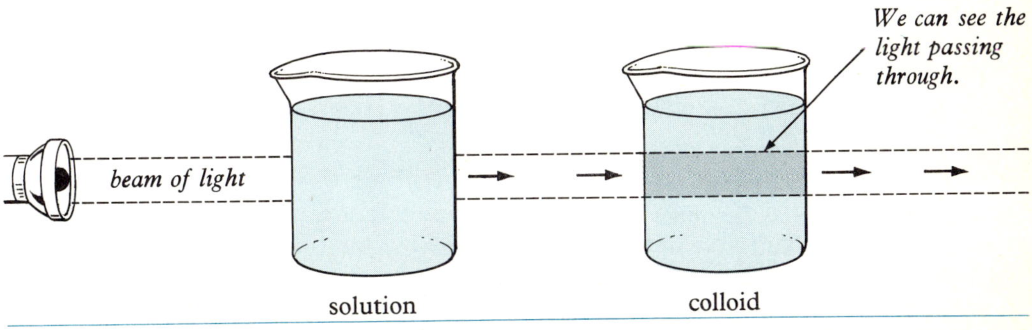

BROWNIAN MOVEMENT. If they are large enough, the particles of colloidal dispersion may be seen under a microscope. What do they look like? The principal feature that can be seen is that the particles move, not smoothly, but in an erratic fashion. This erratic motion of colloidal particles is called *Brownian movement,* after its discoverer, Robert Brown (1773–1858). Brown suggested that the colloidal particles might be alive, but we now know that the dancing motion is caused by the small colloidal particles responding to collisions with the even smaller molecules of the medium. (See Figure 6-10.)

Figure 6-10. Brownian movement is the erratic motion of a colloidal particle.

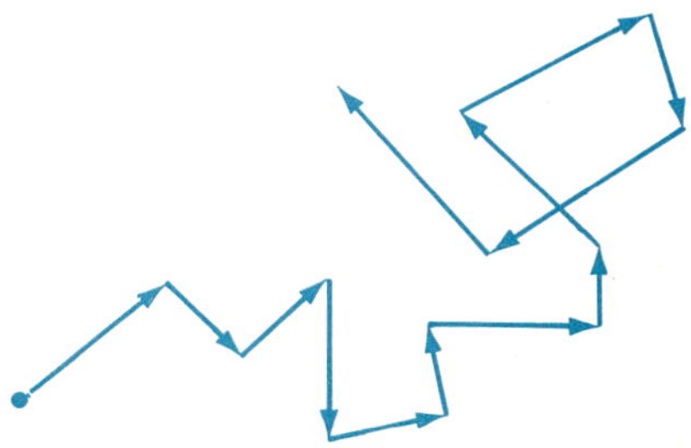

The fact that colloidal particles are small enough to be buffeted about by molecular collisions is one reason that the particles do not settle out of the dispersion medium. The collisions exert a greater force on the particles than does gravity.

DIALYSIS

Plant and animal membranes are similar to osmotic semipermeable membranes in that they allow selective flow of substances from one side to another. A cell membrane is more permeable than an osmotic membrane because small molecules and ions must be able to flow through the membrane. For example, the membrane around a plant or animal cell allows water and nutrients to enter and allows water and waste products to escape.

A membrane that allows selective passage of ions and small molecules (as well as water) is called a *dialyzing membrane.* The process of passage of such substances through a dialyzing membrane is called *dialysis.* Dialysis is very similar to osmosis, and again we find a tendency toward concentration equalization. For example, ions pass through a dialyzing membrane from the more concentrated side to the less concentrated side to equalize concentration.

This tendency of concentration equalization helps an animal to take in oxygen and lose carbon dioxide, which is a waste product. When blood is exposed to the oxygen-rich and carbon dioxide-poor environment of the lungs, it loses its CO_2 (which is then exhaled) and picks up O_2, which is carried to the cells. There, in a reverse process, O_2 is released by the blood to the cells and CO_2 is taken in from the cells. The blood then returns to the lungs to repeat the process. This description is, of course, oversimplified because substances in red blood cells and cell membranes actively help transport O_2, CO_2, and other nutrients and wastes. In addition, CO_2 dissolved in the blood helps maintain the proper acidity of the organism. In other words, CO_2 is not only a waste product, but also a necessary component of an organism.

USES OF DIALYSIS. Colloidal particles adsorb (attract to their surface) ions present in the dispersion medium. Depending on the type of colloid, the particles attract either positive ions or negative ions, but not both. Thus, each colloidal particle is blanketed in ions and repels other colloidal particles. This is a principal reason that colloidal particles do not aggregate and precipitate.

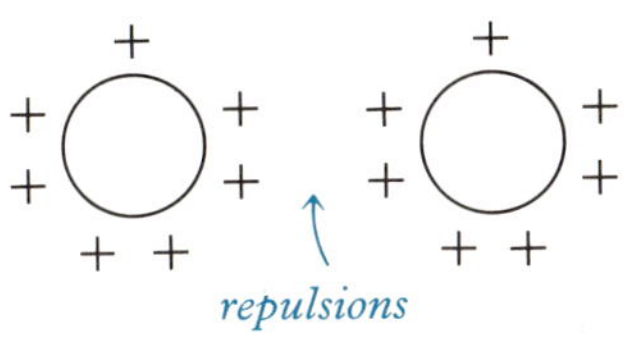

If the dispersion medium has a high ionic concentration, the ionic blanket on the colloidal particles is attracted away, and the colloid may precipitate. In order to stabilize a colloid, it is therefore often necessary to remove some of the ions from the dispersion medium. Dialysis may be used for this purpose.

A dialyzing membrane is sufficiently porous that water and ions can pass through it, but not the colloidal particles, which are too large. Cellophane may be used as a dialyzing membrane. The colloid is placed on one side of the membrane and water on the other. Ions leave the dispersion medium, while the larger colloidal particles are left behind. Usually, running water is used in a dialysis operation to speed up the process of dialysis by removing the ions. (See Figure 6-11.)

Biochemists, who frequently work with macromolecules (that is, giant molecules), find dialysis a useful procedure. For example, inorganic salts may be removed from protein molecules by this technique. Also, with a membrane of suitable porosity, relatively small protein molecules may be separated from larger proteins. Deductions may be made about the sizes and shapes of macromolecules by their behavior in dialysis experiments.

Figure 6–11. Dialysis. The small ions can pass through the dialyzing membrane; the larger particles cannot.

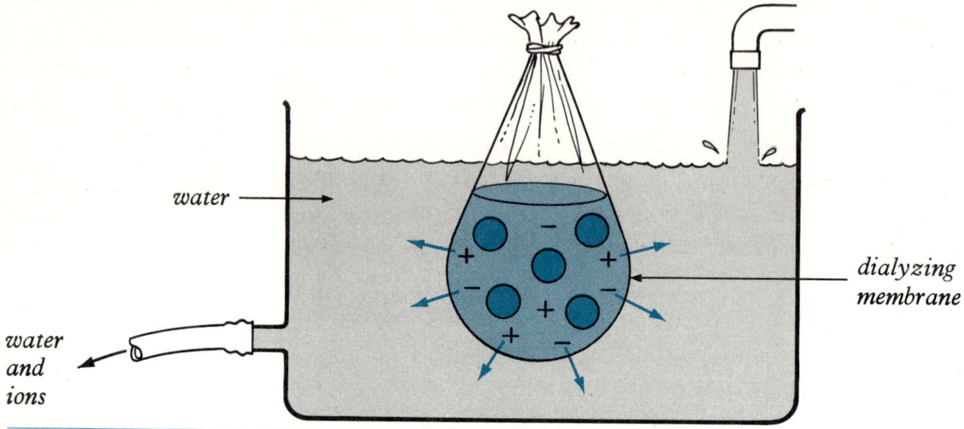

Kidney machines are another example of how dialysis is used. The kidneys remove waste products from the system by dialysis; these products are then eliminated in the urine. If the kidneys are not functioning properly, the waste products build up in the system and eventually poison it.

The development of the kidney machine has given us a means to remove toxic waste products from the blood artificially. The blood of the patient is circulated through tubes of just the right porosity to let the molecules of toxic waste material pass, but to hold back the larger cells and molecules in the blood that are necessary to the system. Individuals who would once have been doomed because of kidney failure may now be able to live relatively normal lives as long as they pay periodic visits to the kidney machine. Unfortunately, the process is still quite expensive, especially for a patient who needs treatment over a long period of time.

STUDY QUESTIONS

6-1. A chemist heated the following mixture in a beaker: 2.0 g sugar, 100 ml H_2O, and 10.0 g coffee grounds. Tell which is the solvent and which is the solute or solutes. Tell which of the ingredients would be considered soluble and which only partially soluble.

6-2. Designate each of the following sets of liquids as miscible or immiscible:
(a) salad oil and vinegar
(b) wine and water
(c) kerosene and water
(d) mercury and water

6-3. The following mixtures have been heated, then cooled to 20°C. Tell whether each is a saturated or unsaturated solution (see Table 6-2).
(a) 38.0 g NaCl in 100 g H_2O
(b) 60.0 g $CuSO_4$ in 200 ml H_2O
(c) 10.3 g NaOH in 100 g H_2O

6-4. Is it possible to supersaturate a solution by simply adding excess solute to the solvent?

6-5. If the pressure above a solution of carbon dioxide in water is changed from three atmospheres to one atmosphere, what is the effect (semiquantitatively) on the solubility of the carbon dioxide?

6-6. What volume of pure ethyl alcohol would be needed in the preparation of 100 ml of a 1% (v/v) solution?

6-7. Find the percent by volume of ethyl alcohol in an aqueous solution prepared by diluting 15 ml of pure ethyl alcohol to 70 ml.

6-8. A chemist dissolved 5.00 ml methyl alcohol in 90.00 ml water and then diluted the solution with enough water to make 100.00 ml. For this solution, give the following concentrations. (The specific gravity of methyl alcohol at 20°C is 0.791.)
(a) % (volume/volume)
(b) % (weight/volume)

6-9. Tell how you would prepare the following solutions:
(a) 200 ml of a 35% (volume/volume) solution of ethyl alcohol in water
(b) 50 ml of a 1% (weight/volume) solution of KOH in ethyl alcohol
(c) 10 g of a 5% (weight/weight) solution of NaOH in water

6-10. An experiment calls for 150 g of a 1% (w/w) solution of phenolphthalein solution. How many grams of phenolphthalein are needed?

6-11. How many grams of $NaHCO_3$ are present in 250 g of a 7.00% (w/w) solution?

6-12. How many grams of $NaHCO_3$ are present in 250 ml of a 7.00% (w/v) solution?

6-13. If the effluent from an industrial plant contains 3.0 ppm of sulfide ions and if the plant discharges 1000 gallons of effluent per second, how many pounds of sulfide ions will be released in a week?

6-14. How many grams of NaCl are contained in 2.0 ml of a 0.5% (weight/volume) solution in water?

6-15. What weight of morphine sulfate is needed to prepare 20 ml of a 10% (w/v) solution of morphine sulfate in water?

6-16. Calculate the number of grams of NaOH that would undergo reaction with
(a) 1.0 g of HCl (b) 3.65 g of HCl
(c) 36.5 g of HCl

6-17. Calculate the molarity of 1000 ml of a solution that contains 2.0 moles of NaOH.

6-18. Calculate the number of moles of $CuSO_4$ present in 10.0 g.

6-19. How many moles of $CuSO_4$ are present in 333 ml of a 0.500 *M* solution?

6-20. Calculate the molarity of 250 ml of solution that contains 2.5 g $CuSO_4$.

6-21. How many grams of KOH would be needed to prepare 100 ml of a 2.00 *M* solution?

6-22. What is the molarity of each of the following solutions?
(a) 20.0 g of $FeCl_3$ in 100 ml of H_2O solution
(b) 1.00 g of $CuSO_4$ in 1050 ml of H_2O solution
(c) 0.050 g of KCl in 500 ml of H_2O solution

6-23. Calculate the number of moles of solute in each of the following solutions:
(a) 100 ml of a 0.5 *M* solution of $CdCl_2$
(b) 50 ml of a 1.5 *M* solution of $Na_2Cr_2O_7$
(c) 30 ml of a 6.0 *M* solution of H_2SO_4

6-24. How many milliliters of each of the following solutions would be required to give 1.5 moles?
(a) 12 *M* H_2SO_4 (b) 0.05 *M* NH_4Cl
(c) 1.0 *M* CH_3CO_2H (acetic acid)

6-25. Give the number of equivalents present in each of the following samples:
(a) 8.0 moles H_2SO_4
(b) 2.0 moles HCl
(c) 0.4 moles $Ca(OH)_2$
(d) 1.5 moles $Ba(OH)_2$

6-26. Give the equivalent weight for each of the following compounds:
(a) NaOH (b) $Ba(OH)_2$
(c) H_2SO_4 (d) H_3PO_4

6-27. Calculate the equivalent weight of each of the following compounds:
(a) HNO_3 (b) $NaHSO_4$
(c) $Al(OH)_3$ (d) NaH_2PO_4

6-28. What is the normality of a 2 *M* H_2SO_4 solution?

6-29. Calculate the normality of 250 ml of a solution that was prepared from 49 g pure H_2SO_4 and water.

6-30. How many (1) grams, (2) moles, (3) equivalents are present in each of the following?
(a) 100 ml of 1.0 *N* NaOH
(b) 50.0 ml of 1.5 *N* H_2SO_4
(c) 10.0 ml of 0.75 *N* H_3PO_4

6-31. It is desired to prepare a 1.0% (weight/volume) solution of NaCl from 25 ml of a 2.5% (weight/volume) solution. How much water should be added to the first solution (assuming the volumes are additive)?

6-32. A chemist diluted 100 ml of 6 *N* H_2SO_4 to 600 ml with water. What is the normality of the new solution?

6-33. For each of the following solutions, how many ml of H_2O should be added to obtain a solution that has a concentration of 1.5 *M* (assuming that the volumes are additive)?
(a) 25 ml of 6 *M* HCl
(b) 75 ml of 2 *M* NaOH
(c) 1000 ml of 10 *M* H_2SO_4
(d) 0.5 ml of 1.6 *M* NaCl

6-34. What will be the final concentration of each of the following if the solution is diluted with 10 ml of H_2O (assuming that the volumes are additive)?
(a) 25 ml of 0.7 *M* NH_4OH
(b) 1.5 ml of 0.25 *M* $CoCl_2$
(c) 0.25 ml of 1.5 *M* $CoCl_2$
(d) 1.0 ml of 5 *M* KCl

6-35. Is it possible to dilute 10 ml of a 0.50 *M* NaCl solution to obtain a 1.0 *M* solution? (Solve the problem by the dilution formula and interpret the result.)

6-36. Give the number of particles of solute that would be formed if each of the following samples were dissolved in water:
(a) 0.5 mole ethyl alcohol
(b) 0.25 mole NaCl
(c) 1.2 moles $MgBr_2$
(d) 3.0 moles KBr

6-37. Calculate the freezing points of solutions prepared by dissolving each of the preceding compounds in 500 ml of water.

6-38. How many solute particles would there be in a solution of 1.0 mole of each of the following in 1000 g water?
(a) NH_4Cl
(b) $HOCH_2CH_2OH$ (ethylene glycol)
(c) KBr

6-39. Calculate the freezing points and boiling points for the solutions in the preceding problem.

6-40. Describe what would happen if equal amounts of the following two solutions were placed on each side of a semipermeable membrane: Solution *A* is 10 *M* NaCl; solution *B* is 0.25 *M* NaCl.

6-41. If pure water is placed inside a bag made of a semipermeable membrane and the bag is then suspended in an ionic solution,
(a) in which direction will the water flow?
(b) will the bag expand or contract?
(c) will the ionic solution increase or decrease in ionic strength?

6-42. Identify the dispersed phase (solid, liquid, or gas) and the dispersion medium for each of the following colloids:

(a) gravy
(b) aerosol dessert topping
(c) vinegar-oil salad dressing

6-43. Give two reasons for the ability of colloidal particles to stay suspended in the dispersion medium.

6-44. Explain why the addition of a large quantity of an ionic substance may cause a colloidal suspension to precipitate.

6-45. A colloidal dispersion of flour and water is placed on side 1 of a dialyzing membrane. A concentrated aqueous solution of NaCl is placed on side 2 of the membrane. Describe the movement across the membrane (if any) of each of the following species:

(a) H_2O (b) Na^+ (c) Cl^- (d) flour

CHAPTER 7 IONS IN SOLUTION

OBJECTIVES

Water as a Solvent *Define hydrogen bond, and draw structures of hydrogen-bonded molecules. * Describe the processes of dissociation and solvation in water. * Write the structure of a hydronium ion.*

Types of Electrolyte *Name and give examples of the three types of electrolyte.*

Solubility of Salts in Water *Describe the equilibrium involved when a salt dissolves in water. * List the types of salt that are generally soluble in water and those that are generally insoluble.*

Ionization of Acids in Water *Differentiate between a strong acid and a weak acid. Write the structures of some examples.*

Dissociation of Bases in Water *Differentiate between a strong base and a weak base. Give examples of each.*

How Ionic Reactions Proceed *Write equations to show three types of occurrence that cause an ionic reaction to proceed. * Convert a molecular equation to a net ionic equation.*

How Reversible Reactions Proceed *Describe the equilibrium for the ionization of a weak electrolyte* *and show how the equilibrium can be shifted. (Use equations.)* * *Write the equilibrium expression for the ionization of a weak electrolyte and use the expression to calculate either the equilibrium constant or the concentrations of species in solution.*

Equilibrium Expression for Water *Write the equilibrium expression for the ionization of water.*

The pH Scale *Define pH, and use pH values in calculations.*

Neutralization Reactions *Define neutralization reaction, and give examples.*

Titration *Describe the process of titration and how the concentration of an unknown sample may be determined.*

Dibasic Acids *Write equations for the stepwise dissociation of a dibasic acid, such as* H_2CO_3.

Amphoteric Substances *Define and illustrate amphoteric substances.*

Buffer Solutions *Describe a buffer solution and some buffering reactions in blood. (Use equations.)*

Pure water cannot conduct electricity. You might ask, "Then how can a person be electrocuted in the bathtub?" The answer is that water containing ions is an excellent conductor of electricity. Bath water contains ionic compounds dissolved from the soil, from the plumbing pipes, from the skin, and from soap; therefore, bath water can conduct electricity.

Because a solution of ions in water can conduct electricity, an ionic compound is said to be an *electrolyte*. A solution of the covalently bonded compound sugar does not conduct electricity because sugar does not ionize in water. Sugar and other nonionic compounds are called *nonelectrolytes*.

Some electrolytes: $Na^+ Cl^-$, $Na^+ OH^-$, $K^+ NO_3^-$

Some nonelectrolytes: $C_{12}H_{22}O_{11}$ (sugar), CH_3CH_2OH (ethyl alcohol)

WATER AS A SOLVENT

Water has some unique properties that make it an excellent solvent for a variety of compounds, both ionic and nonionic. A molecule of water is a polar molecule—that is, it is a molecule with an unequal distribution of charge. You will recall from Chapter 4 that the oxygen atom of a water molecule has a greater attraction for the bonded electrons than do the hydrogen atoms. Electrons in the covalent bonds of water are drawn toward the electronegative oxygen. As a result, oxygen carries a partial negative electrical charge. Electrons are pulled away from the hydrogen atoms; thus, the hydrogen atoms in water have a partial positive electrical charge.

Most substances with low molecular weights are gases at room temperature. By virtue of its molecular weight (18), we would expect water to be a gas. (Even CO_2, with a molecular weight of 44, is a gas.) However, the polar character of the water molecule causes strong attractions between the molecules. A partially positive hydrogen is attracted to a partially negative oxygen in another molecule of water. This attraction is called a *hydrogen bond.*

The hydrogen bond is not nearly as strong as a true covalent bond: It takes about five kilocalories of energy per mole to break a hydrogen bond, and more on the order of 50 to 100 kilocalories per mole to break a true covalent bond.

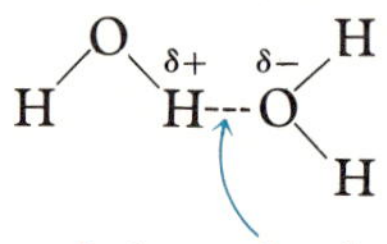

a hydrogen bond

In a cup of water, there are trillions of molecules. Each molecule is attracted to its neighbors by hydrogen bonds, which hold the water molecules together as a liquid rather than letting individual molecules move completely at random as in a gas.

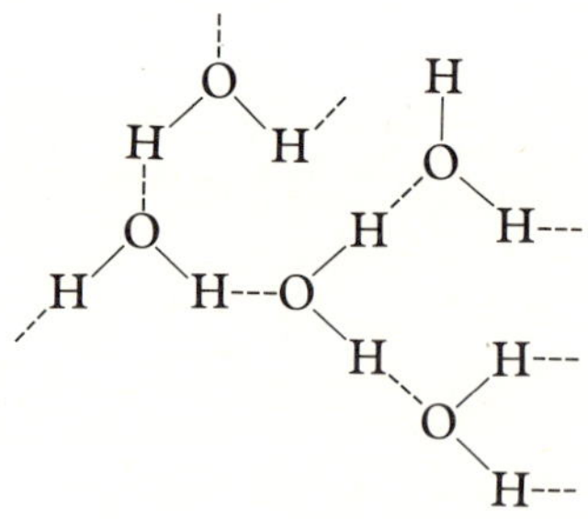

Hydrogen bonds in water.

Polar molecules of water attract not only other water molecules, but other polar molecules or ions. When a crystal of sodium chloride is placed in water, the negative end of a water molecule and a positive sodium ion attract each other. The partially positive hydrogen atom in water and the negative chloride ion also attract each other. Ions are pulled away from the crystal by the attractions of the water. In solution, both the positive cations and the negative anions become surrounded by a group of attracted water molecules. We say the ions are *solvated,* or have undergone *solvation.*

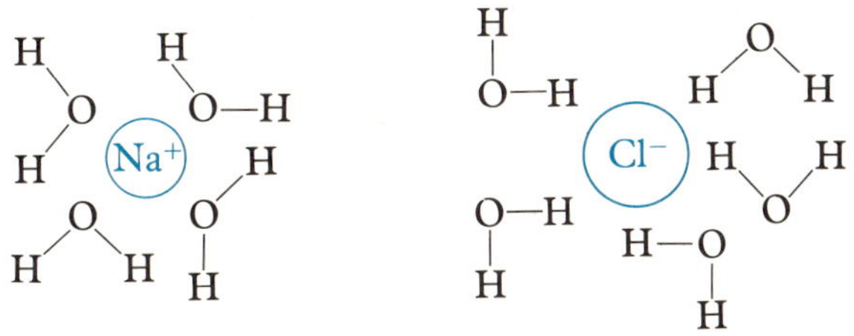

In the process of solvation, water surrounds both the cation and the anion, allowing them to be free from one another in solution.

Because an ionic compound breaks into ions, or dissociates, in solution, the process of its dissolving is called *dissociation.* The dissociation of ionic compounds to form ionic solutions may be exemplified by the following equations.

In water:

$$NaCl_{solid} \longrightarrow Na^+ + Cl^-$$

sodium chloride

$$KOH_{solid} \longrightarrow K^+ + OH^-$$

potassium hydroxide

Because ions are separated from each other in water solution, we often cannot assign a particular cation to a particular anion in a solution of a mixture. For example, blood contains Na^+, Cl^-, and HCO_3^-, along with some organic ions. However, we cannot say that blood really contains sodium bicarbonate ($NaHCO_3$) or sodium chloride (NaCl) as discrete compounds.

Many covalent compounds that are water-soluble do not dissociate. Ethyl alcohol, for example, is water-soluble because it forms hydrogen bonds with water molecules. Alcohols with fewer than four carbon atoms are all soluble in water (and in each other) because of hydrogen bonding. Alcohols with more carbons (for example, $CH_3CH_2CH_2CH_2CH_2OH$) are less soluble in water because of their large, nonpolar, hydrocarbon ends.

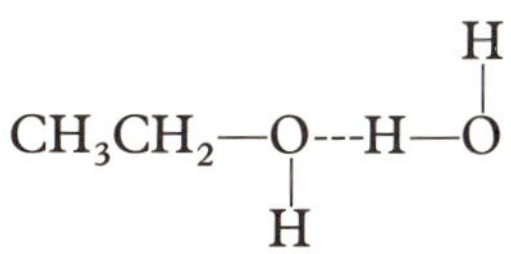

Ethyl alcohol dissolves without ionizing.

Some highly polar covalent compounds ionize (yield ions) in water because the attractions for the water molecules are so great. Hydrogen chloride (HCl) is a covalently bonded gas that undergoes ionization in water. A water solution of hydrogen chloride is called *hydrochloric acid.*

In water:

$$\underset{\text{hydrogen chloride}}{HCl} \longrightarrow \underset{\text{hydrogen ion}}{H^+} + \underset{\text{chloride ion}}{Cl^-}$$

When HCl ionizes in water, it does not actually form a hydrogen ion as we have shown. When a hydrogen ion is formed in water, it is attracted to the partially negative oxygen of water. This oxygen has unshared electrons that form a bond with the H^+ to yield H_3O^+. Figure 7-1 shows the formation of this bond with orbital pictures, but we can represent the formation of the new bond in equations as well.

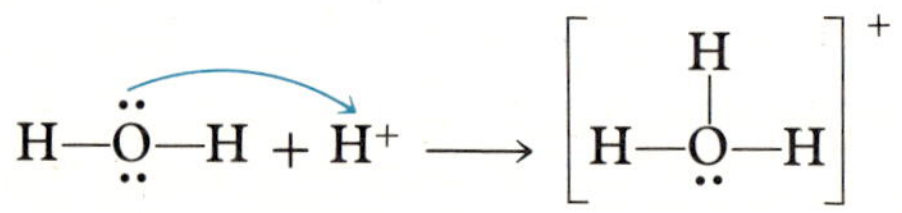

or simply $H_2O + H^+ \longrightarrow H_3O^+$

The H_3O^+ ion is called the *hydronium ion.* In water solution, a hydrogen ion always exists as a hydronium ion rather than as a bare proton. Thus, the ionization of hydrochloric acid, or any other acid (a compound that can give up a hydrogen ion), is best represented as a reaction with water.

$$HCl + H_2O \longrightarrow H_3O^+ + Cl^-$$

$$\underset{\text{acetic acid}}{CH_3CO_2H} + H_2O \rightleftharpoons \underset{\text{acetate ion}}{CH_3CO_2^-} + H_3O^+$$

Figure 7-1. A water molecule forming a bond with a hydrogen ion to yield a hydronium ion.

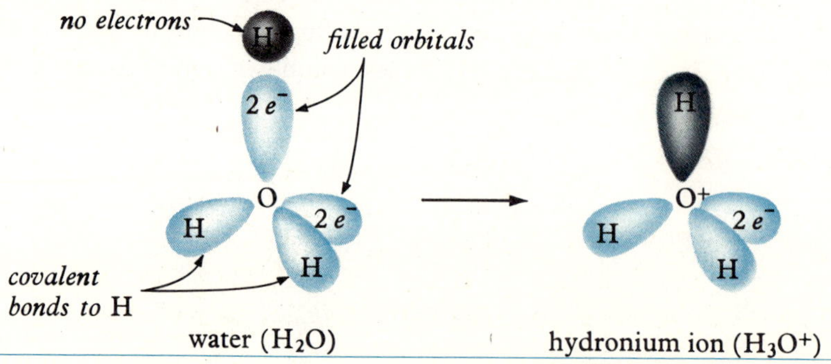

For the sake of simplicity, we will usually represent an acid as ionizing to H^+ instead of H_3O^+, even though it is not strictly correct.

$$HCl \longrightarrow H^+ + Cl^-$$

$$CH_3CO_2H \rightleftharpoons CH_3CO_2^- + H^+$$

One other aspect of water as a solvent should be mentioned. If two water molecules collide with enough energy, a water molecule can be broken into H^+ and OH^-. In other words, water itself can ionize.

$$H{-}\ddot{\underset{\cdot\cdot}{O}}{-}H + \boxed{H}{-}O{-}H \rightleftharpoons H_3O^+ + OH^-$$

or, simplified, $$H_2O \rightleftharpoons H^+ + OH^-$$

All water, no matter how pure, contains a few hydrogen ions (as hydronium ions) and a few hydroxide ions. The concentration of the ions in pure water is far too low to conduct electricity. It has been determined that the hydrogen ion and the hydroxide ion are each present in pure water to the extent of only $1 \times 10^{-7}\,M$, or $0.0000001\,M$.

TYPES OF ELECTROLYTE

The three basic types of electrolyte are *acids, bases,* and *salts.* For the present discussion, we will define an *acid* as a compound that yields hydrogen ions upon ionization.

Some typical acids:

$$HCl \xrightarrow{H_2O} H^+ + Cl^-$$

hydrochloric acid

$$H_2SO_4 \xrightarrow{H_2O} 2\,H^+ + SO_4^{2-}$$

sulfuric acid

$$CH_3CO_2H \xrightarrow{H_2O} H^+ + CH_3CO_2^-$$

acetic acid

One of the properties common to acids is that they have a sour, tart taste. The sour taste of lemons and grapefruits is due to the presence of citric acid. Vinegar contains about 4 percent acetic acid. Gastric juices contain very dilute hydrochloric acid.

Another property of acids is that they undergo reaction with hydroxides and carbonates.

$$2\,H^+ + Ca(OH)_2 \longrightarrow Ca^{2+} + 2\,H_2O$$

calcium hydroxide
(lime)

$$2\,H^+ + CaCO_3 \longrightarrow Ca^{2+} + H_2O + CO_2$$

calcium carbonate
(limestone)

For the present, we will define a *base* as a compound that yields hydroxide ions (OH^-) upon dissociation.

Some typical bases:

$$NaOH \longrightarrow Na^+ + OH^-$$

sodium hydroxide
(lye)

$$Ca(OH)_2 \longrightarrow Ca^{2+} + 2\,OH^-$$

calcium hydroxide

$$NH_3 + H_2O \longrightarrow NH_4^+ + OH^-$$

ammonia

Bases react with large molecules of grease, fats, and oils and break them into smaller, soluble molecules (Chapter 17). Lye (sodium hydroxide) is used to unclog drain pipes because of its ability to dissolve grease and fats. Ammonia is added to household cleaners because it too helps cut grease.

A *salt* is an ionic compound formed from the cation of a base and the anion of an acid; a salt yields neither a hydrogen ion nor a hydroxide ion upon dissociation.

Some typical salts:

$$NaCl \longrightarrow Na^+ + Cl^-$$

sodium chloride

$$KNO_3 \longrightarrow K^+ + NO_3^-$$

potassium nitrate

SOLUBILITY OF SALTS IN WATER

Salts differ from one another in their solubilities in water. In general, a very soluble salt is one in which the attraction of the ions for water molecules (and water molecules for the ions) is greater than the attraction of the two ions for each other. A few generalizations about salt solubilities are listed in Table 7-1.

Table 7-1. Solubilities of Salts

Salts containing these cations are generally soluble	Salts containing these anions are generally soluble	Salts containing these anions are generally not soluble (unless the cation is Na^+, K^+, NH_4^+)
Na^+	NO_3^-	OH^-
K^+	$CH_3CO_2^-$	CO_3^{2-}
NH_4^+	Cl^- (except for AgCl, $PbCl_2$)	S^{2-}
	Br^-	
	I^-	
	SO_4^{2-} (except for $CaSO_4$, $SrSO_4$, $BaSO_4$, and $PbSO_4$)	

Sample Problem: From Table 7-1, predict whether the following salts would be soluble or insoluble:
(a) KNO_3, (b) Na_2CO_3, (c) $MgCl_2$, (d) FeS

Solution: (a) Contains K^+ and NO_3^- and is soluble, (b) is a salt of Na^+ and is soluble, (c) is a salt of Cl^- and is soluble, (d) is a salt of S^{2-} and is insoluble.

Let us take a salt that is soluble in water and prepare a saturated solution with some solid salt still in contact with the solution. As we have said in Chapter 6, such a solution is in equilibrium. Some of the ions are going into solution, while others are coming out, or precipitating.

$$\underset{\text{sodium fluoride}}{NaF_{\text{solid}}} \xrightarrow{\text{dissolving}} Na^+ + F^-$$

$$Na^+ + F^- \xrightarrow{\text{precipitating}} NaF_{\text{solid}}$$

Since the reaction for dissociation is a reversible reaction, we may use just one equation with a double arrow, instead of two separate equations. The solid material, the precipitate, may be indicated in the equation by a down-pointing arrow (↓).

$$\underset{\textit{precipitate}}{NaF\downarrow} \rightleftharpoons \underset{\textit{in solution}}{Na^+ + F^-}$$

Barium sulfate ($BaSO_4$) is an insoluble salt that is opaque to *x*-rays. It is sometimes administered as an enema prior to *x*-rays of the gastrointestinal tract. Because of the opacity of $BaSO_4$, the intestines show up clearly on the *x*-ray film.

Any insoluble salt has a slight solubility. In the case of $BaSO_4$, 0.0002 grams dissolve in 100 grams of water. We may represent this slight solubility by an equation with arrows of different lengths to show in which direction the equilibrium lies. In the case of sodium chloride, we may draw the forward arrow as the longer of the two.

$$BaSO_4 \downarrow \;\rightleftharpoons\; Ba^{2+} + SO_4^{2-}$$

The long arrow shows that there are not many Ba^{2+} and SO_4^{2-} ions in solution.

The long arrow shows that NaCl is readily soluble.

$$NaCl \;\rightleftharpoons\; Na^+ + Cl^-$$

IONIZATION OF ACIDS IN WATER

When a salt dissolves in water, it is generally completely dissociated. Some acids, like salts, are almost completely ionized in water. We call these acids *strong acids.* Hydrochloric acid (HCl), sulfuric acid (H_2SO_4), and nitric acid (HNO_3) are all strong acids. (Sulfuric acid, which can give up two hydrogen ions, will be discussed later in this chapter.) Since they do not contain carbon, these acids are referred to as *mineral acids.*

Strong acids are highly ionized:

$$HCl \longrightarrow H^+ + Cl^-$$
$$H_2SO_4 \longrightarrow H^+ + HSO_4^-$$
$$HNO_3 \longrightarrow H^+ + NO_3^-$$

We dare not taste a 1.0 *M* solution of a strong acid like HCl. The concentration of hydrogen ions would be high enough that the solution would destroy the tissues of the mouth. Yet, vinegar and lemon juice contain acids (acetic acid and citric acid, respectively) in almost that same concentration. Why can a person eat pickles or drink lemonade and suffer no harm? The reason is that, unlike HCl, acetic acid and citric acid do not ionize to a large degree in solution. These acids are called *weak acids.*

Let us use acetic acid as an example of a weak acid. In a water solution of acetic acid, we find a few hydrogen ions and a few acetate ions, but mainly we find dissolved, nonionized acetic acid molecules.

$$CH_3\overset{\overset{\displaystyle O}{\|}}{C}OH \;\rightleftharpoons\; CH_3\overset{\overset{\displaystyle O}{\|}}{C}O^- + H^+$$

acetic acid — acetate ion

a weak acid

In other words, a weak acid is an acid that is only *partially ionized.* Because there are only a few ions in solution to conduct electricity, a partially ionized substance is called a *weak electrolyte,* compared to a strong electrolyte (almost completely dissociated into ions) or a nonelectrolyte (no appreciable ionization).

Table 7-2. Some Typical Strong and Weak Acids

Strong Acids	Weak Acids
Hydrochloric, HCl	Acetic, CH_3CO_2H[a]
Nitric, HNO_3	Boric, H_3BO_3
Sulfuric, H_2SO_4	Carbonic, H_2CO_3
	Citric, $C_3H_5O(CO_2H)_3$

[a]Any organic compound containing the $—CO_2H$ group is acidic.

Keep in mind that the terms *strong acid* and *weak acid* refer to the extent of ionization of an acid and not to its concentration. We may talk about a very dilute solution of a strong acid. Our gastric juices are about 0.5 percent of the strong acid hydrochloric acid. We may also speak of a highly concentrated solution of a weak acid. See Table 7-2 for some typical strong and weak acids.

DISSOCIATION OF BASES IN WATER

Bases, like acids, can be categorized as *strong bases* or *weak bases.* Sodium hydroxide and potassium hydroxide are examples of strong bases, completely dissociated in water solution.

$$\underset{\substack{\text{potassium hydroxide}\\ \textit{a strong base}}}{KOH} \longrightarrow K^+ + OH^-$$

Calcium hydroxide and magnesium hydroxide are not very soluble in water, but the portion that does dissolve is completely dissociated. These compounds are also strong bases.

$$\underset{\text{calcium hydroxide}}{Ca(OH)_2} \longrightarrow Ca^{2+} + 2\,OH^-$$

Ammonia gas (NH_3) dissolves in water to form a basic, or alkaline, solution. This reaction is somewhat different from simple dissociation in that ammonia actually undergoes a reaction with water to form an ammonium ion and a hydroxide ion.

A molecule of ammonia is composed of a nitrogen atom and three hydrogen atoms, and the nitrogen atom has an orbital filled with two unshared electrons. It is these two electrons that are donated to a hydrogen ion from water to form the ammonium ion. (See Figure 7-2.) Because only a small percentage of the ammonia

Figure 7–2. The formation of ammonium ions and hydroxide ions from ammonia and water.

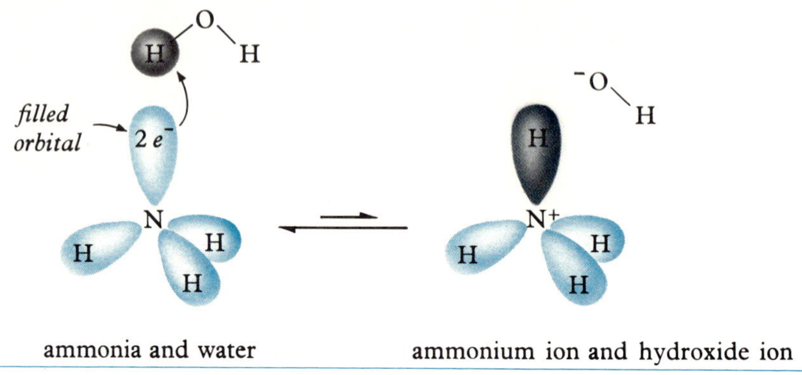

molecules in solution undergo this reaction, an aqueous solution of ammonia is considered a *weak base*. See Table 7–3 for a listing of some strong and weak bases.

$$\underset{\substack{\text{ammonia} \\ \textit{a weak base}}}{NH_3} + H_2O \rightleftharpoons NH_4^+ + OH^-$$

Table 7–3. Typical Strong and Weak Bases

Strong Bases	Weak Bases
Sodium hydroxide, $NaOH$	Ammonia, NH_3
Potassium hydroxide, KOH	(often called ammonium hydroxide, NH_4OH)
Calcium hydroxide, $Ca(OH)_2$	Methylamine, CH_3NH_2
Magnesium hydroxide, $Mg(OH)_2$	

HOW IONIC REACTIONS PROCEED

What happens if sodium chloride ($NaCl$) and potassium nitrate (KNO_3) are dissolved in water? The $NaCl$ dissociates into Na^+ and Cl^- ions, and the KNO_3 dissociates into K^+ and NO_3^- ions. The result is a mixture of four ions: Na^+, K^+, Cl^-, and NO_3^-. In this case, we say that *no reaction* takes place.

Now let us mix a solution of $NaCl$ with a solution of silver nitrate ($AgNO_3$). When we mix these two solutions, we observe a heavy white precipitate. What has happened? The situation is very similar to the previous example, but with one important difference. When Na^+, K^+, Cl^-, and NO_3^- are in solution, no insoluble salt can be formed. $NaCl$, $NaNO_3$, KCl, and KNO_3 are all soluble.

When Ag^+ collides with Cl^-, however, silver chloride (AgCl) is formed. Silver chloride is insoluble (Table 7-1) and does not redissociate. Instead, the ions come out of solution, and we observe a white precipitate of AgCl. The Na^+ and NO_3^- ions remain behind in solution. We may write the following equation to represent this reaction:

$$Na^+ + Ag^+ + NO_3^- + Cl^- \longrightarrow \underset{\substack{\textit{a white}\\ \textit{precipitate}}}{AgCl\downarrow} + Na^+ + NO_3^-$$

The Na^+ and NO_3^- ions are common to both sides of the equations; therefore, we may cancel them.

$$\textit{Cancellation:}\quad \cancel{Na}^+ + Ag^+ + \cancel{NO}_3^- + Cl^- \longrightarrow AgCl\downarrow + \cancel{Na}^+ + \cancel{NO}_3^-$$

$$\textit{Net ionic equation:}\quad Ag^+ + Cl^- \longrightarrow AgCl\downarrow$$

The portion of the equation that is left is called a *net ionic equation.* This equation tells us that any solution that contains Ag^+ ions and Cl^- ions will give a precipitate of AgCl. The other ions that might be present are of no importance in this ionic reaction. We could have started with KCl or $CaCl_2$ instead of NaCl; however, the net ionic equation would be the same.

A *molecular equation,* by contrast, shows us the complete molecules rather than just the ions in the reaction mixture. Throughout this book, we will use both net ionic and molecular equations.

$$\textit{Molecular equation:}\quad NaCl + AgNO_3 \longrightarrow AgCl\downarrow + NaNO_3$$

There must be a driving force to make an ionic reaction yield a product, or *proceed to completion.* In our last example, the precipitation of an insoluble salt resulted in an ionic reaction. Precipitation is just one of three occurrences that can drive an ionic reaction to completion. The three occurrences are (1) the precipitation of an insoluble salt, (2) the evolution of a gas, or (3) the formation of a nonionized substance.

1. The *precipitation of an insoluble salt* as a driving force has been exemplified with the precipitation of silver chloride. Any mixture of ions that leads to an insoluble salt will lead to an ionic reaction. For example, if we mix solutions of sodium sulfide, Na_2S, and lead(II) nitrate, $Pb(NO_3)_2$, the insoluble salt lead(II) sulfide, PbS, will be formed.

$$2\,Na^+ + S^{2-} + Pb^{2+} + 2\,NO_3^- \longrightarrow PbS\downarrow + 2\,Na^+ + 2\,NO_3^-$$

$$\textit{Net ionic equation:}\quad Pb^{2+} + S^{2-} \longrightarrow PbS\downarrow$$

2. The *evolution of a gas* also will drive an ionic reaction to completion. In an equation, we represent a gas coming out of a solution by an up-pointing arrow ($\uparrow$).

We have already mentioned that carbonic acid (H_2CO_3) yields H_2O and the gas CO_2. Therefore, if a solution of sulfuric acid (H_2SO_4) is mixed with a solution of sodium carbonate (Na_2CO_3), the product will be H_2CO_3, which, in turn, gives carbon dioxide, a gas.

$$2\,H^+ + SO_4^{2-} + 2\,Na^+ + CO_3^{2-} \longrightarrow SO_4^{2-} + 2\,Na^+ + H_2O + CO_2\uparrow$$

Net ionic equation: $2\,H^+ + CO_3^{2-} \longrightarrow H_2O + CO_2\uparrow$

3. The *formation of a nonionized compound* will also drive a reaction to completion. For example, we have seen that a hydrogen ion will react with a hydroxide ion to yield water, which is essentially a nonionized compound. If we mix a solution of potassium hydroxide (KOH) with nitric acid (HNO_3), water is formed and the reaction proceeds to completion.

$$H^+ + NO_3^- + K^+ + OH^- \longrightarrow H_2O + NO_3^- + K^+$$

Net ionic equation: $H^+ + OH^- \longrightarrow H_2O$

Sample Problem: Write the net ionic equation for the reaction that occurs in the stomach (which contains dilute HCl) when a person takes the antacid sodium bicarbonate ($Na^+\ {}^-HCO_3$).

Solution: *Step 1:* Write the formulas for the reactant ions.

$$H^+ + Cl^- + Na^+ + HCO_3^- \longrightarrow$$

Mentally, figure the possible combinations of ions and decide if an insoluble salt, a gas, or a nonionized compound can be formed from these ions. In this case, H_2CO_3 can be formed. We have already seen that H_2CO_3 yields water (H_2O) and carbon dioxide gas (CO_2).

Step 2: Write the formulas for the products.

$$\longrightarrow H_2O + CO_2 + Cl^- + Na^+$$

Step 3: Cancel the ions that are common to both sides of the equation.

Net ionic equation: $H^+ + HCO_3^- \longrightarrow H_2O + CO_2$

HOW REVERSIBLE REACTIONS PROCEED

Let us now turn our attention to weakly ionized substances—the weak acids and bases that exist in solution in both ionized and nonionized form. For the purposes of discussion, we will use acetic acid as a typical example of a weak acid and ammonia as a typical example of a weak base. Both acetic acid and ammonia are ionized in water to the extent of about 1.3 percent. In either an acetic acid solution or an ammonia solution, the vast majority (about 98.7 percent) of the molecules in water solution are not ionized.

IONIZATION OF A WEAK ACID. The ionization of acetic acid is a *reversible reaction.* When we add pure acetic acid to water, some of the molecules of acetic acid ionize. Eventually, a point is reached at which the concentrations of hydrogen ion and acetate ion are sufficiently great that they, in turn, start to undergo reaction with each other to go back to nonionized acetic acid molecules.

$$\textit{Ionizing:}\quad \underset{\text{acetic acid}}{CH_3CO_2H} \longrightarrow \underset{\text{acetate ion}}{CH_3CO_2^-} + H^+$$

$$\textit{Recombining:}\quad CH_3CO_2^- + H^+ \longrightarrow CH_3CO_2H$$

At a particular point, the concentrations of hydrogen ion, acetate ion, and acetic acid all become constant, even though the molecules and ions are shifting back and forth from ionic to nonionic states. Thus, there is an equilibrium in the solution between the nonionized acetic acid and the two ions, hydrogen ion and acetate ion. Let us combine the two preceding equations into one equation that is more descriptive of the equilibrium.

$$\textit{Combined equation:}\quad CH_3CO_2H \rightleftharpoons CH_3CO_2^- + H^+$$

Two facets of this equilibrium are worthy of note:

1. If acetic acid is the only compound in the solution, the concentration of acetate ion must be equal to the concentration of hydrogen ion. In other words, for every hydrogen ion that comes from an acetic acid molecule, one acetate ion must also be formed.

   When acetic acid ionizes:
   Concentration of H^+ = Concentration of $CH_3CO_2^-$

2. If the concentration of one of the ions (H^+ or $CH_3CO_2^-$) is increased or decreased by addition or removal by an outside source, the equilibrium will *shift*. To illustrate this statement, let us add a source of H^+, such as HCl, to an acetic acid solution. Now there are more hydrogen ions than before in the solution, and there are more hydrogen ions that can combine with the acetate ions to form acetic acid. The equilibrium shifts to the left. The addition of extra hydrogen ions results in fewer acetate ions and more acetic acid molecules in solution than before the addition.

$$CH_3CO_2H \rightleftharpoons CH_3CO_2^- + \overset{\text{add}}{H^+}$$

$\longleftarrow$
Add H^+; the equilibrium shifts to the left.

If we add a substance that *removes* hydrogen ions from the solution, such as NaOH, there will be *fewer* hydrogen ions to combine with acetate ions. The equilibrium will shift to the *right*. We will find more acetate ions and fewer undissociated acetic acid molecules in the solution than before the addition of NaOH.

$$CH_3CO_2H \rightleftharpoons CH_3CO_2^- + \underset{\text{remove}}{H^+}$$

$$\xrightarrow{\hspace{4cm}}$$

Remove H⁺; the equilibrium shifts to the right.

In the reversible reaction for the ionization of acetic acid in water, the concentrations of acetic acid, acetate ion, and hydrogen ion are all interdependent. This interdependency may be expressed by a mathematical equation.

$$\underset{\textit{reactant}}{CH_3CO_2H} \rightleftharpoons \underset{\textit{products}}{CH_3CO_2^- + H^+}$$

$$K = \frac{[H^+][CH_3CO_2^-]}{[CH_3CO_2H]}$$

where

$$\begin{aligned} K &= \text{The equilibrium constant} \\ [H^+] &= \text{Molarity of } H^+ \\ [CH_3CO_2^-] &= \text{Molarity of } CH_3CO_2^- \\ [CH_3CO_2H] &= \text{Molarity of undissociated } CH_3CO_2H \end{aligned}$$

This equation is called an *equilibrium expression*. It shows that the result of the concentrations (in molarity) of the reaction products multiplied together, then divided by the concentration of the reactant, is a constant. This constant is called the *equilibrium constant* and is represented as K.

Experimentally, it has been found that a 0.100 M solution of acetic acid has a hydrogen ion concentration of 0.00135 M. The concentration of acetate ion must be the same as the hydrogen ion concentration since one molecule of acetic acid ionizes into one acetate ion and one hydrogen ion. Since all of the acetate ions must have come from the original acetic acid, the final concentration of undissociated acetic acid is (0.100 − 0.00135) M, or 0.0986 M (rounded).

$$\underset{0.0986\,M}{CH_3CO_2H} \rightleftharpoons \underset{0.00135\,M}{CH_3CO_2^-} + \underset{0.00135\,M}{H^+}$$

Substituting the numbers into the equilibrium expression and solving for K gives a value of 1.85×10^{-5} for the equilibrium constant for the ionization of acetic acid. (A review of the use of exponents may be found in the Appendix.)

$$K = \frac{[H^+][CH_3CO_2^-]}{[CH_3CO_2H]}$$
$$= \frac{(0.00135)(0.00135)}{0.0986}$$
$$= 1.85 \times 10^{-5} \text{ for acetic acid}$$

For *any* dilute solution of acetic acid (whether it be 0.1 M, 0.02 M, or 0.0832 M), the value for K will calculate to be the same value, 1.85×10^{-5}, even though the concentrations of H^+, $CH_3CO_2^-$, and CH_3CO_2H vary. (Note that K is constant only for dilute solutions. In more concentrated solutions, ionic attractions become important, and the relationship is no longer valid.)

Sample Problem: A 0.020 M solution of acetic acid has a hydrogen ion concentration of 6.0×10^{-4} M. Calculate the value for K.

Solution: *Step 1:* Write the equation and the equilibrium expression.

$$CH_3CO_2H \rightleftharpoons CH_3CO_2^- + H^+$$
$$K = \frac{[H^+][CH_3CO_2^-]}{[CH_3CO_2H]}$$

Step 2: Before substituting in the equilibrium expression, list the known values.

$$K = ?$$
$$[H^+] = 6.0 \times 10^{-4}\,M$$
$$[CH_3CO_2^-] = 6.0 \times 10^{-4}\,M$$
$$[CH_3CO_2H] = .020 - 0.0006 = 0.0194$$

Step 3: Substitute the values in the equilibrium expression and solve the equation.

$$K = \frac{(6.0 \times 10^{-4})(6.0 \times 10^{-4})}{0.0194}$$
$$= 1.8 \times 10^{-5}$$

IONIZATION OF A WEAK BASE. The reaction of ammonia with water is also a reversible reaction, with ammonia and water molecules forming ions, and the ions going back to ammonia and water. This reaction, too, has an equilibrium expression.

$$\underset{\textit{reactants}}{NH_3 + H_2O} \rightleftharpoons \underset{\textit{products}}{NH_4^+ + OH^-}$$

$$K' = \frac{[NH_4^+][OH^-]}{[NH_3][H_2O]} \text{ for ammonia in water}$$

Since the concentration of water is large and relatively constant compared to that of the other species in solution, it is included in the value for K. (For dilute solutions, it is a common practice to include the concentration of water as part of K in this fashion.)

$$\underbrace{(K')[H_2O]}_{\text{constant}} = \frac{[NH_4^+][OH^-]}{[NH_3]}$$

$$K = \frac{[NH_4^+][OH^-]}{[NH_3]} \text{ for ammonia in water}$$

CALCULATIONS WITH EQUILIBRIUM CONSTANTS. Any reversible reaction has its own equilibrium constant K.

$$K = \frac{[\text{Products}]}{[\text{Reactants}]}$$

Table 7-4 shows the constants for the ionization of a few weak acids and bases. The larger the value for K, the greater the extent of ionization. Thus, HCN (hydrocyanic acid, or prussic acid) with $K = 4.9 \times 10^{-10}$ is a much weaker acid than acetic acid.

Table 7-4. Equilibrium Constants for Some Weak Acids and Weak Bases

Name	Formula	Equilibrium Constant, K
Acetic acid	CH_3CO_2H	1.85×10^{-5}
Boric acid	H_3BO_3	7.3×10^{-10}
Hydrocyanic acid	HCN	4.9×10^{-10}
Ammonia	NH_3	1.8×10^{-5}
Lead(II) hydroxide	$Pb(OH)_2$	9.6×10^{-4}

Sample Problem: What is the hydrogen ion concentration of a 0.50 M solution of HCN?

Solution: *Step 1:* Write the equation for the ionization and the equilibrium expression.

$$HCN \rightleftharpoons H^+ + CN^-$$

$$K = \frac{[H^+][CN^-]}{[HCN]}$$

Step 2: List the known values.

$$K = 4.9 \times 10^{-10} \text{ (from Table 7-4)}$$
$$[H^+] = ?$$
$$[CN^-] = ?$$
$$[HCN] = ?$$

We want to find $[H^+]$; let us call that value x. However, since one HCN molecule dissociates to give equal numbers of H^+ and CN^-, $[CN^-]$ must also be x. The original concentration of HCN is 0.5 *M*, but a tiny amount of this concentration is lost as H^+ and CN^-. For most weak acids and bases, the proportional amount of ionized substance is very small compared to the amount of nonionized substance. The concentration of nonionized HCN is still essentially 0.5 *M*. (It is as if we had 50 billion ants and stepped on five. We can still say we have approximately 50 billion ants.)

Now we can fill in the values.

$$K = 4.9 \times 10^{-10}$$
$$[H^+] = x$$
$$[CN^-] = x$$
$$[HCN] = 0.5\,M$$

Step 3: Substitute into the equation and solve.

$$4.9 \times 10^{-10} = \frac{(x)(x)}{0.5}$$
$$x^2 = (4.9 \times 10^{-10})(0.5)$$
$$= 2.45 \times 10^{-10}$$
$$x = \sqrt{2.45 \times 10^{-10}}$$
$$= \sqrt{2.45} \times 10^{-5}$$
$$= 1.6 \times 10^{-5}$$
$$[H^+] = 1.6 \times 10^{-5}\,M$$

Sample Problem: What is the concentration of NH_3 in a solution that has a hydroxide ion concentration of $1.8 \times 10^{-4}\,M$?

Solution: *Step 1:* Write the equation and the equilibrium expression.

$$NH_3 + H_2O \rightleftharpoons NH_4^+ + OH^-$$
$$K = \frac{[NH_4^+][OH^-]}{[NH_3]}$$

Step 2: List the known values,

$$K = 1.8 \times 10^{-5} \text{ (from Table 7-4)}$$
$$[NH_4^+] = 1.8 \times 10^{-4} \text{ (same as } OH^-)$$
$$[OH^-] = 1.8 \times 10^{-4}\,M$$
$$[NH_3] = ?$$

Step 3: Substitute and solve.

$$1.8 \times 10^{-5} = \frac{(1.8 \times 10^{-4})(1.8 \times 10^{-4})}{[NH_3]}$$
$$[NH_3] = \frac{(1.8 \times 10^{-4})(1.8 \times 10^{-4})}{1.8 \times 10^{-5}}$$
$$= 1.8 \times 10^{-3}\,M$$

EQUILIBRIUM EXPRESSION FOR WATER

We have already mentioned that water ionizes to a small extent into H^+ and OH^-. Just as for any ionization of a weakly ionized substance, we can express this equilibrium by an equilibrium expression.

$$H_2O \rightleftharpoons H^+ + OH^-$$

$$K = \frac{[H^+][OH^-]}{[H_2O]}$$

Again, the concentration of water, $[H_2O]$, is grouped with the constant K. The resulting expression, $K\,[H_2O]$, is called K_w. This constant (K_w) has been found to be 1.0×10^{-14} at 25°C.

$$K_w = [H^+][OH^-] = 1.0 \times 10^{-14}$$

In pure water, the concentration of H^+ must be equal to the concentration of OH^-. Calculating the concentrations of H^+ and OH^- from the equilibrium expression, we find that both H^+ and OH^- are present in pure water to the extent of $1 \times 10^{-7}\,M$.

$$\begin{aligned}[H^+] = [OH^-] &= x \\ [H^+][OH^-] &= 1 \times 10^{-14} \\ x^2 &= 1 \times 10^{-14} \\ x &= 1 \times 10^{-7}\,M\end{aligned}$$

Not only for pure water, but for *any* aqueous solution (such as a solution of NaCl) that has equal concentrations of H^+ ions and OH^- ions, the concentrations of both H^+ and OH^- equal $1 \times 10^{-7}\,M$. A solution in which the concentrations of H^+ and OH^- are equal is called a *neutral solution.* A neutral solution is neither acidic nor alkaline. A solution that contains more hydrogen ions than hydroxide ions is said to be an *acidic solution.* A solution with more hydroxide ions than hydrogen ions is said to be a *basic solution,* or an *alkaline solution.*

The multiplication product of the concentrations of H^+ ions and OH^- ions in an acidic solution still must equal 1×10^{-14}. Consequently, even an acidic solution contains some OH^- ions. For example, if the concentration of H^+ is $1 \times 10^{-2}\,M$, then the concentration of OH^- ions is $1 \times 10^{-12}\,M$.

THE pH SCALE

It is often inconvenient to say that hydrogen ion concentration is, for example, $1 \times 10^{-6}\,M$, because most people prefer not to work with exponents. Therefore, a scale has been set up using just the single number that appears in the exponent. This number, the negative exponent of the hydrogen ion concentration, is called the pH of a solution.

$$1 \times 10^{-⑥}\ M$$ ← *used for the pH scale*

Reworded, we say that the pH of a solution is *the negative log of the hydrogen ion concentration.*

$$\text{pH} = -\log [\text{H}^+]$$

Table 7-5 shows some examples of hydrogen ion concentrations and the corresponding pH values. Note that the higher the concentration of hydrogen ion, the smaller is the pH value.

Table 7-5. Hydrogen Ion Concentration and pH Values

$[\text{H}^+]$ Expressed as a Decimal	$[\text{H}^+]$ Expressed with an Exponent	pH Value
10 *M*	$1 \times 10^{1}\ M$	−1
0.1 *M*	$1 \times 10^{-1}\ M$	1
0.001 *M*	$1 \times 10^{-3}\ M$	3
0.00001 *M*	$1 \times 10^{-5}\ M$	5
0.0000001 *M*	$1 \times 10^{-7}\ M$	7
0.00000001 *M*	$1 \times 10^{-8}\ M$	8

Sample Problem: What is the pH of a solution that is 0.01 *M* in hydrogen ion?

Solution:

$$[\text{H}^+] = 0.01\ M = 1 \times 10^{-2}\ M$$

$$\text{pH} = -\log [\text{H}^+] = -\log (1 \times 10^{-2}) = -(-2) = +2$$

FRACTIONAL pH VALUES. Hydrogen ion concentrations that fall *between* 0.01 *M* and 0.1 *M*, for example, may be expressed by using fractional pH values.

Sample Problem: What is the pH of a solution in which $[\text{H}^+] = 0.035\ M$?

Solution:

$$[\text{H}^+] = 3.5 \times 10^{-2}\ M$$

$$\begin{aligned}\text{pH} &= -\log [\text{H}^+] \\ &= -\log (3.5 \times 10^{-2}) \\ &= -\log 3.5 - \log (10^{-2}) \\ &= -0.54 + 2 \\ &= 1.46\end{aligned}$$

(−0.54: *from a log table, slide rule, or electronic calculator*)

As a check on your answer, it is wise to ask yourself if the pH value is reasonable. In this case, a hydrogen ion concentration of 0.035 M lies *between* 0.01 M and 0.1 M. Therefore, the pH value must lie between +2 and +1, which indeed, it does.

USING pH VALUES. How can we use pH values? A few rules of thumb may be helpful:

1. A neutral solution, neither acidic nor basic, has pH 7.
2. Acidic solutions have pH values *less than 7.*
3. Alkaline solutions have pH values *greater than 7.*

−2 −1 0 1 2 3 4 5 6 ⑦ 8 9 10 11 12 13 14

← Increasing acidity ↑ Increasing basicity →

Neutral

Sample Problem: Normal human blood has a pH range of 7.3 to 7.5. Is human blood acidic or basic?

Solution: The pH value is slightly greater than 7; therefore, human blood is slightly basic.

We can calculate either the hydrogen ion concentration or the hydroxide ion concentration from the pH of a solution by means of the following formulas.

$$pH = -\log [H^+] \qquad [H^+][OH^-] = 1 \times 10^{-14}$$

Sample Problem: What is the hydroxide ion concentration of a solution with pH 8?

Solution: *Step 1:* Find $[H^+]$.

$$\begin{aligned} -\log [H^+] &= pH \\ &= 8 \\ \log [H^+] &= -8 \\ [H^+] &= 1 \times 10^{-8}\,M \end{aligned}$$

Step 2: Find $[OH^-]$.

$$\begin{aligned} [H^+][OH^-] &= 1 \times 10^{-14} \\ (1 \times 10^{-8})[OH^-] &= 1 \times 10^{-14} \\ [OH^-] &= \frac{1 \times 10^{-14}}{1 \times 10^{-8}} \\ &= 1 \times 10^{-6}\,M \end{aligned}$$

NEUTRALIZATION REACTIONS

In an ionic solution, the formation of a nonionized product causes a reaction to proceed to completion. Water is essentially a nonionized product produced by the reaction of an acid and a base. If we mix equivalent quantities of an acid and a base, we will end up with a neutral solution, neither acidic nor basic. Thus, a reaction between an acid and a base is called a *neutralization reaction.*

$$\underset{\text{an acid}}{H^+} + \underset{\text{a base}}{:\ddot{\underset{..}{O}}H^-} \longrightarrow \underset{\text{water}}{H_2\ddot{O}:}$$

We can add a base to a weak acid, as well, and neutralize the acid by shifting the equilibrium until the weak acid has undergone complete reaction.

$$\underset{\text{a weak acid}}{CH_3CO_2\textcircled{H}} + \underset{\text{a base}}{:\ddot{\underset{..}{O}}H^-} \longrightarrow CH_3CO_2^- + \underset{\text{water}}{H_2\ddot{O}:}$$

Compounds other than hydroxides can act as bases in neutralization reactions. If a neutral solution is the result, we still have a neutralization reaction.

$$\underset{\text{an acid}}{2\,H^+} + \underset{\text{a base}}{CO_3^{2-}} \longrightarrow \underset{\text{water}}{H_2O} + CO_2\uparrow$$

Our earlier definition of a base being a substance that forms hydroxide ions is therefore too restrictive. We must redefine the term *base* to mean *any substance that accepts hydrogen ions.* Hydroxides react with hydrogen ions—they are but one type of base. Carbonates react with hydrogen ion—they are bases also.

What of the other ions in solution? We cannot simply mix up plain hydrogen ions and plain hydroxide ions, but must start with electrically neutral compounds. Let us rewrite our neutralization equations to show the nonreacting ions as well as the reacting ones.

$$\underbrace{H^+ + Cl^-}_{\text{hydrochloric acid}} + \underbrace{Na^+ + OH^-}_{\text{sodium hydroxide}} \longrightarrow H_2O + \underbrace{Na^+ + Cl^-}_{\text{sodium chloride}}$$

$$\underset{\text{acetic acid}}{CH_3CO_2H} + \underbrace{Na^+ + OH^-}_{\text{sodium hydroxide}} \longrightarrow H_2O + \underbrace{Na^+ + CH_3CO_2^-}_{\text{sodium acetate}}$$

$$\underbrace{2\,H^+ + SO_4^{2-}}_{\text{sulfuric acid}} + \underbrace{2\,Na^+ + CO_3^{2-}}_{\text{sodium carbonate}} \longrightarrow H_2O + CO_2\uparrow + \underbrace{2\,Na^+ + SO_4^{2-}}_{\text{sodium sulfate}}$$

In each case, we have a *salt* left over in the solution. A reaction of the strong, corrosive base NaOH with the strong, corrosive acid HCl results simply in an aqueous solution of NaCl, or ordinary salt water. Thus, *a neutralization reaction is the reaction of an acid and a base to yield water and a salt.*

TITRATION

A common laboratory problem is determining the quantity of an acid or a base in a given sample. The procedure that is used to solve this problem is called *titration.* Titration involves measuring the volume of the sample solution and then slowly adding to it another solution of known concentration until a neutralization reaction just reaches completion.

As an example, let us say we have 25.00 ml of a solution of a base of unknown concentration. We add to it a *standardized solution* (a solution of a known concentration) of an acid, such as HCl, until the unknown solution is exactly neutral. The standardized solution is added from a specially calibrated tube called a *buret.* The volume of the standardized solution needed to react completely with the unknown sample is read directly from the buret. (See Figure 7-3.)

In order to carry out a titration, the operator must know when the reaction has reached completion. An acid-base reaction gives no outward appearance that the reaction is complete, and we need something to tell us STOP when the neutral point is reached. As a consequence, an *indicator* is added to the mixture. An indicator is an organic compound that is one color in acidic solution and another color in alkaline solution. At the point of neutrality, or *end point,* the indicator changes color—a very effective way of saying "Stop titrating. You have added just enough acid to neutralize the base."

The use of indicators is widespread in laboratory work. In titrations of acids and bases, *phenolphthalein* is a common indicator. This compound is colorless in acidic solution and pink in alkaline solution. For cruder work, like a quick check on whether a solution is acidic and basic, *litmus paper* is commonly used. Blue litmus paper turns pink when dipped into acidic solution, whereas pink litmus paper turns blue in basic solution. Also available to the laboratory worker is *pH paper.* This is a paper impregnated with a variety of indicators that can turn one of a variety of colors depending upon the pH. Yet another way of determining the pH of a solution is with a *pH meter,* an instrument that measures the pH of a solution electrically.

In a titration, we carry out a neutralization reaction just to the end point. At this point, the number of equivalents of the acid (or base) added just equals the number of equivalents of base (or acid) in the unknown solution.

You will recall that the normality (N) of a solution is equal to the number of equivalents of a substance in one liter of solution and that the normality may be

Figure 7-3. A titration apparatus.

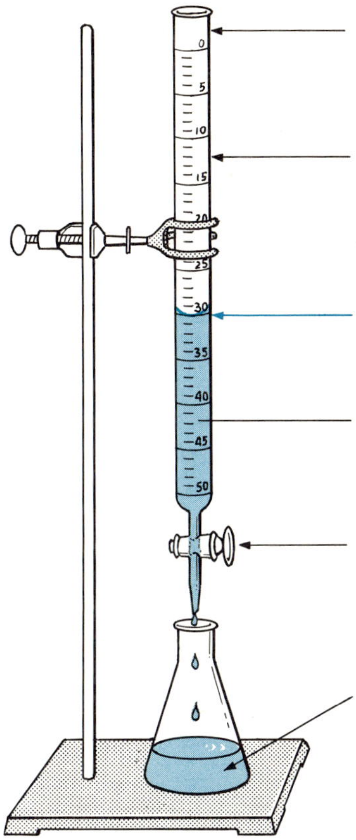

calculated by the following equation:

$$N = \frac{\text{Number of equivalents of solute}}{\text{Number of liters of solution}}$$

Let us rearrange this equation to give the number of equivalents:

$$\begin{aligned}\text{Number of equivalents} &= (N)\ (\text{Number of liters}) \\ &= NV\end{aligned}$$

In a neutralization reaction, the number of equivalents of the acid equals the number of equivalents of the base. Therefore, we can set up a convenient equation with which to make calculations:

$$\text{Number of equivalents of acid} = \text{Number of equivalents of base}$$

or

$$N_{\text{acid}}V_{\text{acid}} = N_{\text{base}}V_{\text{base}}$$

In this equation, we may use milliliters as well as liters, as long as we are consistent. The use of this equation is shown in the following sample problem.

Sample Problem: A solution of 15.00 ml of a base requires 30.00 ml of 0.2000 *N* HCl to be neutralized. What is the normality of the basic solution?

Solution:

$$N_{acid}V_{acid} = N_{base}V_{base}$$

$$(0.2000\ N)(30.00\ \text{ml}) = (N_{base})(15.00\ \text{ml})$$

$$N_{base} = \frac{(0.2000\ N)(30.00\ \text{ml})}{(15.00\ \text{ml})}$$

$$= 0.4000$$

DIBASIC ACIDS

Certain acids, such as H_2SO_4, contain two hydrogen ions that can ionize. These acids are called *dibasic*, or *diprotic*, acids. (The term *dibasic* comes from the property that such an acid can undergo reaction with *two* equivalents of *base*.) A *tribasic acid* is one with *three* ionizable hydrogens.

H_2SO_4	H_2CO_3	H_3PO_4
sulfuric acid	carbonic acid	phosphoric acid
dibasic acids		*a tribasic acid*

When these acids dissociate in water, they do so in steps. First, one hydrogen ion ionizes; then the second one. The second dissociation always occurs to a lesser extent than the first.

Let us use sulfuric acid as an example. The first hydrogen ion separates completely from the sulfuric acid molecule in water solution. Sulfuric acid is a strong acid.

single arrow shows complete dissociation

Step 1: $H_2SO_4 \longrightarrow H^+ + HSO_4^-$

sulfuric acid (*a strong acid*) → hydrogen ion + bisulfate ion

The bisulfate, or hydrogen sulfate, ion (HSO_4^-) that is formed can lose yet another hydrogen ion, but this second ionization is incomplete. The bisulfate ion is a weak acid.

double arrow shows incomplete dissociation

$$\textit{Step 2:}\quad \underset{\textit{a weak acid}}{HSO_4^-} \rightleftharpoons H^+ + \underset{\text{sulfate ion}}{SO_4^{2-}}$$

Carbonic acid is a weak acid that ionizes only partially to hydrogen ion and bicarbonate ion (HCO_3^-).

$$\textit{Step 1:}\quad \underset{\text{carbonic acid}}{H_2CO_3} \rightleftharpoons H^+ + \underset{\text{bicarbonate ion}}{HCO_3^-}$$

a weak acid

The bicarbonate ion is a weaker acid yet and ionizes scarcely at all. In fact, as we shall see, the bicarbonate ion generally acts like a base, not an acid.

$$\textit{Step 2:}\quad HCO_3^- \rightleftharpoons H^+ + \underset{\text{carbonate ion}}{CO_3^{2-}}$$

AMPHOTERIC SUBSTANCES

Certain compounds or ions can act as either acids or bases. These are called *amphoteric compounds* or *amphoteric ions*. Sodium bicarbonate ($NaHCO_3$) is an amphoteric compound. $NaHCO_3$ can act as a base, a proton acceptor, in its reaction with HCl.

Sodium bicarbonate as a base:

Molecular equation: $HCl + NaHCO_3 \longrightarrow NaCl + H_2CO_3$

Net ionic equation: $H^+ + HCO_3^- \longrightarrow H_2CO_3$ (which, in turn, goes to $H_2O + CO_2$)

However, the bicarbonate ion can also *lose* a proton, or act as an acid, as in the following reaction with NaOH.

Sodium bicarbonate as an acid:

Molecular equation: $NaOH + NaHCO_3 \longrightarrow Na_2CO_3 + H_2O$

Net ionic equation: $OH^- + HCO_3^- \longrightarrow CO_3^{2-} + H_2O$

Hydrogen phosphate salts, such as Na_2HPO_4 and NaH_2PO_4, are also amphoteric because the hydrogen phosphate ions can either gain or lose hydrogen ions.

$$\underset{\text{an acid}}{HPO_4^{2-}} + OH^- \longrightarrow PO_4^{3-} + H_2O$$

$$\underset{\text{a base}}{HPO_4^{2-}} + H^+ \longrightarrow H_2PO_4^-$$

BUFFER SOLUTIONS

A compound or a group of compounds that helps a solution resist changes in pH is called a *buffer*. Solutions of buffers are called *buffer solutions*.

Blood is an example of a buffer solution. The buffering action in blood that maintains the pH at a constant value is caused primarily by the bicarbonate ion and carbonic acid. (Certain proteins and, to a small extent, hydrogen phosphate ions also help buffer the blood.)

If a base is added to the bloodstream, it is neutralized by the carbonic acid.

$$OH^- + H_2CO_3 \rightleftharpoons HCO_3^- + H_2O$$

If, on the other hand, excess acid is added to the bloodstream, it is neutralized by the bicarbonate ion.

$$H^+ + HCO_3^- \rightleftharpoons H_2CO_3$$

Carbon dioxide is formed as a waste product by cells. Excess carbon dioxide is eliminated by way of the lungs. If the blood does not contain enough carbonic acid for buffering action, however, more of the carbon dioxide from the cells is retained in the blood.

$$H_2O + CO_2 \rightleftharpoons H_2CO_3$$

Any buffer system can be overwhelmed by the addition of large amounts of acid or base because the acid-base equilibria would then be shifted to one side. For example, if large amounts of acid were added to blood, the blood would become deficient in bicarbonate ions and less able to dispose of the excess acid.

$$\underset{\text{excess}}{H^+} + HCO_3^- \longrightarrow H_2CO_3$$

The kidneys help prevent excessive acid in the blood by the excretion of H^+ from carbonic acid by way of the urine and the return of bicarbonate ions to the bloodstream. (See Chapter 29.)

STUDY QUESTIONS

7-1. Which of the following compounds would you expect to be electrolytes?
(a) $MgSO_4$ (b) H_2O
(c) CH_4 (d) H_2SO_4

7-2. What is the distinction between a hydrogen ion and a hydronium ion?

7-3. Classify each of the following electrolytes as an acid, a base, or a salt:
(a) KCl (b) NaH_2PO_4
(c) $NaHSO_4$ (d) $Al(OH)_3$

7-4. Dilute hydrochloric acid is often used to remove rust (Fe_2O_3) from metals and from sinks. Write the equation for the chemical reaction that takes place.

7-5. Using Table 7-1, predict the solubilities of the following compounds in water:
(a) NH_4Cl (b) Na_2CO_3
(c) $MgCO_3$ (d) $Sr(NO_3)_2$
(e) $CoCl_2$

7-6. Is it correct to say that a very concentrated solution of acetic acid is a strong acid?

7-7. Write the equation for the reaction that occurs when NH_3 dissolves in water.

7-8. Write (and balance) the following equations as net ionic equations:
(a) $MgCl_2 + NaOH \longrightarrow Mg(OH)_2\downarrow + NaCl$
(b) $NaHCO_3 + HCl \longrightarrow CO_2\uparrow + H_2O + NaCl$
(c) $Na_2S + CoCl_2 \longrightarrow CoS\downarrow + NaCl$
(d) $H_3PO_4 + FeCl_3 \longrightarrow FePO_4\downarrow + HCl$

7-9. Write the net ionic equation for the reaction that would occur when solutions of sodium sulfide (Na_2S) and copper(II) chloride ($CuCl_2$) are mixed.

7-10. In the reaction $HCN \rightleftharpoons H^+ + CN^-$, determine how the equilibrium will be affected by the addition of
(a) NaCN (b) NaOH
(c) HCN (d) HCl

7-11. Set up the equilibrium expression for each of the following reactions:
(a) $HCN \rightleftharpoons H^+ + CN^-$
(b) $AgCl \rightleftharpoons Ag^+ + Cl^-$
(c) $CH_3CH_2CO_2H \rightleftharpoons CH_3CH_2CO_2^- + H^+$
(d) $CH_3NH_2 + H_2O \rightleftharpoons CH_3NH_3^+ + OH^-$

7-12. A 0.500 M solution of chloroacetic acid ($ClCH_2CO_2H$) ($K = 140 \times 10^{-5}$) would have a hydrogen ion concentration of ______ M.

7-13. What would be the hydrogen ion concentration of a 0.250 M solution of trichloroacetic acid (Cl_3CCO_2H) ($K = 1.3 \times 10^{-1}$)?

7-14. Calculate the hydroxide ion concentration of a solution if the hydrogen ion concentration is
(a) $7.0 \times 10^{-7}\ M$ (b) $4.0 \times 10^{-3}\ M$
(c) $3.0 \times 10^{-10}\ M$ (d) $1.5 \times 10^{-5}\ M$

7-15. Given the following hydrogen ion concentrations, calculate the pH values:
(a) $3.5 \times 10^{-5}\ M$ (b) $6.0 \times 10^{-2}\ M$
(c) $8.0 \times 10^{-13}\ M$ (d) $0.50 \times 10^{-1}\ M$
(e) $6.0 \times 10^{-4}\ M$ (f) $4.4 \times 10^{-10}\ M$
(g) 1.0 M (h) 5.0 M

7-16. Calculate the hydrogen ion concentration if the pH is
(a) 6.0 (b) 5.5 (c) 1.2 (d) 8.6
(e) −1.0 (f) 0 (g) 7.2 (h) 10.8

7-17. State whether solutions with the following pH values are acidic or basic:
(a) 6.9 (b) 1.0 (c) 8.0 (d) 7.1

7-18. Someone states that the pH of a solution is 15. Is this statement reasonable?

7-19. What are the hydroxide ion concentrations of the following solutions?
(a) pH 2 (b) pH 13
(c) pH 4.5 (d) pH 7.2

7-20. The pH of urine is usually about 6.0.
(a) Is urine acidic or alkaline?
(b) What is the hydrogen ion concentration of urine?
(c) What is the hydroxide ion concentration?

7-21. Which of the following equations represent neutralization reactions?

(a) $2\,HNO_3 + K_2CO_3 \longrightarrow CO_2 + H_2O + 2\,KNO_3$

(b) $AgNO_3 + NaCl \longrightarrow AgCl + NaNO_3$

(c) $NaCN + NaHSO_4 \longrightarrow HCN + Na_2SO_4$

(d) $Na_2S + CuCl_2 \longrightarrow 2\,NaCl + CuS$

7-22. If a solution of 25.00 ml of an acid requires 15.00 ml of 0.500 *N* sodium hydroxide solution for neutralization, what is the normality of the original acid solution?

7-23. How many ml of a 0.100 *N* NaOH solution would be required to neutralize each of the following?

(a) 5.00 ml of a 0.500 *N* HCl

(b) 10.00 ml of 0.250 *N* H_2SO_4

(c) 10.00 ml of 0.250 *M* H_2SO_4

(d) 30.00 ml of a solution that has a pH of 2.00

7-24. An aqueous solution of sodium bicarbonate is not neutral, but is alkaline. Suggest a reason for this.

7-25. One of the acids produced when fat is broken down by the cells is acetoacetic acid. (The production of this acid is one of the factors that leads to *acidosis,* excessive acid in the blood.) Show by an equation how this acid can be neutralized by the bicarbonate ion.

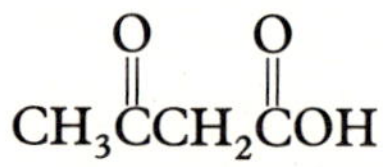

acetoacetic acid

CHAPTER 8 OXIDATION AND REDUCTION REACTIONS

OBJECTIVES

Oxidation and Reduction *Define oxidation and reduction, and write equations for examples of each.*

Determining an Oxidation Number *List the rules used in the calculations of oxidation numbers of atoms.*

Balancing Simple Oxidation-Reduction Equations *Balance an oxidation-reduction equation using half-reactions.*

Oxidation and Reduction of Organic Compounds *Describe with examples how we determine whether an organic compound is oxidized or reduced.*

Oxidation and Reduction in Our World *Define and illustrate complete combustion and incomplete combustion.* * *Give equations for the formation and the decomposition of ozone in the upper atmosphere.*

OXIDATION AND REDUCTION

The burning of gasoline in an automobile engine, the digestion of food, the bleaching of clothes, and the corrosion of metals—all of these reactions and many others involve oxidation and reduction reactions.

$$2\,C_8H_{18} + 25\,O_2 \longrightarrow 16\,CO_2\uparrow + 18\,H_2O\uparrow + \text{Energy}$$

octane
in gasoline

$$C_6H_{12}O_6 + 6\,O_2 \longrightarrow 6\,CO_2\uparrow + 6\,H_2O + \text{Energy}$$

glucose
from carbohydrates in the diet

Before we consider these reactions, we must discuss a few of the conventions that have been devised to handle oxidation and reduction reactions. Recall that the *oxidation number* is the electrical charge on an atom, a charge that results from an unequal number of protons and electrons in the atom. An atom of a substance in the elemental state has no electrical charge, and the oxidation number is zero. For example, a neutral atom of calcium metal contains 20 protons and 20 electrons; the electrical charge and the oxidation number are both zero.

oxidation number of 0 → Ca^0

20 protons
20 electrons

oxidation number of +2 → Ca^{2+}

20 protons
18 electrons

oxidation number of 0 → Cl^0

17 protons
17 electrons

oxidation number of −1 → Cl^-

17 protons
18 electrons

oxidation number of 0 → $Cl_2{}^0$

a total of 34 protons and 34 electrons for the two chlorine atoms

Oxidation numbers can be changed in chemical reactions. An *increase in the oxidation number* of an atom is defined as *oxidation.* Oxidation of an atom occurs by the loss of electrons. Thus, when Ca^0 loses two electrons to become Ca^{2+}, its oxidation number increases from zero to +2, and an oxidation reaction takes place. (Note that the oxidation number here is the same as the valence.)

an increase from 0 to +2 in oxidation number

$$Ca^0 \longrightarrow Ca^{2+} + 2\,e^-$$

An oxidation reaction: a loss of electrons.

Reduction is defined as a *decrease in the oxidation number* of an atom. In other words, for an atom to be reduced, it must *gain* electrons.

a decrease in oxidation number from 0 to −1

$$Cl_2{}^0 + 2\,e^- \longrightarrow 2\,Cl^-$$

A reduction reaction: a gain of electrons.

An *oxidation-reduction reaction* is one in which there is *electron transfer.* Any time we observe an oxidation, there must be a corresponding reduction of something else. If one substance gains electrons, another substance must have lost the same number of electrons. One of the keys to understanding oxidation-reduction reactions is remembering that an oxidation reaction cannot occur without a reduction reaction also occurring and that electrons are neither lost nor gained in the overall combined reaction.

$$Ca + Cl_2 \longrightarrow CaCl_2$$

Ca is oxidized to Ca^{2+} (two electrons lost).
Cl_2 is reduced to two Cl^- (two electrons gained).

In the reaction of calcium metal with chlorine, we see that one calcium atom loses two electrons and is oxidized. Each chlorine atom gains one electron and is reduced. The number of electrons lost by one calcium atom is the same as the total number of electrons gained by the two chlorine atoms.

In the following reaction, we also find that both an oxidation reaction and a reduction reaction occur and that all the electrons in the combined reaction can be accounted for.

$$2\,Na + Br_2 \longrightarrow 2\,NaBr$$

Two Na atoms are oxidized to two Na^+ (two electrons lost).
Br_2 is reduced to two Br^- (two electrons gained).

Before we can proceed with our discussion of oxidation and reduction, two more conventions must be defined. These are the terms "oxidizing agent" and "reducing agent." An *oxidizing agent* is a substance that causes the oxidation of something else. While it is causing an oxidation, the oxidizing agent is reduced. In the following equation, bromine causes the oxidation of sodium. Therefore, bromine acts as an oxidizing agent and, in the process, is reduced.

$$2\,Na \quad + \quad Br_2 \longrightarrow 2\,NaBr$$

reducing agent — 2 Na; *oxidizing agent* — Br_2; in 2 NaBr, Na is *oxidized* and Br is *reduced*.

A *reducing agent* is a substance that causes the reduction of something else. In the process, the reducing agent is oxidized. The definitions and conventions used in oxidation and reduction are summarized in Table 8-1.

Table 8-1. Conventions Used for Oxidation and Reduction Reactions

$$2\,Na^0 + Br_2{}^0 \longrightarrow 2\,Na^+\,Br^-$$

$2\,Na^0 \longrightarrow 2\,Na^+ + 2\,e^-$	$Br_2{}^0 + 2\,e^- \longrightarrow 2\,Br^-$
Oxidation Reaction	Reduction Reaction
Loss of electrons	Gain of electrons
Increase in oxidation number	Decrease in oxidation number
Na is oxidized	Br_2 is reduced
Na acts as a reducing agent	Br_2 acts as an oxidizing agent

DETERMINING AN OXIDATION NUMBER

The determination of oxidation numbers in simple ionic compounds is relatively easy. But, if a molecule contains covalent bonds, the oxidation numbers of the atoms in the molecule may not be immediately apparent. For example, what is the oxidation number of nitrogen in potassium nitrate ($K^+\,NO_3{}^-$)?

Most complex ions that we deal with, such as $NO_3{}^-$, contain oxygen. In its compounds, oxygen almost always has an oxidation number of -2. (See Table 8-2.) We can therefore calculate the oxidation number of nitrogen in the nitrate ion, $NO_3{}^-$.

First, we determine the sum of the oxidation numbers in the ion or molecule. The sum of the oxidation numbers in $NO_3{}^-$ is -1 (because this ion has an overall

Table 8-2. Some Typical Oxidation Numbers of Atoms

Oxidation Number	Where Found
0	Any element, such as Fe or O_2
$+1$	H^+, Li^+, Na^+, K^+
-1	F^-, Cl^-, Br^-, I^-
$+2$	Mg^{2+}, Ca^{2+}
-2	O^{2-} or in complex ions

ionic charge of −1). In a neutral molecule, such as KNO_3 or Fe_2O_3, the sum of the oxidation numbers is zero (because these molecules carry no overall ionic charge).

The oxidation number of nitrogen in NO_3^-:

$$\begin{aligned} \text{Nitrogen} + 3(\text{Oxygen}) &= -1 \\ \text{Nitrogen} + 3(-2) &= -1 \\ \text{Nitrogen} &= -1 + 6 \\ \text{The oxidation number of N} &= +5 \end{aligned}$$

The oxidation number of Fe in Fe_2O_3:

$$\begin{aligned} 2(\text{Fe}) + 3(\text{O}) &= 0 \\ 2(\text{Fe}) + 3(-2) &= 0 \\ 2(\text{Fe}) &= 6 \\ \text{Fe} &= +3 \end{aligned}$$

Sample Problem: What is the oxidation number of Mn in $KMnO_4$?

Solution: $K = +1 \qquad Mn = ? \qquad O = -2$

The sum of the oxidation numbers is zero:

$$\begin{aligned} +1 + \text{Mn} + 4(-2) &= 0 \\ +1 + \text{Mn} - 8 &= 0 \end{aligned}$$

The oxidation number of Mn = +7

BALANCING SIMPLE OXIDATION-REDUCTION EQUATIONS

The balancing of oxidation-reduction equations often cannot be done quickly by the inspection method. Therefore, techniques have been developed to aid us in balancing these equations. Any such technique must take into account the balancing of electrons and ionic charges as well as the balancing of atoms. The technique we will present is called the *half-reaction method* of balancing oxidation-reduction equations.

A *half-reaction* is either the oxidation portion or else the reduction portion of the overall reaction. It is just what its name implies—half of a reaction. All oxidation-reduction reactions have two half-reactions: the *oxidation half-reaction* and the *reduction half-reaction*. The *net reaction* is the sum of two half-reactions. (When the two half-reactions have been added, the electrons on each side of the equation for the net reaction may be cancelled.)

Oxidation half-reaction: $2\,Na \longrightarrow 2\,Na^+ + 2\,e^-$

Reduction half-reaction: $Cl_2 + 2\,e^- \longrightarrow 2\,Cl^-$

Net reaction: $2\,Na + Cl_2 + \cancel{2e^-} \longrightarrow 2\,Na^+ + 2\,Cl^- + \cancel{2e^-}$

To use the half-reaction method to balance an oxidation-reduction equation, we must proceed by a series of steps in a specific order.

1. Determine the oxidation numbers of the participating atoms.
2. Decide which atoms are oxidized and which atoms are reduced.
3. Write the two half-reactions—the one for the oxidation and the one for the reduction. In the oxidation half-reaction put the number of electrons lost on the right side of the equation. In the reduction half-reaction, put the electrons gained on the left. (The number of electrons is the difference between the two oxidation numbers of the atom involved in the half-reaction.)
4. Balance each of the half-reactions by inspection. If the reaction is run in water, it may be necessary to add water, hydrogen ion (if the reaction is run in acidic solution), or hydroxide ion (if the reaction is run in alkaline solution). Both the atoms and the electrical charges of each half-reaction must be balanced before proceeding.
5. The electrons lost in the oxidation half-reaction must equal the electrons gained in the reduction half-reaction. If necessary, multiply either or both half-reactions by appropriate whole numbers to make the numbers of electrons equal. When multiplying, multiply *all* the members of a particular half-reaction by the same number.
6. Add the two half-reactions, and cancel the electrons and atoms that appear equally on both sides of the net equation.
7. Check the answer by counting the different atoms on each side of the equation to see that they are equal.

Sample Problem: The biological oxidation of food is not a simple reaction, but a series of reactions in which many compounds are oxidized and others are reduced. The final step, the reduction of oxygen in a cell, is accomplished by a compound called a *cytochrome,* which contains the iron(II) ion. Balance the following equation, which shows the results of this reaction:

$$Fe^{2+} + O_2 + H^+ \longrightarrow Fe^{3+} + H_2O$$

Solution: *Step 1:* Determine the oxidation numbers.

$$\underset{(+2)}{Fe^{2+}} + \underset{(0)}{O_2} + \underset{(+1)}{H^+} \longrightarrow \underset{(+3)}{Fe^{3+}} + \underset{(+1)\,(-2)}{H_2O}$$

Step 2: Fe^{2+} is oxidized; O_2 is reduced.

Steps 3 and 4: Oxidation: $Fe^{2+} \longrightarrow Fe^{3+} + e^-$

Reduction: $O_2 + 4\,H^+ + 4\,e^- \longrightarrow 2\,H_2O$

Step 5: Multiply the oxidation half-reaction by 4.

$$4\,Fe^{2+} \longrightarrow 4\,Fe^{3+} + 4\,e^-$$

Step 6: Add the two half-reaction equations and cancel.

$$4\,Fe^{2+} + O_2 + 4\,H^+ \longrightarrow 4\,Fe^{3+} + 2\,H_2O$$

OXIDATION AND REDUCTION OF ORGANIC COMPOUNDS

In the later chapters of this book, we will discuss how the oxidation of nutrients, which are organic compounds, provides energy for organisms. Generally, we say that an organic compound is oxidized if it *gains oxygen* or *loses hydrogen* (or both).

Oxidation of organic compounds:

$$\underset{\text{methane}}{CH_4} + 2\,O_2 \xrightarrow{\text{spark}} \underset{\text{carbon dioxide}}{CO_2} + 2\,H_2O$$

$$\underset{\text{ethyl alcohol}}{CH_3CH_2OH} + O_2 \xrightarrow{\text{enzymes}} \underset{\text{acetic acid (vinegar)}}{CH_3\overset{\overset{O}{\|}}{C}OH} + H_2O$$

Reduction is the loss of oxygen or the gain of hydrogen.

Reduction of organic compounds:

$$\underset{\text{ethylene}}{CH_2{=}CH_2} + H_2 \xrightarrow{\text{catalyst}} \underset{\text{ethane}}{CH_3CH_3}$$

$$\underset{\text{acetone}}{CH_3\overset{\overset{O}{\|}}{C}CH_3} + H_2 \xrightarrow[\text{heat, pressure}]{\text{catalyst}} \underset{\text{isopropyl alcohol}}{CH_3\overset{\overset{OH}{|}}{C}HCH_3}$$

Sample Problem: Label each of the following conversions as the oxidation or reduction of the organic reactant.

(a) $CH_3CH_2OH \longrightarrow CH_3\overset{\overset{O}{\|}}{C}H$

(b) $CH{\equiv}CH \longrightarrow CH_2{=}CH_2$

(c) $HO\overset{\overset{O}{\|}}{C}CH{=}CH\overset{\overset{O}{\|}}{C}OH \longrightarrow HO\overset{\overset{O}{\|}}{C}CH_2CH_2\overset{\overset{O}{\|}}{C}OH$

Solution: (a) is an oxidation (loss of hydrogen). (b) and (c) are reductions (gain of hydrogen).

OXIDATION AND REDUCTION IN OUR WORLD

COMBUSTION. Oxygen gas is a potent oxidizing agent, as we might guess from the name *oxidizing* agent. The burning of a substance in air, called *combustion,* is the rapid oxidation of that substance by oxygen. During combustion, energy is given off in the form of heat and light, which we often harness for power and heat.

A substance that can undergo combustion is said to be *combustible, flammable,* or *inflammable.* Although the prefix *in-* usually means *not,* the word *inflammable* means the same as the word *flammable*—the substance can burn. The term *nonflammable* is used to describe a substance that does not burn.

$$\underset{\substack{\text{methane} \\ \textit{in natural gas}}}{CH_4} + 2\,O_2 \longrightarrow CO_2 + 2\,H_2O + \text{Energy}$$

$$\underset{\substack{\text{magnesium} \\ \textit{used in flares}}}{2\,Mg} + O_2 \longrightarrow 2\,MgO + \text{Energy}$$

INCOMPLETE COMBUSTION. Incomplete combustion is the incomplete reaction of a substance (generally a carbon compound) with oxygen. The complete combustion of carbon or a carbon compound gives carbon dioxide as a product. Carbon cannot be oxidized any further than carbon dioxide.

Complete combustion:

$$CH_4 + 2\,O_2 \longrightarrow CO_2 + 2\,H_2O$$
$$C + O_2 \longrightarrow CO_2$$

Incomplete combustion leading to carbon monoxide:

$$2\,CH_4 + 3\,O_2 \longrightarrow 2\,CO + 4\,H_2O$$
$$2\,C + O_2 \longrightarrow 2\,CO$$

Incomplete combustion gives carbon products with carbon in a lower oxidation state than $+4$. The formation of the poisonous gas carbon monoxide is an example of incomplete combustion. Virtually any device that uses carbon fuels, such as automobile engines and gas furnaces, produces some carbon monoxide. If the oxygen supply is inadequate, incomplete combustion is an especially prevalent reaction.

Another product of incomplete combustion may be carbon itself. Black, sooty smoke is smoke that contains particles of carbon. Compounds that contain aromatic pi clouds often undergo incomplete combustion to yield carbon. An inadequate oxygen supply can also lead to the formation of carbon.

OXIDIZING AGENTS OTHER THAN OXYGEN GAS. In *hydrogen peroxide* (H_2O_2, or H—O—O—H), each oxygen has an oxidation number of -1. A 3 percent solution of hydrogen peroxide is used as a mild antiseptic for minor wounds or as a mouthwash.

A 10 percent solution is used to bleach hair. In more concentrated solutions, hydrogen peroxide is likely to explode in contact with material that can be oxidized.

Ozone gas (O_3) is an oxidizing agent that we hear about in connection with smog. Ozone is formed in a complex series of reactions by the action of sunlight on air pollutants. Such reactions are called *photochemical* reactions (from the Greek *photos* meaning light). A human breathing 2 ppm of ozone in air experiences respiratory symptoms and loss of coordination. The usual levels of ozone encountered in smog (as, for example, in Los Angeles) are under 0.3 ppm. Ozone, especially in combination with other pollutants, is also damaging to plant life. When properly used, ozone is a useful gas. It is used commercially as a bleach for oils, waxes, fabric, and starch. Ozone, like chlorine, is used by some cities to kill bacteria in drinking water.

Besides being formed from the reaction of sunlight on air pollutants, ozone is produced by the action of electrical discharges on oxygen gas. We can often detect the sharp, irritating odor of ozone near an electrical device or after a lightning storm.

In the upper atmosphere, powerful ultraviolet (uv) rays from the sun provide the energy to cause an exothermic reaction in which oxygen is converted to ozone. This ozone can go back to oxygen; energy in the form of heat is released in this reaction.

Formation of ozone:

$$\textit{Step 1:}\quad O_2 + \text{uv from sun} \longrightarrow O + O$$
$$\textit{Step 2:}\quad O + O_2 \longrightarrow O_3 + \text{Heat}$$

Decomposition of ozone:

$$2\,O_3 \longrightarrow 3\,O_2 + \text{Heat}$$

The formation and decomposition of ozone in the upper atmosphere serve a very useful purpose. This cycle protects living organisms on the surface of the earth by removing the most powerful of the ultraviolet radiation from the sun's rays and, in doing so, helps keep the earth warm.

Chlorine is an oxidizing agent that is used by many cities to kill bacteria in water supplies. The chlorine laundry bleaches are made by dissolving chlorine gas in dilute, aqueous sodium hydroxide solution. Chlorine undergoes reaction with the sodium hydroxide to form sodium hypochlorite, NaOCl, in the following reversible reaction:

$$Cl_2 + 2\,NaOH \rightleftharpoons \underset{\substack{\text{sodium}\\ \text{hypochlorite}\\ \textit{an oxidizing agent}}}{NaOCl} + NaCl + H_2O$$

In neutral or acidic solution, the equilibrium between hypochlorite ion and chlorine gas is shifted to the left, and the toxic chlorine gas is given off. For this reason, chlorine bleaches should never be used in conjunction with acidic household cleaners, such as toilet bowl cleaners, which usually contain the acidic salt sodium bisulfate ($NaHSO_4$).

STUDY QUESTIONS

8-1. What is the oxidation number of each atom in the following species?
(a) Fe^{3+} (b) Cl_2 (c) Cu (d) $Hg_2{}^{2+}$

8-2. Tell which of the following are oxidations and which are reductions, and add the proper number of electrons to the proper side of each equation.
(a) $Fe^{3+} \longrightarrow Fe^{2+}$ (b) $O_2 \longrightarrow 2\,O^{2-}$
(c) $Na \longrightarrow Na^+$ (d) $Cu^+ \longrightarrow Cu^{2+}$

8-3. In the following equations, which atoms are oxidized and which are reduced?
(a) $CuCl_2 + Fe \longrightarrow FeCl_2 + Cu$
(b) $Cl_2 + 2\,Br^- \longrightarrow Br_2 + 2\,Cl^-$
(c) $2\,Na + 2\,H_2O \longrightarrow 2\,NaOH + H_2$
(d) $CH_4 + 2\,O_2 \longrightarrow CO_2 + 2\,H_2O$

8-4. For each of the equations in the preceding problem, give the number of electrons that have been transferred.

8-5. State which is the reducing agent and which is the oxidizing agent in the preceding equations.

8-6. What is the oxidation number in each of the following examples?
(a) S in $Na_2S_2O_8$ (b) Cr in $Na_2Cr_2O_7$
(c) N in N_2O_4 (d) Cl in $HClO_4$

8-7. Complete and balance the following equations:
(a) $Br_2 + SO_2 \longrightarrow H^+ + Br^- + SO_4{}^{2-}$
(b) $H_2S + H_2O_2 \longrightarrow S\downarrow + H_2O$
(c) $HBrO \longrightarrow H^+ + Br^- + O_2\uparrow$
(d) $H_2SO_4 + HBr \longrightarrow SO_2 + Br_2 + H_2O$

8-8. Three percent hydrogen peroxide often effervesces (gives off gas) when it is poured from a bottle. What is the gas? Write an equation for its formation.

8-9. When metals are extracted from ores containing sulfur compounds, some of the sulfur is converted to the toxic gas sulfur dioxide (SO_2). In the presence of sunlight and O_2 (from the air), SO_2 can be further oxidized to sulfur trioxide (SO_3). Write a balanced equation for this oxidation of SO_2.

8-10. SO_3 reacts with the water in moist or humid air to form the very corrosive sulfuric acid (H_2SO_4).
(a) Write the equation for this reaction.
(b) Is this an oxidation-reduction reaction?

8-11. Silver metal (Ag) is not readily oxidized by air. Silver does react with hydrogen sulfide, H_2S (from sulfur-containing foods such as eggs), and oxygen (O_2) to yield silver sulfide (AgS), which we observe as *tarnish.* Write a balanced equation for the tarnishing of silver.

8-12. Research has been aimed at the development of fuel cells to provide the energy for the operation of automobiles. One such fuel cell would use the reaction between lithium metal and chlorine. Write the half-reactions for this fuel cell.

8-13. The following reactions all occur in animal cells. (The first requires several steps.) Tell whether each is an oxidation or reduction of the organic reactant, or neither.

(a) $\underset{\text{glucose}}{C_6H_{12}O_6} \longrightarrow 6\,CO_2 + 6\,H_2O$

(b) $HOCH_2\overset{O}{\overset{\|}{C}}CH_2OH \longrightarrow HOCH_2\overset{OH}{\overset{|}{C}}H\text{—}\overset{O}{\overset{\|}{C}}H$

(c) $HOCH_2\overset{OH}{\overset{|}{C}}H\text{—}\overset{O}{\overset{\|}{C}}H \longrightarrow HOCH_2\overset{OH}{\overset{|}{C}}H\text{—}\overset{O}{\overset{\|}{C}}OH$

(d) $\underset{\text{pyruvic acid}}{CH_3\overset{O}{\overset{\|}{C}}\text{—}\overset{O}{\overset{\|}{C}}OH} \longrightarrow \underset{\text{lactic acid}}{CH_3\overset{OH}{\overset{|}{C}}H\text{—}\overset{O}{\overset{\|}{C}}OH}$

(e) $\underset{\text{carbonic acid}}{HO\overset{O}{\overset{\|}{C}}OH} \longrightarrow CO_2 + H_2O$

CHAPTER 9 THE CHEMISTRY OF THE ELEMENTS

OBJECTIVES

Properties of the Metals *List the physical properties generally common to the metals.* * *Describe the principal chemical property common to most metals.*

The Alkali Metals *Describe (using equations) the chemical features common to the alkali metals.*

The Alkaline Earth Metals *Describe the chemical features common to the alkaline earth metals. Describe the occurrences of calcium compounds in nature.* * *Define hard water and write an equation for its reaction with soap.* * *Define strontium-90 and describe why this isotope is of concern in regards to human health.*

The Transition Metals *List some properties and uses of copper, mercury, and iron.* * *Define and give examples of complex compounds and chelates.*

Aluminum *Describe the chemical features of aluminum.*

Lead and Silicon *List some uses of lead; list some causes and symptoms of lead poisoning.* * *Describe the occurrence of silicon.*

Properties of the Nonmetals *Describe and illustrate the types of bonds formed by the nonmetals.*

Hydrogen *Describe the properties and occurrence of* H_2.

Nitrogen and Phosphorus *Draw and describe the nitrogen cycle.* * *Write structures of phosphoric acid and its anhydrides.*

Oxygen and Sulfur *Give examples of oxygen and sulfur compounds.* * *Draw and describe the oxygen cycle.*

The Halogens *Describe the reactivity of the halogens and the properties of the hydrogen halides.*

The Inert Gases *Describe the properties of the inert gases.*

Almost all elements can be classified into two main divisions—the metals and the nonmetals. These two classifications do not have a sharp dividing line. A few elements, called the *metalloids,* such as silicon (Si) and germanium (Ge), have some of the properties of metals and some of the properties of nonmetals. In this chapter, we will briefly consider the properties of some of the important elements, beginning with the metals, and some of their compounds.

PROPERTIES OF THE METALS

Although approximately three-fourths of the elements are metals, the ancient Greeks were familiar with only seven of them: gold, silver, iron, copper, tin, lead, and mercury. These metals either are found in the elemental state in nature or are easily obtained from their naturally occurring compounds, for example, by heating. (Note that the atomic symbols for these "ancient" metals are all derived from their Latin names—for example, the symbol Au for gold is from *aureum.*) Today we use almost all of the metals in the Periodic Table in commerce and industry.

The importance of metals to mankind is due to the properties of metallic elements. Let us look briefly at some of the key properties:

1. Metals are *conductors* of heat and electricity.
2. Metals are *malleable*; that is, they can be hammered into a shape. Metals can also be molded and flattened into sheets. Moreover, most metals are *ductile*—they can be drawn into wires.
3. Metals have *luster*; they are shiny, or can be made so by polishing.
4. Metals are generally *denser* than nonmetals (although the heaviest nonmetal, iodine, is denser than some of the lighter metals).

These properties are what make metals so important to us. The fact that metals can be molded into useful shapes and the fact that they are hard and tough allow us to use them to make tools with which we can shape our world.

The physical properties of metals are a result of their atomic structures and bonding. Recall that metals are formed from atoms with low electronegativities, a property determined by the size of the atom and the number of electrons in the outer shell. (This topic was discussed in more detail in Chapter 4. A review of that section may be helpful.) An atom of a metal can lose its outer electrons fairly easily. The structure of a piece of metal is a group of cations awash in a sea of bonding electrons. The strong three-dimensional attractions between the cations and the electrons give rise to the strength of metals.

The distinguishing chemical feature of the metals is their ability to be oxidized to the cationic state. The alkali metals, such as sodium (Na), react spontaneously with water or air to form hydroxides or oxides. The alkaline earth metals, such as calcium (Ca) and magnesium (Mg), are not quite as reactive. The transition metals (with higher electronegativities) are less reactive yet. The transition metals can form covalent compounds as well as ionic compounds, and a few, such as copper (Cu) and gold (Au), may be found in their elemental states in nature.

Metallic ions bear little resemblance to the free metals. Sodium ion (along with chloride ion) is essential to the ionic balance of body fluids. Calcium ions, in $Ca_3(PO_4)_2$, form bones and teeth. Iron(II) ions are essential in hemoglobin (p. 187), the transporter of oxygen in the blood stream, and in the oxidation of foodstuffs. (See Chapter 25.) Magnesium ions are an indispensable portion of chlorophyll (p. 188), the green material in plants that makes possible the utilization of the energy of sunlight for the production of carbohydrates. Magnesium ions are also indispensable to many of the reactions in animal systems. Table 9-1 shows the relative abundance of some metal ions in human blood plasma.

Table 9–1. Relative Abundance of Some of the Ions in Sea Water and in Human Blood Plasma

Ion	Percent by Weight in Sea Water	Percent by Weight in Human Blood Plasma
Cations:		
Na^+	1.1	0.32
Mg^{2+}	0.13	0.003
Ca^{2+}	0.04	0.02
K^+	0.04	0.015
Anions:		
Cl^-	1.9	0.36
Br^-	0.006	. . .

THE ALKALI METALS

The metals in Group IA of the Periodic Table are known as the *alkali metals.* (See Figure 9-1.) The hydroxides of these metals are strongly basic, or *alkaline,* hence the term *alkali* metals. The metals in this group, of which lithium (Li), sodium (Na), and potassium (K) are the common ones, are noted primarily for their chemical reactivity. All the Group IA metals have one electron in the outer shell, and this electron is easily lost ($K \longrightarrow K^+ + e^-$). Therefore, these metals are easily oxidized and are good reducing agents.

Figure 9-1. The alkali metals: Group IA.

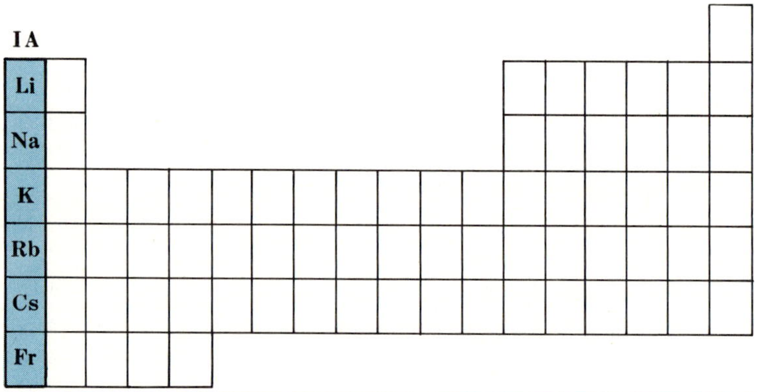

The alkali metals react vigorously, sometimes explosively, with water and also with oxygen. Thus, they will be oxidized if exposed to air or moisture.

$$\underset{\text{sodium}}{4\,Na} + O_2 \longrightarrow \underset{\text{sodium oxide}}{2\,Na_2O}$$

$$2\,Na + 2\,H_2O \longrightarrow \underset{\text{sodium hydroxide}}{2\,NaOH} + H_2\uparrow$$

The alkali metals are never found in the metallic form in nature, as they are far too reactive; however, salts of sodium and potassium are common. In animal systems, the sodium ion is the principal cation of extracellular fluids, while the potassium ion is the principal cation within the cells. (These cations are found with inorganic anions, such as the chloride ion, the sulfate ion, and the phosphate ion, or with organic anions, such as the acetate ion and other more complex ions.)

THE ALKALINE EARTH METALS

The alkaline earth metals are the metals found in Group IIA of the Periodic Table. (See Figure 9-2.) Calcium (Ca) and magnesium (Mg) are the most common metals in this group.

Because the alkaline earth metals all contain two electrons in the outer shell, all alkaline earth metals readily lose two electrons and become ions in the +2 oxidation state ($Ca \longrightarrow Ca^{2+} + 2\,e^-$).

Figure 9-2. The alkaline earth metals: Group IIA.

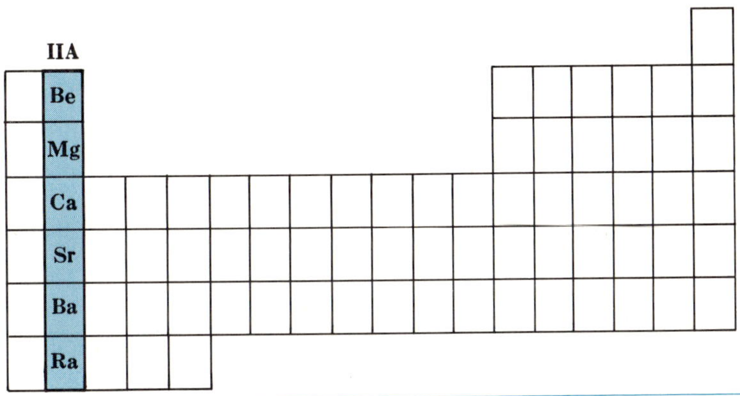

The metals in this group are not as reactive as the alkali metals, but all will react slowly with water to form strong, but fairly insoluble, hydroxides.

$$\underset{\text{calcium}}{Ca} + 2\,H_2O \xrightarrow{\text{not vigorous}} \underset{\text{calcium hydroxide}}{Ca(OH)_2} + H_2\uparrow$$

$$\underset{\text{magnesium}}{Mg} + 2\,H_2O \xrightarrow{\text{heat}} \underset{\text{magnesium hydroxide}}{Mg(OH)_2} + H_2\uparrow$$

Calcium is found in the earth's crust in limestone and marble, which are both composed primarily of calcium carbonate ($CaCO_3$). Pearls are also composed of calcium carbonate, while chalk, coral, eggshells, and the shells of many sea animals are composed of mixtures of calcium and magnesium carbonates. Our bones and teeth contain calcium in the form of calcium phosphate, $Ca_3(PO_4)_2$.

Chemically, strontium ions (Sr^{2+}) are very similar to calcium ions (Ca^{2+}). Our body processes cannot distinguish between the two and will deposit strontium phosphate in our bones as readily as they will calcium phosphate. In the 1940's and 1950's, atmospheric testing of atomic bombs produced radioactive strontium ions (strontium-90, an isotope of normal strontium), which were washed to the ground by rain. Once on the ground, the strontium-90 ions were taken up by plants and animals and eventually by humans. In certain areas of the world, the buildup of strontium-90 was so great that its presence could be detected in the bones of humans. This unforeseen problem was one of the reasons that atmospheric testing of atomic devices has been banned.

CALCIUM ION IN HARD WATER. Hard water is water that contains ions of metals other than the alkali metals. Most hard water contains soluble calcium salts, such as calcium bicarbonate, $Ca(HCO_3)_2$. Calcium ions in solution form a precipitate with the anions of soap; we see this precipitate as a scum in the bath water or as a ring around the tub. Detergents are synthesized from anions that do not precipitate with calcium ions.

$$\underset{\textit{from soap}}{Ca^{2+} + 2\,CH_3(CH_2)_{16}CO_2^-} \longrightarrow [CH_3(CH_2)_{16}CO_2]_2Ca\downarrow$$

The softening of water involves the removal of the metal ions (primarily calcium) that precipitate with soap. Water may be softened in a number of ways. The most common method is by treating the water in a water softener that contains sodium

Figure 9-3. Diagram of a water softener.

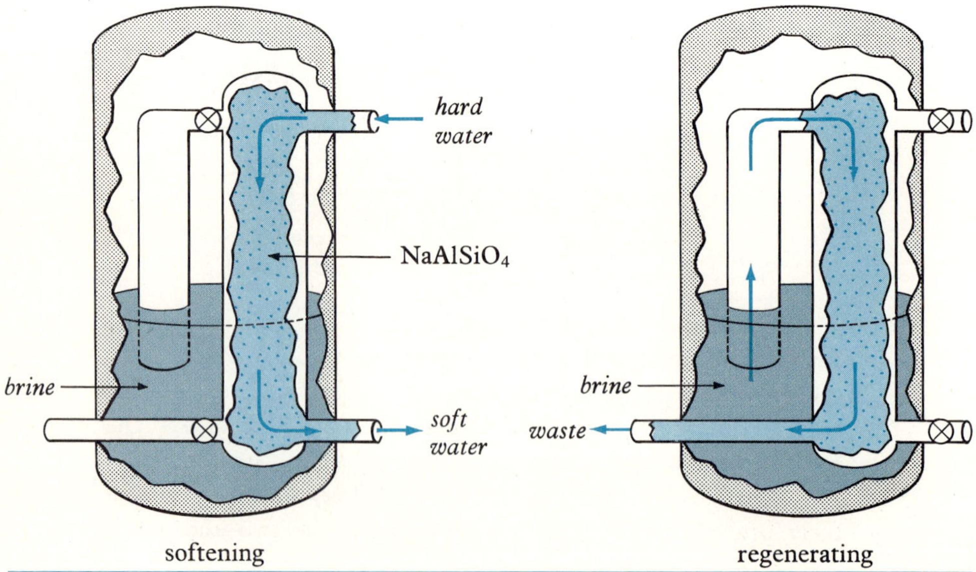

aluminum silicate (Zeolite), a compound that forms a precipitate with the calcium ions and releases sodium ions, which do not precipitate with soap. This type of reaction is called an *ion exchange reaction.*

Softening the water:

$$2\,NaAlSiO_4 + Ca^{2+} \longrightarrow Ca(AlSiO_4)_2\downarrow + 2\,Na^+$$

removed from the water (Ca^{2+}); *released into the water* (Na^+)

When the silicate salt has been used up, it may be regenerated by being flooded with brine (sodium chloride solution). The calcium ions are then flushed away as soluble calcium chloride. (See Figure 9-3.)

Regenerating:

$$Ca(AlSiO_4)_2 + 2\,NaCl \longrightarrow 2\,NaAlSiO_4 + CaCl_2$$

flushed away ($CaCl_2$)

THE TRANSITION METALS

No discussion of metals would be complete without a mention of the highly important *transition metals*, a group that includes iron (Fe), copper (Cu), silver (Ag), and mercury (Hg), each of importance since ancient times. (See Figure 9-4.)

The transition metals scandium (Sc) through zinc (Zn) are those metals with electrons in the fourth shell, but with an incomplete third shell. In the next row of the Periodic Table are another group of transition metals, yttrium (Y) through cadmium (Cd), which have electrons in the fifth shell, but have an incomplete fourth shell.

Figure 9-4. The transition metals.

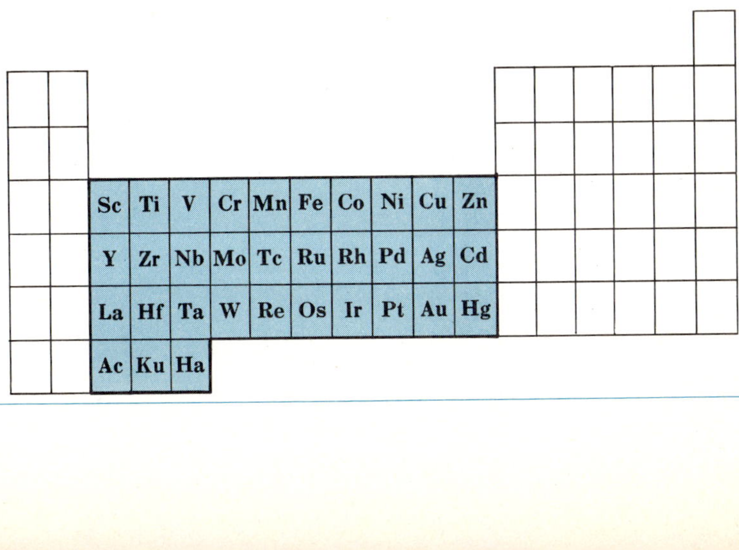

Some transition metals are found in compounds in two or more oxidation states (for example, Fe^{2+} and Fe^{3+}). This is possible because electrons can be lost from the two outermost electron shells.

Many transition metals are as important commercially today as they were in ancient times. *Copper* is one of the few metals that is occasionally found as a pure metal in nature. For this reason, it was probably the first metal to be used for utensils. Today, copper is second only to iron as an important metal. We use copper for electrical wiring, cooking utensils, water pipes, gas tubing, and automobile parts. As an alloying agent, copper is used in brass, German silver, and coins. (An *alloy* is a mixture of two or more metals that resembles a solid solution.)

Mercury is liquid at room temperature. For this reason, mercury is sometimes called *quicksilver*. Mercury is certainly one of the more useful metals. Because it is liquid over a wide temperature range (melting point $-38.9\,^{\circ}C$; boiling point $356.6\,^{\circ}C$), mercury is a common liquid in thermometers. Because it has a high density, mercury is used in barometers. Mercury is also used for seals, for electrical contacts, and in mercury vapor lamps, which are commonly used as street lights.

Metallic mercury is a cumulative poison. A particularly insidious cause of mercury poisoning results from inhalation of the vapors. Although the vapor pressure of mercury is not high at room temperature, over a period of time vapors from mercury spills in the laboratory or elsewhere still constitute a hazard. Mercury compounds, like the compounds of most transition metals, are also poisonous. The mercury(II) ion interferes with the ability of the kidney to function normally in removing waste products from the blood. It is interesting to note that some diuretics (drugs that increase kidney activity and cause fluid loss) are organomercury compounds.

Iron metal is abundant, cheap, and strong. It may be alloyed with other substances to make steel. Well over 100 million tons of steel are produced yearly in the United States. Much of this steel comes from the mining and smelting of iron ore; a growing proportion comes from scrap cars and other scrap iron. As we use up useful iron deposits, we will have to recycle more and more scrap iron. Fortunately, there are a number of large deposits of iron ore in the world. At the present time, most of the iron mined in the United States comes from huge deposits of iron(III) oxide (Fe_2O_3) in the Lake Superior area.

COMPLEX COMPOUNDS. The transition metals can form stable *complexes*, or *coordination compounds*, with anions or with neutral molecules that have unshared electrons to donate. (Other metals may also form complexes, but they are generally not as stable as those of the transition metals.)

Some complex ions:

$$Ag(NH_3)_2{}^{+} \qquad Fe(CN)_6{}^{3-} \qquad MnF_6{}^{3-}$$

In complex ion formation, the unshared electrons of the *ligand* (the anion or neutral molecule) are donated to the metal ion. Ammonia and related organic compounds are excellent ligands because they have unshared electrons.

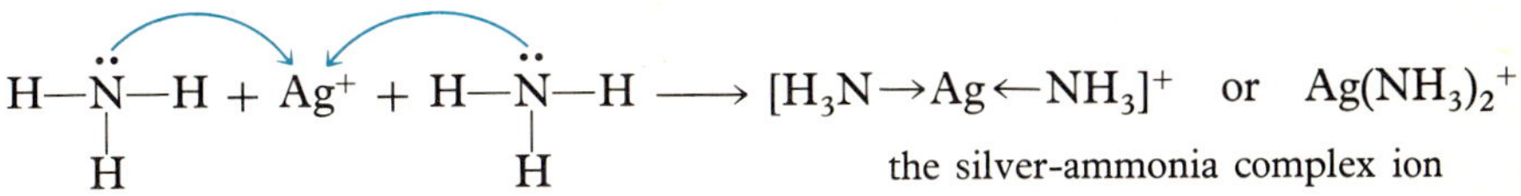

An interesting type of complex may be formed when two or more of the ligand atoms are in the same molecule. In the following example, two molecules of ethylenediamine ($H_2NCH_2CH_2NH_2$) have added to a copper(II) ion:

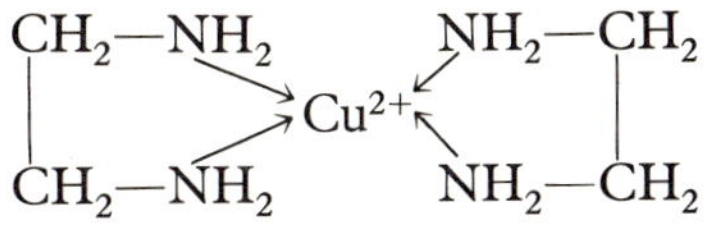

This type of complex is called a *chelate,* from the Greek word *kela* (crab's claw)—the imagery is that of a cation being caught in a pincer-like claw. A very effective chelating agent, often used as an antidote in heavy metal poisoning, is ethylenediaminetetraacetic acid (more conveniently called EDTA), an organic compound with several atoms that have unshared electrons.

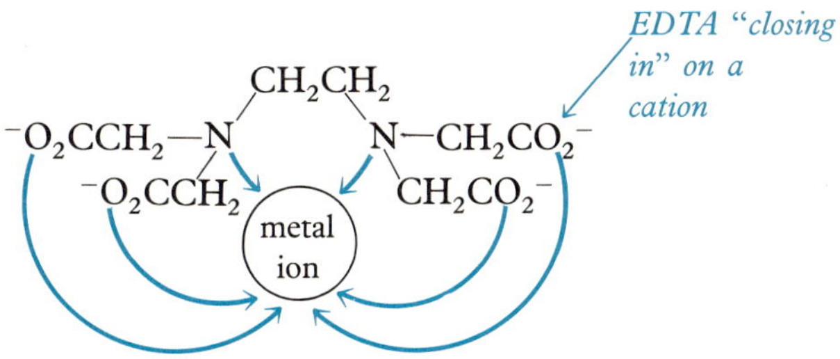

Many indispensable natural compounds are chelates—for example, heme (the oxygen-carrier portion of hemoglobin) and chlorophyll. (See Figure 9-5.)

Note the similarities in the organic portions of heme and chlorophyll. The ring system, called the *porphyrin* ring system, is ideally structured for the chelation of metal ions. Also, note that, as complicated as these structures may appear, they are primarily metal ions bonded to nitrogen atoms:

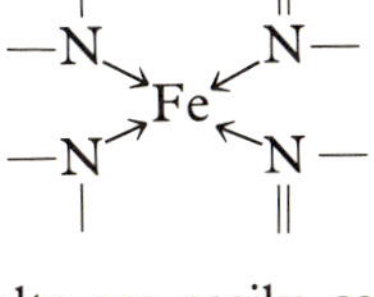

While most soluble iron salts are easily converted by air and water to the insoluble iron(III) hydroxide, $Fe(OH)_3$, the stable iron complexes do not undergo this

Figure 9–5. Two important natural chelates.

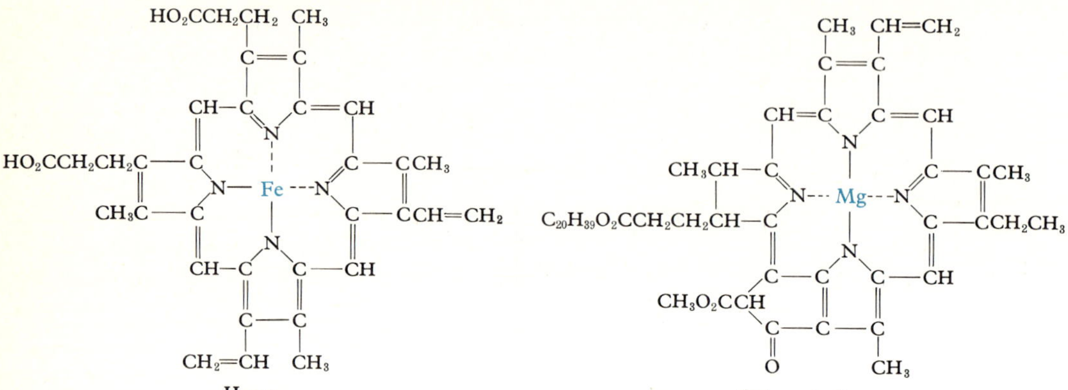

reaction readily. Many plants and trees, such as citrus trees and azalea plants, need a supply of soluble iron salts. Insoluble iron salts are useless to them. Therefore, certain of the complexes of iron are valuable in plant food.

In living systems, iron is found in its complexed form, usually in structures related to heme. Seventy percent of the iron in the human body is contained in hemoglobin itself. About 3 percent is contained in *myoglobins,* compounds similar to hemoglobin but found in muscle cells rather than in the blood. Most of the remainder of the iron in the body is stored in the spleen and liver in a form called *ferritin,* which is composed primarily of protein material bound to ferric hydroxides and phosphates.

ALUMINUM

The element in Group IIIA that needs concern us is aluminum, Al. (See Figure 9-6.) In the form of complex silicates, aluminum is the most abundant metal in the earth's crust. The other Group IIIA elements are quite rare.

Aluminum metal is familiar to us as a lightweight metal useful for making foil, pots, pans, lawn chairs, and recyclable beer cans. Because of its light weight and strength, it is also useful in alloys and is the metal of choice for the manufacture of airplanes.

Aluminum has three electrons in the outer shell. It tends to lose these three electrons and usually forms compounds in the +3 oxidation state:

$$Al \longrightarrow Al^{3+} + 3\,e^-$$

Pure aluminum is fairly reactive and in the presence of air forms aluminum oxide:

$$4\,Al + 3\,O_2 \longrightarrow 2\,Al_2O_3$$

Figure 9-6. The boron and aluminum group of metals: Group IIIA.

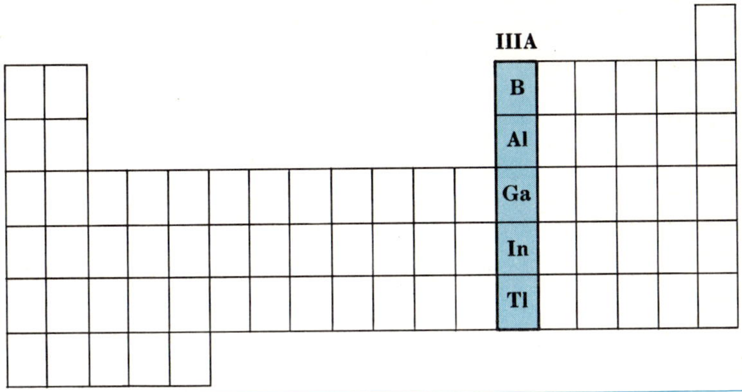

The reason that aluminum metal does not slowly rust away as iron does is that Al_2O_3 forms a tight coating on aluminum metal and protects the metal from further corrosion. (It is interesting to note that sapphires and rubies are also composed of aluminum oxide; the colors of these gems are imparted by impurities.)

SILICON AND LEAD

Group IVA of the Periodic Table (Figure 9-7) contains the important nonmetal carbon (C), the metalloid silicon (Si), and the metal lead (Pb).

Carbon is the element that makes life as we know it possible. Most of the remainder of this book will be devoted to organic chemistry, the chemistry of carbon

Figure 9-7. Group IVA.

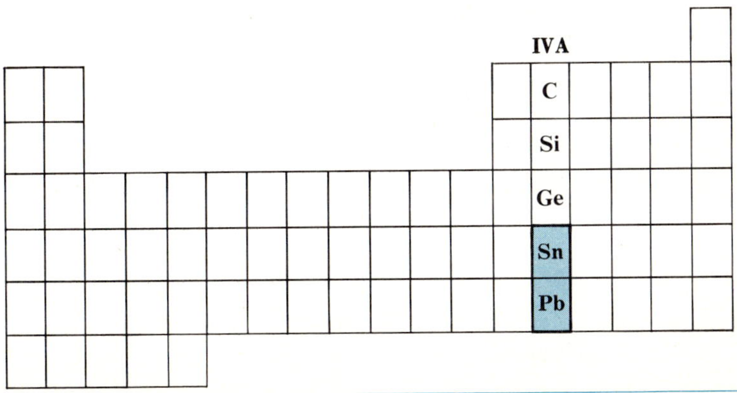

compounds, and biochemistry, the chemistry of living systems. We have mentioned that carbon forms covalent bonds and not ionic bonds. Silicon is similar to carbon in this respect. Both carbon and silicon have four outer electrons and form four covalent bonds in compounds.

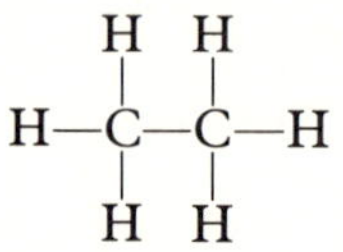

ethane
a compound of carbon

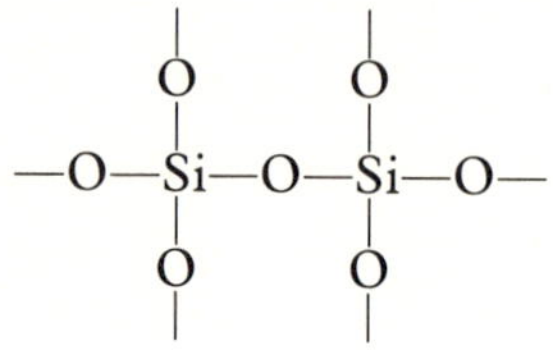

a portion of a silicate molecule

Silicon, in combination with other elements (mainly oxygen) comprises one fourth of the earth's crust. Silicon-oxygen compounds, called *silicates,* form sand and quartz, for example. In combination with potassium, aluminum, and other metals, silicon and oxygen form clay. Emeralds are primarily silicon and oxygen.

It is said that we eat a peck of dirt before we die. Indeed, silicon compounds are so omnipresent that it is practically impossible to raise an animal in a silicon-free environment. Consequently, there is still controversy whether higher animals actually require trace amounts of silicon compounds or not. It is known that certain plants (some spruces, some grasses, horsetails, rice, and so forth) and certain lower animals (notably diatoms) do require silicon compounds as nutrients.

Lead is a useful metal for alloying and is used in plumbers' solder, type metal, pewter, shot, and bullets. (Lead pencils contain no lead, but rather graphite, a form of carbon.)

Lead forms compounds in the $+2$ and $+4$ oxidation states ($Pb \longrightarrow Pb^{2+} + 2\,e^-$; $Pb \longrightarrow Pb^{4+} + 4\,e^-$).

In the presence of air or moisture, lead metal, like aluminum, forms a protective coating of lead(II) oxide, PbO, or of lead(II) hydroxide, $Pb(OH)_2$. The coating protects the metal from further corrosion. Consequently, lead has found wide use in the manufacture of drain pipes.

Lead piping should not be used for carrying drinking water. When water courses through a pipe with a coating of lead(II) hydroxide, some of the lead salt is carried along with it. Lead compounds are absorbed slowly by the body. Over a period of time, lead ions accumulate and give rise to the symptoms of lead poisoning: headache, discomfort, and a general feeling of illness. In more advanced cases, severe abdominal cramps (called "lead colic") are noted. The ancient Romans used lead pipes to transport their drinking water. Unfortunately, we will never know how this practice affected the course of the Roman Empire. We do know that many Romans suffered from chronic lead poisoning. Of a certainty, lead poisoning shortened the lives, and possibly the tempers, of many prominent Roman leaders.

Cases of lead poisoning still appear in our society. "White lead," really a mixed salt of $Pb(OH)_2$ and $PbCO_3$, has been used in paints from 400 B.C. up until the present time. Manufacturers no longer use lead compounds in house paints, but older homes present a problem. Even though lead paint may be the fourth layer of paint down from the surface, lead compounds may be present in bits of paint that flake off walls or ceilings. Lead poisoning has been observed in adults because these flakes have contaminated their food, and in children who may simply eat the flakes of paint or may teethe on window sills. Other sources of present-day lead poisoning are unglazed pottery, which sometimes contains lead salts, and bootleg whiskey, which may have been distilled through an automobile radiator containing lead metal.

PROPERTIES OF THE NONMETALS

Although most of the elements are metallic, the nonmetallic elements play the leading role in the world around us. Animal systems are composed almost entirely of compounds of the nonmetals carbon, hydrogen, oxygen, and nitrogen. Plants, our atmosphere, the oceans and their inhabitants—all are composed mainly of compounds of the nonmetallic elements.

Metallic elements have many physical and chemical properties in common. They are almost all lustrous solids, able to conduct electricity, and easily oxidized to cations. Unlike the metals, the nonmetals vary widely in both physical and chemical properties. Some are gases, some are liquids, and some are solids. Some form ionic compounds, and some do not.

One common thread in the chemistry of nonmetals results from the high electron affinity of their atoms. Many of the nonmetals form anions, and almost all the nonmetals are capable of forming covalent compounds. The halogens, with their high electron affinities, are reactive in their elemental states and readily form ions. Oxygen and sulfur are not so reactive—both are found in the elemental state in nature; however, their atoms will form anions. Nitrogen is even less reactive than oxygen and forms covalent bonds rather than ionic bonds.

Decreasing electron affinity ↓

$$\underset{\text{chlorine}}{Cl_2} + 2\,e^- \longrightarrow 2\,Cl^-$$

$$\underset{\text{oxygen}}{O_2} + 4\,e^- \longrightarrow 2\,O^{2-}$$

$$\underset{\text{nitrogen}}{N_2} \longrightarrow \text{No anion formation}$$

HYDROGEN

Although hydrogen is placed above lithium and sodium in the Periodic Table (Figure 9-8), it does not behave like these alkali metals. Instead of forming cations, hydrogen atoms share electrons and form covalent bonds with other elements. (Even though we may write the structure of a hydrogen ion as H^+, the hydrogen ion is truly bonded to another atom, as in the hydronium ion, H_3O^+.)

In the universe, hydrogen is by far the most abundant element. Free hydrogen atoms are found in outer space and hydrogen atoms compose most of the mass of the sun, where they produce prodigious amounts of energy by undergoing nuclear reactions leading to helium and higher elements. On the earth, elemental hydrogen is not abundant; most of the earth's hydrogen is bound up in compounds such as water.

With a molecular weight of only two, hydrogen (H_2) is the lightest element and the least dense of all gases. Because of its light weight, hydrogen gas was once used to fill dirigibles and blimps. However, hydrogen is a potent reducing agent; if a mixture of hydrogen and oxygen is ignited by a spark, the two elements unite explosively to form water. The explosion of the German dirigible *Hindenburg* in New Jersey in 1937 was the end of the practice of filling lighter-than-air craft with hydrogen. The inert gas helium, although not as light as hydrogen and far more expensive, is now used for filling blimps, circus balloons, and the like.

$$\underset{\text{hydrogen}}{2\,H_2} + \underset{\text{oxygen}}{O_2} \longrightarrow \underset{\text{water}}{2\,H_2O} + \text{Energy}$$

Figure 9-8. Hydrogen in the Periodic Table.

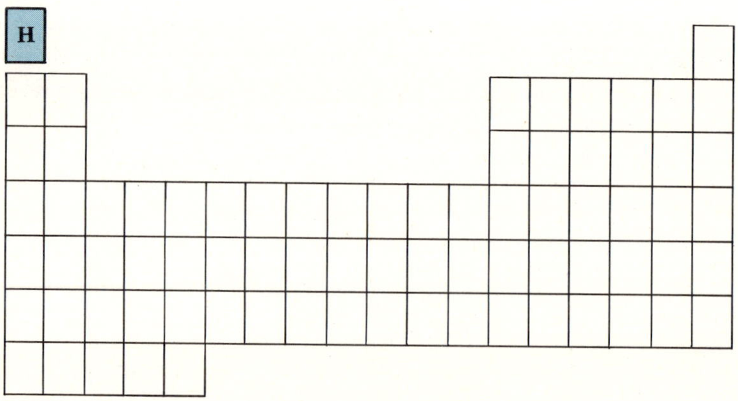

NITROGEN AND PHOSPHORUS

In Group VA of the Periodic Table, we find two typical nonmetals, nitrogen (N) and phosphorus (P). (See Figure 9-9.) In the elemental state, nitrogen is found as nitrogen gas (N_2), a relatively inert gas that comprises 78 percent of our atmosphere. Elemental phosphorus is a highly reactive solid. These two dissimilar nonmetals are both biologically important. The proteins that form skin and muscles are organic polymers containing covalently bonded nitrogen atoms. The compounds that form bones, teeth, nerve cells, and genes are all rich in covalently bonded phosphorus atoms.

Both nitrogen and phosphorus can form compounds with three covalent bonds. Phosphorus can also form compounds with five bonds; we will discuss such a compound, phosphoric acid, shortly.

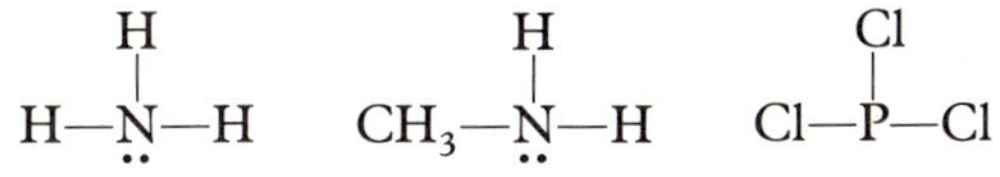

ammonia   methylamine   phosphorus trichloride

THE NITROGEN CYCLE. Carbon, hydrogen, and oxygen are all available to plants from air and water. The element next most important to plants is nitrogen, which is an indispensable part of protein molecules and of nucleic acids, as well as of chlorophyll. Unfortunately, plants cannot use atmospheric nitrogen directly; it is too inert. The conversion of atmospheric nitrogen to nitrogen compounds that can be used by plants is called *nitrogen fixation.*

One way that nitrogen is fixed is by lightning during thunderstorms. The electrical discharge supplies the energy for nitrogen to react with oxygen in the air and

Figure 9-9. Nitrogen and phosphorus: Group VA.

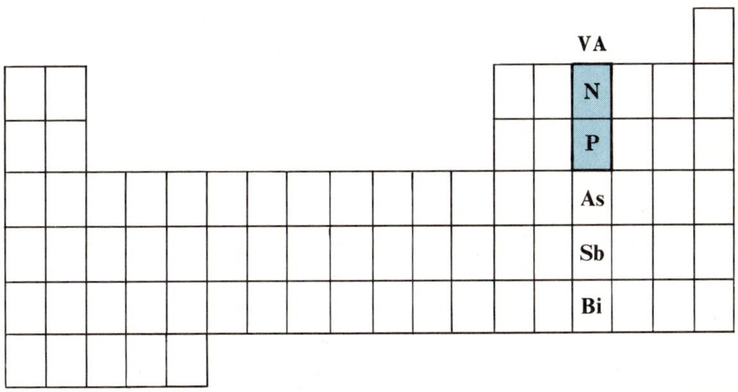

form the gas nitric oxide (NO), which spontaneously undergoes further reaction with oxygen to yield nitrogen dioxide (NO_2), another gas.

$$N_2 + O_2 \xrightarrow{\text{lightning}} \underset{\text{nitric oxide}}{2\,NO}$$

$$2\,NO + O_2 \longrightarrow \underset{\text{nitrogen dioxide}}{2\,NO_2}$$

Rain washes the nitrogen dioxide from the air to the ground, where the nitrogen dioxide is converted to the nitrate ion (NO_3^-), which can be readily utilized by plants.

Another process of nitrogen fixation is carried out by legumes—a type of plant of which clover, beans, and alfalfa are representative. These plants have nodules on their roots that harbor nitrogen-fixing bacteria. Legumes, therefore, indirectly increase the amount of nitrogenous material in the soil. A crop of legumes may be plowed under and a crop of nonlegumes planted; the new crop will be fertilized by the nitrate and nitrite (NO_2^-) salts left over by the legumes. Alternatively, mixed crops of legumes and nonlegumes may be planted; the nonlegume partakes of the nitrates formed by the bacteria at the roots of the growing legume. A lawn planted with nitrogen-fixing clover and nonnitrogen-fixing grass does not require the application of as much fertilizer as does a plain grass lawn.

Plants eventually die. Animals that have eaten plants also die. When plants and animals decay, the nitrogen in their systems returns to the soil and air in the form of nitrate ions, ammonia gas, and nitrogen gas, and the cycle is completed.

The nitrogen cycle:

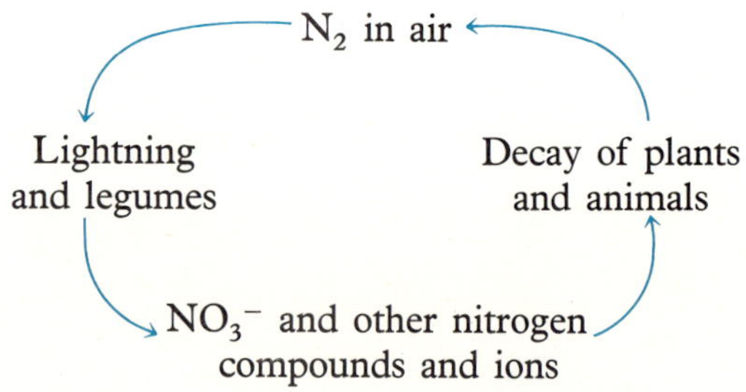

It might be of interest to note that the process of nitrogen fixation is not always a beneficial one. Nitrogen dioxide may be formed in automobile engines and in furnaces by the oxidation of atmospheric nitrogen. The action of sunlight on NO_2 can cause the nitrogen dioxide to cleave and undergo reaction with oxygen to form ozone.

$$NO_2 + uv \longrightarrow NO + O$$

$$O + O_2 \longrightarrow O_3$$

The action of sunlight on NO_2 in conjunction with unburned gasoline or other fuel can lead to the formation of peroxyacyl nitrates (PAN), which are highly toxic eye irritants. An example of one of these compounds is peroxyacetyl nitrate:

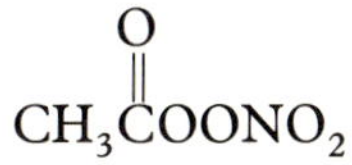

PHOSPHORIC ACID. Phosphoric acid (more correctly, orthophosphoric acid) (H_3PO_4) is useful both in the laboratory and in the production of fertilizers. Compounds containing phosphoric-acid structural units are found in the genes and are used in the synthesis of proteins and in the conversion of food to energy.

Phosphoric acid is a tribasic acid; it can form a series of salts by losing one, two, or three hydrogen ions.

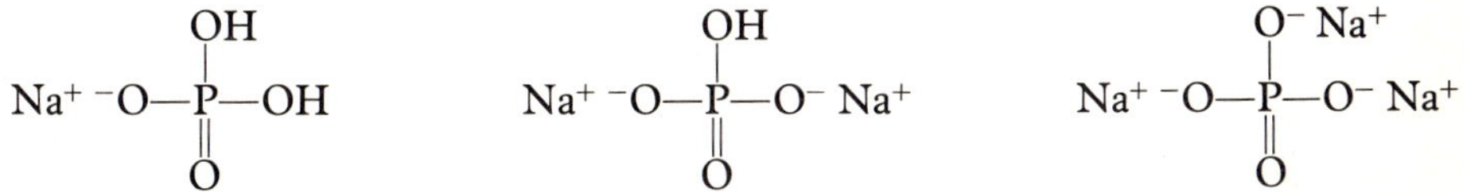

monosodium phosphate (NaH_2PO_4) — disodium phosphate (Na_2HPO_4) — trisodium phosphate (Na_3PO_4)

Phosphoric acid can undergo dehydration to form a series of *anhydrides,* which are compounds formed by the loss of water. When phosphoric acid is heated to 250°C, two molecules of phosphoric acid combine, and one molecule of water is lost. The result is an anhydride of phosphoric acid called *pyrophosphoric acid.* (The prefix *pyro-*, like the term "funeral *pyre,*" comes from the Greek word *pyros,* fire. In this case, *pyro-* is used to mean *heat.* Pyrophosphoric acid arises from heated phosphoric acid.)

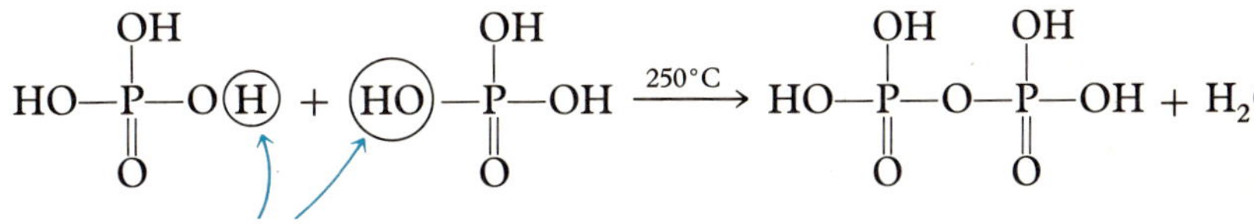

the molecules combine with the loss of H_2O

pyrophosphoric acid

an anhydride

Another molecule of phosphoric acid may link up with pyrophosphoric acid. The product is called triphosphoric acid because it is formed from three units of phosphoric acid.

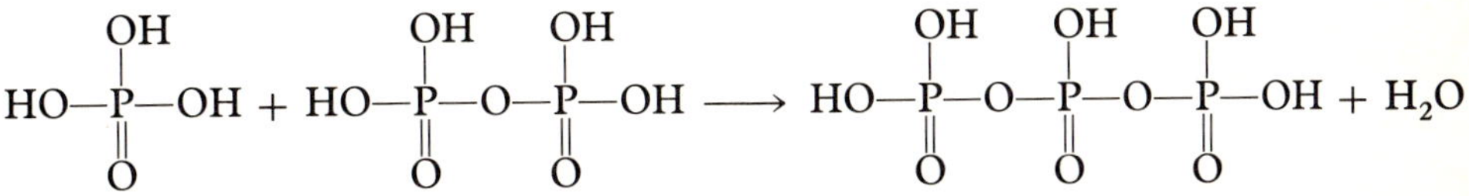

triphosphoric acid

another anhydride

In living systems, a derivative of triphosphoric acid is used to store the energy obtained from food. This anhydride is named adenosine triphosphate (ATP); its complete structural formula is found in Chapter 25. The hydrolysis of ATP to adenosine diphosphate (ADP) supplies energy to the living system when it is needed.

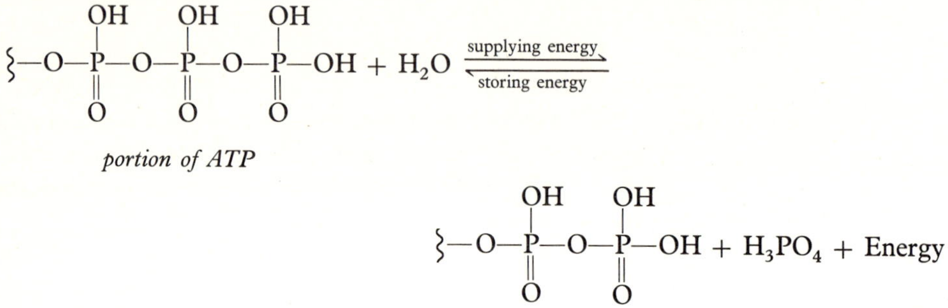

portion of ATP

portion of ADP

OXYGEN AND SULFUR

In Group VIA of the Periodic Table, we find the important nonmetals, oxygen (O) and sulfur (S). (See Figure 9-10.) Oxygen is the most abundant element on the earth. By weight, oxygen as O_2 is 21 percent of our atmosphere. Oxygen also comprises about 50 percent of the earth's crust, where it is found in combination with other elements. By weight, oxygen also comprises 89 percent of water.

An oxygen atom contains six electrons in its outer shell. Thus, we would expect an oxygen atom to tend to gain two electrons by electron transfer or by sharing. In

Figure 9-10. Oxygen and sulfur: Group VIA.

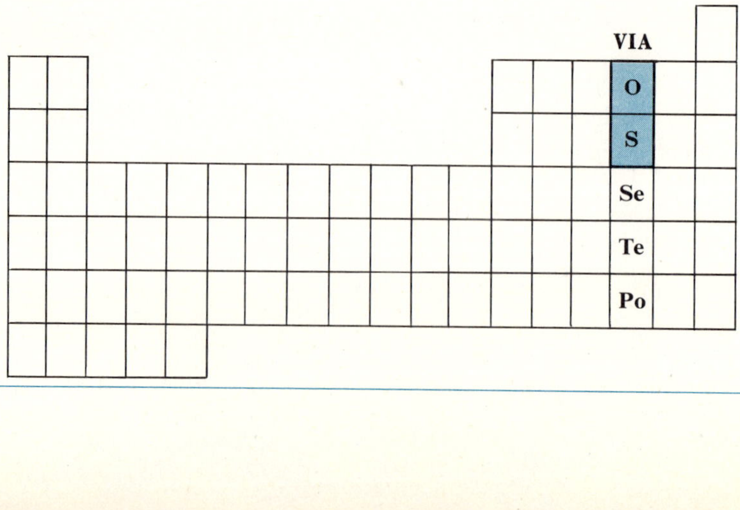

compound formation, oxygen generally behaves as predicted. In each of the following compounds, an oxygen atom has gained two electrons either by electron transfer or by sharing.

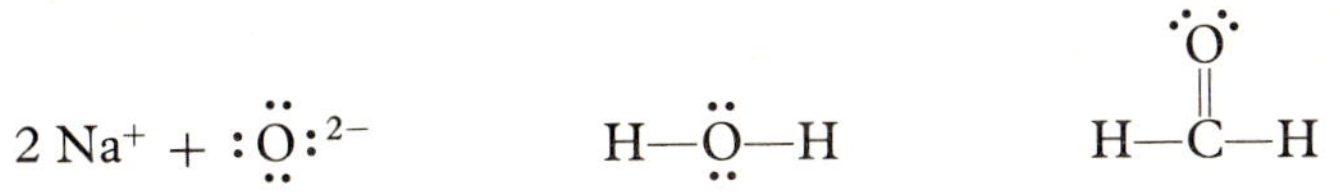

sodium oxide (Na_2O) — water — formaldehyde

The element sulfur is a bright yellow solid that was known to the ancients; it is sometimes found around the mouths of volcanoes. In antiquity, sulfur was called by the picturesque name *brimstone.*

Because of the similarities in electronic structure between the sulfur and oxygen atoms, sulfur undergoes many reactions that are similar to those of oxygen. Sulfur tends to gain two electrons and form a sulfide ion (S^{2-}) or to form compounds in which it has two covalent bonds to other elements. Because there are more orbitals available in the third shell, sulfur can form more than two bonds, as in sulfuric acid.

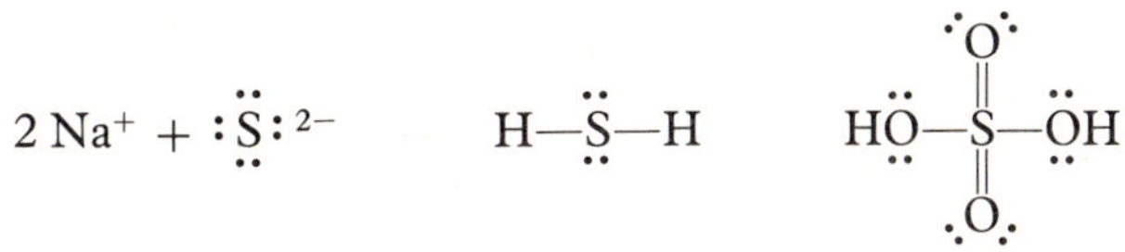

sodium sulfide (Na_2S) — hydrogen sulfide (H_2S) — sulfuric acid (H_2SO_4)

Hydrogen sulfide (H_2S) is worthy of special mention. This compound is a foul-smelling gas with the odor of rotten eggs. Hydrogen sulfide is produced in small quantities when sulfur-containing animal matter (such as eggs) decays. It is also given off in many industrial processes. We can detect the odor of H_2S when it is present in the air in unbelievably small quantities. In a way, this is fortunate. In higher concentrations, H_2S is lethal; its toxicity is about the same as that of hydrogen cyanide (HCN).

THE OXYGEN CYCLE. Oxygen is necessary for human life. It is used to oxidize foodstuffs and produce the energy that supports the living system. Carbon dioxide is a product of these oxidation reactions.

$$\underset{\substack{\text{glucose} \\ \textit{from food}}}{C_6H_{12}O_6} + 6\,O_2 \longrightarrow 6\,CO_2 + 6\,H_2O + \text{Energy}$$

When wood, gasoline, natural gas, or any other carbonaceous fuel is burned, oxygen is used for the combustion, and carbon dioxide is again produced. When plant or animal material decays, the complex compounds of the once-living systems are oxidized; again, oxygen is used, and carbon dioxide is produced. Therefore, we have three major types of oxygen consumers, each one a producer of carbon dioxide.

At this rate, it would seem that we would run out of oxygen and be smothered by an ever-increasing amount of carbon dioxide in the atmosphere. However, thanks to the earth's plant life, there is no real danger in this respect. In the process of photosynthesis, plants utilize atmospheric carbon dioxide to produce carbohydrates; oxygen is produced in this reaction. On the earth, the oxygen consumed and the oxygen produced by plants are in a state of almost perfect balance, even in this present period of rapid consumption of fossil fuels. (It is interesting to note that more than half of the oxygen produced by plants comes from microscopic aquatic plants, rather than from the more noticeable plants of the forests and prairies.)

$$CO_2 + H_2O + \text{Sunlight} \longrightarrow C_6H_{12}O_6 + O_2\uparrow$$

We call the consumption and subsequent regeneration of oxygen the *oxygen cycle*.

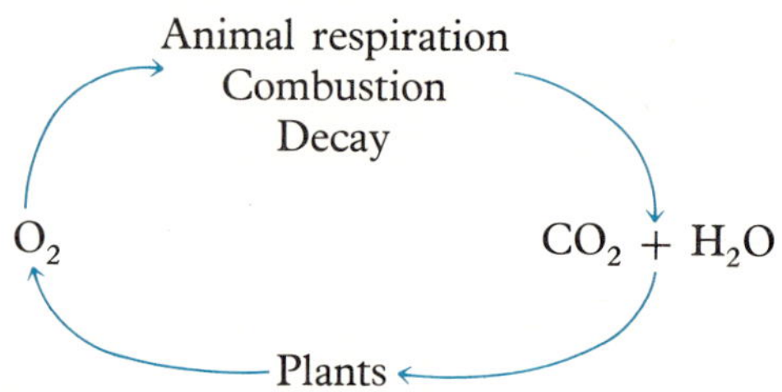

THE HALOGENS

The elements in Group VIIA of the Periodic Table (F, Cl, Br, and I) are called the *halogens*, a word that means *salt formers*. (See Figure 9–11.) A halogen atom has an outer shell containing seven electrons. In the elemental state, each of the halogens is a diatomic molecule with the two atoms connected by a covalent bond. (See Chapter 4.)

The halogens, as a group, exhibit a series of gradated properties. They vary from gas to solid, from lightly colored to dark, from vigorous oxidizing agent to mild oxidizing agent. All of the halogens are too reactive to be found in nature in the elemental state, and all the halogens are toxic.

F_2	Cl_2	Br_2	I_2
gas	gas	liquid	solid
(pale yellow)	(yellow-green)	(red-brown)	(violet-black)

Decreasing oxidizing power ⟶

The halogens are reactive oxidizing agents because a halogen atom has a high electron affinity and has a tendency to obtain an eighth electron from another atom in order to complete its outer shell. In nature, halogens are almost always found in the -1 oxidation state, as *halide ions.*

$$\underset{\text{chlorine}}{Cl{-}Cl} + 2\,e^- \longrightarrow \underset{\substack{\text{chloride ion} \\ \textit{a halide ion}}}{2\,Cl^-}$$

Chloride ions are the principal inorganic anions of living systems, where they help to maintain the ionic, or electrolytic, balance. This balance is necessary for maintenance of the proper osmotic activity between cellular fluids and other body fluids.

Traces of iodide ion are necessary in human nutrition for the proper functioning of the thyroid gland. Lack of iodides results in *goiter,* an enlarged thyroid gland. In coastal areas, iodides are found in the soil and in the fruits and vegetables growing there. Salt-water fish also contain iodides. However, over the eons, most of the iodides in the soil of inland areas have been leached out by rain water, and natural food from these inland areas does not contain a sufficient amount of iodides to keep the body in balance. The introduction and widespread use of *iodized salt,* salt to which a trace of sodium iodide or potassium iodide has been added, has alleviated the problem of iodide deficiency in inland areas.

Figure 9-11. The halogens: Group VIIA.

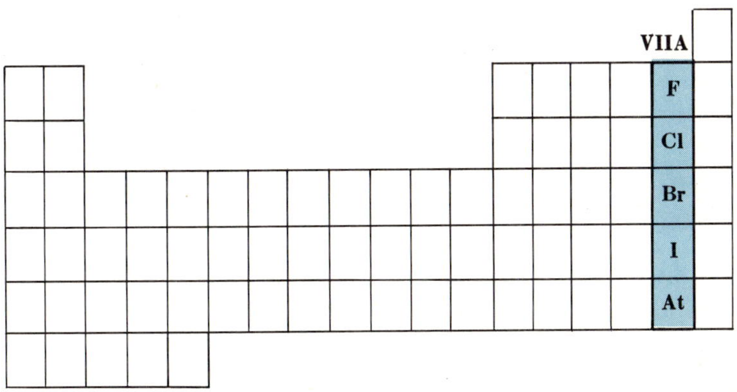

THE HYDROGEN HALIDES. All the halogens can combine with hydrogen to form *hydrogen halides:* HF, HCl, HBr, and HI. Although all of the hydrogen halides are gases, the boiling point of hydrogen fluoride (19 °C) is near room temperature, and it is sometimes considered as a liquid.

The high boiling point of HF, compared to the lower boiling points of the other hydrogen halides, is attributed to hydrogen bonding between molecules of HF. The hydrogen bonds are a result of the high electronegativity of the fluorine atom.

$$\mathrm{H{-}F \cdots H{-}F \cdots H{-}F \cdots H{-}F}$$

hydrogen bonding in hydrogen fluoride

In the gaseous state, the hydrogen atom in each of the hydrogen halides is attached to the halogen atom by a polar covalent bond.

$$\overset{\delta+}{\mathrm{H}}{-}\overset{\delta-}{\mathrm{F}} \qquad \overset{\delta+}{\mathrm{H}}{-}\overset{\delta-}{\mathrm{Cl}} \qquad \overset{\delta+}{\mathrm{H}}{-}\overset{\delta-}{\mathrm{Br}} \qquad \overset{\delta+}{\mathrm{H}}{-}\overset{\delta-}{\mathrm{I}}$$

polar covalent bonds in the hydrogen halides

In water solution, we find that hydrogen fluoride is a weak acid, while the other hydrogen halides are strong acids.

A weak acid: $\underset{\text{hydrofluoric acid}}{HF} \rightleftharpoons H^+ + F^-$

A strong acid: $\underset{\text{hydrochloric acid}}{HCl} \longrightarrow H^+ + Cl^-$

Water solutions of hydrogen halides are called *hydrohalic acids:* hydrofluoric acid, hydrochloric acid, hydrobromic acid, and hydroiodic acid. Of the hydrohalic acids, hydrochloric acid is the cheapest and the one used most extensively. This acid, often called *muriatic acid* in commerce, is used in the production of metal chlorides, dyes, and other materials, and for removing metal oxides from the surfaces of metals that are to be coated with paint or another metal. Hydrochloric acid is also one of the most commonly used sources of hydrogen ion in the laboratory.

Reaction with a metal oxide:

$$\underset{\textit{insoluble}}{CuO} + 2\,HCl \longrightarrow \underset{\textit{soluble}}{CuCl_2} + H_2O$$

THE INERT GASES

The *inert gases,* also called the *rare gases,* or the *noble gases,* comprise Group 0 of the Periodic Table. (See Figure 9-12.) These gases are helium (He), so named because it was first discovered in the sun (Greek *helios,* sun); argon (Ar) (Greek *argon,* inert); krypton (Kr) (Greek *krypton,* hidden); xenon (Xe) (Greek *xenos,* strange); and radon (Rn), a name coined because this gas is radioactive. The outer shells of atoms of the inert gases are filled. This is the principal reason that these elements are nonreactive.

Figure 9-12. The inert gases: Group 0.

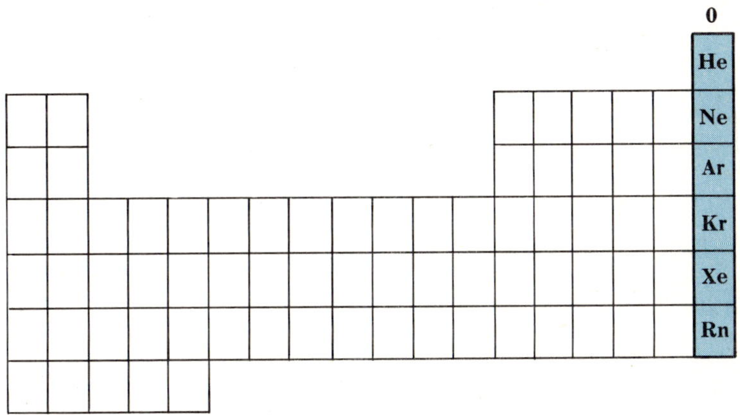

It was not until 1962 that the first compound of an inert gas was prepared. This was a yellow solid, $Xe(PtF_6)$. Shortly after this synthesis, xenon tetrafluoride (XeF_4) was prepared. Since these first two compounds were made, other compounds of some of the inert gases have also been synthesized in the laboratory. (The bonding in these compounds involves the unfilled inner orbitals.) However, we are still correct in calling these gases inert because they will undergo reactions only under unusual circumstances.

STUDY QUESTIONS

9-1. List the principal physical properties that characterize metals.

9-2. Which would you expect to be the more reactive metal?
(a) Ca or Al (b) Mg or Fe (c) Li or Sn

9-3. Why are the metals in Group IA called the *alkali* metals?

9-4. Why is potassium metal usually stored in mineral oil? Could it be stored in some other organic solvent such as ethyl alcohol (CH_3CH_2OH)?

9-5. Predict the products.
(a) $Cs + H_2O \longrightarrow$
(b) $Rb + O_2 \longrightarrow$
(c) $Li + CH_3OH \longrightarrow$
(d) $Na + HCl \longrightarrow$

9-6. What is the common oxidation number of the alkali metals in compounds? The alkaline earth metals?

9-7. What is the difference between hard water and soft water?

9-8. In which of the following liquids would you expect to be able to dissolve a pearl?
(a) ethyl alcohol (b) water
(c) vinegar (d) 6 *N* HCl

9-9. Write the equation for the principal reaction that occurs when milk of magnesia (magnesium hydroxide) is swallowed.

9-10. Both aluminum and magnesium undergo reaction readily with oxygen to form oxides. Yet, both of these metals are used extensively in society. Explain.

9-11. What is meant by the term *transition metal?*

9-12. Which of the following contain complexed metal ions?
(a) NH_4Cl (b) $Ag(NH_3)_2Cl$
(c) $Na_3Fe(CN)_6$ (d) H_2SO_4

9-13. Silver chloride (AgCl) is insoluble in water but soluble in a dilute aqueous solution of ammonia. Explain.

9-14. Which of the following elements would you expect to have a relatively high electron affinity?
(a) Mg (b) I (c) Se (d) Sr

9-15. What is the oxidation number of phosphorus in each of the following:
(a) phosphorus pentoxide (P_2O_5)
(b) phosphoric acid
(c) pyrophosphoric acid
(d) ATP

9-16. Predict the product: $CH_3\ddot{N}H_2 + H^+ \longrightarrow$

9-17. Write a chemical equation that illustrates nitrogen fixation.

9-18. Outline the oxygen cycle.

9-19. Give the oxidation number for the sulfur atom in each of the following:
(a) S_8 (b) SO_2 (c) H_2SO_4
(d) H_2S (e) SO_3 (f) $NaHSO_3$

9-20. Using dots to represent electrons in the outer energy level, show the electronic structure of F_2.

9-21. Bromine (Br_2) can be obtained from sea water (which contains both Cl^- and Br^- ions) by treatment with chlorine gas. Give the equation for the chemical reaction.

9-22. What is the difference between helium and xenon that allows $Xe(PtF_6)$ or XeF_4 to be prepared, but not the corresponding helium compounds?

CHAPTER 10 NUCLEAR CHEMISTRY AND RADIOACTIVITY

OBJECTIVES

Types of Radioactivity *Write the symbols for alpha particles, beta particles, and gamma rays, and describe their properties.*

Why Nuclei Decay *Describe two factors that affect the stability of a nucleus.*

Rate of Radioactive Decay *Define half-life.* * *Describe the technique of dating objects by the use of radioactive isotopes.*

Artificial Transmutations *Define transuranium element.*

Nuclear Fission *Describe the chain reaction involved in the disintegration of uranium-235.* * *Define mass defect and describe why nuclear fission produces energy.* * *Describe a breeder reactor.*

Nuclear Fusion *Write an equation that shows the origin of the energy liberated in a nuclear fusion reaction.*

Using Radioactive Isotopes *Describe some uses of radioactive isotopes in medicine.*

The Dangers of Exposure to Radiation *Describe radiation sickness.*

How We Measure Radiation *Define curie, roentgen, and rad.*

The alchemists of the Middle Ages hoped to discover all the secrets of the universe. They were not really chemists, but would-be magicians who dabbled in chemistry. Their writings were mystical and secretive, often in code. Because many alchemists were charlatans and cheats, the semi-science of alchemy has not been revered in more recent times.

One of the dreams of the alchemists was to find the *essence of all things,* which they also called the *philosophers' stone* and which reputedly could cure all disease and change lead into gold. The alchemists extracted, distilled, mixed, and heated all manner of chemicals in order to find this magical substance. Undoubtedly, had the writings of the alchemists been more intelligible, the early chemists who followed the alchemists would have had a larger background of chemical knowledge upon which to draw.

The alchemists never did find the philosophers' stone. It was not until the 20th century that the dream of *transmutation of the elements,* changing one element into another, came true. However, the modern-day "philosophers' stones" are nuclear-particle accelerators and nuclear reactors, not a magical substance at all.

In the late 1800's, a French physicist, Henri Becquerel, stored some photographic film in a drawer where he had also stored a uranium compound. Imagine his surprise to find later that the carefully wrapped film had darkened. Investigation by Becquerel showed him that all uranium compounds were capable of affecting film in this way. Even today, special "film badges" are worn by workers that might be exposed to radioactivity. Darkening of the film is used as a measure of the amount of exposure to radiation by the wearer of the badge.

Radioactivity is the emission of high-energy electromagnetic waves or subnuclear particles, or both. In Becquerel's time, one form of radioactive emission, the *x*-ray, was already known. The causes and effects of *x*-rays were not yet well understood, but Becquerel correctly reasoned that his uranium compound was emitting some type of radiation.

Interest in the phenomenon of radiation became high in that era as it is now. After Becquerel's discovery, Marie Curie and her husband Pierre Curie undertook to investigate and isolate radioactive elements, that is, elements that emit radiation. This famous team discovered polonium (named after Madame Curie's native country, Poland), then radium, so named because of its very high level of radioactivity. The Curies' dedication to their research is attested to by the fact that to get 0.1 gram of radium, they had to work up painstakingly one ton of ore. Becquerel and the Curies received a joint Nobel prize in 1903 for their discoveries, but unfortunately only after Pierre Curie was killed in an accident.

TYPES OF RADIOACTIVITY

From the work of these early chemists and physicists and their successors, we now know that natural radioactivity arises from the breakup of unstable nuclei in atoms. This breakup is called *radioactive decay.*

ALPHA RADIATION. There are several types of radiation that arise from radioactive decay. The first is the *alpha* (α) *particle.* An alpha particle is a helium nucleus, a particle containing two protons and two neutrons. Alpha rays travel at about one tenth the speed of light. However, because of their relatively large size and positive charge, they are not very penetrating and cannot get past the molecules in the first layer of human skin or through a piece of cardboard.

One nuclear reaction in which alpha rays are formed is the radioactive disintegration of uranium-238.

$$\underset{\text{uranium-238}}{{}^{238}_{92}\text{U}} \longrightarrow \underset{\text{thorium-234}}{{}^{234}_{90}\text{Th}} + \underset{\text{alpha particle}}{{}^{4}_{2}\text{He}}$$

We must pause in our discussion of radiation for some explanation of the way we write nuclear equations. Since we are dealing with changes in the nuclei of atoms, we need a notation that will help us keep track of the nuclear particles. Therefore, we include the *atomic mass* (the number of protons plus the number of neutrons) at the top left of the atomic symbol. The *atomic number* (the number of protons) is at the lower left of the atomic symbol.

atomic mass ↘

$${}^{238}_{92}\text{U}$$

atomic number ↗

In a nuclear equation, like any other equation, the particles, or *nucleons,* on the left side must balance with the nucleons on the right side. We might call this statement the "Conservation of Nucleons." In the preceding equation for the decomposition of uranium-238, we can see that the atomic masses on the right (234 + 4) equal those on the left (238), and that the number of protons on the right (90 + 2) is the same as on the left (92).

Sample Problem: Describe in words the meanings of (a) ${}^{234}_{90}\text{Th}$, (b) ${}^{4}_{2}\text{He}$, and (c) ${}^{14}_{6}\text{C}$.

Solution:

(a) An atom of thorium (atomic number 90) with an atomic mass of 234 amu.

(b) An atom of helium (atomic number 2) with an atomic mass of 4 amu.

(c) An atom of carbon (atomic number 6) with an atomic mass of 12 amu.

Sample Problem: Rewrite the following equation, filling in the missing numbers.

$$^{14}_{7}\text{N} + ^{4}_{2}\text{He} \longrightarrow ^{?}_{?}\text{O} + ^{?}_{?}\text{H}$$

nitrogen-14 helium-4 oxygen-? hydrogen-1

Solution:

$$^{14}_{7}\text{N} + ^{4}_{2}\text{He} \longrightarrow ^{17}_{8}\text{O} + ^{1}_{1}\text{H}$$

oxygen-17

BETA RADIATION. A second type of radiation, called *beta* (β) *radiation,* is the emission of beta particles. Beta particles are high-energy electrons that travel at an exceedingly high speed, up to nine-tenths the speed of light. Because of their high speed and small size, beta particles are more penetrating than alpha particles and can penetrate a slight distance into the skin. For this reason, beta radiation can cause burns. An example of a nuclear reaction that yields beta particles follows.

$$^{234}_{90}\text{Th} \longrightarrow ^{234}_{91}\text{Pa} + ^{0}_{-1}e$$

thorium-234 protactinium-234 beta particle

The symbol we use for an electron in nuclear equations follows the same scheme as those for nuclei. The atomic mass of an electron is essentially 0, and the "atomic number" is -1.

The electron that is emitted in this reaction does not come from electronic shells around the nucleus, but from the nucleus itself. A neutron in the nucleus loses the electron and becomes a proton. (Figure 10–1 illustrates this phenomenon.)

$$^{1}_{0}n \longrightarrow ^{1}_{1}p + ^{0}_{-1}e$$

neutron	proton
atomic mass 1	*atomic mass 1*
atomic number 0	*atomic number 1*

The atomic mass of protactinium-234 is the same as that of thorium-234, but the atomic number has been raised by one. In the equation for the decay of thorium-234, the particles on the left still balance with those on the right. On the right,

Figure 10–1. A depiction of the decay of thorium-234, in which a neutron emits a beta particle and becomes a proton.

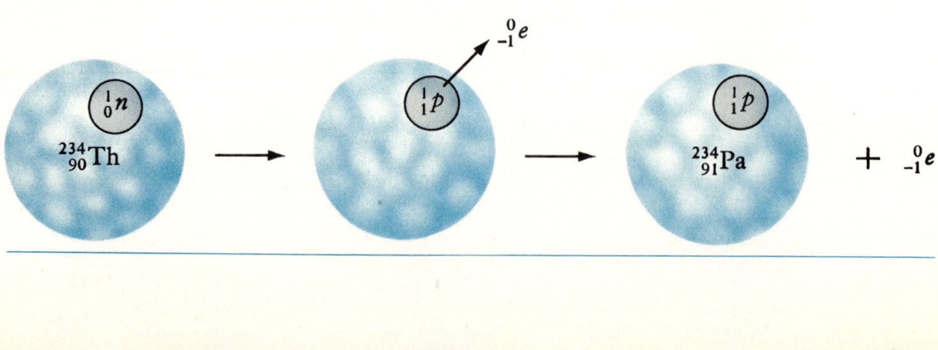

we have a total atomic mass of 234 (234 + 0) and a total number of protons of 90 (91 − 1), the same as on the left.

GAMMA RAYS. A third type of radiation is the *gamma* (γ) *ray*. Gamma rays are electromagnetic waves; they have no mass or electrical charge. Their speed is the speed of light. Gamma rays are extremely penetrating and can pass through several feet of concrete or dirt. These rays may or may not accompany an emission of alpha or beta radiation. For example, the decay of thorium-234 produces gamma radiation as well as beta radiation. Since gamma rays have no mass or charge, the emission of this type of radiation does not alter the atomic mass or atomic number of an atom. See Table 10-1 for a summary of the types of radiation.

$$^{234}_{90}\text{Th} \longrightarrow {}^{234}_{91}\text{Pa} + {}^{0}_{-1}e + \underset{\text{gamma ray}}{\gamma}$$

Table 10-1. Types of Radiation from Nuclear Decay

Radiation	Symbol	Atomic Mass	Charge	Speed
Alpha (α)	$^{4}_{2}\text{He}$	4	+2	0.1 speed of light
Beta (β)	$^{0}_{-1}e$	0	−1	0.9 speed of light
Gamma (γ)	—	0	0	Speed of light

COSMIC RAYS. Cosmic rays are subnuclear particles from the sun and outer space that enter our atmosphere. Their intensity on the earth is the highest during periods of solar flares. These rays consist mainly of high-speed protons, but they also contain beta particles, alpha particles, and the nuclei of other elements.

When cosmic rays enter our atmosphere, they collide and undergo reaction with the gaseous molecules in the atmosphere. One result of these reactions is the *aurora borealis* (northern lights). Another is the formation of radioactive carbon, which is used in carbon dating and which we shall discuss shortly. The reactions of cosmic rays with our atmosphere also produce other particles (electrons, protons, neutrons, and so forth) that we can detect on the ground with appropriate instruments. Cosmic rays cause a portion of the *background radiation*, or natural radiation, to which the inhabitants of earth are constantly exposed.

WHY NUCLEI DECAY

The forces that hold a nucleus together are not entirely understood. We do know that there are other particles besides neutrons and protons in the nucleus, and it is believed that they aid in binding the neutrons and protons together.

It has been observed that the nuclei of the natural, stable isotopes of most of the elements of lower molecular weight contain equal numbers of neutrons and protons.

The nucleus of carbon-12, for example, contains six protons and six neutrons; that of nitrogen-14 contains seven protons and seven neutrons.

The nuclei of the natural, stable isotopes of the elements of higher molecular weight contain a greater proportion of neutrons than do the lower-molecular-weight elements. For example, the nucleus of bromine-79 contains 35 protons and 44 neutrons.

If the ratio of neutrons to protons is changed dramatically from the normal, the result is an unstable, or radioactive, isotope that will undergo decay. We find that carbon-14 with six protons and eight neutrons is an unstable isotope.

$$\underset{\text{carbon-14}}{{}^{14}_{6}\mathrm{C}} \longrightarrow \underset{\text{nitrogen-14}}{{}^{14}_{7}\mathrm{N}} + {}^{0}_{-1}e$$

It is apparent then that the *ratio of neutrons to protons* is one factor in the stability of a nucleus. (See Figure 10–2.) Another factor that affects the stability of a nucleus is the *size of the nucleus*. Every known isotope of every element with an atomic number greater than 82 is radioactive. In other words, lead is stable, but bismuth and each heavier element is unstable and undergoes radioactive decay.

Figure 10–2. The number of neutrons *versus* the number of protons in stable isotopes.

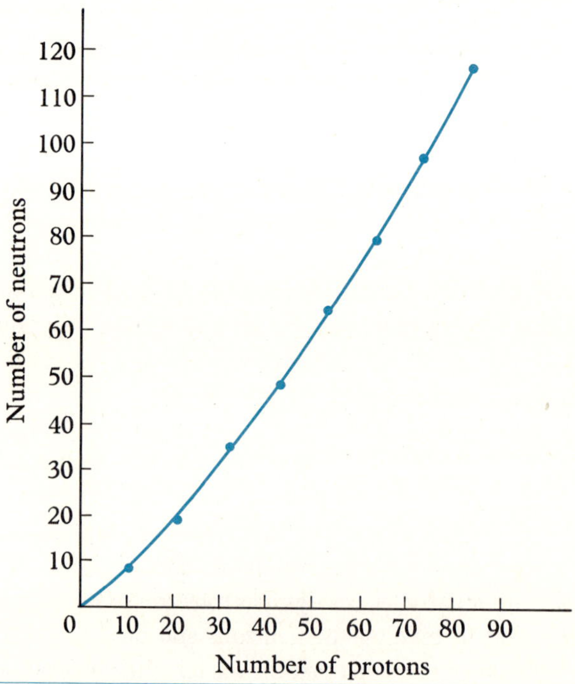

Keep in mind that many elements do have more than one stable isotope. For example, ^1_1H and ^2_1H are both stable, while ^3_1H is radioactive.

RATE OF RADIOACTIVE DECAY

When uranium-238 undergoes radioactive decay, it yields thorium-234, which is also radioactive and undergoes further decay. The nucleus continues its decay until a stable nucleus is formed. In the case of uranium-238, the final product is lead-206, which is a stable isotope of lead.

The process of disintegration of uranium-238 does not occur quickly. In fact, one gram of uranium-238 would produce 0.4326 grams of lead-206 after 4.5 billion years! One half (0.5000 g) of the original uranium-238 would still be left at that time. In another 4.5 billion years, half of the remaining uranium-238 would decay and 0.25 grams would be left from our original 1.0 gram. In another 4.5 billion years, 0.125 grams of uranium-238 would remain from that original 1.0 gram. (See Figure 10–3.)

The time it takes for one half of a given amount of radioactive isotope to decay is defined as the *half-life* of that isotope. For uranium-238, the half-life is 4.5 billion years; for radon-222, the half-life is four days; while for protactinium-234, the half-life is only 69 seconds. The half-life of uranium or any other radioactive isotope is unaffected by temperature changes or any chemical change. There is no known method by which we can change the half-life of a radioactive isotope.

Because the rate of decay of an element is constant, the relative amounts of uranium-238 and lead-206 found in uranium ore may be used to estimate the age of

Figure 10–3. In each 4.5-billion-year period, one half of the uranium-238 decays.

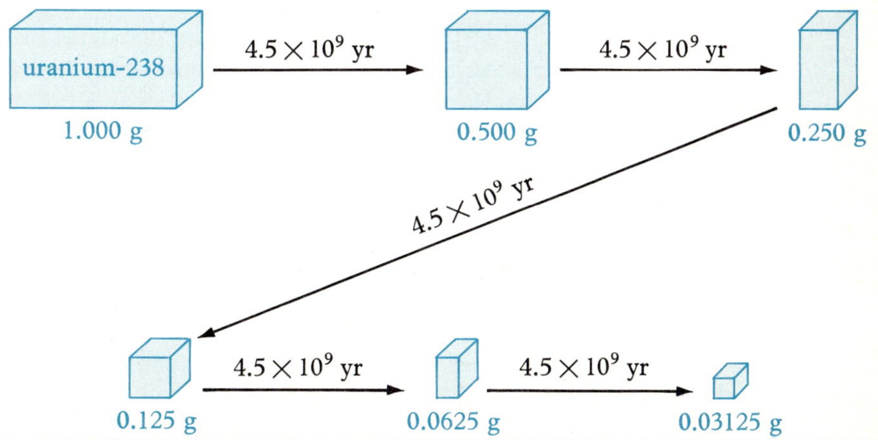

the earth's crust. By this technique, scientists estimate that the earth's crust is at least 2.6 billion years old. Meteorites have been found that are estimated to be 4.5 billion years old.

CARBON DATING. The radioactive isotope carbon-14 is formed in the atmosphere by the action of cosmic rays on nitrogen.

$$\textit{Formation of carbon-14:}\quad {}^{14}_{7}N + {}^{1}_{0}n \longrightarrow {}^{14}_{6}C + {}^{1}_{1}p$$

$$\textit{Decay of carbon-14:}\quad {}^{14}_{6}C \longrightarrow {}^{14}_{7}N + {}^{0}_{-1}e$$

As soon as it is formed, the carbon-14 begins to decay; its half-life is 5570 years. A living plant, which absorbs carbon dioxide containing both carbon-12 and carbon-14, contains the same ratio of carbon-12 and carbon-14 as is found in the atmosphere. When the plant dies, it ceases to absorb carbon dioxide; however, the carbon-14 absorbed by the plant when alive continues to decay. Thus, the age of a wood object such as a carved figurine may be estimated from its carbon-14 content. If the wood contains half the carbon-14 that today's trees do, we estimate that the tree from which the object was made died 5570 years ago. Since living animals eat living plants, the technique can be used to date animal material also.

The disadvantage of carbon-14 dating is that the half-life of this isotope is relatively short. After 50,000 years, so little carbon-14 is left in an object that it is difficult to make accurate measurements. New dating techniques using radioactive potassium (which has a half-life of 1 billion years) have been developed.

ARTIFICIAL TRANSMUTATIONS

In 1919, Ernst Rutherford tried bombarding nitrogen gas with the alpha particles that were emitted from radium. His experiment was a milestone in the history of science. From the nitrogen and the alpha particles, Rutherford obtained small amounts of oxygen-17 and hydrogen. Rutherford was the first to effect an artificial transmutation of the elements. Since Rutherford's experiment, hundreds of transmutations have been made.

$$\underset{\text{nitrogen-14}}{{}^{14}_{7}N} \quad + \quad \underset{\text{alpha particle}}{{}^{4}_{2}He} \quad \longrightarrow \quad \underset{\text{oxygen-17}}{{}^{17}_{8}O} \quad + \quad \underset{\text{hydrogen-1}}{{}^{1}_{1}H}$$

TRANSURANIUM ELEMENTS. Until 1940, uranium was the heaviest element known. It is apparently the heaviest *natural* element. In 1940, McMillan and Abelson synthesized the first *transuranium element,* an element with an atomic mass greater than that of uranium. The new element, which was named neptunium, was formed by bombarding uranium-238 with high-energy *deuterons,* nuclei of the hydrogen-2 isotope, or $^{2}_{1}H$. This bombardment produced uranium-239, which decayed to neptunium-239. Since 1940, elements 94 through 106 have been synthesized.

$$^{238}_{92}U + \underset{\text{deuteron}}{^{2}_{1}H} \longrightarrow {}^{239}_{92}U + {}^{1}_{1}H$$

$$^{239}_{92}U \longrightarrow \underset{\text{neptunium-239}}{^{239}_{93}Np} + {}^{0}_{-1}e$$

NUCLEAR FISSION

After Rutherford's first transmutation, other scientists became interested in bombarding uranium with particles and in creating new, heavier elements. As we have seen, this goal has been accomplished. However, in the late 1930's Otto Hahn in Germany and others elsewhere noticed an unusual occurrence. When uranium was bombarded with neutrons, small amounts of barium and other elements from the middle of the Periodic Table were produced in the target samples. Was the bombardment of uranium actually causing uranium atoms to split in two? This, indeed, was the case. It was then shown that the isotope that was breaking up, or undergoing *nuclear fission,* was uranium-235, which comprises less than 1 percent of naturally occurring uranium (mainly uranium-238).

$$^{235}_{92}U + {}^{1}_{0}n \longrightarrow {}^{236}_{92}U \longrightarrow {}^{141}_{56}Ba + {}^{92}_{36}Kr + 3\,{}^{1}_{0}n + \gamma + \text{Energy}$$

The discovery of nuclear fission was the dawn of the age of atomic energy. A few years after this discovery, the United States became involved in World War II and initiated the most fantastic (and expensive) problem-solving project that the world had ever seen—the development of the atomic bomb.

When uranium-235 captures one neutron, it undergoes fission and produces *three* neutrons. When small quantities of uranium-235 undergo fission, these newly produced neutrons escape from the sample into the surroundings. When a large enough mass of uranium-235 is present, the neutrons are contained and collide with other nuclei, causing more fissions and the production of more neutrons. Consider this process similar to a chain letter. You write three letters. Each of the recipients writes three more. Each of those recipients does the same. If this were kept up, the world would soon be inundated with letters. In a similar way, the nuclei of uranium-235 are bombarded with more and more neutrons. (See Figure 10–4.)

A self-sustaining reaction such as the fission of uranium-235 is called a *chain reaction.* The mass of uranium-235 necessary to capture enough neutrons and undergo a chain reaction is called the *critical mass.* For pure uranium-235, the critical mass is

Figure 10–4. Each fission of a uranium-235 nucleus requires one neutron and produces three neutrons.

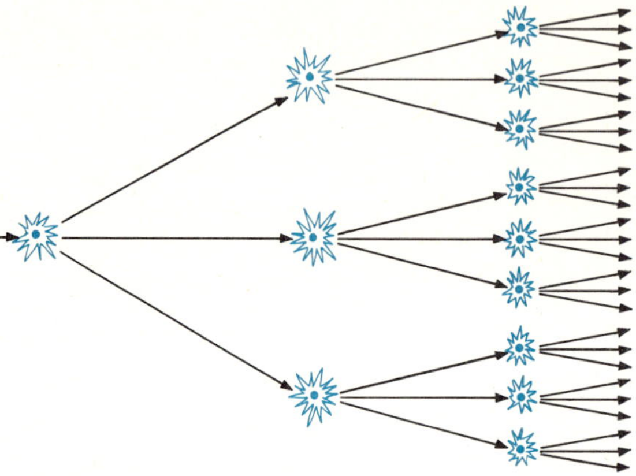

a little smaller than a football. In an atomic bomb, small masses of uranium-235 are kept separate within the bomb. When the small masses are combined to produce a critical mass, the result is a nuclear explosion.

WHERE DOES THE ENERGY COME FROM? When an atom such as uranium-235 undergoes fission, tremendous amounts of energy are released. Where does this energy come from?

Let us take a simple example—a helium atom. It has been determined that a helium atom has an atomic mass of 4.002604. A helium atom consists of two protons, two neutrons, and two electrons. One would expect that if we added the masses of two protons, two electrons, and two neutrons, we would arrive at the mass of the helium atom; however, this is not the case. The sum of the masses of two protons, two neutrons, and two electrons is 4.032980. A helium atom weighs *less* than the sum of its parts.

The difference between the actual atomic mass and the expected atomic mass is called the *mass defect*. If we could stick two protons, two electrons, and two neutrons together to create a helium atom, mass would be lost, and it would be lost as energy. This energy is called the *binding energy* of the atom. The binding energies of atoms vary with the type of atom. The greater the binding energy per nucleon, the more stable the nucleus of an atom. It has been found that middleweight elements (atomic numbers 30 to 63) have the greatest binding energies per nucleon and are the most stable.

Early in the 1900's, Albert Einstein put forth his famous equation $E = mc^2$. E is the energy in ergs (1 erg $= 2.4 \times 10^{-8}$ cal), m is the mass in grams, and c is the speed of light (30,000,000,000 cm/sec). From Einstein's equation, the amount of

Figure 10–5. In the fission of uranium-235, a small part of its mass is converted to energy.

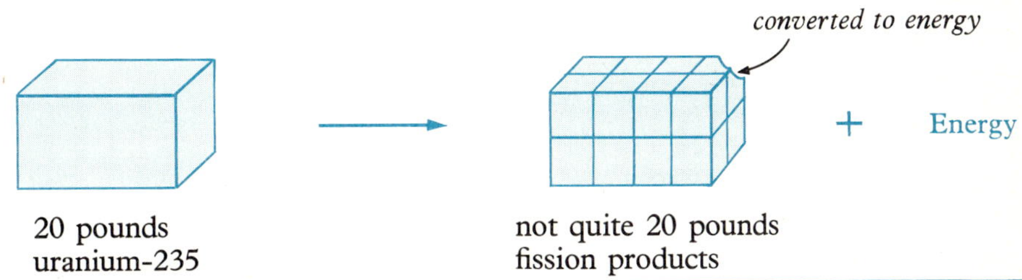

energy formed from a given mass of matter can be calculated. How much energy would be formed if 1.0 gram of matter were completely converted to energy? Almost 20 billion kilocalories! This 1.0 gram of matter could heat hundreds of homes during a long, hard winter.

When twenty pounds of uranium-235 undergo fission, the atoms cleave into more-stable atoms that have a greater mass defect. Even though we get almost twenty pounds of smaller, more-stable atoms out of the reaction, some of the mass is changed to energy. The energy released by twenty pounds of uranium-235 (the approximate size of the atomic bombs dropped in World War II) is equivalent to the energy released by 20,000 tons of TNT. (See Figure 10–5.)

HARNESSING THE ENERGY. Our supply of fossil fuels, such as petroleum and natural gas, is being depleted at a rapid rate. To support the world's population, new sources of energy are needed. Atomic energy seems to be one answer to our fuel needs.

In 1942, Enrico Fermi succeeded in building an *atomic pile,* a device in which a nuclear chain reaction can be controlled. To have a controlled chain reaction, enough neutrons must be supplied to a sample of uranium-235 to keep the fission going, but not so many neutrons that an explosion could result. The present nuclear reactors used for power production are based upon this principle of the controlled nuclear fission of uranium-235. The heat produced by fission in the reactor is trapped and used for the production of electricity. These reactors are designed so that the possibility of an explosion is remote. The radioactive material is well shielded so that workers and nearby residents are not exposed to dangerous radiation.

THE BREEDER REACTOR. The number of elements that undergo fission is limited. Our supply of uranium-235 is also limited. At our present rate of consumption in nuclear power plants, we will be running short of uranium-235 in not too many years.

Present research is therefore aimed at perfecting the *breeder reactor,* a reactor that creates its own fissionable material while it is operating. Uranium-238 is far more abundant than uranium-235. Unlike uranium-235, uranium-238 is not fissionable;

however, when bombarded with neutrons, it can be changed to a fissionable isotope of plutonium. Thus, when some of the neutrons from the fission of uranium-235 are used to bombard uranium-238, new nuclear fuel is generated.

NUCLEAR FUSION

Yet another source of energy may be found in *nuclear fusion reactions,* or *thermonuclear reactions. Fission* means "breaking apart"; *fusion* means "putting together." Nuclear fusion is the putting together of small atoms to make larger ones. The energy of the sun comes from the fusion of hydrogen atoms into helium atoms.

$$4\,{}^{1}_{1}\text{H} \longrightarrow {}^{4}_{2}\text{He} + 2\,{}^{0}_{1}e + \text{Energy}$$

The mass of four hydrogen atoms is greater than the mass of one helium atom. In other words, helium has a greater mass defect than hydrogen. In this fusion reaction, mass is converted to energy. The sun converts 5 million tons of matter into energy each second! In the production of only 1.0 mole (4.0 grams) of helium from hydrogen, 620 million kilocalories of energy are produced.

For a fusion reaction to occur requires temperatures of 1,000,000 °C and higher. These are not temperatures we can normally reach in a laboratory. Therefore, the thought of using nuclear fusion here on the earth would seem an impossible dream, and so it was until the development of the atomic bomb. In a nuclear fission explosion, temperatures become high enough to allow fusion to occur. Thus, the atomic fission bomb was the key to the fusion bomb, also called the *hydrogen bomb,* or *thermonuclear device.* In one type of hydrogen bomb, lithium deuteride is placed next to the fissionable material in an atomic bomb. When the fissionable material explodes, the lithium compound is cleaved to helium and isotopes of hydrogen. The isotopes of hydrogen then undergo fusion and release huge quantities of energy.

$$\underset{\text{lithium deuteride}}{{}^{6}_{3}\text{Li}\,{}^{2}_{1}\text{H}} + {}^{1}_{0}n \longrightarrow {}^{3}_{1}\text{H} + {}^{4}_{2}\text{He} + {}^{2}_{1}\text{H}$$

$${}^{3}_{1}\text{H} + {}^{2}_{1}\text{H} \longrightarrow {}^{4}_{2}\text{He} + {}^{1}_{0}n + \text{Energy}$$

USING RADIOACTIVE ISOTOPES

Radioactive isotopes are different from stable isotopes in that we can detect their presence with such instruments as Geiger counters and scintillation counters. These instruments pick up and *count* the radiation; that is, they can tell us how much radiation is being given off by a substance.

Chemically, a radioactive isotope behaves like a stable isotope. In the body or in a laboratory reaction, sodium-24 chloride behaves just like the normal, nonradioactive sodium-23 chloride. Therefore, radioactive isotopes can be used as *tracers* in the human body and are a useful diagnostic tool. For example, the sodium-24 isotope is used to detect for faulty circulation of the blood. A measured dose of sodium-24 chloride solution is injected into the bloodstream and its path followed with a counter. If it takes longer for the isotope to show up at a particular part of the body than it should, it is assumed that the circulation is impaired at that spot.

Radioactive sodium ions can even be used to determine the *volume* of blood in a patient by a technique called *isotopic dilution.* A measured amount of sodium-24 chloride is injected intravenously, then its concentration in the blood is determined. From this concentration, the volume of blood can be calculated.

Iodine-131 (in the form of sodium iodide) is used to test thyroid function. Iodides become concentrated in the thyroid gland at a certain rate. A patient is given a measured dose of radioactive sodium iodide in a glass of water, a radioactive cocktail. After a given period of time, the amount of iodine-131 concentrated in the thyroid gland of the patient is counted, and the rate of concentration compared with the normal rate.

Radioactive isotopes are also used therapeutically because they can destroy living tissue. (We will discuss how radiation affects cells shortly.) One of the best-known therapeutic uses of radiation is in the treatment of cancer.

A cancer patient may be subjected to an external source of radiation, which is aimed at the tumor or other tissue to be destroyed. Sometimes the radiation source is actually implanted in a tumor. Another technique of killing certain cells selectively is to administer a radioactive isotope that will become concentrated at the site of malignancy. Iodine-131 thus finds use in treating cancer of the thyroid gland. Phosphorus is concentrated in the bones; therefore, phosphorus-32 may be used to treat a bone malignancy.

THE DANGERS OF EXPOSURE TO RADIATION

We are continuously exposed to radiation in many forms. The ultraviolet light from the sun is a form of radiation that helps plants to grow and gives us a healthy-looking tan. But ultraviolet light can also cause sunburn and skin cancer. *X*-rays are used to detect tumors in the body and cavities in the teeth. However, too great or too prolonged exposure may cause burns and cell damage. Having an occasional diagnostic *x*-ray is not likely to cause harm, but a dentist or *x*-ray technician must take precautions to avoid overexposure. Lead is one of our denser metals and an effective shield against radiation. Hence, lead aprons and lead walls are commonly used to protect humans against radiation exposure.

It is likely that many of the early workers in the field of nuclear chemistry died as a result of radiation poisoning. The symptoms of radiation sickness are nausea and a drop in the numbers of red and white blood cells. There may be loss of appetite, general feeling of illness, diarrhea, and hair loss. A cancer patient undergoing radiation treatment may experience some of these symptoms and may not be able to fight off infections, such as pneumonia, as a normal person could. Of course, a large dose of radiation or prolonged exposure may itself cause death, without the complications of secondary infections.

It is thought that radiation sickness is caused by the ionizing properties of radiation. Radiation can cause molecules to break apart into ions and free radicals (particles with unpaired electrons). The free radicals can lead to chain reactions that result in the destruction of biomolecules and the eventual death of cells. It has also been found that radiation can damage the molecules that constitute the genes. The effects of this type of damage may not show up for generations.

HOW WE MEASURE RADIATION

Because we use radioactive materials in the laboratory, in medicine, and in industry, we need units for measuring radiation. By necessity, several types of measurements, and therefore several units of measure, have been devised. We will briefly look at some of these.

RATE OF RADIOACTIVE DECAY. The rate at which a radioactive substance decays is dependent upon the number of atoms in a sample that undergo disintegration in a given period of time. This rate is measured in *curies* (Ci), where 1.0 curie is defined as 3.7×10^{10} nuclear disintegrations per second. When smaller units are needed, the *millicurie*, the *microcurie*, or the *micromicrocurie* (also called the *picocurie*) is used.

$$\begin{aligned} 1.0 \text{ curie (Ci)} &= 3.7 \times 10^{10} \text{ disintegrations/sec} \\ 1.0 \text{ millicurie (mCi)} &= 10^{-3} \text{ Ci} \\ 1.0 \text{ microcurie } (\mu\text{Ci}) &= 10^{-6} \text{ Ci} \\ 1.0 \text{ micromicrocurie } (\mu\mu\text{Ci}) &= 10^{-12} \text{ Ci} \end{aligned}$$

EXPOSURE TO RADIATION. Besides knowing how fast a radioactive sample is decaying, we need a value that can tell us the intensity of a given exposure to radiation. The unit of exposure is the roentgen (R). By definition, the roentgen applies only to the very penetrating x-rays and gamma rays.

1.0 roentgen (R) = the amount of x-radiation or γ radiation that will produce 2.1×10^9 units of ionic electrical charge in 1.0 cc of dry air at standard temperature and pressure.

From a practical standpoint, knowing the number of roentgens to which a person is exposed will tell us what symptoms that person may be expected to develop. An exposure of 600 roentgens, for example, is very likely to be fatal.

ABSORPTION OF RADIATION. A measure of exposure to radiation does not tell us how much radiation is actually absorbed by a system. As a unit of absorption, we use the *rad,* which stands for *radiation, absorbed dose.*

1.0 rad = 100 ergs of energy absorbed by 1.0 g of tissue

An absorbed dose of 200 rads of gamma radiation will lead to symptoms of radiation poisoning. A dose of 500 rads is, of course, much more serious. Statistically, about half of the recipients would die from such a dose of radiation.

STUDY QUESTIONS

10-1. Match the following terms:

(a) α particle	(1) electromagnetic waves
(b) β particle	(2) mass of 4 amu
(c) γ ray	(3) mass of 0 amu (or approximately so)
	(4) helium nucleus
	(5) electron

10-2. Define the following terms:
(a) ${}^{40}_{19}K$ (b) ${}^{0}_{-1}e$ (c) ${}^{6}_{3}Li$ ${}^{2}_{1}H$

10-3. Complete the following equations:
(a) ${}^{214}_{83}Bi \longrightarrow {}^{214}_{84}Po + \_\_\_\_$
(b) ${}^{226}_{88}Ra \longrightarrow \_\_\_\_ + {}^{4}_{2}He$

10-4. Which of the following would you expect to be a stable isotope?
(a) carbon-14 (b) nitrogen-14
(c) sodium-24 (d) oxygen-18

10-5. Tritium (${}^{3}_{1}H$) is prepared by bombarding lithium-6 with a neutron. Write the equation for this reaction.

10-6. Write balanced nuclear equations for each of the following:
(a) loss of a β particle by ${}^{115}_{48}Cd$
(b) loss of an α particle by ${}^{220}_{86}Rn$

10-7. The half-life for radon-222 is four days. Starting with 10.0 g of radon, how much is left in 12 days?

10-8. The half-life of cobalt-60 is five years. How long would it take for a 100 g sample to decay to 25 g of cobalt-60?

10-9. An artifact of antiquity contains 6 percent as much carbon-14 as does the atmosphere. What would you estimate to be the age of the artifact?

10-10. Complete the following equations:
(a) ${}^{14}_{7}N + {}^{1}_{1}H \longrightarrow \_\_\_\_ + {}^{4}_{2}He$
(b) ${}^{7}_{3}Li + \_\_\_\_ \longrightarrow 2\,{}^{4}_{2}He$

10-11. Write the nuclear symbol for each of the following terms:
(a) deuteron
(b) uranium-235
(c) sodium-24 deuteride
(d) uranium-238

10-12. The radioactivity of a sample is measured and it is found that 500 disintegrations occur per second. What is the rate of decay in microcuries?

CHAPTER 11 INTRODUCTION TO ORGANIC CHEMISTRY

OBJECTIVES

A Short Review of Covalent Bonding *Define and describe with formulas the following: bond distance, bond angle, sigma bond, pi bond, single bond, double bond, triple bond, and aromatic pi cloud.* * *Tell how many bonds a carbon atom forms in compounds.*

Formulas for Organic Compounds *Compare the advantages of structural formulas with those of molecular formulas in organic chemistry.* * *Define and show examples of structural isomerism.* * *Recognize the difference between formulas for isomers and those for the same compound.* * *Convert structural formulas for cyclic compounds to "polygon" formulas.*

Functional Groups *Define functional group and list the structures and names of some common ones.*

Organic chemistry is the chemistry of compounds containing carbon. The term "organic" comes to us from an early misconception concerning organic compounds. At one time, it was believed that organic compounds could come only from living, or *organic,* systems. This belief was called the *Vital Force Theory.* The demise of this theory occurred when it was shown that organic compounds can be prepared from inorganic compounds in the laboratory. An experiment performed by the German chemist, Friedrich Wöhler, is generally cited as the dawn of modern organic chemistry. Wöhler was able to show that the organic compound urea could be prepared by heating the inorganic compound ammonium cyanate.

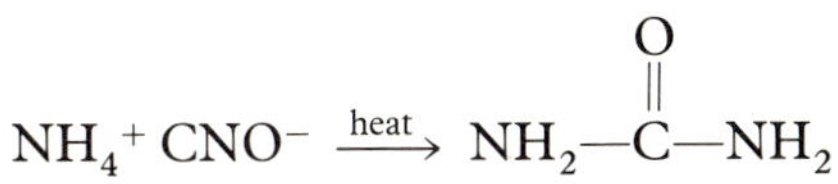

ammonium cyanate urea

What is so unique about carbon compounds that they justify a major division in the study of chemistry? The answer lies in the ability of carbon atoms to form stable covalent bonds with other carbon atoms. Carbon atoms can join together to form an almost unlimited number of types of molecules, such as short chains, long chains, branched chains, small rings, large rings, and connected rings. In addition, carbon can form stable covalent bonds with other elements, notably hydrogen, oxygen, and nitrogen. The number of different ways that these elements can be combined is beyond comprehension!

Some compounds of carbon and hydrogen:

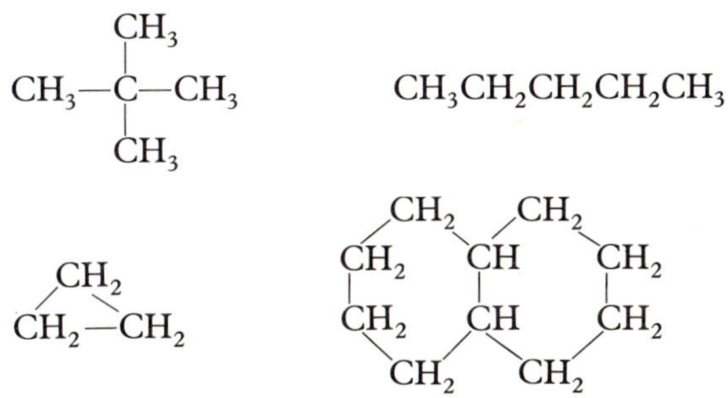

Living systems are composed primarily of organic compounds and water. Biochemistry, in many respects, is organic chemistry applied to living systems. For this reason, we must survey the structures and properties of some compounds of carbon before we can discuss the chemistry of living systems.

A SHORT REVIEW OF COVALENT BONDING

Since the chemistry of carbon compounds is the chemistry of covalently bonded compounds, we will summarize some of the important features of covalent bonds here. A review of Chapter 4, which covers covalent bonds in more detail, is also suggested.

1. Atoms held together by covalent bonds are not usually free to dissociate in solution, as are atoms joined by ionic bonds.
2. Atoms that are joined by covalent bonds are held at fixed *distances* from each other and at fixed *angles* in relation to each other.
3. There are two basic types of covalent bond: the *sigma bond* and the *pi bond.*
   (a) A single bond is a sigma bond. (See Figure 4-14.)
   (b) A double bond is composed of one sigma bond and *one* pi bond. (See Figure 4-18.)
   (c) A triple bond is composed of one sigma bond and *two* pi bonds. (See Figure 4-22.)

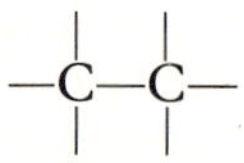

single bond

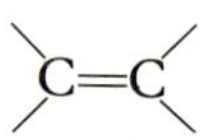

double bond

triple bond

4. Electrons in pi bonds can be delocalized. Complete delocalization results in an aromatic pi cloud. (See Figure 19-1.)

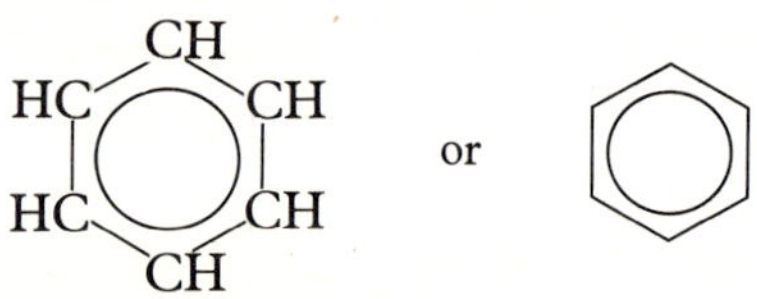

benzene

(*a compound with an aromatic pi cloud*)

5. *Carbon always forms four bonds.*

FORMULAS FOR ORGANIC COMPOUNDS

The various types of formula—molecular, structural, and condensed structural—are of great importance in even the most prosaic of organic reactions. Therefore, we will take a look at these different types of formula in relation to organic molecules.

A *molecular formula*, as you will recall, tells how many of each type of atom are in a molecule. Although molecular formulas, such as Na_2SO_4 or H_2SO_4, are used extensively in inorganic chemistry, when we enter the realm of organic chemistry we find that molecular formulas are not very useful at all. Consider the formula C_4H_{10}. This formula tells us that a molecule contains four carbon atoms and ten hydrogen atoms. The unwary might think that the formula C_4H_{10} refers to a single compound. However, there are actually *two* compounds that have the formula C_4H_{10}. In order to see this, we have to expand the formula and show the attachment of each atom to the others—that is, the *structures* of the molecules.

A *structural formula* shows the order of attachment of the atoms in a molecule. There are two common types of structural formula. The *expanded structural formula* shows all the bonds. The *condensed structural formula* gives this same information, but in an abbreviated manner.

Let us consider the structural formulas for C_4H_{10}. We can write a structural formula or a condensed structural formula for a molecule in which four carbon atoms are attached in a continuous chain. The name of this compound is butane—a familiar name because of its use as a fuel and in butane lighters.

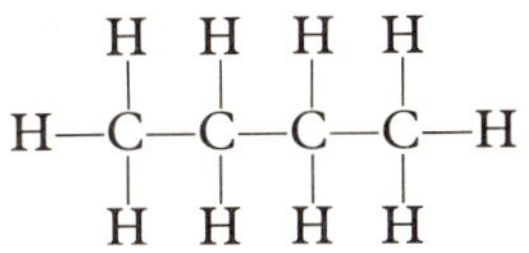

or

$CH_3CH_2CH_2CH_3$

expanded structural formula *condensed structural formula*

butane (C_4H_{10})

We can also arrange the carbon atoms so that the structure has a branched carbon chain, rather than a continuous chain. The name of this branched compound is methylpropane, named from propane ($CH_3CH_2CH_3$) with a methyl group ($—CH_3$) at the second carbon.

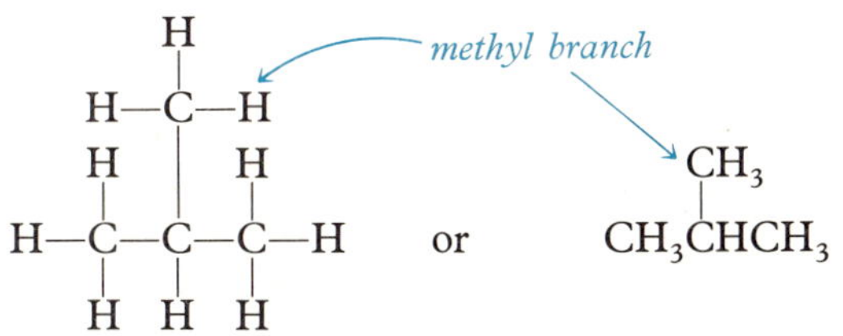

expanded structural formula *condensed structural formula*

methylpropane (C_4H_{10})
(also called isobutane)

ISOMERISM. Why is the order of attachment of atoms to each other so important? The reason is that two molecules that have the same number and types of atoms, but have the atoms attached in different orders, are different compounds with different physical and chemical properties. For example, the boiling point of butane is $-0.5\,°C$, while that of methylpropane is $-12\,°C$.

Compounds that have the same molecular formula but differ in the arrangement of their atoms are called *structural isomers* of each other. Butane and methylpropane are structural isomers. The number of isomers for a given molecular formula is dependent upon the number of atoms in the molecule. In a compound of only carbon and hydrogen with one, two, or three carbon atoms, there is only one possible way the atoms can be attached and retain the same molecular formula. There are no structural isomers for these compounds.

CH_4	CH_3CH_3	$CH_3CH_2CH_3$
methane	ethane	propane

As we have seen, the formula C_4H_{10} can represent two structural isomers. The formula C_5H_{12} represents three structural isomers.

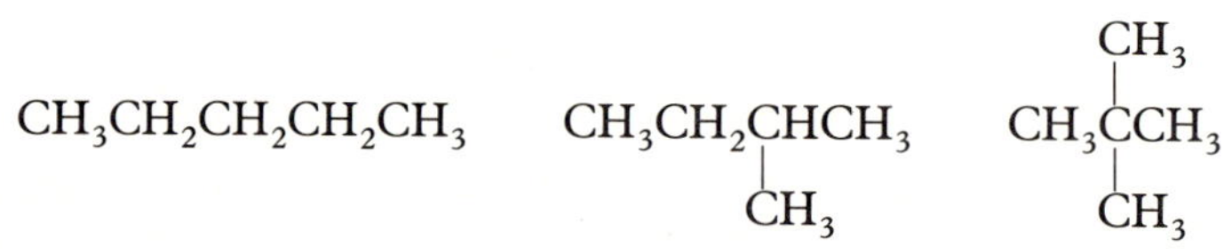

As the number of carbon atoms in a molecule increases, we find that the number of possible isomers increases at a fantastic rate. There are 75 isomers for $C_{10}H_{22}$ and 62,500,000,000,000 isomers for $C_{40}H_{82}$! The presence of an atom besides carbon (such as oxygen or nitrogen) in a molecule increases the number of isomers even more.

MOLECULES IN THREE DIMENSIONS. The structure of butane may be drawn on paper in a number of different ways. Are these differently drawn structures isomers? No, they are different projections of the same compound. In each of the three different projections, the four carbon atoms are connected to each other in a continuous chain. The order of attachment of the atoms is not changed. The projections only appear different on paper.

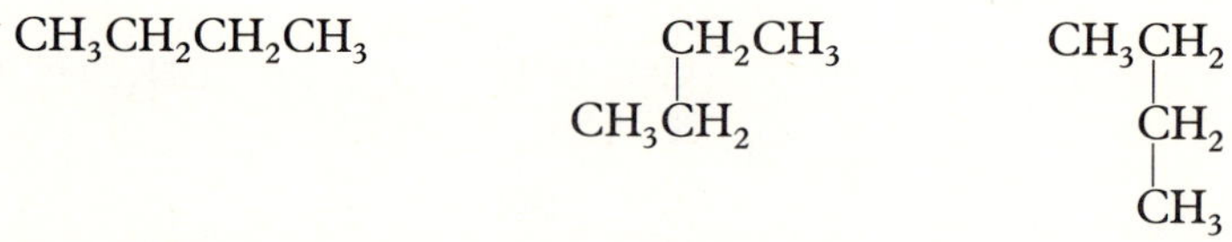

Although these structural formulas look different from each other, any one of them can be a correct representation of butane. This is because groups of atoms attached by sigma bonds can rotate with respect to each other. Two groups joined by a sigma bond might be compared to two wheels attached to an axle. The axle keeps the two wheels at a fixed distance from each other and parallel to each other, but does not prevent the wheels from rotating. The same may be said for a sigma bond—it keeps the atoms at a fixed distance from each other and the bonds at fixed angles, but does not prevent the atoms and groups from rotating around the bond. (See Figure 11-1.)

Figure 11–1. A group can rotate around a sigma bond just as a wheel can rotate around an axle.

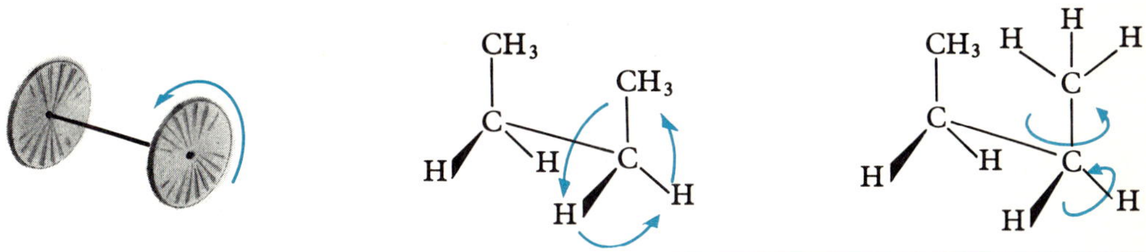

The butane molecule can move around in space, and each sigma bond in the molecule is capable of rotation. The butane molecule may assume a variety of shapes. Therefore, the point to remember in determining whether or not one structure is a structural isomer of another structure is the *order in which the atoms are joined.* If a bond must be broken to convert one structure to the other, the structures are isomers. If rotation around bonds or moving the structure as a whole is all that is needed to convert one paper structure to the other, they are not isomers.

Sample Problem: Do the following structural formulas represent the same compound or a pair of structural isomers?

```
  Cl              Cl
  |               |
ClCHCH3        CH3CH
                  |
                  Cl
```

Solution: The order of attachment of the atoms is the same. The formulas represent the same compound.

Sample Problem: In the following set of formulas, three formulas represent the same compound, while one represents an isomer. Which is the isomer?

```
                                OH
                                |
(a) CH3CHCH3           (b) CH3CHCH3
       |
       OH

(c) CH3CHOH            (d) CH3CH2CH2OH
       |
       CH3
```

Solution: In (a), (b), and (c), the —OH group is attached to the central carbon. In (d), the —OH group is attached to the end carbon. Therefore, (d) is the isomer.

CYCLIC COMPOUNDS. In drawing the formulas for cyclic organic compounds, we use a special shorthand symbol. Instead of writing the atomic symbols for the carbon atoms and hydrogen atoms connected in a ring, we use polygons. Thus, the formula for cyclohexane becomes a hexagon with each corner of the hexagon being understood to represent a carbon atom with its hydrogens. The sides of the hexagon represent the covalent bonds.

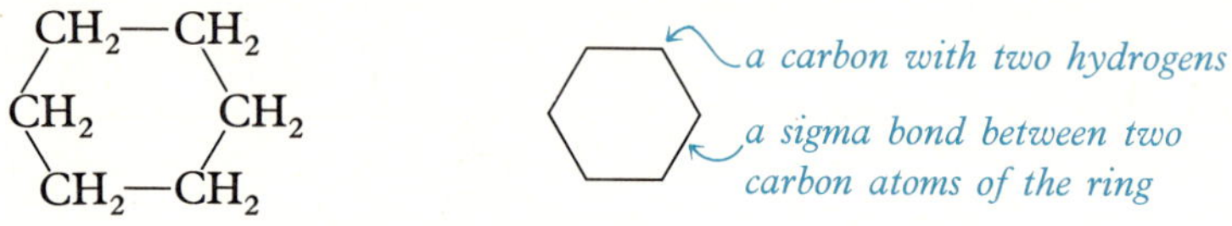

cyclohexane

This type of shorthand formula may also be used when there are other groups attached to a ring carbon. (Note that the number of hydrogens on a ring carbon with a substituent is decreased so that each carbon atom still has four bonds.)

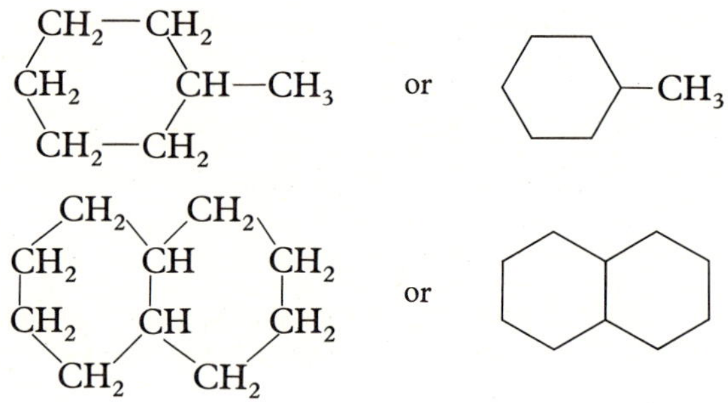

We have encountered benzene before in the discussion of pi bonds and the aromatic pi cloud. Benzene contains six carbons in a ring joined together by sigma bonds *and* the aromatic pi cloud. The aromatic pi cloud is one of the four bonds of each carbon atom of the ring. The shorthand symbol of benzene looks very similar to the shorthand symbol of cyclohexane. The two are distinguished by the fact that the symbol for benzene has a circle representing the aromatic pi cloud in the hexagon.

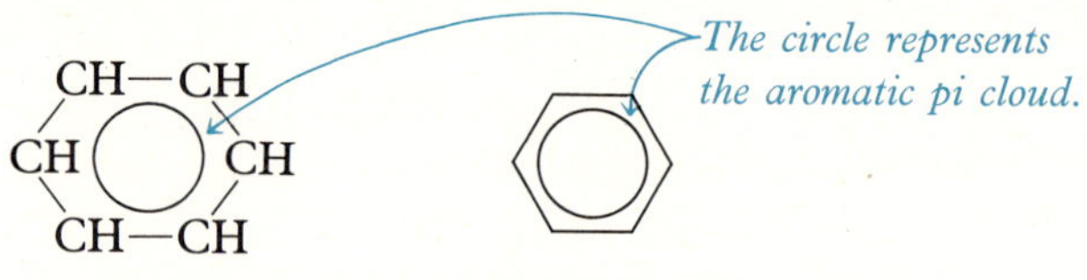

Sample Problem: What is *wrong* with the following structure?

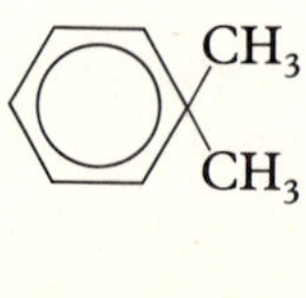

Solution: This is an impossible structure because it shows a carbon atom with five bonds: two to the —CH_3 groups, two to adjacent ring carbons, and one by its participation in the aromatic pi cloud. Remember that carbon can form only *four* bonds.

FUNCTIONAL GROUPS

Our presentation of organic chemistry will be organized according to the functional groups found in certain compounds. A *functional group* is the *site of chemical reactivity* in an organic molecule. An organic molecule may be large or small, but its functional group behaves the same as the same functional group in another molecule. When armed with a vocabulary of functional-group reactions, we will be better able to understand the diversity and similarities of the chemistry of carbon compounds. (See Table 11-1.)

These compounds contain the —OH functional group and behave similarly in chemical reactions:

CH_3OH $CH_3CH_2CH_2OH$ 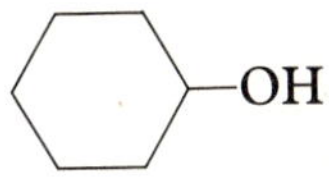

These compounds contain no functional group; they are all comparatively unreactive:

$CH_3CH_2CH_3$ 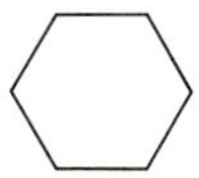 $CH_3CH_2CH_2CH(CH_3)CH_3$

Table 11-1. Some Common Functional Groups

Name of Group or Compound Class	Structure	Example
Double bond	$>C{=}C<$	$CH_3CH{=}CH_2$
Triple bond	$—C{\equiv}C—$	$CH_3C{\equiv}CH$
Chloro-	—Cl	$CH_3CH_2CH_2Cl$
Hydroxy-	—OH	$CH_3CH_2CH_2OH$
Amino-	—NH_2	$CH_3CH_2CH_2NH_2$
Carboxylic acid	$—C({=}O)—OH$	$CH_3CH_2C({=}O)OH$

STUDY QUESTIONS

11-1. What is the principal difference between an ionic bond and a covalent bond? Give an example of each.

11-2. Differentiate between a sigma bond and a pi bond.

11-3. Describe the three types of hybridization of orbitals that carbon undergoes when it forms covalent bonds with other atoms. Given an example of a compound that illustrates each case.

11-4. Which of the following compounds have aromatic pi clouds?

(a) 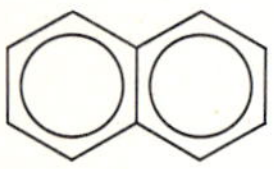 (b) 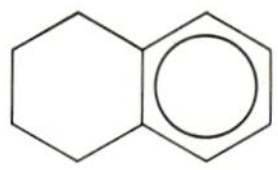

(c) 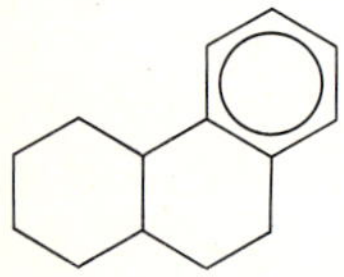 (d) 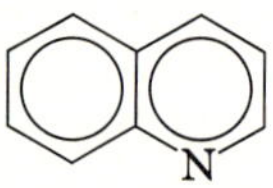

11-5. Write each of the following as a molecular formula, an expanded structural formula, and a condensed structural formula (whichever are not shown).

(a) 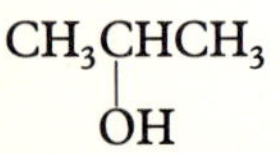 (b) 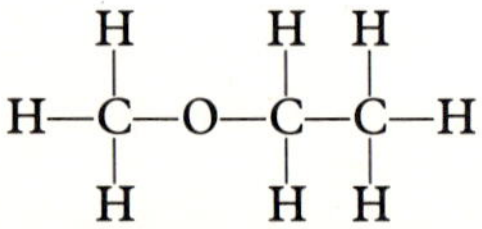

(c) $CH_3CH_2CH_2OH$ (d) C_3H_8O

11-6. Draw one structural isomer for each of the following compounds. (In each case there may be more than one possible isomer.)

(a) CH_3OCH_3 (b) 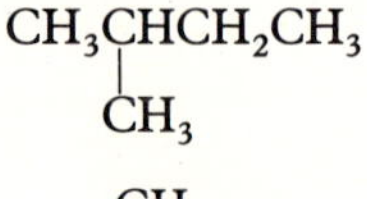

(c) CH_3CHCl_2 (d) 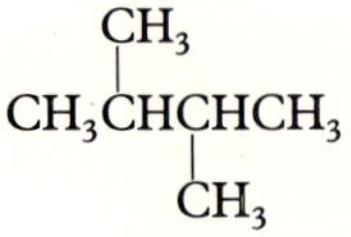

11-7. Which of the following compounds has no structural isomers?

(a) $CH_3CH_2CH_3$ (b) $CH_3CHClCH_3$

(c) CH_3OH (d) 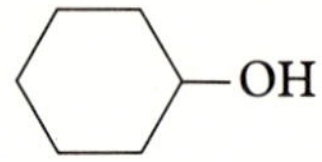

11-8. Which of the following pairs are structural isomers and which are different representations of the same compound?

(a) 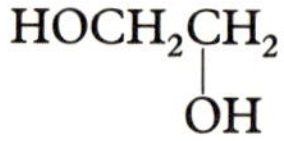 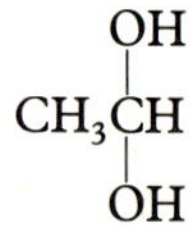

(b) 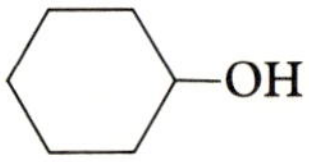 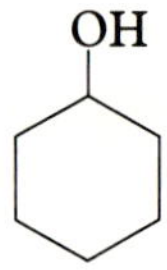

(c) 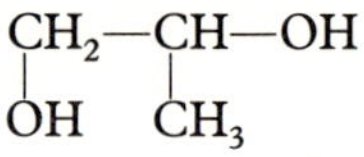 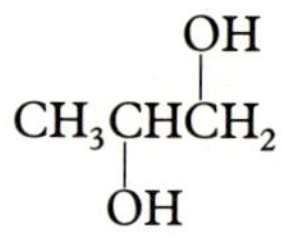

(d) 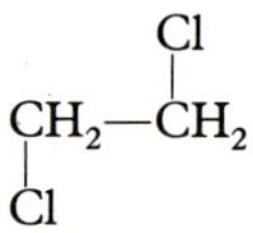 $ClCH_2CH_2Cl$

11-9. In the following set of formulas, three formulas represent the same compound, while one represents a structural isomer. Which is the isomer?

(a) 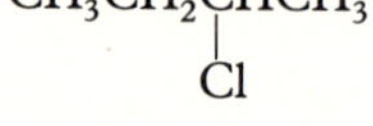 (b) 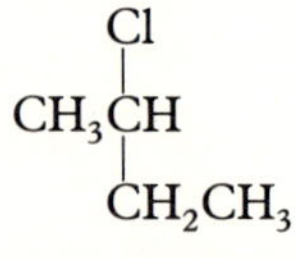

(c) 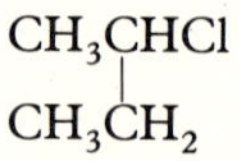 (d) 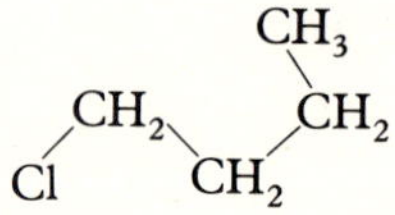

11-10. Draw the shorthand symbol or the complete condensed structural formula (whichever is not given) for each of the following cyclic structures:

(a) 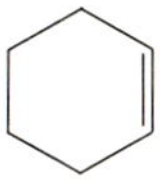

(b) 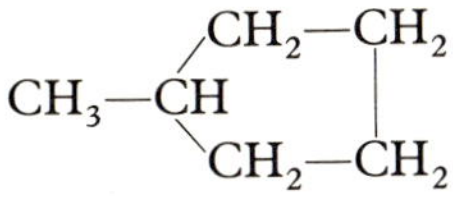

(c) 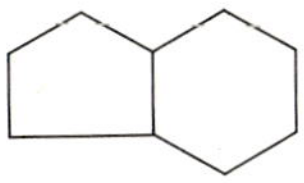

(d) 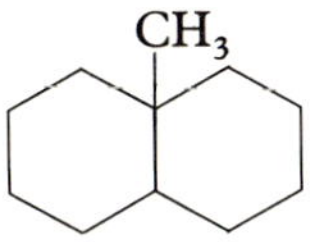

(e) 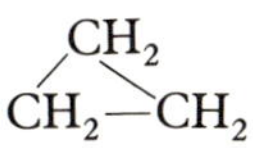

(f) 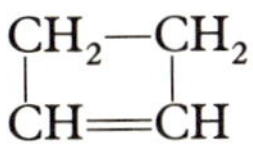

(g) 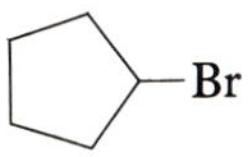

(h) 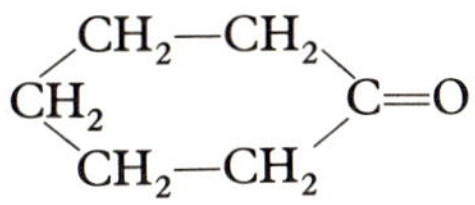

(i) 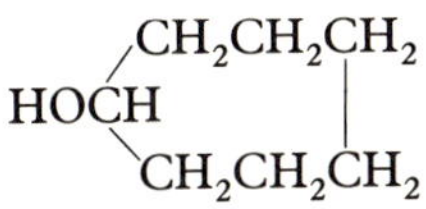

11-11. What is *wrong* with each of the following structures?

(a) 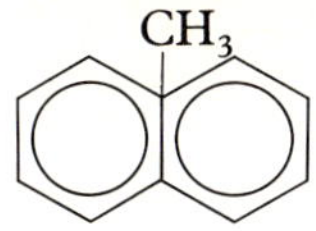

(b) 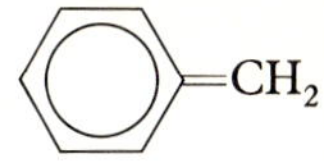

(c) 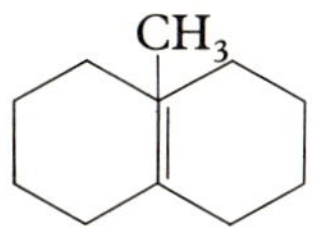

11-12. Identify the functional group in each of the following compounds:

(a) 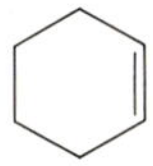

(b) $CH_3CH_2\overset{O}{\overset{\|}{C}}OH$

(c) 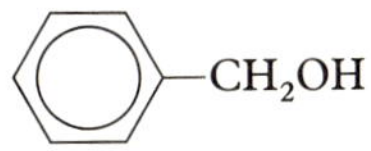

(d) $CH_3CH_2\overset{NH_2}{\overset{|}{C}}HCH_2CH_3$

11-13. Use a four-carbon compound to illustrate each of the common functional groups given in Table 11-1 (page 225).

11-14. Keeping the same functional group, give the structure of a structural isomer of each of your examples in the preceding problem.

CHAPTER 12 ALKANES

OBJECTIVES

Hydrocarbons *Write structural formulas to exemplify a hydrocarbon, a saturated compound, an unsaturated compound, a cycloalkane, and an aliphatic compound.*

Naming Alkanes and Cycloalkanes *List the names for the continuous-chain alkanes containing one through ten carbon atoms.* * *Name the cycloalkanes containing three through ten carbon atoms in their rings.* * *Name an alkane that has attached a continuous-chain alkyl group, a common branched alkyl group, and more than one branch.*

Reactions of the Alkanes *Describe the reactivity of alkanes toward common laboratory reagents, toward halogens in the presence of ultraviolet light, and toward oxygen when ignited. Write equations where appropriate.*

Petroleum *Describe briefly the processes involved in refining crude oil.* * *Define compression ratio, antiknock fuel, and octane number.* * *Define cracking.*

Natural Gas *List the various sources of methane, both commercial and experimental.*

HYDROCARBONS

The *hydrocarbons* are compounds that contain carbon and hydrogen and no other element. Hydrocarbons may be classified as *saturated* and *unsaturated* hydrocarbons. A saturated hydrocarbon is one that contains only single bonds and no pi bonds or aromatic pi clouds.

Typical saturated hydrocarbons:

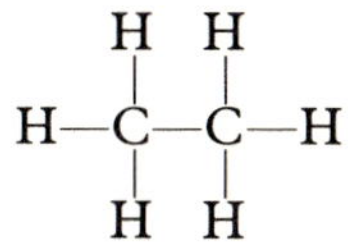

ethane

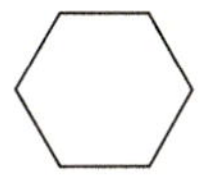

cyclohexane

Compounds that contain pi bonds or aromatic pi clouds are said to be *unsaturated.* An unsaturated hydrocarbon has a functional group—that is, the double bond, the triple bond, or the aromatic pi cloud.

Typical unsaturated hydrocarbons:

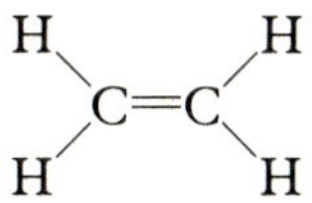

ethylene

$H{-}C{\equiv}C{-}H$

acetylene

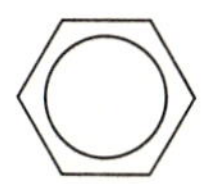

benzene
an aromatic hydrocarbon

The term "saturated" is derived from the idea that a compound is *saturated with hydrogen.* That is, a saturated hydrocarbon will not react with hydrogen. An unsaturated hydrocarbon can react with hydrogen to become saturated.

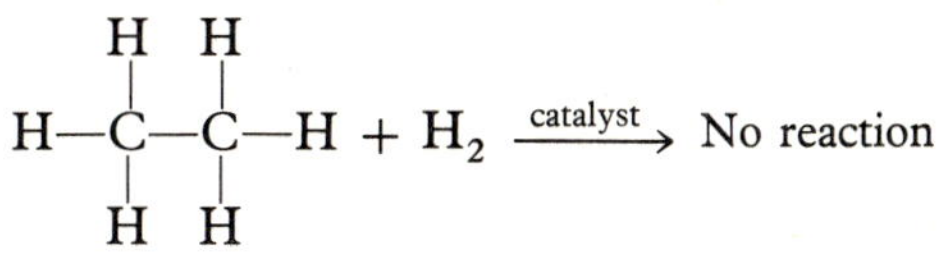

ethane
a saturated hydrocarbon

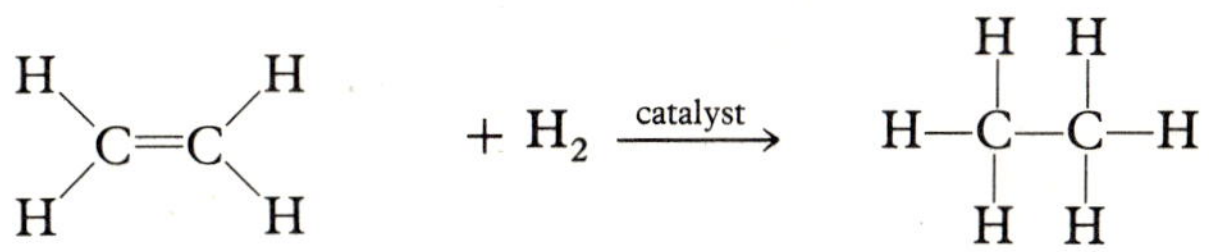

ethylene
an unsaturated hydrocarbon

ethane
a saturated hydrocarbon

In this chapter, we will discuss the *saturated hydrocarbons*. This group may be further divided into compounds in which the carbon atoms are connected together in chains, called *alkanes*, and those in which the carbon atoms are connected together to form rings, called *cycloalkanes*.

Some alkanes:

$CH_3CH_2CH_2CH_3$ $\qquad$ CH_4

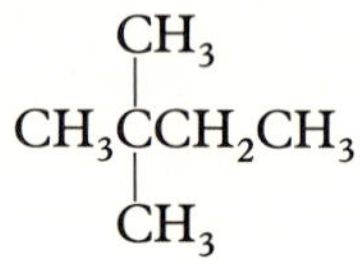

Some cycloalkanes:

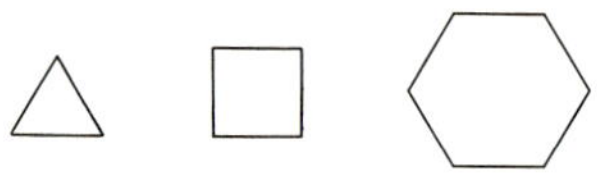

NAMING ALKANES AND CYCLOALKANES

In the 19th century, before chemists had learned the structures of many compounds, a new organic compound was named after its source, after some chemical or physical property, or even after a friend or relative. Such names are called *trivial names*.

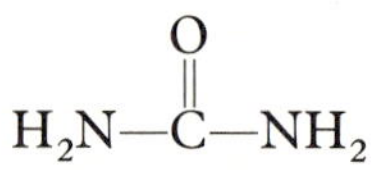

urea
found in urine

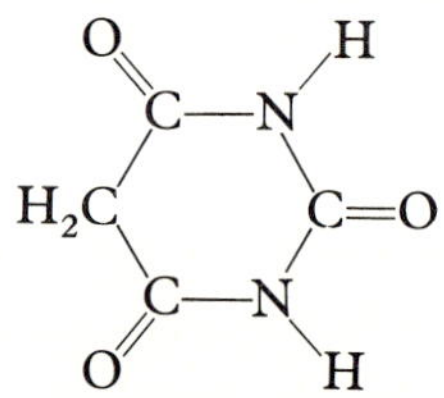

barbituric acid
from "Barbara"

Before 1900, chemists realized that some system of nomenclature should be devised to give the thousands of organic compounds names based upon molecular structure. The idea was that each organic compound must have an unambiguous name derived from its structure, and that the system should be designed so that any new compound could also be given a name. The nomenclature system that was developed is now known as the *IUPAC system*, after the International Union of Pure and Applied Chemistry, the group that is responsible for the continual upgrading and revising of the system. Although chemists still use many trivial names (and we will also use them in this book), the IUPAC system forms the foundation for almost all organic nomenclature.

The IUPAC system is based on the premise that each organic compound has a basic skeletal structure, called the *parent*, to which hydrocarbon branches and functional groups have been added. In the following compound, the branches and functional group are encircled. The parent structure is in a box. In naming a compound, the branches and functional groups are specified by prefixes and suffixes added to the name of the parent structure.

Branches and functional groups on a parent skeleton:

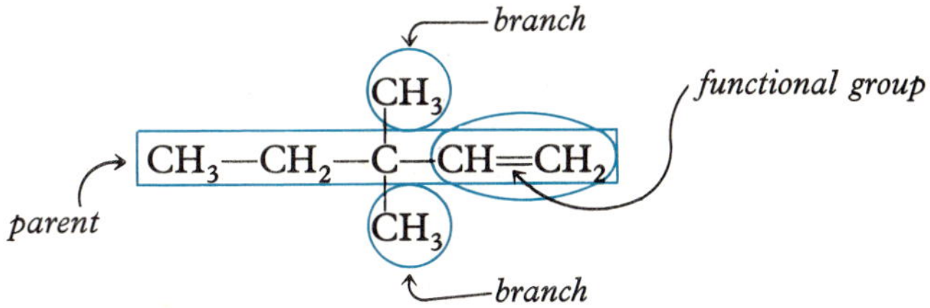

Sample Problem: Circle and label the functional groups and branches in the following structure and draw a box around the parent.

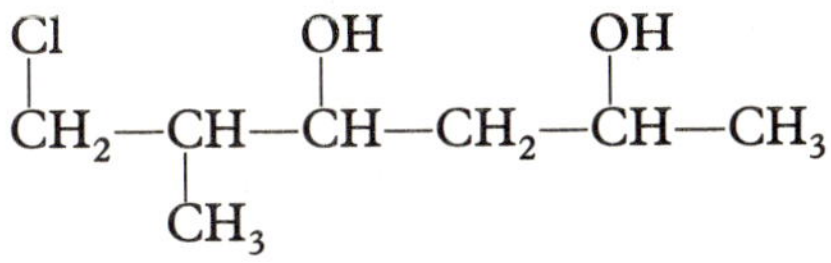

Solution:

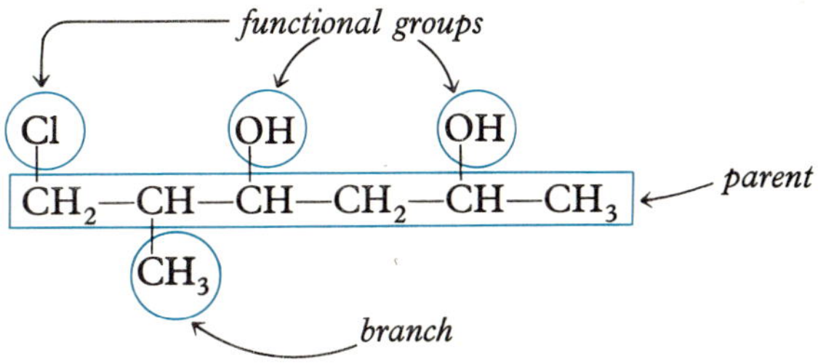

The names of the continuous-chain alkanes are used as parent names for nonaromatic compounds, whether or not they contain branches or functional groups. The names for the first ten continuous-chain alkanes are given in Table 12-1.

The names of the alkanes are composed of two parts. The first part of the name tells us the number of carbon atoms in the chain; the second part tells us that the compound is saturated. For example, the name *pentane* starts with the parent *pent-*, indicating the presence of five carbon atoms, and ends with the suffix *-ane*, indicating that the compound is a saturated hydrocarbon, or an alkane. (The names of unsaturated hydrocarbons have different endings.)

Table 12–1. Names and Structures of the First Ten Continuous-Chain Alkanes

Name	Number of Carbon Atoms in Chain	Structure
Methane	1	CH_4
Ethane	2	CH_3CH_3
Propane	3	$CH_3CH_2CH_3$
Butane	4	$CH_3CH_2CH_2CH_3$ or $CH_3(CH_2)_2CH_3$
Pentane	5	$CH_3CH_2CH_2CH_2CH_3$ or $CH_3(CH_2)_3CH_3$
Hexane	6	$CH_3(CH_2)_4CH_3$
Heptane	7	$CH_3(CH_2)_5CH_3$
Octane	8	$CH_3(CH_2)_6CH_3$
Nonane	9	$CH_3(CH_2)_7CH_3$
Decane	10	$CH_3(CH_2)_8CH_3$

CYCLOALKANES. The names of the ring compounds are taken directly from the names of the continuous-chain parents with the prefix *cyclo-* added. The number of carbon atoms in the ring determines the parent name. For example, cyclohexane is a cycloalkane with six carbons in a ring. Cyclooctane is a cycloalkane with eight carbons in a ring.

cyclohexane

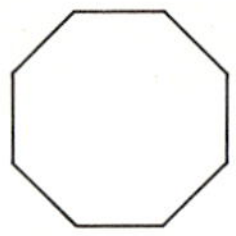

cyclooctane

BRANCHED ALKANES. The parent of a branched-chain alkane is the *longest continuous chain.* The branches are designated by prefixes added to the parent name. When naming a branched-chain alkane, (a) *find and name the longest continuous chain of carbon atoms,* and then (b) *number that chain starting at the end closer to the branch.*

Let us illustrate these rules with a specific alkane, $CH_3CH_2CH_2CH(CH_3)_2$. This structure has a total of six carbons, but only five of the carbon atoms are in one continuous chain. The structure is named as a pentane.

Now the five-carbon chain must be numbered. We start at the end closer to the branch.

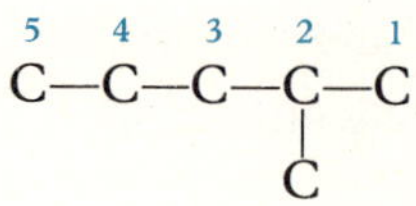

not

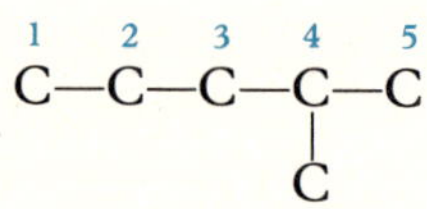

The one-carbon branch at carbon number two is called a methyl group (*meth-*, indicating one carbon; *-yl*, indicating that it is attached to something else). The complete name of this compound is 2-methylpentane. The number 2 designates the position of the branch; *methyl* designates what group is at that position; and *pentane* is the parent alkane. The position number is separated by a hyphen from the rest of the name; "methylpentane" is one word.

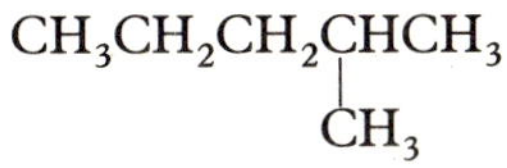

2-methylpentane

The names of the continuous-chain branches that are commonly encountered are listed in Table 12-2. In each case, the name of the group is taken directly from the name of the continuous-chain parent hydrocarbon with the *-ane* ending of the alkane replaced by *-yl*. These groups are referred to as *alkyl groups*.

Table 12-2. Names of Continuous-Chain Alkyl Groups

Name	Structure
Methyl	CH_3—
Ethyl	CH_3CH_2—
Propyl	$CH_3CH_2CH_2$—
Butyl	$CH_3CH_2CH_2CH_2$—
Pentyl	$CH_3CH_2CH_2CH_2CH_2$—
Hexyl	$CH_3(CH_2)_4CH_2$—

BRANCHED ALKYL GROUPS. Alkyl substituents may have branching within themselves. A few branched alkyl groups are so common that they have specific names. (See Table 12-3.) The commonly used prefixes for these branched alkyl groups are *iso-*, *secondary-* (abbreviated *sec-*), and *tertiary-* (abbreviated *t-*). Frequently, to differentiate a continuous-chain alkyl group from a branched one, the prefix *normal* (abbreviated *n-*) is used to name a continuous-chain alkyl group. For example, *n*-hexyl means $CH_3CH_2CH_2CH_2CH_2CH_2$—. (A small *n* must be used and not a capital N. Capital N denotes nitrogen.) The prefix *n-* is actually redundant since the absence of any other prefix implies a continuous chain.

Sample Problem: Draw the structures for (a) 4-propylheptane (which could also be called 4-*n*-propylheptane), and (b) isobutylcyclopentane.

Solution:

(a) 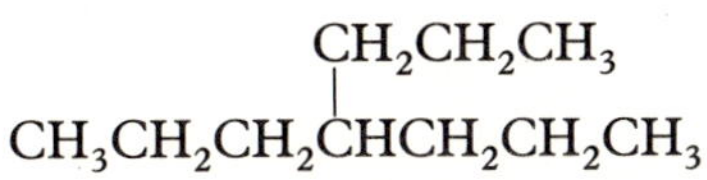 (b) 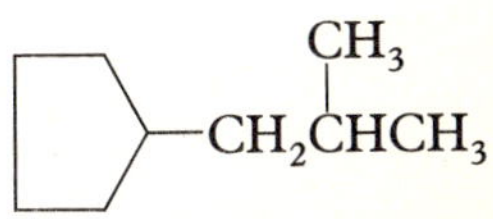

Table 12–3. Names of Common Branched Alkyl Groups

Name	Structure	Example
Isopropyl	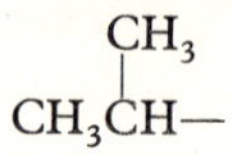	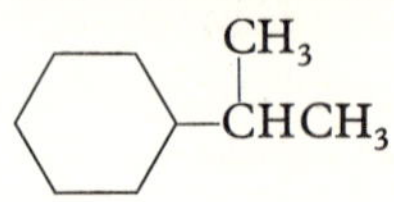 Isopropylcyclohexane
Isobutyl	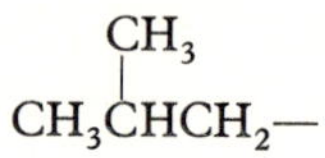	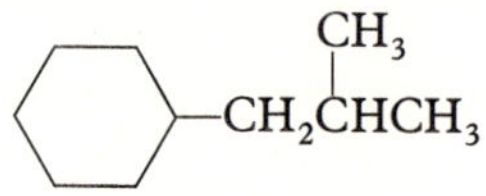 Isobutylcyclohexane
sec-Butyl	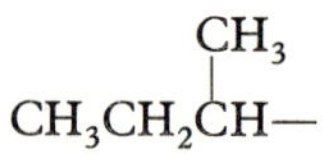	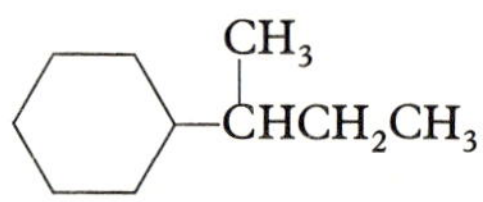 *sec*-Butylcyclohexane
t-Butyl	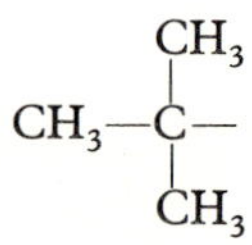	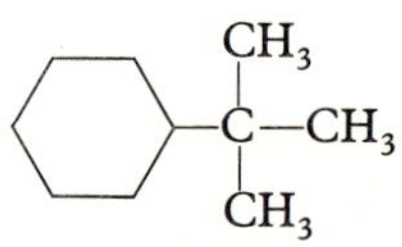 *t*-Butylcyclohexane

COMPOUNDS WITH MORE THAN ONE SUBSTITUENT. When a compound contains two or more of the same substituent, the prefix is preceded by *di-*, *tri-*, and so forth. This additional prefix tells us the number of times that a substituent is found in the molecule. For example, 2,3-dimethylbutane is a four-carbon continuous chain (butane) that has two (*di*) methyl groups located at carbons 2 and 3, respectively. (See Table 12–4.)

Table 12–4. Some Prefixes Used to Indicate the Number of Times a Substituent Is Found

Total Number of the Same Group	Prefix Used to Indicate that Number
Two	Di-
Three	Tri-
Four	Tetra-
Five	Penta-
Six	Hexa-

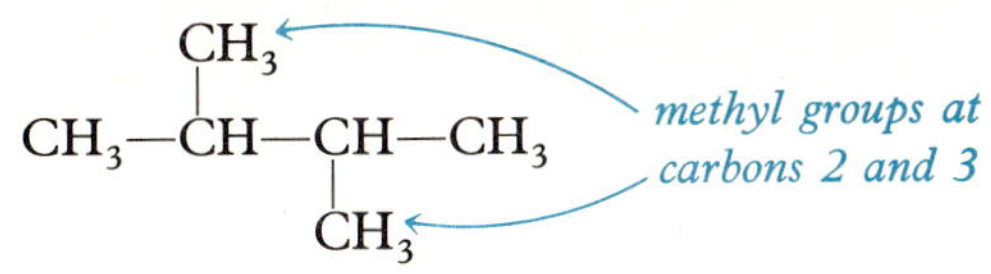

2,3-dimethylbutane

In naming complex alkanes, (a) *be careful to find the longest continuous chain* (the chain may or may not be written in a straight line); (b) *number the chain, starting at the end that will give the smallest prefix numbers;* and (c) *group more than one of the same prefix together.*

Sample Problem: Give the IUPAC name for the following:

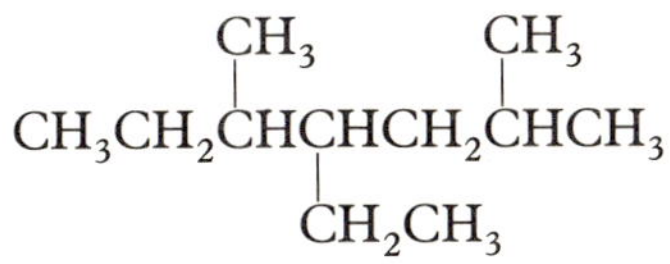

Solution: *Step 1:* Number the longest continuous chain from the end that will give the lower prefix numbers.

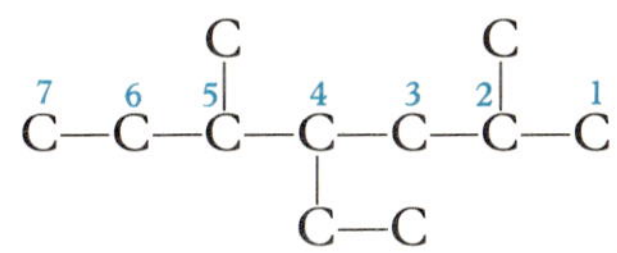

Step 2: Name the parent.

heptane

Step 3: Add the branches, grouping like branches together.

4-ethyl-2,5-dimethylheptane

Sample Problem: Write the structure for 1-isopropyl-3-methylcyclopentane.

Solution: *Step 1:* Write the structure of the parent (cyclopentane), and number the carbon atoms.

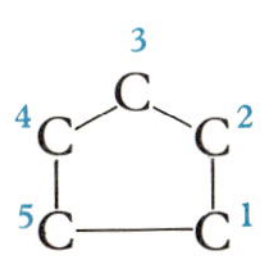

Step 2: Insert the branches.

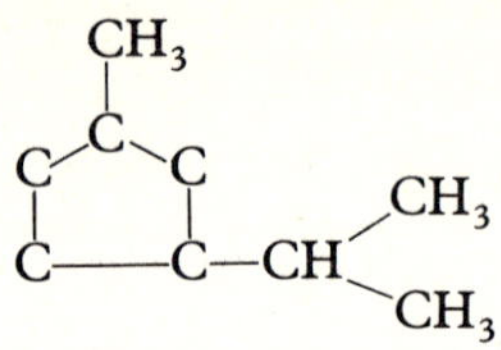

Step 3: Insert the hydrogens so that each carbon has four bonds.

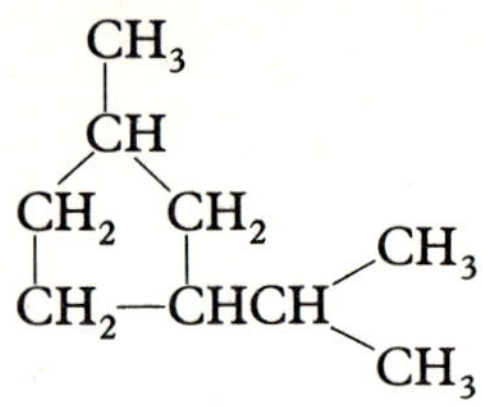

REACTIONS OF THE ALKANES

For the most part, alkanes are chemically unreactive. They do not react with acids, bases, reducing agents, and many oxidizing agents.

$$CH_3CH_2CH_2CH_2CH_3 + H_2SO_4 \longrightarrow \text{No reaction}$$
$$CH_3CH_2CH_2CH_2CH_3 + NaOH \longrightarrow \text{No reaction}$$
$$CH_3CH_2CH_2CH_2CH_3 + H_2 \longrightarrow \text{No reaction}$$
$$CH_3CH_2CH_2CH_2CH_3 + Na \longrightarrow \text{No reaction}$$
$$CH_3CH_2CH_2CH_2CH_3 + KMnO_4 \longrightarrow \text{No reaction}$$

Because the alkanes are immiscible with water and are nonreactive, they are widely used as protective coatings. Lubricating oils, among other purposes, help protect automobile parts from rusting. Petroleum jelly is used to protect skin from moisture. Paraffin wax is used to seal homemade jams and jellies. All these substances are mixtures of alkanes.

Alkanes will react with the halogens. These reactions lead to a mixture of products. We will discuss this type of halogenation reaction further in Chapter 14.

$$CH_4 + Cl_2 \xrightarrow{\text{ultraviolet light}} CH_3Cl + CH_2Cl_2 + CHCl_3 + CCl_4 + HCl$$

Although alkanes will burn in air, they do so only if ignited.

$$\underset{\substack{\text{methane} \\ \textit{in natural gas}}}{CH_4} + 2\,O_2 \xrightarrow{\text{spark}} CO_2 + 2\,H_2O + \text{Energy}$$

PETROLEUM

Our modern society runs on energy, and we obtain much of the energy that we need from petroleum fuels. Petroleum has been formed by the anaerobic decay (decay without air) of organic materials originally buried in sediment, but now found in underground deposits. Crude oil, or petroleum, as it comes from oil wells is a fantastically complex mixture of organic compounds: straight-chain and branched-chain hydrocarbons, saturated and unsaturated hydrocarbons, cyclic hydrocarbons, and aromatic hydrocarbons. In addition, crude oil contains sulfur-, nitrogen-, and oxygen-containing organic compounds. The relative amounts of all the different compounds in crude oil vary widely and depend upon the source of the oil.

Because it is a mixture of so many components, crude oil as it comes from the ground is not particularly useful. To obtain usable products, the crude oil is subjected to a process called *refining,* in which the crude oil is separated by distillation into fractions of different boiling ranges. After distillation, the fractions may be subjected to further distillation, extraction, or chemical modification in order to enrich the proportions of the more useful components.

Direct distillation of crude oil is called *straight-run distillation.* The principal fractions from this distillation are gases, gasoline, kerosene, gas oil, and the residue. The residue is distilled under vacuum to yield lubricating oils and petroleum waxes. The residue of the vacuum distillation is used primarily as asphalt.

GASOLINE. Gasoline is a complex mixture of hydrocarbons boiling in the range of 0° to 200°C. Because of the nature of our gasoline engines, commercial gasoline is often modified chemically, and additives, such as tetraethyllead, $(CH_3CH_2)_4Pb$, are added in order to increase the quality of the gasoline.

The efficiency of a gasoline engine increases with an increase in the *compression ratio* of the engine, that is, the ratio of the initial volume of the gas-air mixture as it enters a piston to the final volume to which it is compressed by the piston. At low compression ratios (4 to 1), gasoline burns evenly. At higher compression ratios (7.5–10 to 1), gasoline burns evenly at first, but then the rate speeds up toward the end of the burn. The increased rate of burn is accompanied by an increased rate of pressure increase, which causes shock waves. These shock waves we hear as knocking in an automobile engine. Knocking decreases the efficiency of an engine, and also increases wear and tear on the engine. Thus, a high compression ratio means a more efficient engine, but also means more knocking.

Increases in the compression ratios of gasoline engines have been made practical by the development of *antiknock* fuel. This is a gasoline that has been modified to burn evenly, even in an engine with a high compression ratio. The term we use in comparing the antiknock properties of gasolines is the *octane number.*

The octane number of gasoline is based upon an empirical scale. Heptane, a continuous-chain alkane, has very poor burning characteristics and causes a substantial amount of knocking. This compound has been assigned an octane number of 0. At one time, the branched-chain alkane 2,2,4-trimethylpentane (trivial name: isooctane) was the best antiknock fuel known. This compound was assigned an octane number of 100.

$CH_3(CH_2)_5CH_3$

heptane

octane number 0

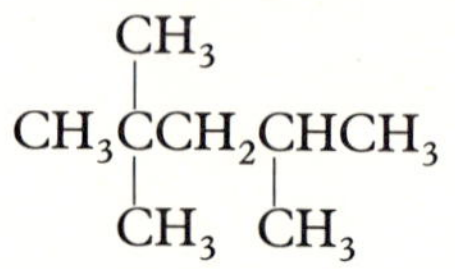

2,2,4-trimethylpentane (isooctane)

octane number 100

To rate an ordinary gasoline, its burning characteristics are compared to mixtures of these two compounds. A gasoline that has an octane number of 80 has the same knocking characteristics as a mixture of 20 percent heptane and 80 percent isooctane. Straight-run gasolines without any additives have octane numbers that vary between 24 and 80, depending upon the source of the crude oil.

Straight-run distillation of crude petroleum does not yield enough of the gasoline fractions to supply the needs of society. A large part of the distillate is composed of less useful, higher-boiling components of too high a molecular weight to use as gasoline. However, petroleum chemists have discovered methods to cleave these larger, less useful molecules into smaller molecules. These processes are called *cracking* processes.

A cracking reaction:

$$\underset{\text{heptane}}{CH_3(CH_2)_5CH_3} \xrightarrow[\text{catalyst}]{\text{heat, pressure}} \underset{\text{pentane}}{CH_3(CH_2)_3CH_3} + \underset{\text{ethylene}}{CH_2{=}CH_2}$$

NATURAL GAS

Natural gas, which is primarily methane (CH_4), is used to heat a large percentage of homes in the United States. Like petroleum, natural gas has been formed by the anaerobic decay of organic materials and is often found with petroleum deposits. Because the world's supply of natural gas is limited, other sources of gaseous fuels are being investigated. In some communities, sewage treatment plants are powered by methane from the anaerobic decomposition of the sewage. Similarly, some experimental farms are heated by methane from the decomposition of manure.

Combustible gases can also be obtained from coal (primarily carbon) in processes called *coal gasification*. The reaction of coal and steam yields a low-quality fuel containing the noncombustible gas carbon dioxide plus the combustible gases carbon monoxide (CO) and hydrogen (H_2). Methane, a higher quality fuel, can be obtained from this mixture by the catalytic combination of carbon monoxide and hydrogen.

$$\underset{\text{steam}}{\text{coal} + H_2O} \longrightarrow CO_2 + \underbrace{CO + H_2}_{\xrightarrow{\text{catalyst}} CH_4 + H_2O}$$

Alternatively, coal can be treated directly with hydrogen to yield methane.

$$\text{coal} + H_2 \longrightarrow CH_4$$

Unfortunately, all these sources of gaseous fuel are expensive. Current research is aimed at finding an *economical* technique of converting coal to methane.

STUDY QUESTIONS

12-1. Classify each of the following hydrocarbons as an alkane, a cycloalkane, a nonaromatic but unsaturated hydrocarbon, or an aromatic hydrocarbon:

(a) 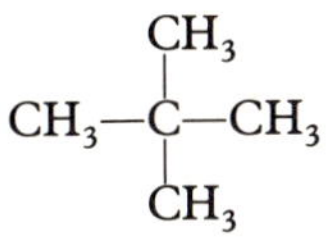 (b) 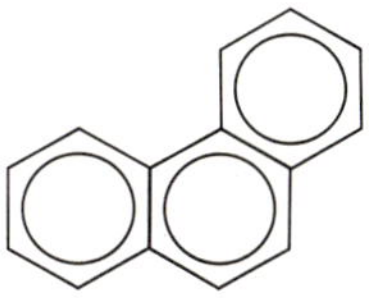

(c)  (d) 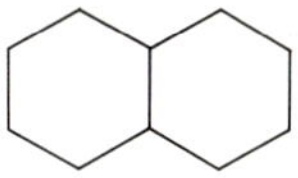

12-2. Using six carbon atoms, give structures for (a) a saturated hydrocarbon, (b) an unsaturated hydrocarbon, and (c) an aromatic hydrocarbon.

12-3. Would you expect paraffin wax to undergo reaction with hydrogen in the presence of a catalyst?

12-4. Circle the functional groups and draw boxes around the parent skeletons:

(a) 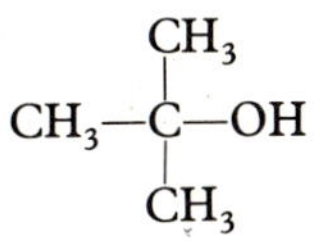 (b) 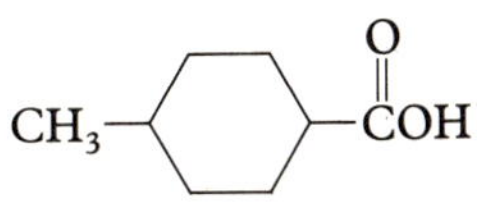

(c) $CH_3CH_2CH_2NH_2$ (d) 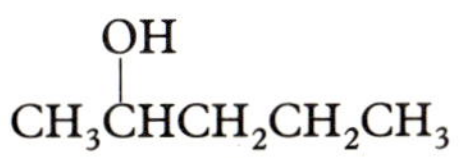

12-5. From memory, write the structure and name for each of the first ten straight-chain hydrocarbons.

12-6. Name the following cycloalkanes:

(a) 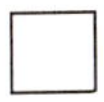 (b) 

(c) 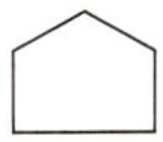 (d) 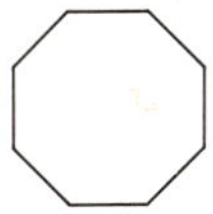

12-7. In each of the following carbon skeletons, give the number of carbon atoms in the longest continuous chain:

(a) 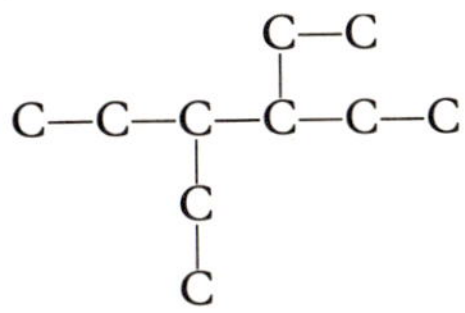

(b) 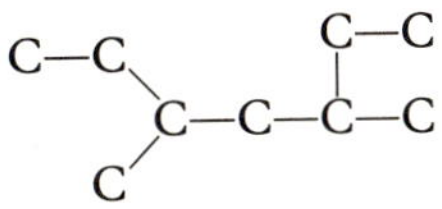

(c) 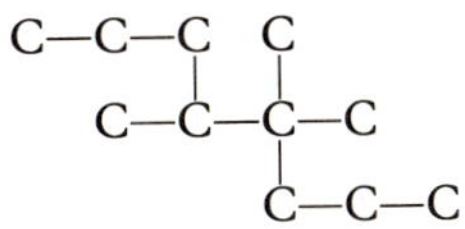

(d) 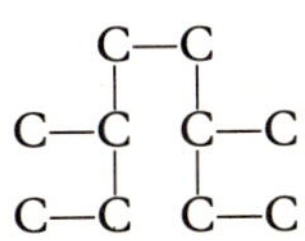

12-8. Which of the following structures contains an "ethyl" in its name?

(a) 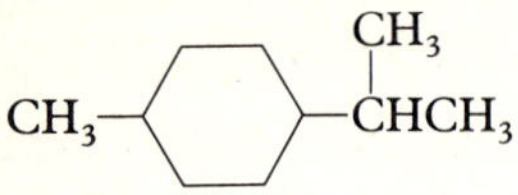

(b) 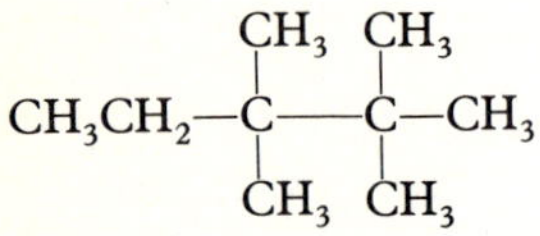

(c) 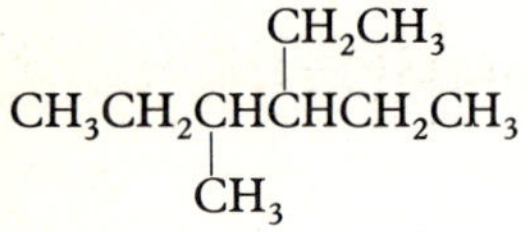

(d) 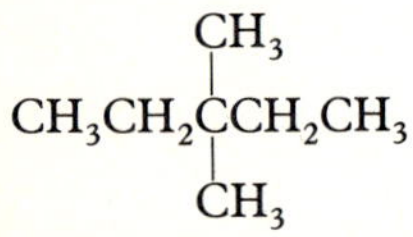

12-9. Name the branched alkyl group attached to each of the following ring systems:

(a) 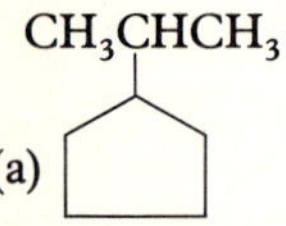

(b) 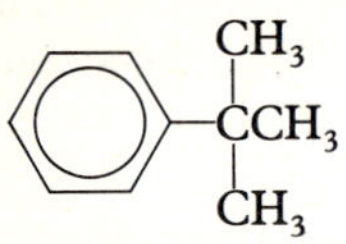

(c) 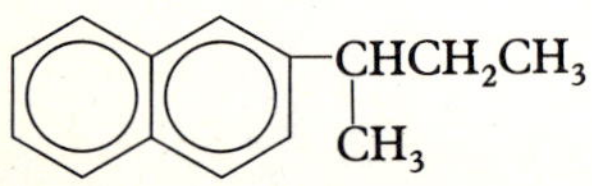

(d) 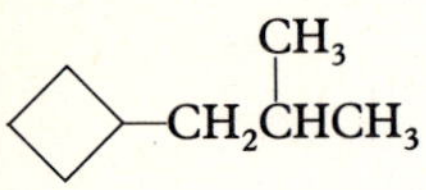

12-10. Give the IUPAC name for each of the following compounds:

(a) 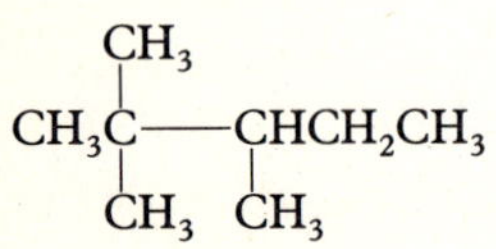

(b) 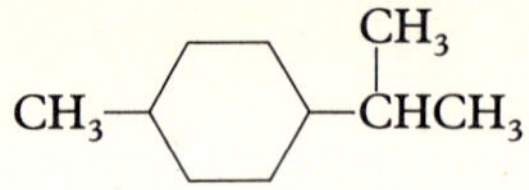

(c) 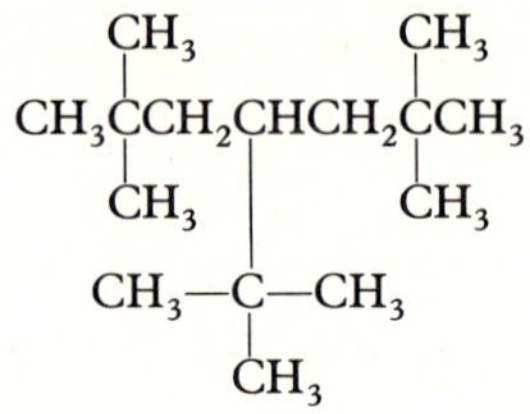

(d) 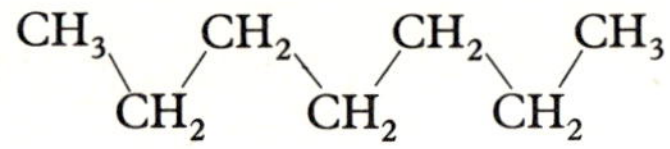

(e) 

12-11. Draw the structure for each of the following:

(a) 1,2,3-tri-*t*-butylcyclobutane
(b) 2,2,3,3-tetramethylbutane
(c) *n*-hexylcyclohexane
(d) 3-methyl-4-propyloctane
(e) 2-cyclobutylheptane

12-12. Complete and balance the following equations (if no reaction occurs, write *No Reaction*):

(a) $CH_4 + H_2SO_4 \longrightarrow$

(b)  $+ H_2 \xrightarrow{\text{catalyst}}$

(c) $CHCl_3 + Cl_2 \xrightarrow{\text{ultraviolet light}}$

(d) $CH_3CH_2CH_2CH_3 + O_2 \xrightarrow{\text{spark}}$

(e)  $+ H_2 \xrightarrow{\text{catalyst}}$

CHAPTER 13 ALKENES AND ALKYNES

OBJECTIVES

Naming the Alkenes and Alkynes *Name alkenes and alkynes by the IUPAC system.*

Geometric Isomerism *Explain why the presence of a pi bond or a ring creates the possibility of geometric isomerism.* * *Give examples of* cis- *and* trans-*isomers.*

Reactions of the Alkenes and Alkynes *Give equations for the additions of hydrogen, halogens, hydrogen halides, and water to alkenes.* * *Predict the products of unsymmetrical addition reactions.* * *Give equations for the oxidation of alkenes with mild oxidizing agents (such as cold $KMnO_4$ solution)* and stronger oxidizing agents (*such as hot $KMnO_4$ solution*). * *Define polymer, describe the naming of polymers, and give equations for the preparation of some polymers.*

Unsaturation in an organic molecule is a site of reactivity in the molecule; that is, a double bond or a triple bond is a functional group, the first of many that will be covered in this and subsequent chapters. Compounds containing double bonds are found so universally in nature and are used so widely in industrial processes that it would be impossible to list the useful and interesting compounds that contain this functional group. A thumbing through of the remainder of the book will show that many, many compounds with double bonds are found in plants and animals—for example, vegetable oils and steroids (Chapter 24) and vitamins and drugs (Chapter 30). The sex attractant of the housefly is an example of a biologically active hydrocarbon containing a double bond as the only functional group.

$$CH_3(CH_2)_{12}CH{=}CH(CH_2)_7CH_3$$

sex attractant of the housefly

Hydrocarbons that contain a double bond as the only functional group are called *alkenes,* or *olefins,* while those that contain a triple bond as the sole functionality are called *alkynes,* or *acetylenes.* Since the reactivity of these compounds depends for the most part on the reactivity of the pi bond, these two classes of compounds have many reactions in common and thus will be discussed together.

NAMING THE ALKENES AND ALKYNES

The nomenclature of the alkenes and alkynes is a direct extension of the nomenclature for the alkanes. In an alkene, the ending *-ane* of the alkane name is changed to *-ene.* Ethene is the IUPAC name for a two-carbon compound containing a double bond.

An alkyne is named by changing the *-ane* ending of the alkane name to *-yne.* Ethyne is the systematic name for the two-carbon compound with a triple bond.

CH_3CH_3	$CH_2{=}CH_2$	$CH{\equiv}CH$
ethane	ethene	ethyne
an alkane	*an alkene*	*an alkyne*
$CH_3CH_2CH_3$	$CH_3CH{=}CH_2$	$CH_3C{\equiv}CH$
propane	propene	propyne
an alkane	*an alkene*	*an alkyne*

In chains longer than three carbon atoms, there is more than one possibility for the position of unsaturation. To distinguish between the possible structural isomers, a number preceding the name is used to tell where the double or triple bond is located in the parent skeleton. The carbon atoms of the parent skeleton are numbered starting from the end *closer to the double or triple bond* in order to give the lowest prefix numbers. The number is separated from the parent name by a hyphen.

$$\overset{1}{C}H_3\overset{2}{C}H{=}\overset{3}{C}H\overset{4}{C}H_2\overset{5}{C}H_2\overset{6}{C}H_3 \quad \text{and not} \quad \overset{6}{C}H_3\overset{5}{C}H{=}\overset{4}{C}H\overset{3}{C}H_2\overset{2}{C}H_2\overset{1}{C}H_3$$

2-hexene

Note that, although the double bond (or triple bond) connects *two* carbons, the prefix number is the number of the carbon at which the double bond *begins*.

$$\overset{1}{C}H_3\overset{2}{C}H_2\overset{3}{C}H{=}\overset{4}{C}H\overset{5}{C}H_2\overset{6}{C}H_2\overset{7}{C}H_2\overset{8}{C}H_3$$

double bond joins carbons 3 and 4

3-octene

A compound with two double bonds is called a *diene*, while one with three double bonds is called a *triene*. In naming one of these compounds, *di-* or *tri-* is used in the name just before the *-ene* ending.

$$\overset{1}{C}H_2{=}\overset{2}{C}H{-}\overset{3}{C}H{=}\overset{4}{C}H_2$$

1,3-butadiene

The first double bond starts at carbon 1, and the second starts at carbon 3.

two double bonds

four carbons

When branches are attached to the parent skeleton, they are handled in the same way as they were in the case of the alkanes. The parent chain of a branched alkene or alkyne is the *longest continuous chain that contains the functional group*. The numbers and names of any branches go before the entire alkene or alkyne name, with hyphens separating any numbers from the word parts of the name.

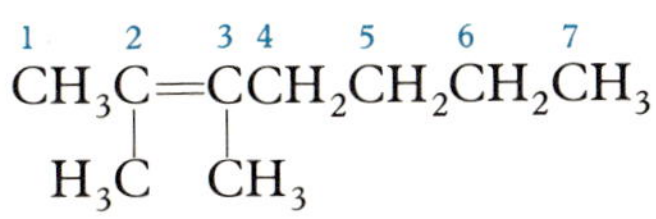

2,3-dimethyl-2-heptene

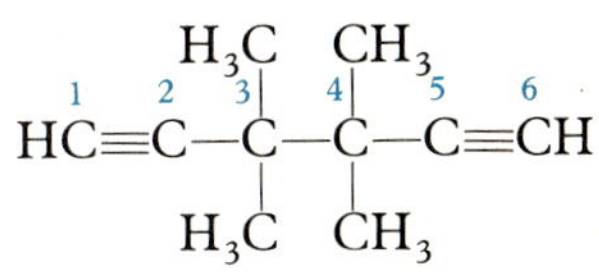

3,3,4,4-tetramethyl-1,5-hexadiyne

Sample Problem: Name the following alkene:

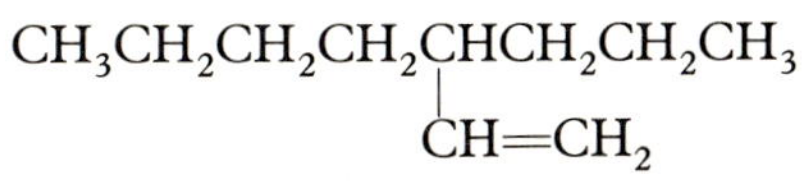

Solution: *Step 1:* Find the longest continuous chain that contains the double bond, and number the chain, starting at the end closer to the double bond.

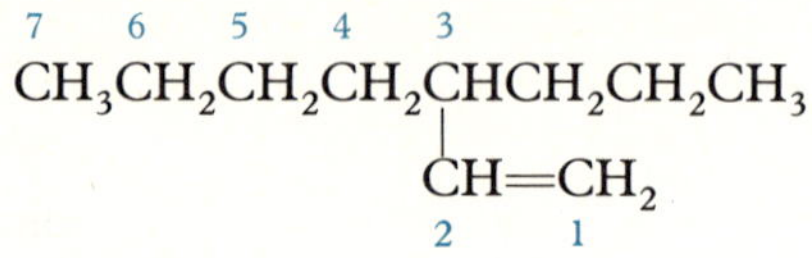

Step 2: Name the parent alkene and insert the number of the position of the double bond.

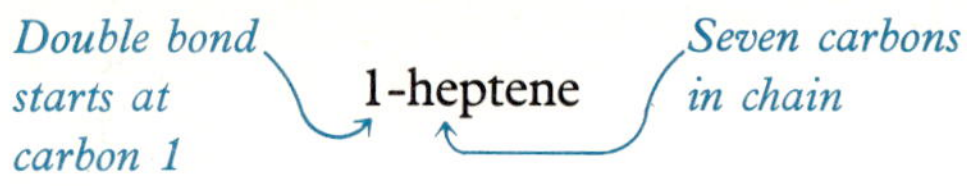

Step 3: Add the name and number for the branch.

3-propyl-1-heptene

Sample Problem: What is the structure of 3-ethyl-4-methyl-1,3-pentadiene?

Solution: *Step 1:* Write and number the parent chain—five carbons, in this case.

$$\overset{1}{C}-\overset{2}{C}-\overset{3}{C}-\overset{4}{C}-\overset{5}{C}$$

Step 2: Insert the double bonds at the proper positions.

$$\overset{1}{C}=\overset{2}{C}-\overset{3}{C}=\overset{4}{C}-\overset{5}{C}$$

Step 3: Add the branches.

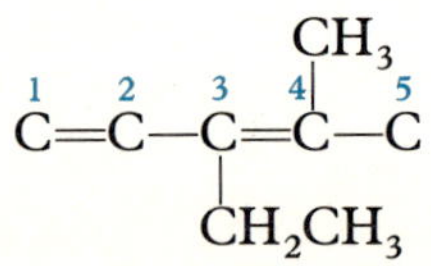

Step 4: Fill in the hydrogens so that each carbon atom has a total of four bonds.

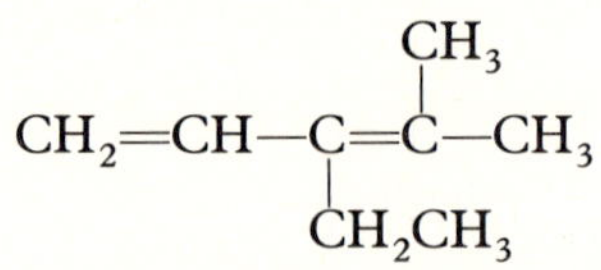

In many cases, the trivial names of alkenes and alkynes are the ones usually encountered. Although any chemist would know that ethene is $CH_2{=}CH_2$, this compound is usually called ethylene. The commonly used name for $CH{\equiv}CH$ is acetylene, not ethyne.

GEOMETRIC ISOMERISM

Recall from Chapter 4 that the carbon atoms involved in a pi bond are in the sp^2-hybrid state. Each of the two carbons has three sigma bonds and a p orbital that is used in the pi bond. The three sigma bonds of each carbon atom form a plane, and the p orbital is at a right angle to this plane. The pi bond then is formed from the side-to-side overlap of the two p orbitals. This overlap is contrasted to that of the sigma bonds, which are formed from the end-to-end overlap of orbitals.

Carbons that are bonded by sigma bonds are free to rotate around the bond. Rotation of groups connected by pi bonds, however, is not possible. If the groups were to rotate, the pi bond would break. (See Figure 13-1.) Under normal circumstances, there is not enough energy available to a molecule for its pi bonds to break. Since groups bonded by a pi bond cannot rotate, the pi-bonded portion of the molecule is held rigidly in position.

In the following two structures, note that, in the first case, the two hydrogens are on the same side of the pi bond. In the second case, the hydrogens are on opposite sides.

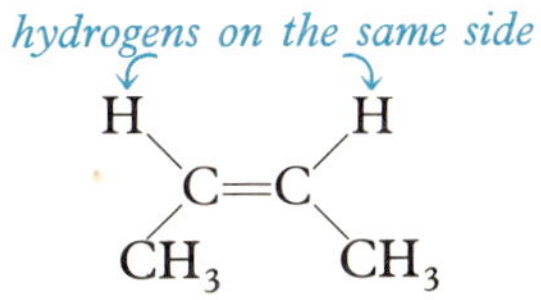

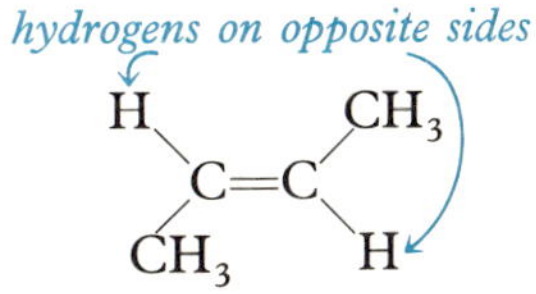

Because the groups adjacent to the pi bond are held rigidly, the pair of compounds cannot be interconverted. The interconversion would involve rotation about the pi bond and its breakage. Thus, the two compounds are isomers of each other. Because the *order* of attachment of the atoms is the same, the two compounds are not structural isomers. However, the geometric relationship of the groups around the pi bond is different. This pair of isomers is thus called a pair of *geometric isomers.*

Geometric isomers of simple alkenes are differentiated from each other by the prefixes *cis-* and *trans-*. The prefix *cis* means that hydrogen atoms are on the *same* side of the pi bond. The prefix *trans* means that they are on *opposite* sides. (Of course, when

Figure 13-1. Rotation around a pi bond would result in the breaking of the pi bond.

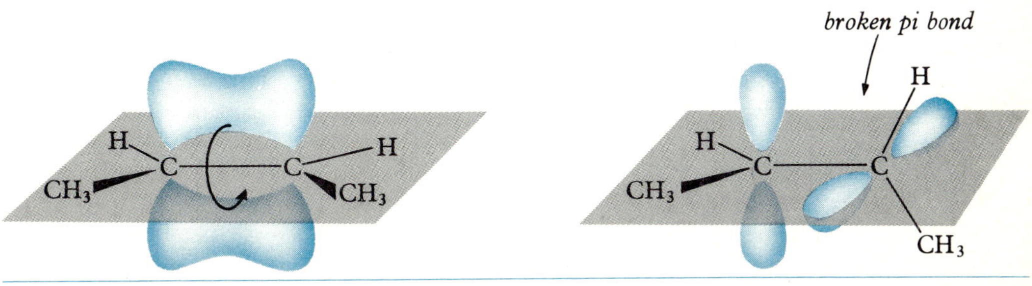

two hydrogen atoms are *cis* to each other, the other substituents, such as methyl groups, are also *cis* to each other. When the two hydrogen atoms are *trans,* the other substituents are also *trans.*)

cis means hydrogens on the same side of the pi bond:

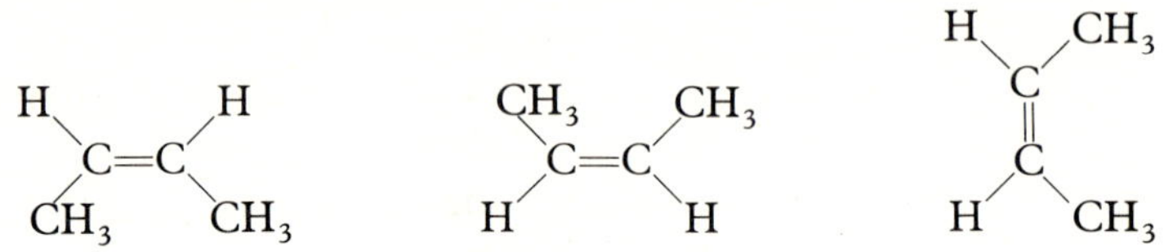

cis-2-butene

bp 3.8°C

trans means hydrogens on opposite sides of the pi bond:

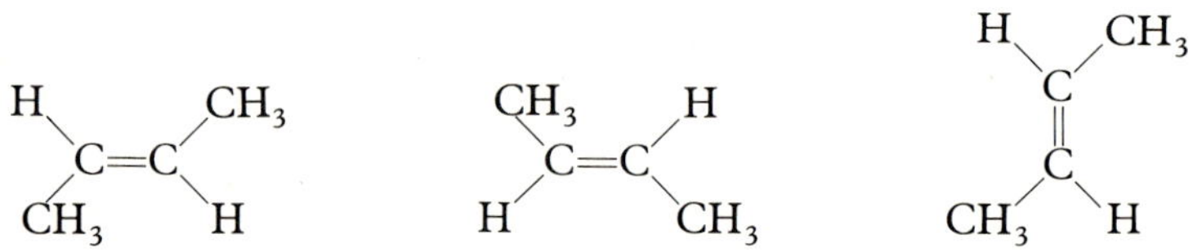

trans-2-butene

bp 0.9°C

Sample Problem: Write structural formulas and label the geometric isomers of 2-hexene.

Solution:

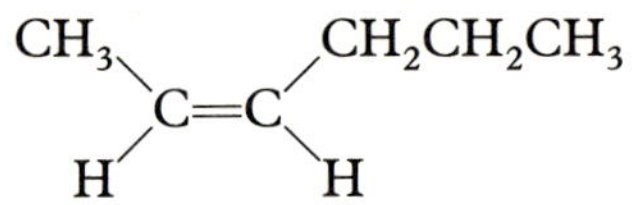

cis-2-hexene

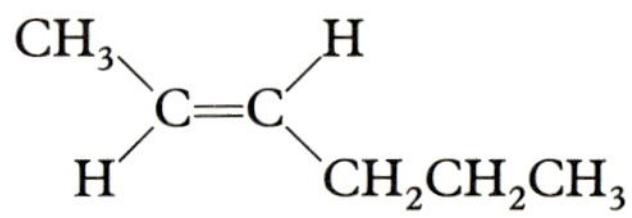

trans-2-hexene

Not all alkenes are capable of geometric isomerism. In order to have geometric isomers, the two carbons of the double bond must each have two different groups attached.

These two structures are the same compound, and not cis- and trans-isomers, because one carbon of the double bond has two groups (hydrogen) that are the same.

Sample Problem: Which of the following compounds have geometric isomers?

(a) 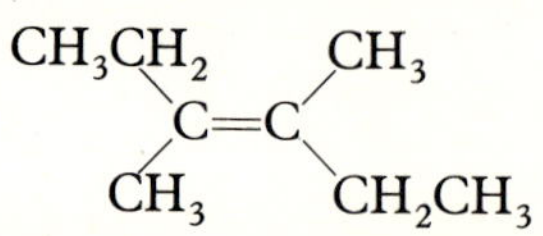 (b) 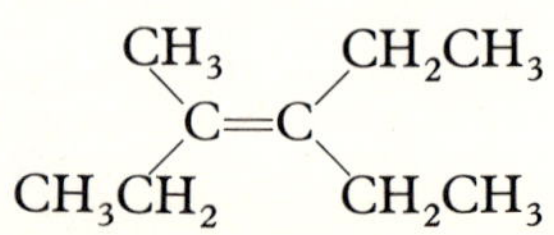 (c) 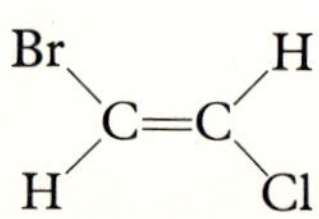

Solution: In (a) and (c), each pi-bonded carbon has two different groups attached; these are capable of geometric isomerism. In (b), one of the pi-bonded carbons has two ethyl groups attached; this compound has no geometric isomer.

One other class of organic compounds that can have *cis*- and *trans*-isomers is the cyclic compounds. In a cycloalkane such as cyclopropane, the ring carbons cannot rotate around their sigma bonds without breaking the ring. Therefore, when two groups are attached to two different ring carbon atoms, they can be on the *same* side of the ring (*cis*-) or on *opposite* sides of the ring (*trans*-). (In the following structures, the dashed lines represent bonds beneath the plane of the ring; solid lines represent those above the plane.)

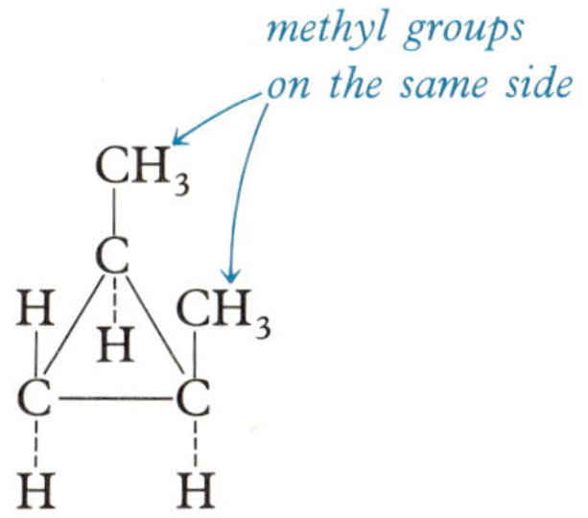

cis-1,2-dimethylcyclopropane

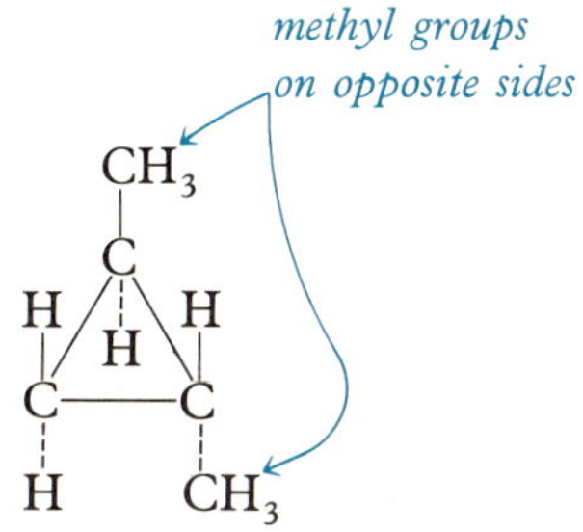

trans-1,2-dimethylcyclopropane

Table 13-1 gives some other examples of *cis*- and *trans*-isomers of cycloalkanes. In these examples, the hydrogen atoms are not shown, but are merely implied.

Table 13-1. Examples of Geometric Isomerism of Cyclic Compounds

cis-Isomer	*trans*-Isomer
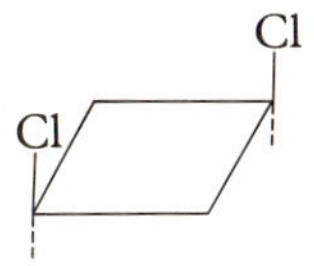	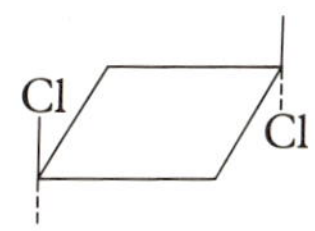
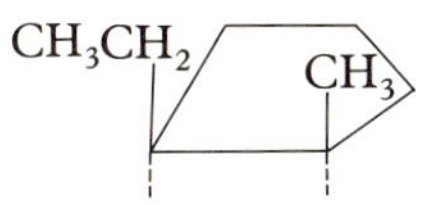	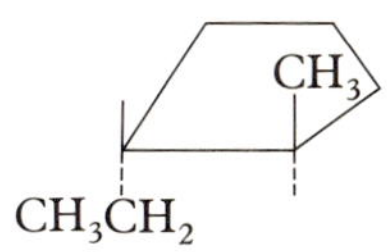
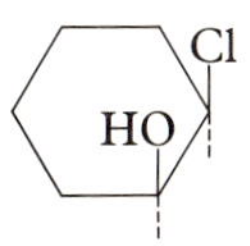	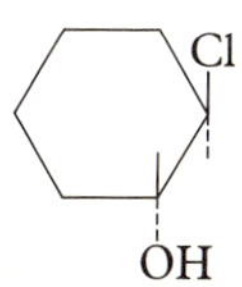

Sample Problem: Label each of the following compounds as being either a *cis*-isomer or a *trans*-isomer:

(a) 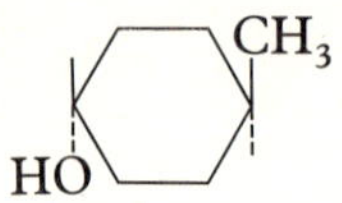

(b) 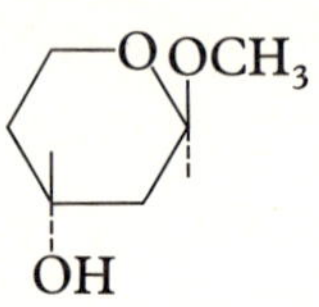

(c) 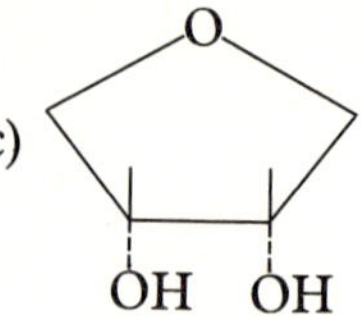

Solution: (a) *trans,* (b) *trans,* (c) *cis*

Geometric isomerism plays an important role in many biochemical processes. One interesting example involves reception of light by the retina of the eye. The retina contains a compound called *rhodopsin,* or *visual purple,* which is responsible for black-and-white vision and for vision in dim light.

Rhodopsin consists of a relatively small molecule (11-*cis*-retinal) that just fits in a pocket of a larger protein molecule (opsin). When light hits the 11-*cis*-retinal molecule in rhodopsin, the 11-*cis* double bond is converted to a *trans* double bond. This conversion changes the shape of the retinal molecule so that it can no longer fit in the opsin pocket. The all-*trans*-retinal is therefore released from the opsin and, in the process, triggers a nerve impulse that results in our being able to see.

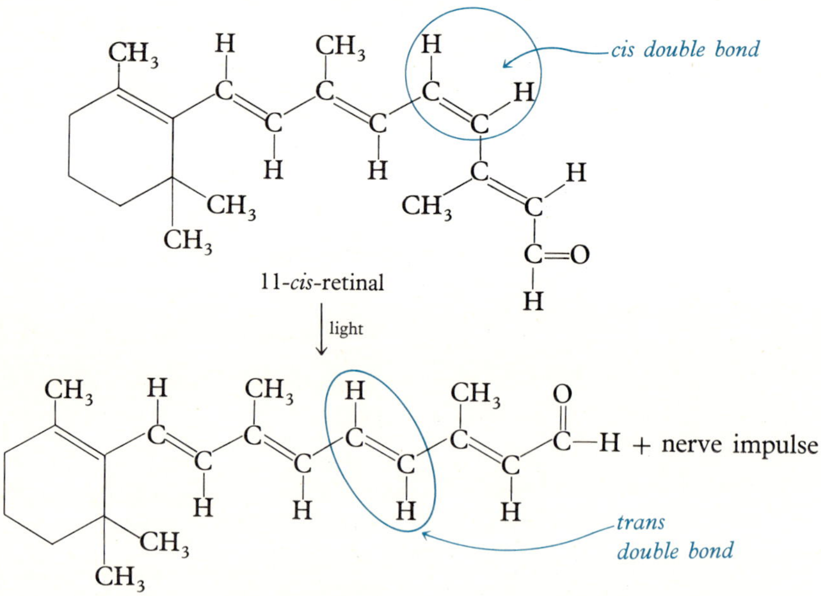

all-*trans*-retinal

Retinal is synthesized in the eye from vitamin A (see page 476). One of the symptoms of vitamin A deficiency is night blindness, which results from a lack of rhodopsin.

REACTIONS OF THE ALKENES AND ALKYNES

The characteristic chemical property of a compound containing a carbon-carbon pi bond is that reagents add across the pi bond. The pi bond is broken in the reaction, and two new sigma bonds, one to each carbon of the pi bond, are formed.

HYDROGENATION. Hydrogen gas can add across a pi bond. The reaction, called *hydrogenation*, is an oxidation-reduction reaction. The alkene is reduced to an alkane. Hydrogen is acting as a reducing agent and is oxidized in the reaction.

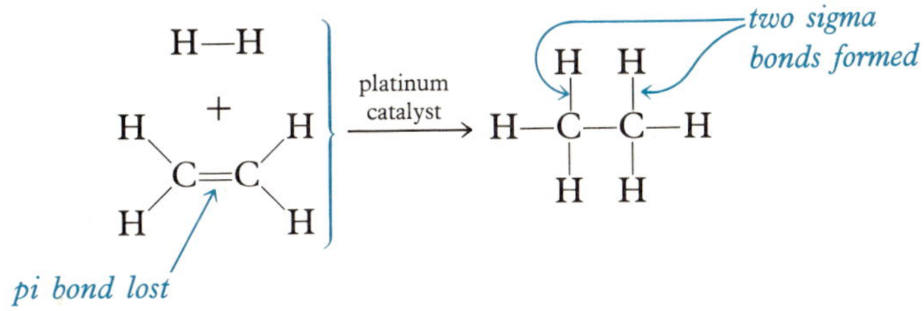

Hydrogenation reactions are general for alkenes:

$$CH_3CH{=}CHCH_3 + H_2 \xrightarrow{Pt} CH_3CH_2CH_2CH_3$$

2-butene (*cis* or *trans*) — butane

$$CH_3CH{=}CHCH{=}CHCH_3 + 2\,H_2 \xrightarrow{Pt} CH_3CH_2CH_2CH_2CH_2CH_3$$

2,4-hexadiene — hexane

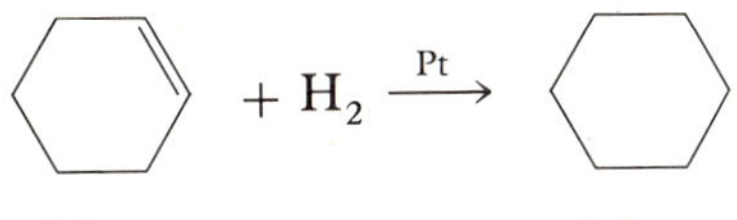

cyclohexene — cyclohexane

Alkynes contain two pi bonds. Two moles of hydrogen gas per mole of alkyne are needed for complete hydrogenation.

$$CH{\equiv}CH + 2\,H_2 \xrightarrow{Pt} CH_3CH_3$$

acetylene — ethane

A hydrogenation reaction does not proceed without the aid of a catalyst. Platinum (Pt) is a very effective catalyst for the reaction of alkenes and alkynes with hydrogen. The catalyst functions by adsorbing the hydrogen gas onto its surface. The association between H_2 and the metal catalyst weakens the hydrogen-hydrogen sigma bond and lowers the energy of activation necessary for the reaction with the alkene.

REACTION WITH THE HALOGENS. Bromine and chlorine readily add across the pi bond of an alkene or alkyne. No catalyst is needed in this reaction.

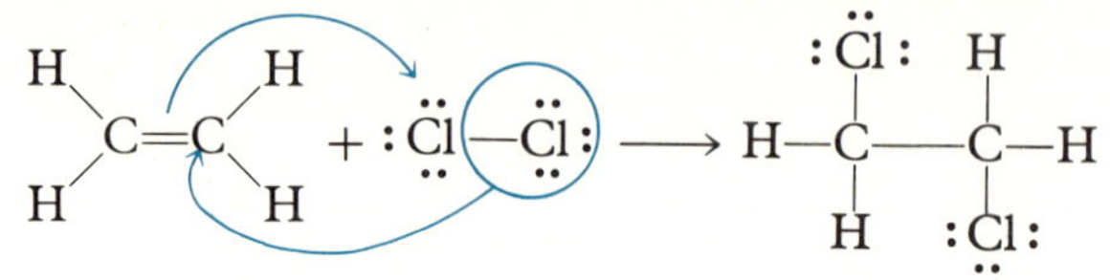

This reaction is general for alkenes and alkynes:

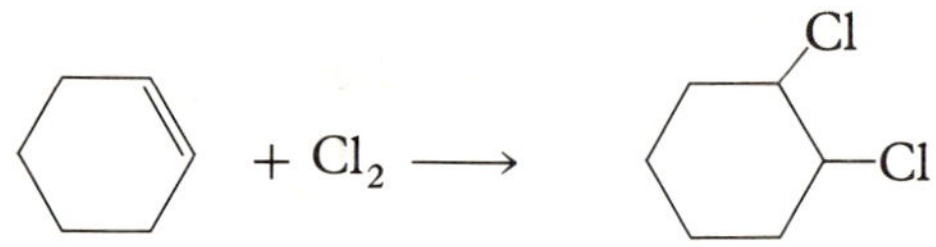

cyclohexene 1,2-dichlorocyclohexane

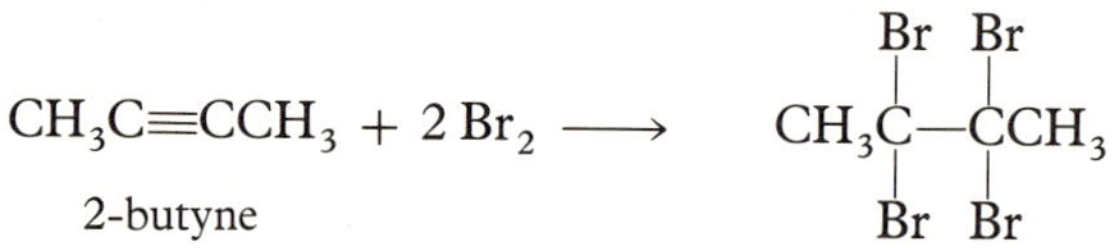

2-butyne 2,2,3,3-tetrabromobutane

Sample Problem: Predict the product of each of the following reactions:

(a) $CH_3CH_2CH{=}CH_2 + H_2 \xrightarrow{Pt}$

(b) 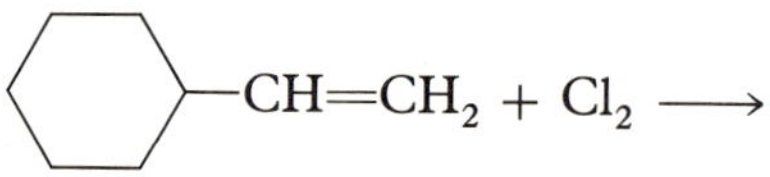

Solution:

(a) $CH_3CH_2CH_2CH_3$ (b) 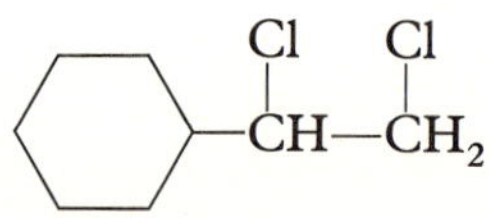

REACTION WITH THE HYDROGEN HALIDES. The hydrogen halides (HCl, HBr, HI) also add across the pi bonds of alkenes or alkynes. Again, no catalyst is required.

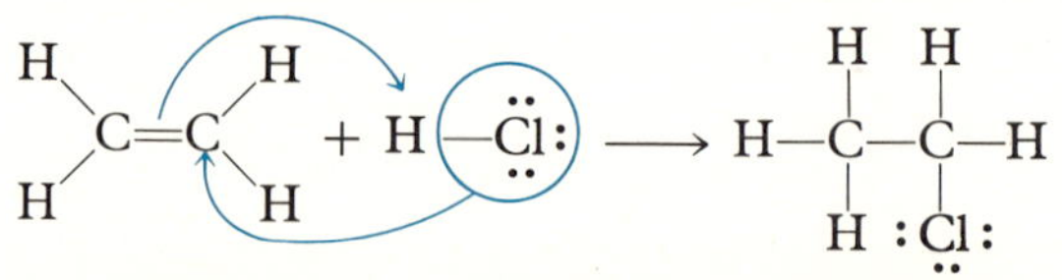

ethylene chloroethane

This reaction, too, is general for alkenes and alkynes:

$$CH_3CH{=}CHCH_3 + HBr \longrightarrow CH_3CH_2\underset{\displaystyle Br}{\underset{|}{C}}HCH_3$$

2-butene (*cis* or *trans*) — 2-bromobutane

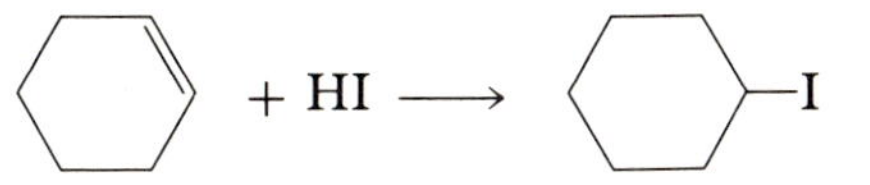

cyclohexene — iodocyclohexane

The hydrogen halides are different in structure from the other two reagents that we have just studied in that they are *unsymmetrical* reagents, while hydrogen gas and the halogens are *symmetrical* reagents. A symmetrical reagent is one in which the two halves that add to the pi bond are identical. The result of addition of H_2 is the addition of two identical H atoms. An unsymmetrical reagent is one in which the two halves are different; the addition of HCl results in two different atoms (H and Cl) being added to a pi bond.

Symmetrical reagents:

H—H Br—Br Cl—Cl *two halves are identical*

Unsymmetrical reagents:

H—Cl H—Br H—I *two halves are different*

Alkenes and alkynes may also be classified as being symmetrical and unsymmetrical. If we draw a line through the double or triple bond, we can see that the parts of the molecule on either side of this line will be either identical or different.

Sample Problem: Classify each of the following alkenes as symmetrical or as unsymmetrical: (a) 2-butene, (b) 1-butene, (c) cyclohexene, (d) 3-methyl-1-cyclohexene.

Solution: Compounds (a) and (c) are symmetrical.

$$CH_3CH{=}CHCH_3$$

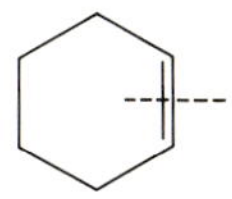

Compounds (b) and (d) are unsymmetrical.

$$CH_3CH_2CH{=}CH_2$$

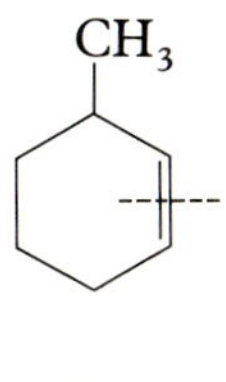

When either reactant is symmetrical, only one addition product is possible. However, if neither the alkene nor the reagent is symmetrical, there are *two* possible products. The reason that there could be two products is that the unsymmetrical reagent could add to the unsymmetrical alkene in two different ways.

One possibility:

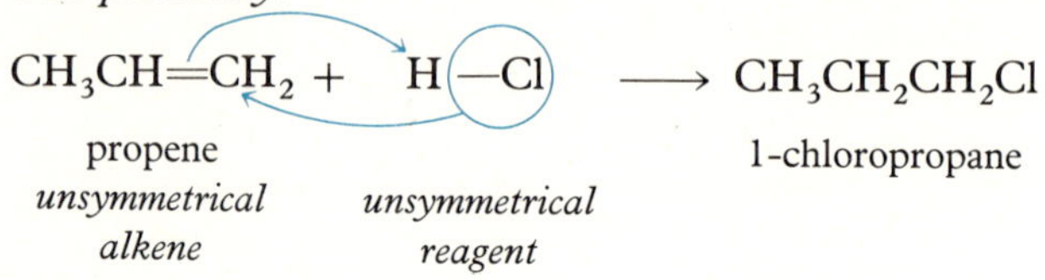

The other possibility:

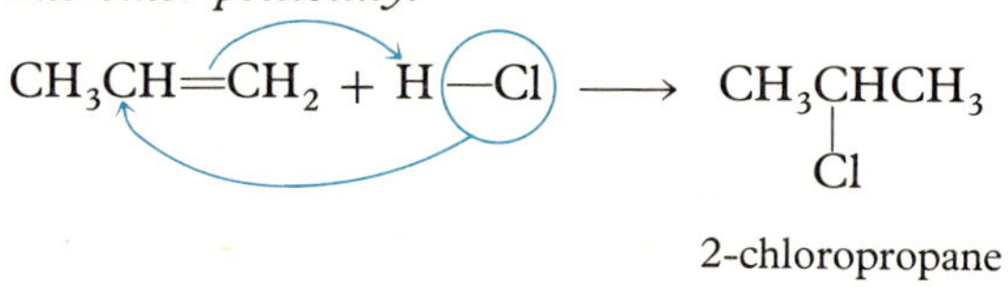

Even though there is the possibility of forming two products, the two products are *not* formed in equal abundance. Organic chemists have developed a rule of thumb to predict the major product of an unsymmetrical addition reaction. *The hydrogen of the reagent goes to the carbon of the double bond that has the greater number of hydrogens already.* (We might say that *the rich get richer.*)

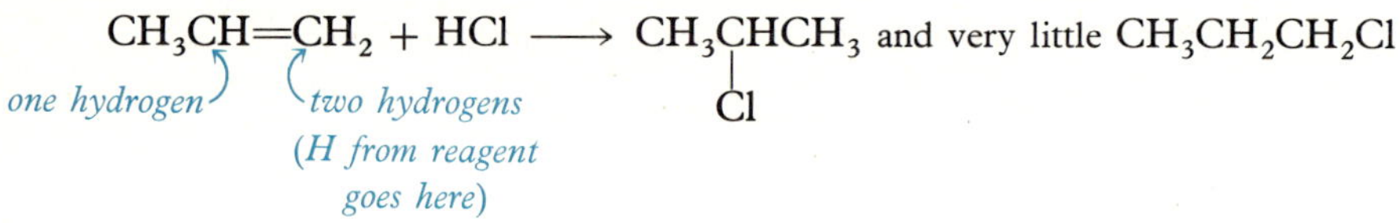

Sample Problem: Predict the product of the reaction of methylpropene with HCl.

Solution:

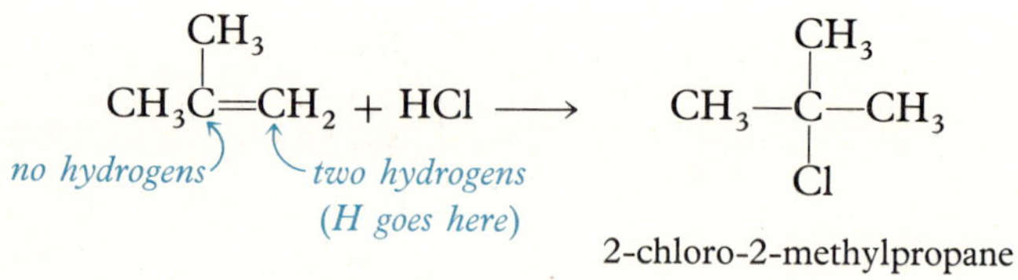

REACTION WITH WATER. Alkenes undergo reaction with water provided that an acidic catalyst is present.

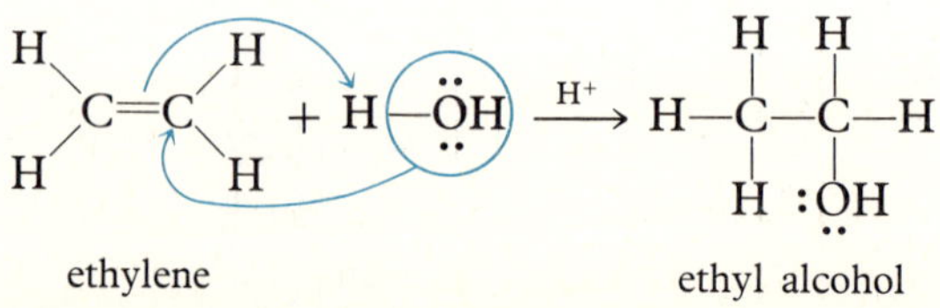

In reactions with double bonds, water acts as an unsymmetrical reagent because H^+ goes to one carbon of the double bond and OH^- goes to the other carbon. In the reaction with an unsymmetrical alkene, the hydrogen from the water goes to the carbon of alkene that has the greater number of hydrogens.

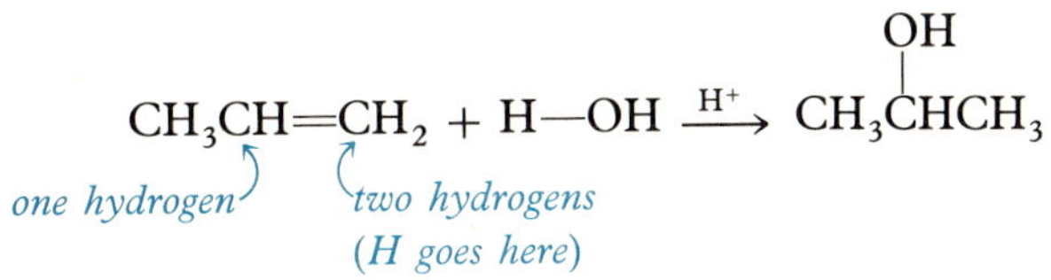

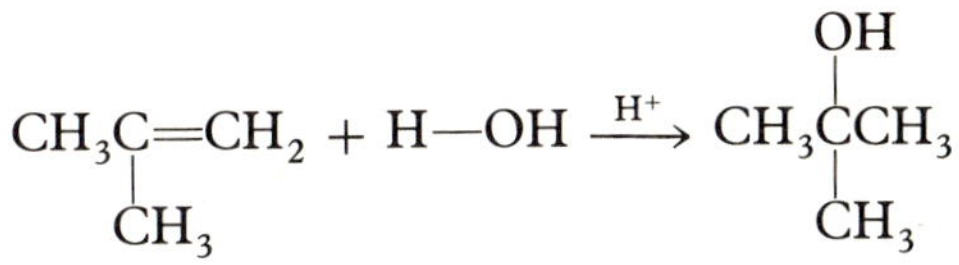

OXIDATION. The pi bond is a point of weakness, or chemical reactivity, within a molecule. An organic compound can be selectively oxidized at the site of the pi bond.

Cold potassium permanganate solution ($K^+\ MnO_4^-$) is a reagent that oxidizes the pi bond of an alkene by the addition of oxygen atoms to each side of the pi bond. The product of this oxidation is an organic compound that has two hydroxyl (—OH) groups on adjacent carbon atoms. The name for this class of organic compounds is *glycol*. One member of this class, ethylene glycol, is used as an antifreeze agent in automobile radiators.

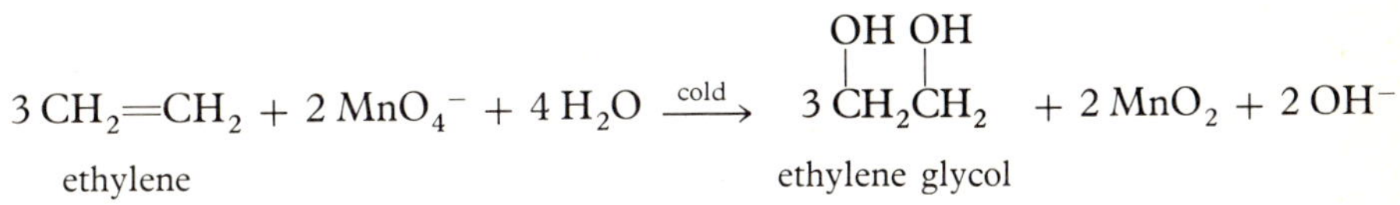

If a stronger oxidizing agent than cold $KMnO_4$ is used to treat an alkene, cleavage occurs. Once the oxidation starts, it continues and results in breakage of *both* the pi bond and the sigma bond. Strong oxidizing reagents, such as hot potassium permanganate solution, are used to carry out this type of reaction.

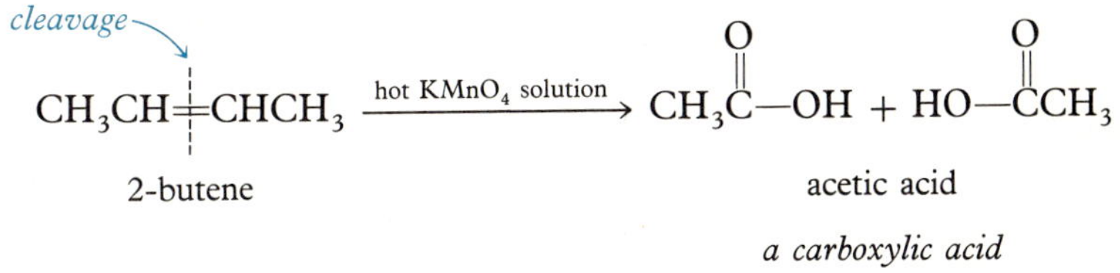

In a case, such as the preceding example, where there is a hydrogen on the pi-bond carbon, a carboxylic acid results from the cleavage.

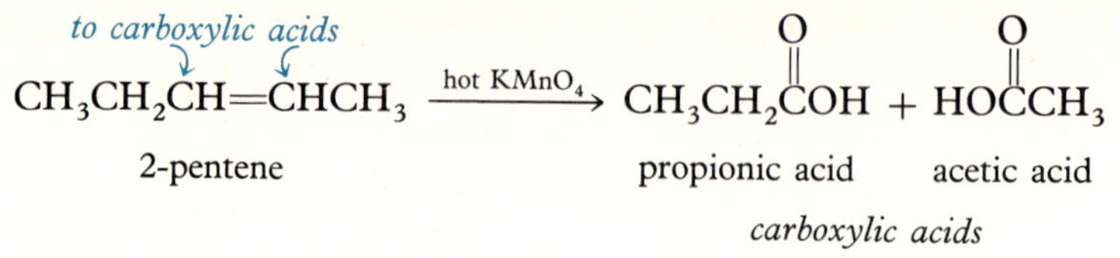

If the pi-bond carbons of the starting alkene do *not* have hydrogen atoms attached, ketones (compounds containing the >C=O group) are the result of the cleavage.

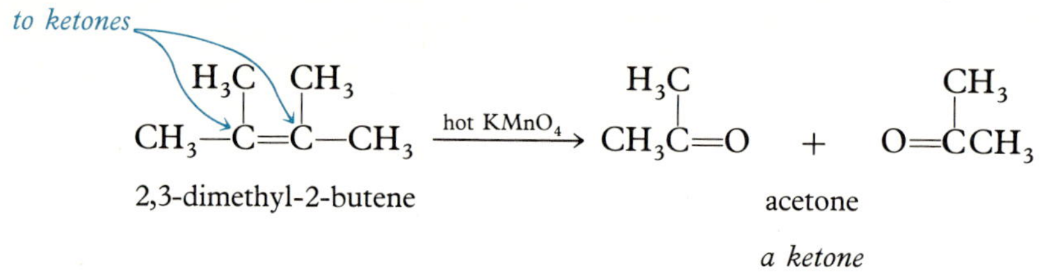

An alkene may have the two types of pi-bond carbons, one with a hydrogen and one without, in one molecule. In this case, we would predict a carboxylic acid and a ketone as the products.

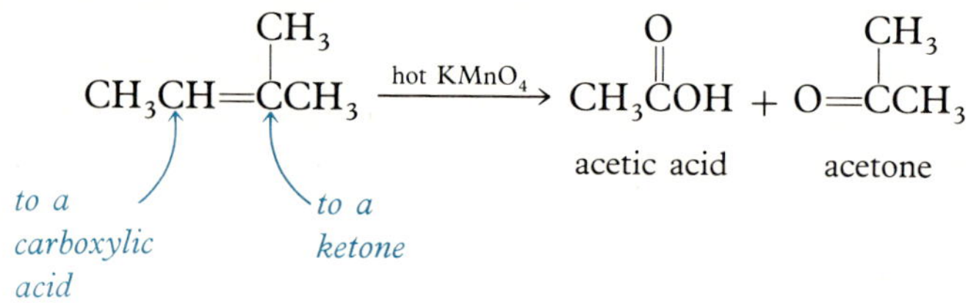

Sample Problem: What are the products of the reaction of 2-methyl-2-pentene with (a) cold $KMnO_4$ solution and (b) hot $KMnO_4$ solution?

Solution:

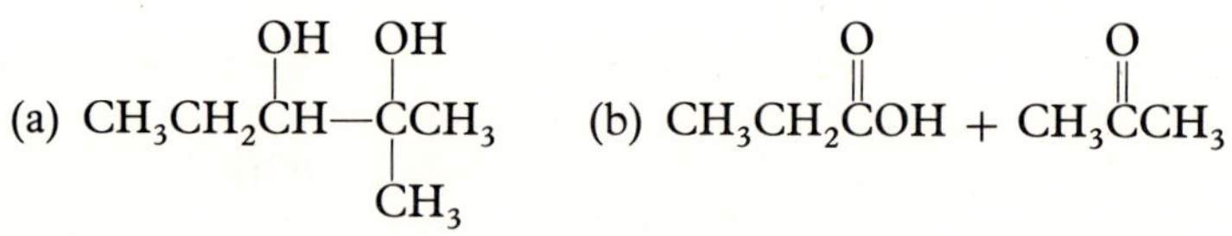

POLYMERIZATION REACTIONS. The word "polymer" is derived from the Greek *poly* (many) and *mer* (units). A polymer is a giant molecule that contains many repeating units. Polymeric molecules form nerve cells, skin, hair, muscles, and chromosomes. Plant fibers are also composed of polymers. We will discuss some of these natural polymers in Chapters 22, 23, and 28.

Synthetic polymers are tremendously useful in every walk of life; plastics, synthetic rubbers, and synthetic fibers are all synthetic polymers. We sit on polymer-covered chairs and write on polymer-coated table tops. Synthetic polymers are used to make fibers for the rugs on our polymeric floors, and for the clothes that we wear. Indeed, the age we live in might well be called the Polymer Age. In this chapter, we will consider polymers formed from alkenes, a group of polymers that includes polyethylene, polypropylene, acrylic fibers, and synthetic rubber.

The compound used to make a polymer is called the *monomer* (*mono-*, one). For polyethylene, the monomer is ethylene. Most polymers are named after their monomer with the prefix *poly-* added. Under the influence of a catalyst, two ethylene molecules combine to form a *dimer* (*di-*, two). Then another molecule of ethylene is added to form the *trimer* (*tri-*, three). This process continues until a giant molecule has been formed from thousands of ethylene molecules—this, of course, is the polymer.

$$\text{Many molecules of } \underset{\substack{\text{ethylene} \\ \textit{a monomer}}}{CH_2{=}CH_2} \xrightarrow{\text{catalyst}} \underset{\textit{portion of a polyethylene molecule}}{{-}CH_2CH_2{-}CH_2CH_2{-}\overbrace{CH_2CH_2}^{\textit{the repeating unit}}{-}}$$

For ease in writing a polymerization equation, chemists use the letter x or n to designate a very large number. The end groups of a polymer are not usually written out. Thus, the polymerization of ethylene may be represented by the following abbreviated equation:

$$\underset{\text{ethylene}}{x\,CH_2{=}CH_2} \xrightarrow{\text{catalyst}} \underset{\text{polyethylene}}{(\!\!-CH_2CH_2-\!\!)_x}$$

The catalysts used in polymerization reactions are many and varied; therefore, the reaction mechanisms of polymerization reactions are also varied. In living systems, enzymes are the catalysts for polymerization reactions leading to proteins and other polymers. One type of industrial catalyst attaches itself to a monomer by the pi electrons of the double bond and creates a positive charge on the monomer. This is an unstable situation, and another molecule of monomer is added to alleviate the situation. Another unstable ion is formed, and the process continues.

$$\text{Catalyst} + CH_2{=}CH_2 \longrightarrow \text{Catalyst}{-}CH_2{-}CH_2^+$$

$$\text{Catalyst}{-}CH_2{-}CH_2^+ + CH_2{=}CH_2 \longrightarrow \text{Catalyst}{-}CH_2CH_2{-}CH_2CH_2^+$$

and so on

No polymerization reaction continues forever. At some point, the catalyst falls off the growing polymer; a growing chain finds another way than continued polymerization to relieve its unstable situation; or maybe the monomer has been used up. In any case, the polymerization ceases, or *terminates*. Because the length of the chain has a decided effect on the properties of a polymer, different catalysts and different procedures have been developed to control the size of polymeric molecules.

The properties of polymers may also be changed by using alkenes with different substituents. While polyethylene is suitable for plastic squeeze bottles and plastic bags for kitchen use and for protecting clothes from the cleaners, polypropylene is more suitable for indoor-outdoor carpeting. Some examples of some common polymerization reactions follow:

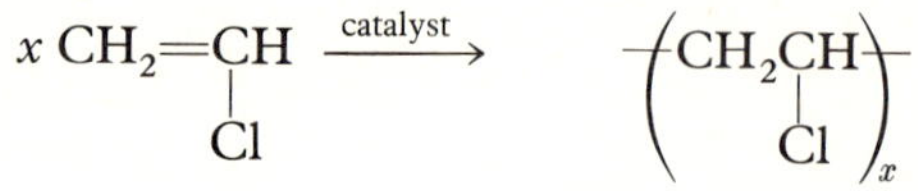

vinyl chloride — polyvinyl chloride (PVC)

used for vinyl floor covering and piping

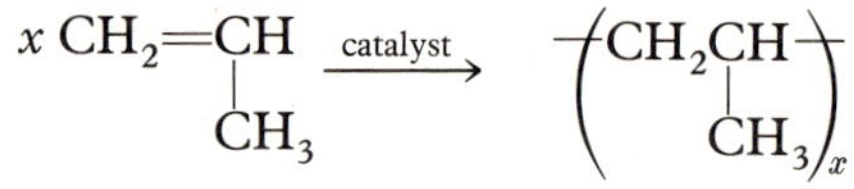

propylene — polypropylene

used for piping and carpeting

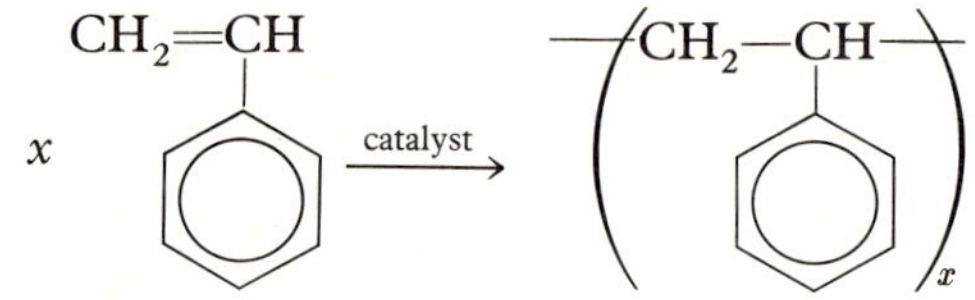

styrene — polystyrene

used for toys, hairbrush handles, and also to make styrofoam

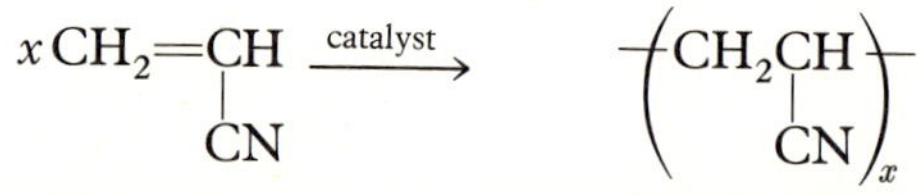

acrylonitrile — polyacrylonitrile

used for fibers for rugs and clothes (Acrilan, Orlon)

Sample Problem: Polytetrafluoroethylene (commonly called *Teflon*) is used as a heat- and corrosive-resistant material for gaskets and valves. Cookware is sometimes coated with this polymer. From its name, deduce the structure of the polymer and write an equation for its formation from its monomer.

Solution:

$$x\,CF_2{=}CF_2 \xrightarrow{\text{catalyst}} {-}\!(CF_2CF_2)\!{-}_x$$

tetrafluoroethylene — polytetrafluoroethylene

Another type of polymer synthesized from alkenes is formed from dienes, compounds with two double bonds. In these cases, each polymer unit retains one of the double bonds. Polymers of this type are generally flexible and find use as synthetic rubbers and the like. Natural rubber is a polymer that contains *cis* bonds within its structure. Below, we show the formation of natural rubber by the polymerization of isoprene. Although polyisoprene can be synthesized this way, natural rubber is synthesized in the rubber tree (*Hevea braziliensis*) in a lengthy and complex series of reactions that do not involve isoprene.

$$x\,CH_2{=}\underset{\displaystyle CH_3}{\underset{|}{C}}{-}CH{=}CH_2 \xrightarrow{\text{catalyst}} \left(CH_2\underset{\displaystyle CH_3}{\underset{|}{C}}{=}CHCH_2\right)_x$$

isoprene — polyisoprene, *natural rubber*

STUDY QUESTIONS

13-1. Name the following hydrocarbons:

(a) $CH_3\underset{\displaystyle CH_3}{\underset{|}{C}}HCH{=}CH_2$

(b) 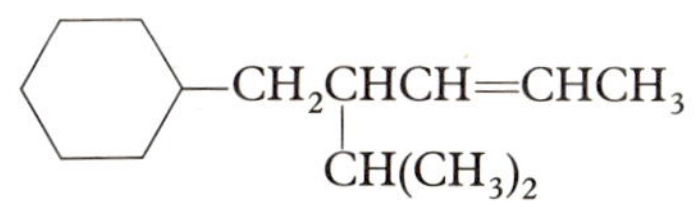

(c) $(CH_3)_2CHC{\equiv}CH$

(d) $CH_2{=}\underset{\displaystyle CH_3}{\underset{|}{C}}CH{=}CHCH_2CH_2CH_3$

(e) $CH_3C{\equiv}\underset{\displaystyle C(CH_3)_3}{\underset{|}{C}}HC{\equiv}CH$

13-2. Draw the condensed structural formula of each of the following compounds:

(a) 2,3,4-trimethyl-2,4-hexadiene
(b) ethylene
(c) 4-methyl-4-ethyl-2-heptyne

13-3. Give the structures for the geometric isomers of the following compounds. Label each isomer as *cis* or *trans*.

(a) $CH_3CH_2CH{=}CHCH_3$
(b) $ClCH{=}CHBr$
(c) $CH_3CH_2CH{=}CHCH_2CH_3$
(d) $BrCH_2CH{=}CHCH_2Br$

13-4. Explain why *trans*-cyclohexene is not a stable compound.

13-5. Which of the following compounds have geometric isomers?

(a) $CH_3C{\equiv}CCH_3$
(b) $Cl_2C{=}CHCl$
(c) $BrCH{=}CHBr$
(d) $CH_3CH_2CH{=}CHCH_2CH_2Cl$

13-6. Give all of the possible geometric isomers of 1,2,3-cyclohexanetriol (OH groups at positions 1, 2 and 3).

13-7. Each of the following structures is represented in the *cis* or *trans* form. State which form it is and draw its geometric isomer.

(a) 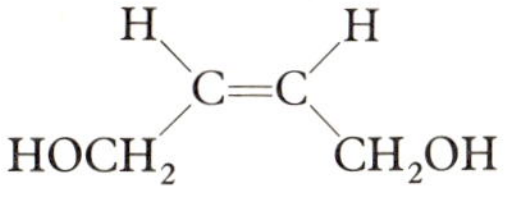

(b) 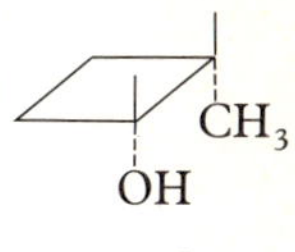

(c) 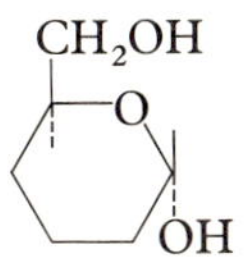

(d) 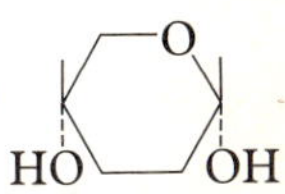

13-8. What would be the product of the complete hydrogenation of each of the following? [In (d), the $—CO_2H$ group is unaffected.]

(a) 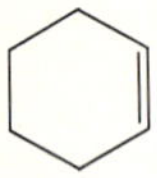

(b) $CH_3C{\equiv}C$— 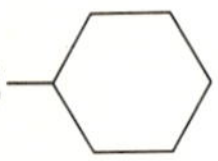

(c) $CH_2{=}CHCH_2CH_2CH{=}CH_2$

(d) $CH_3(CH_2)_7CH{=}CH(CH_2)_7CO_2H$

13-9. Bromine or chlorine was added to a series of alkenes resulting in the products below. What was the structure of the original alkene in each case?

(a) 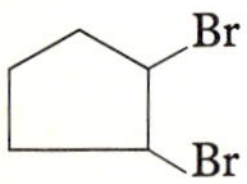

(b) $CH_3CH_2C(CH_3)_2CHClCH_2Cl$

(c) $(CH_3)_2CBrCH_2Br$

(d) 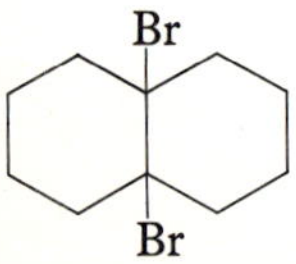

13-10. Can 1,3-dibromopropane be obtained from the addition of bromine to an alkene?

13-11. Classify the following alkenes as symmetrical or unsymmetrical:

(a) $CH_3CH_2CH{=}CHCH_2CH_3$

(b) $CH_2{=}CHCl$

(c) 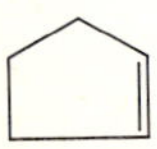

(d) $ClCH_2CH_2CH{=}CHCH_2CH_2OH$

13-12. Complete the following equations:

(a) $CH_3CH_2CH{=}CH_2 + HCl \longrightarrow$

(b) 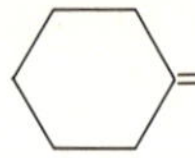 $=CH_2 + H_2O \xrightarrow{H^+}$

(c) $CH_3C{\equiv}CH + 1\ HCl \longrightarrow$

(d) $(CH_3)_2C{=}CH_2 + HBr \longrightarrow$

13-13. Write the complete balanced equation for the combustion of 2-octene.

13-14. Write the equations for (1) the oxidation to a glycol and (2) the oxidative cleavage for each of the following alkenes:

(a) propene

(b) cyclopentene

(c) 1-methyl-1-cyclohexene

(d) 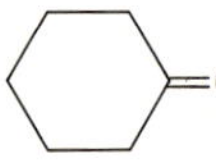 $=CHCH_3$

13-15. Predict the structure of the polymer formed from each of the following monomers:

(a) 1-butene

(b) trichloroethylene

(c) 2-bromo-1,3-butadiene

13-16. Match the following:

(a) 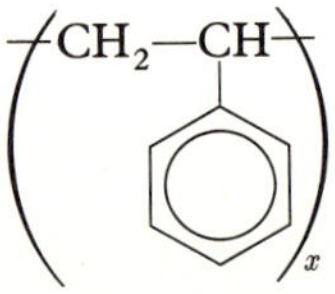 (1) polyvinyl chloride

(b) $\left(CH_2CHCl\right)_x$ (2) polystyrene

(c) $\left(CH_2C(CH_3){=}CHCH_2\right)_x$ (3) polyisoprene

13-17. What is the monomer for each of the following polymers?

(a) $\left(CH_2CH(CH_3)\right)_x$ (b) $\left(CH_2CH(OC(=O)CH_3)\right)_x$

CHAPTER 14 HALOGEN COMPOUNDS

OBJECTIVES

Naming the Alkyl Halides *Give examples of alkyl halides and aryl halides.* * *Name alkyl halides by the IUPAC system and by the functional group system.*

Halogenation and Other Free-Radical Reactions *Define free radical.* * *Write equations for the steps in the free-radical chlorination of an alkane.* * *Describe auto-oxidation.* * *Write the structure of a simple antioxidant.*

Substitution Reactions of Alkyl Halides *Write specific and generalized equations for the substitution reactions of alkyl halides with a variety of bases.*

Elimination Reactions of Alkyl Halides *Write equations for elimination reactions of alkyl halides.*

A hydrocarbon that has one or more halogen atoms attached to a carbon atom is called a *halogenated hydrocarbon*. This class of compound may be divided into the *alkyl halides* and the *aryl halides*. Alkyl halides are those in which the halogen is attached to a carbon in an alkyl group (continuous-chain, branched-chain, or cyclic). An aryl halide is one in which a halogen is attached to a carbon that is part of an aromatic ring.

Some alkyl halides:

CH_3CH_2Cl

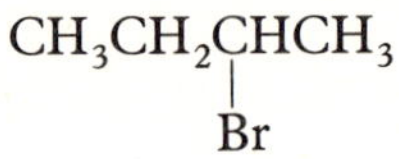

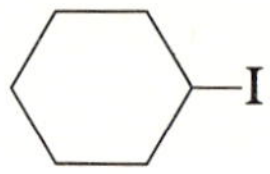

Some aryl halides:

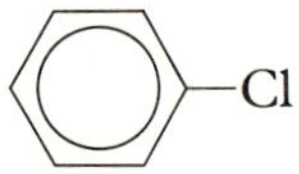

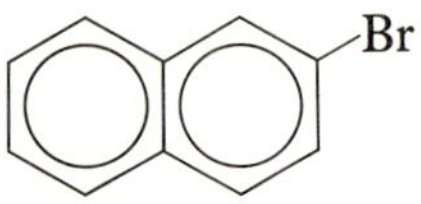

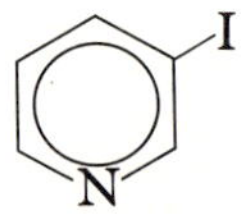

Naturally occurring organohalogen compounds are rather rare. One example is thyroxine, the hormone of the thyroid gland.

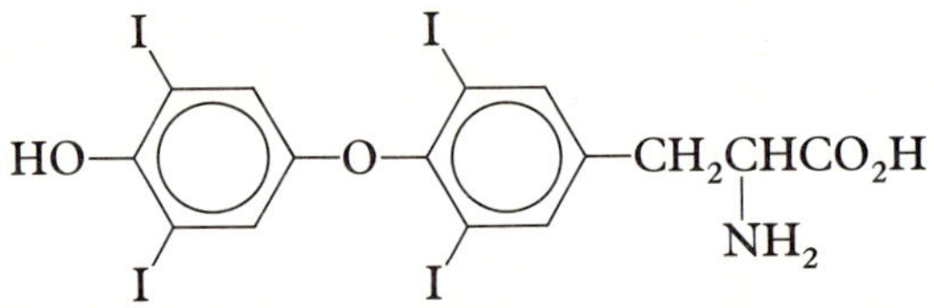

thyroxine

Alkyl halides find use in industry and commerce as solvents. Alkyl halides also can react with a wide variety of reagents to yield an equally wide variety of products. For this reason, alkyl halides are used as intermediates in the preparation of other organic compounds.

Synthetic organohalogen compounds are often toxic and cause liver damage. When working with common halogen-containing solvents such as chloroform ($CHCl_3$) or carbon tetrachloride (CCl_4), one should avoid excessive inhalation or contact.

Insecticides containing halogen atoms are generally quite stable and are only slowly broken down by the environment; therefore, the use of these compounds has declined in recent years. The following is the structure of the insecticide chlordane.

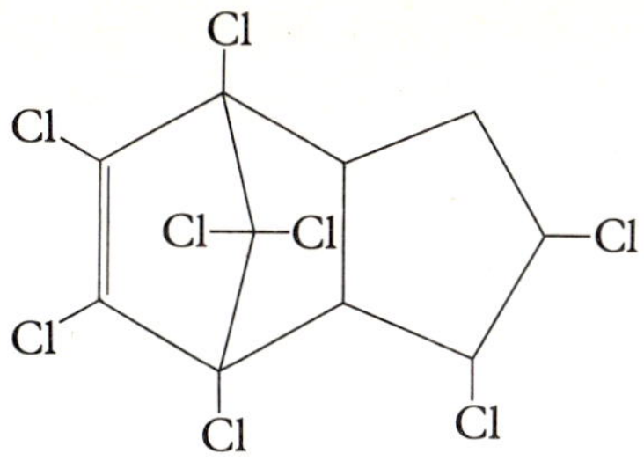

chlordane
a long-lasting insecticide

NAMING THE ALKYL HALIDES

In the IUPAC system, the name of the halogen is a prefix to the parent name. The prefixes that are used are listed in Table 14–1. The rules that apply to the naming of alkenes and compounds with alkyl branches also apply here.

Table 14–1. Prefix Names for the Halogens

Halogen	Prefix Name
F—	Fluoro-
Cl—	Chloro-
Br—	Bromo-
I—	Iodo-

First find the *longest continuous chain containing the functional group* (halogen, in this case). Number the chain from *the end closer to the halogen.* Then name the compound with the proper prefix number and halogen prefix preceding the parent name. Group like substituents together in the name and use the prefixes *di-*, *tri-*, and so forth, where appropriate.

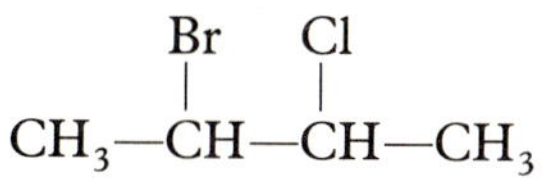

2-bromo-3-chlorobutane

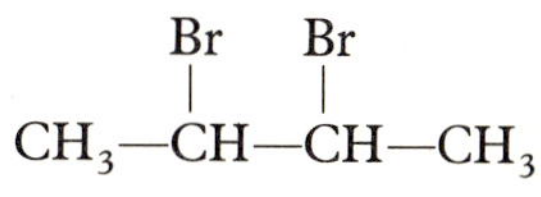

2,3-dibromobutane

Another common way of naming alkyl halides is by using the functional group name. In this nomenclature method, the alkyl halides are called fluorides, chlorides, bromides, or iodides, whichever is applicable. The name of the attached alkyl group is given first as a separate word. This type of name is similar to the names for halide salts (for example, sodium chloride); however, alkyl halides contain a covalent bond, not an ionic bond, to the halogen atom.

CH_3Br	CH_3CH_2Cl
methyl bromide	ethyl chloride

HALOGENATION AND OTHER FREE-RADICAL REACTIONS

A very important class of reactions, both in the laboratory and in biological systems, involves intermediates with unpaired electrons. These intermediates are called *free radicals*. Reactions caused by free radicals are called *free-radical reactions*.

Some free radicals:

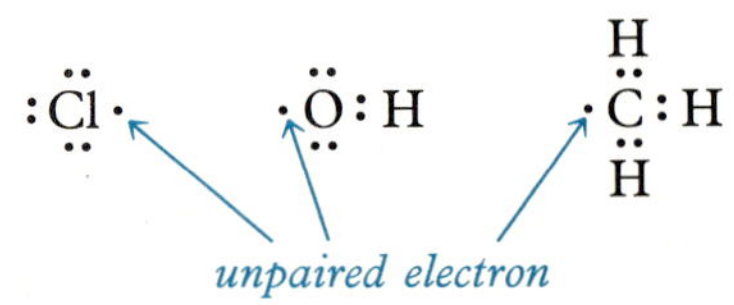

Let us introduce free-radical reactions with the chlorination of methane, a reaction that is catalyzed by ultraviolet light (uv).

$$CH_4 + Cl_2 \xrightarrow{\text{uv}} CH_3Cl + CH_2Cl_2 + CHCl_3 + CCl_4 + HCl$$

methane — *a mixture of products*

The first step in the reaction, called the *initiation step*, is the action of ultraviolet light on the chlorine molecule. The Cl_2 absorbs the energy from the ultraviolet light and cleaves into free radicals (in this case, a pair of Cl atoms, each with one unpaired electron). We represent a free radical by the atomic symbol with one dot to denote the unpaired electron. The other outer electrons are ignored in this symbolism.

Step 1: $Cl:Cl \longrightarrow 2\,Cl\cdot$

two chlorine free radicals

A situation in which an atom has an unpaired electron is not a stable one; most free radicals are reactive and of very high energy. A chlorine free radical reacts with methane by abstracting a hydrogen atom with one electron. In this fashion, Cl· goes to HCl. But, in the formation of HCl, a *methyl free radical* (a CH_3 group with an unpaired electron, or $\cdot CH_3$) is formed.

Step 2: 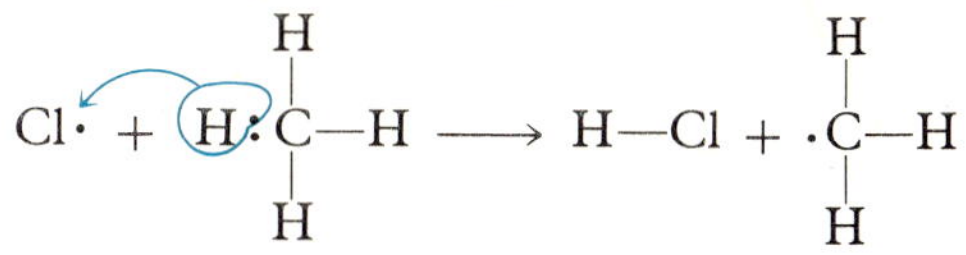

methyl free radical

A methyl free radical is also an unstable, high-energy species. When it collides with a molecule of chlorine, it removes a chlorine atom with one electron. The unpaired electron on the methyl free radical and the electron on the chlorine atom pair off to form a sigma bond, and methyl chloride is formed.

Step 3: 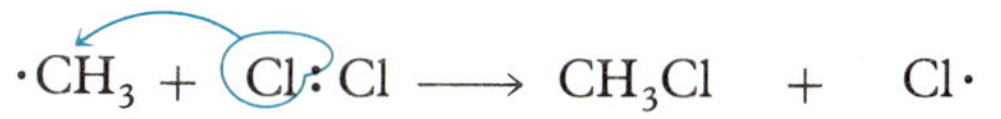

methyl chloride — *a new chlorine free radical*

In the formation of methyl chloride, another chlorine free radical is generated to begin the process anew. Steps 2 and 3 are therefore *propagation* steps in a chain reaction that can continue until the free radicals are terminated or until a reactant is used up.

The reactive chlorine free radical is not selective about which hydrogen it abstracts. As the methane is being converted to methyl chloride, there are fewer and fewer methane molecules in the reaction mixture, and more and more methyl chloride molecules. At some point, the chlorine free radical is more likely to abstract a hydrogen atom from methyl chloride to yield a chloromethyl free radical ($\cdot CH_2Cl$). This chloromethyl free radical, in turn, undergoes reaction with chlorine to yield methylene chloride (CH_2Cl_2) and a new chlorine free radical in another propagation step.

As the process continues, more CH_2Cl_2 is available for reaction with the chlorine free radical, and chloroform ($CHCl_3$) is produced. Chloroform, in turn, is eventually converted to carbon tetrachloride (CCl_4).

Free-radical reactions are terminated by the destruction of reactive free radicals. An example of a termination reaction would be the combination of two free radicals to form a stable product.

$$CH_3\cdot + Cl\cdot \longrightarrow CH_3Cl$$

We have used the chlorination of methane to illustrate the three characteristic processes common to all free-radical reactions: (1) *initiation* (formation of free radicals), (2) *propagation* (reactions in which products and new free radicals are formed), and (3) *termination* (destruction of reactive free radicals).

Sample Problem: Write an equation for each step in the following conversion:

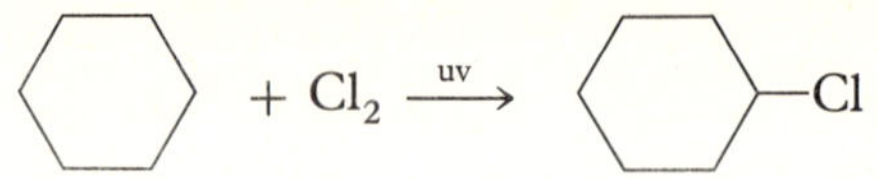

cyclohexane chlorocyclohexane

Solution:

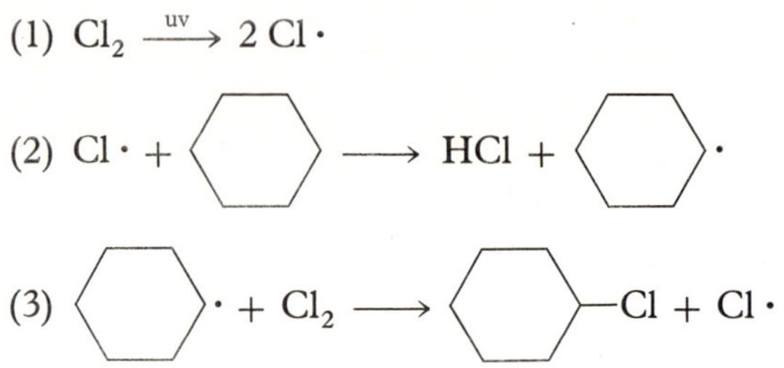

Another type of free-radical reaction is caused by oxygen gas, a molecule that has two unpaired electrons and is called a *diradical.*

$$\cdot O{-}O \cdot$$

molecular oxygen
a stable diradical

Free-radical oxidation by oxygen in the air, called *auto-oxidation,* causes deterioration of organic compounds, such as plastics or rubber. Rancidity of fats and gum formation in gasoline are also partly the results of auto-oxidation. Auto-oxidation leads initially to *hydroperoxides,* which are reactive and often cause cleavage of large molecules into smaller fragments. The unpleasant smell of rancid butter is due primarily to butyric acid, $CH_3CH_2CH_2CO_2H$, which is an auto-oxidation product. (In the following equation, R represents an alkyl group.)

An auto-oxidation reaction:

$$RCH{=}CHCH_2R + O_2 \longrightarrow RCH{=}CH\underset{\displaystyle}{\overset{\displaystyle O{-}OH}{C}}HR \longrightarrow RCH{=}CH\overset{\displaystyle O\cdot}{C}HR + \cdot OH$$

a hydroperoxide

↳ Mixture of aldehydes, ketones, and carboxylic acids

Reagents that prevent auto-oxidation are called *antioxidants, free-radical inhibitors,* or *food preservatives,* depending upon their use. Generally, these substances are phenol derivatives—that is, they contain a hydroxyl group attached to an aromatic ring.

Some antioxidants:

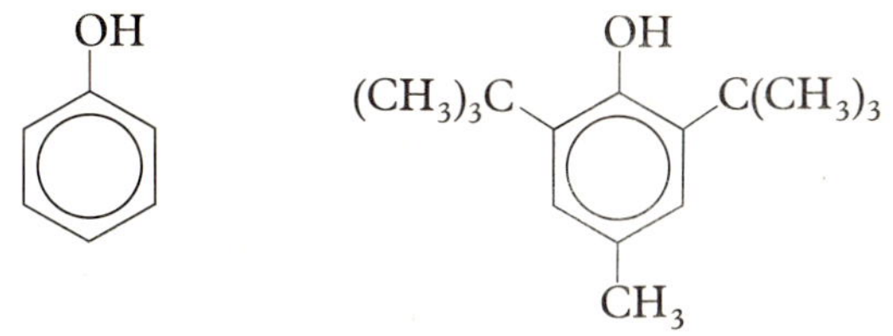

phenol ionol (BHT) hydroquinone

Vegetable oils, such as wheat germ oil, do not auto-oxidize as readily as animal fats and oils because they contain a naturally occurring phenol derivative called α-tocopherol, which acts as a natural antioxidant.

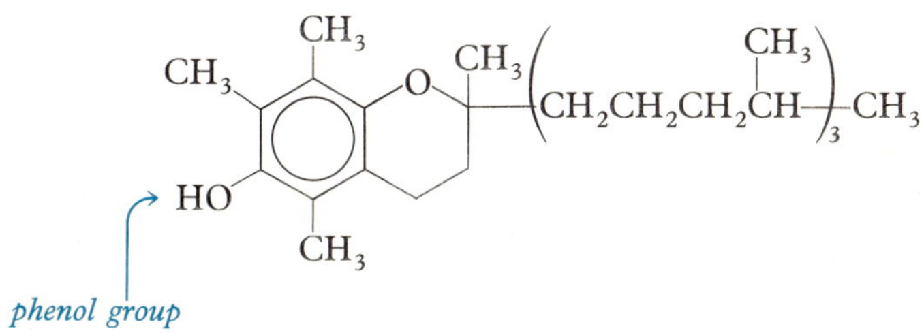

α-tocopherol (vitamin E)
a naturally occurring antioxidant

Phenols serve as inhibitors of auto-oxidation by acting as *free-radical traps.* A phenol reacts readily with a free radical to terminate its further reaction. Although the phenolic compound becomes a free radical in the process, it is more stable and less reactive than most free radicals.

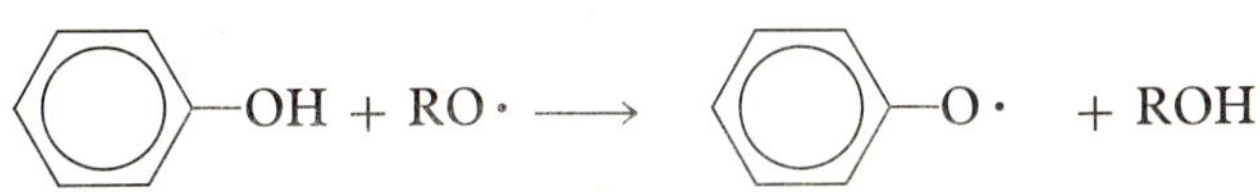

phenol *a stable free radical*

An example of how the stability of phenol free radicals is used to advantage in a living system is the uptake of oxygen by a cell. When hemoglobin delivers an oxygen molecule to the cell of an organism, the cell uses the oxygen to oxidize a Fe^{2+} ion to Fe^{3+}.

$$\tfrac{1}{2}\,O_2 + 2\,H^+ + 2\,Fe^{2+} \xrightarrow{\text{cell}} H_2O + 2\,Fe^{3+}$$

The Fe^{3+} ion then oxidizes a phenolic compound (dihydroubiquinone) by way of a free radical to a dicarbonyl compound called ubiquinone. The hydrocarbon group (R) in ubiquinone varies with the species of plant or animal.

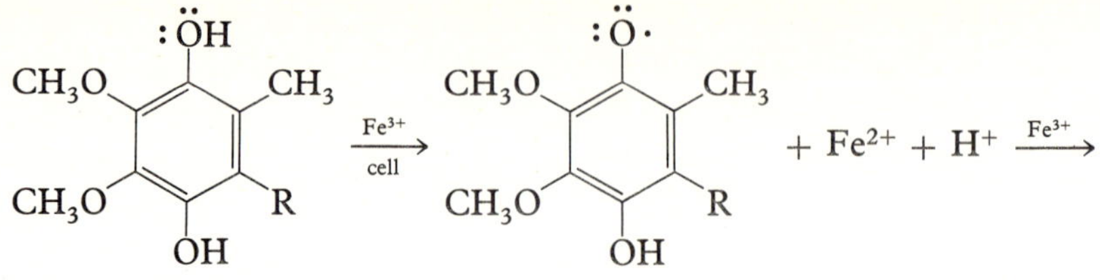

dihydroubiquinone — *a free radical*

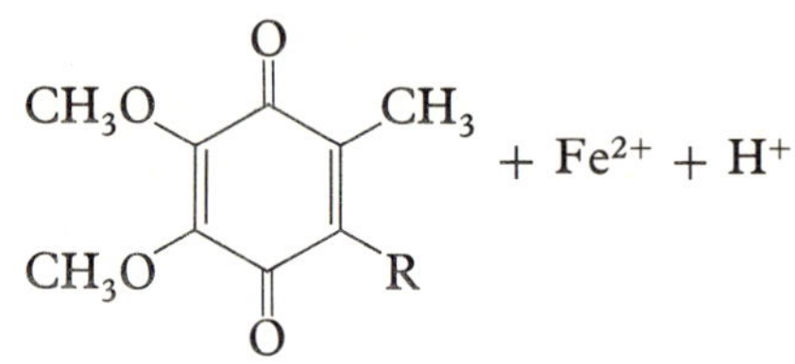

ubiquinone
(coenzyme Q)

In this way, Fe^{2+} is regenerated and can react with more O_2 molecules. Ubiquinone, in turn, is reduced by another reducing agent back to dihydroubiquinone. The new reducing agent is oxidized in this reaction. After a long series of reactions, we finally reach an oxidized compound that takes part in the oxidation of food to yield CO_2, H_2O, and energy. (See Chapters 25 and 26.) However, the whole sequence of the oxidation of nutrients depends on this initial uptake of O_2 by the cell.

SUBSTITUTION REACTIONS OF ALKYL HALIDES

The carbon-halogen bond of an alkyl halide is a polar covalent bond. The halogen atom is more electronegative than the carbon atom and has a greater attraction for the bonding electrons. Therefore, the carbon attached to the halogen atom has a partial positive charge.

$$CH_3\overset{\delta+}{C}H_2\text{—}\overset{\delta-}{Br}$$

This partially positive carbon atom can be attacked by a base. For example, ethyl bromide can undergo reaction with hydroxide ions to yield ethyl alcohol and bromide ions. In this reaction, electrons move in pairs. There is no unpaired electron, as there is in a free-radical reaction.

$$CH_3CH_2\text{—}\ddot{\underset{\cdot\cdot}{Br}}: + :\ddot{\underset{\cdot\cdot}{O}}H^- \longrightarrow CH_3CH_2\text{—}\ddot{\underset{\cdot\cdot}{O}}H + :\ddot{\underset{\cdot\cdot}{Br}}:^-$$

We might say that the hydroxide ion has moved in, attached itself to the carbon atom, and pushed off the bromide ion. Instead, we usually say that the bromine has become *substituted* or *displaced* by the hydroxide ion. A reaction of this type is called a *substitution reaction* or a *displacement reaction.* A variety of alkyl halides undergo this reaction with hydroxide ion.

$$\underset{\text{(isopropyl bromide)}}{\underset{\text{2-bromopropane}}{CH_3\overset{\overset{Br}{|}}{C}HCH_3}} + OH^- \longrightarrow \underset{\text{(isopropyl alcohol)}}{\underset{\text{2-propanol}}{CH_3\overset{\overset{OH}{|}}{C}HCH_3}} + Br^-$$

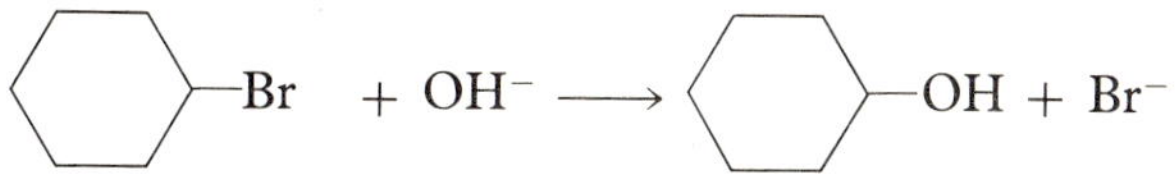

bromocyclohexane cyclohexanol

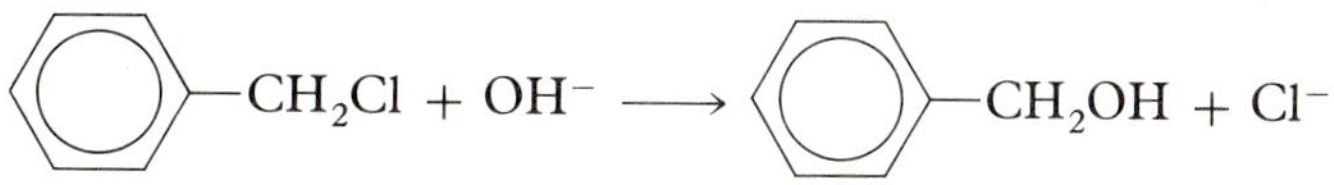

benzyl chloride benzyl alcohol

To show that the reaction of an alkyl halide with hydroxide ions is fairly general, let us use the symbol R to stand for an alkyl group, such as $CH_3—$, $CH_3CH_2—$, $(CH_3)_2CH—$,  , or any of a variety of alkyl groups.

Of the four halogens, three (Cl, Br, and I) behave chemically in a similar fashion. That is, each can be displaced by hydroxide ion. (The carbon-fluorine bond is exceptionally stable, and alkyl fluorides do not tend to behave like the other halogens.) Let us use X to represent Cl, Br, or I. Now we can write the following *general* equation.

Generalized:

$$\underset{\textit{an alkyl halide}}{RX} + OH^- \longrightarrow \underset{\textit{an alcohol}}{ROH} + \underset{\textit{a halide ion}}{X^-}$$

Other anions can also displace a halogen atom from an alkyl halide. A few substitution reactions for alkyl halides are shown in Table 14-2.

Table 14-2. Generalized Substitution Reactions for the Alkyl Halides

$RX + OH^- \longrightarrow R—OH + X^-$
$RX + RO^- \longrightarrow R—OR + X^-$
$RX + HS^- \longrightarrow R—SH + X^-$
$RX + CN^- \longrightarrow R—CN + X^-$

Sample Problem: Complete the following equations for substitution reactions.

(a) $CH_3CH_2I + {}^{-}OCH_3 \longrightarrow$

(b) $CH_3CH_2CH_2Cl + {}^{-}CN \longrightarrow$

Solution: (a) $CH_3CH_2OCH_3 + I^-$

(b) $CH_3CH_2CH_2CN + Cl^-$

ELIMINATION REACTIONS OF ALKYL HALIDES

Another common reaction of the alkyl halides is the loss of the halogen atom along with a hydrogen atom from an adjacent carbon atom. The reaction results in the formation of a pi bond. Since both a hydrogen and a halogen (or HX) are lost from the molecule, the reaction is called a *dehydrohalogenation reaction.* These reactions are also called *elimination reactions,* since part of the molecule is eliminated.

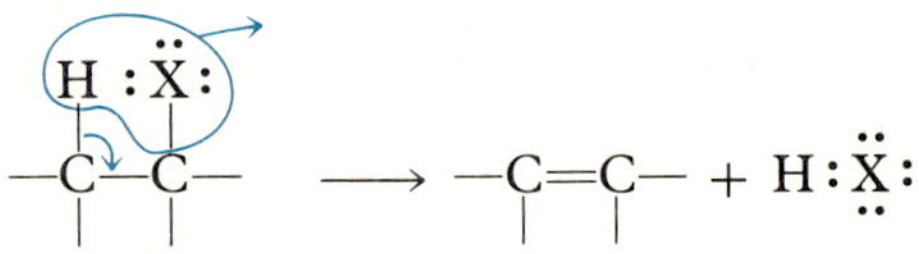

HX is eliminated *an alkene is formed*

A dehydrohalogenation reaction is carried out in base; therefore, H_2O and X^- are the actual products, instead of HX.

$$HX + OH^- \longrightarrow H_2O + X^-$$

from the alkyl halide

For an example of an elimination reaction, let us consider the reaction of *t*-butyl chloride with hydroxide ion.

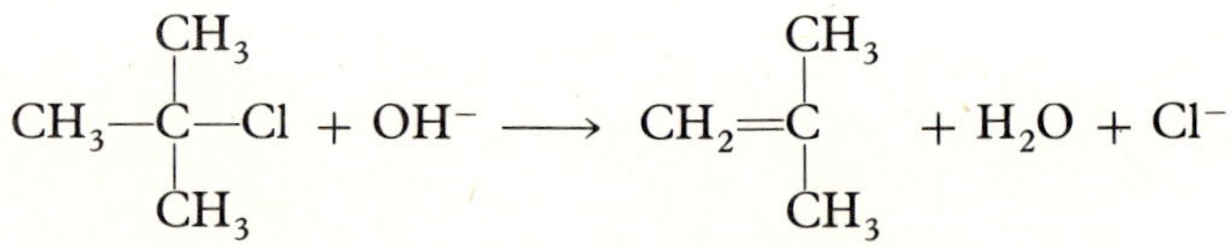

t-butyl chloride methylpropene (isobutylene)

In this reaction, a hydrogen is lost from a carbon adjacent to the halogen carbon, and a pi bond is formed.

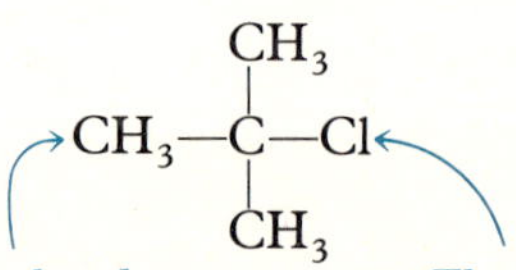

This carbon loses a hydrogen. *The halogen is lost.*

In our example, there are three carbons bonded to the carbon that has the halogen, and together they have a total of nine hydrogens. All of these hydrogens are equivalent. The loss of any one of these nine hydrogens will result in the formation of the same alkene, methylpropene. If these adjacent hydrogens are not all the same, then a mixture of alkenes is formed.

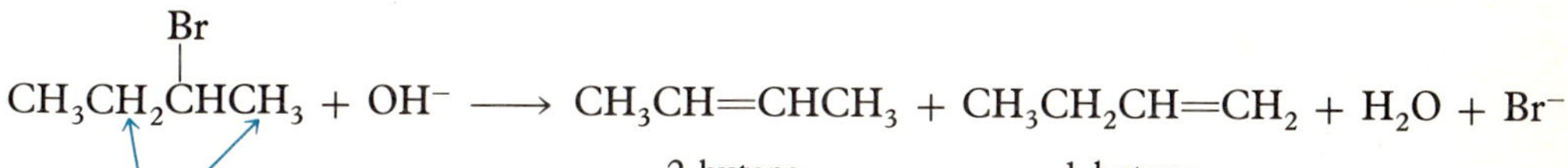

these hydrogens are not equivalent

2-butene *from loss of CH_2 hydrogen*

1-butene *from loss of CH_3 hydrogen*

One alkyl halide can react with a base to yield a substitution product and one or more elimination products, not necessarily in the same amounts. By experimentation, a chemist learns which conditions favor the desired product. For example, heating an alkyl halide and a base usually leads to the alkene. If a substitution product is desired, the chemist would not usually heat the reaction mixture, but would let it stand at room temperature.

Sample Problem: Write equations for (a) the substitution reaction and (b) the elimination reaction of 2-bromo-2-methylbutane with NaOH solution. In (b), show all the alkene products.

Solution: (a) Substitution:

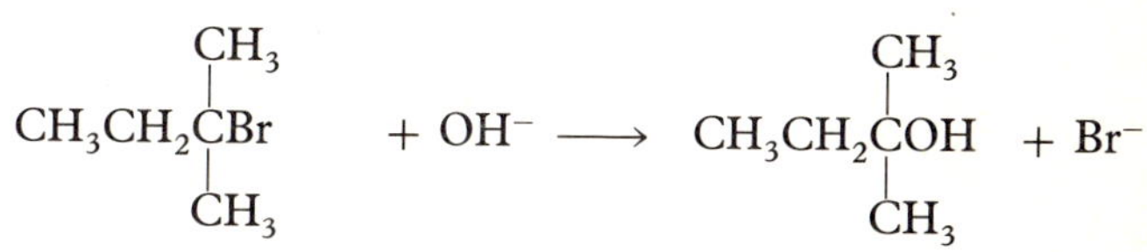

2-bromo-2-methylbutane 2-methyl-2-butanol

(b) Elimination:

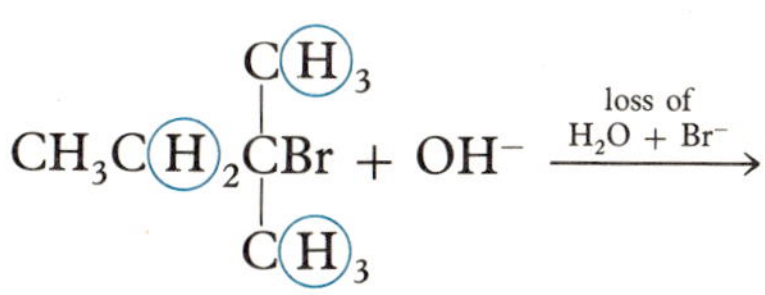

The hydrogen atoms that can be lost are circled.

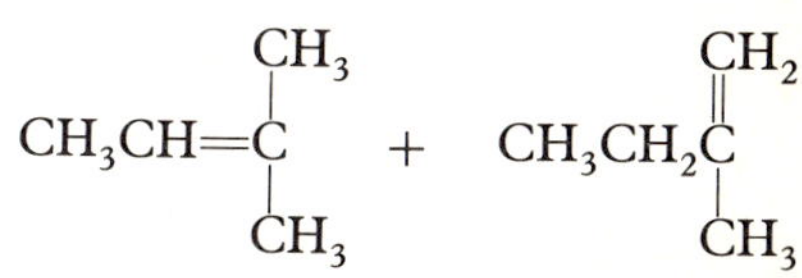

2-methyl-2-butene *from loss of CH_2 hydrogen*

2-methyl-1-butene *from loss of either CH_3 hydrogen*

STUDY QUESTIONS

14-1. Classify each of the following compounds as an aryl halide or an alkyl halide:

(a) 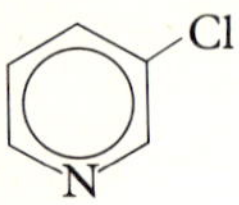 (b) $CH_2{=}CHCH_2CH_2Cl$

(c) 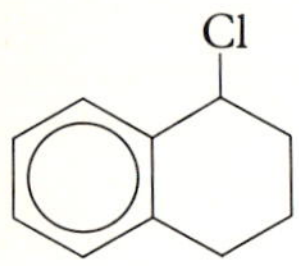

14-2. Write an acceptable name for each of the following compounds:

(a) 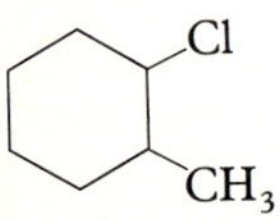 (b) $(CH_3)_2C(CH_2CH_3)CH_2Cl$

(c) $(CH_3CH_2)_2C(CH_3)Cl$

14-3. Give the structure of each of the following compounds:
(a) *t*-butyl fluoride
(b) isopropyl bromide
(c) benzyl chloride
(d) 1-chloro-2,2-dimethylpentane
(e) 5,5-diiodo-1-phenyloctane

14-4. Give the structure of each of the following free radicals:
(a) chlorine free radical
(b) methyl free radical
(c) ethyl free radical
(d) cyclohexyl free radical

14-5. Write equations for (a) the initiation reaction and (b) the propagation reactions for the following free-radical chlorination:

$$Cl_2CHCH_2Cl + 3\,Cl_2 \xrightarrow{uv} Cl_3CCCl_3 + 3\,HCl$$

14-6. Given the following generalized equations, write the specific equations:

Example: $RI + OH^- \longrightarrow ROH + I^-$
where $R = CH_3—$
Answer: $CH_3I + OH^- \longrightarrow CH_3OH + I^-$

(a) $RBr + SH^- \longrightarrow RSH + Br^-$
where $R = n\text{-}C_4H_9—$

(b) $RCl + CH_3O^- \longrightarrow ROCH_3 + Cl^-$
where R = 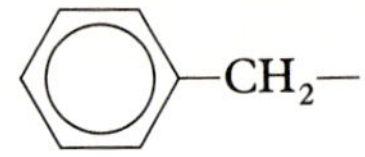

(c) $RCl + CN^- \longrightarrow RCN + Cl^-$
where $R = (CH_3)_2CH—$

14-7. Define and give an illustration for each of the following:
(a) a substitution reaction
(b) an elimination reaction

14-8. Which of the following ions might cause an alkyl halide to undergo a substitution or elimination reaction?
(a) $(CH_3)_3CO^-$ (b) H^+ (c) Na^+
(d) 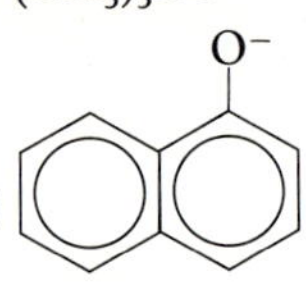 (e) NH_4^+

14-9. Complete the following equations for substitution reactions:

(a) $CH_3I +$ 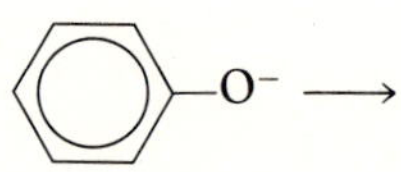

(b) $CH_3S^- + CH_3CH_2Br \longrightarrow$

(c) 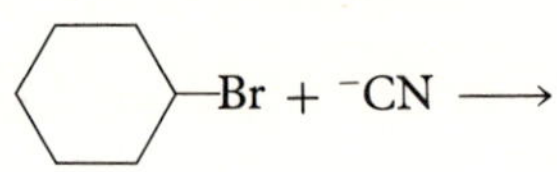

(d) $CH_3CH_2CH(CH_3)Cl + OH^- \longrightarrow$

(e) 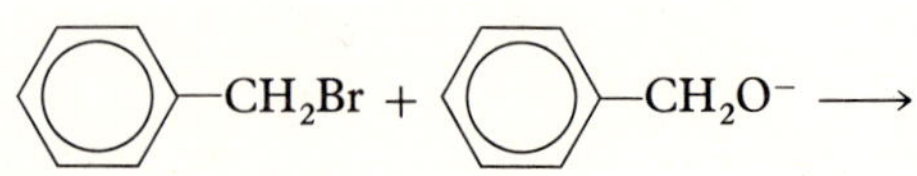

14-10. For each of the following alkyl halides, (1) circle the hydrogens that could be lost in an elimination reaction and (2) designate the *equivalent* adjacent hydrogens:

(a) $(CH_3CH_2)_3CBr$ (b) 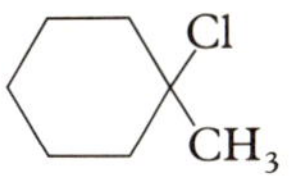

(c) 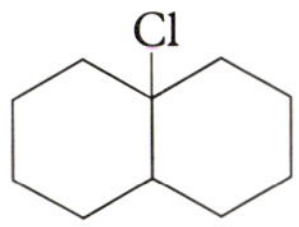

14-11. Write equations for the elimination reactions for the compounds in Problem 14-10. Show all possible isomeric products.

14-12. Write the structures of the substitution and elimination products of the following reactions:

(a) $(CH_3)_2CClCH_2CH_2CH_3 + Na^+ \ ^-CN \longrightarrow$

(b) 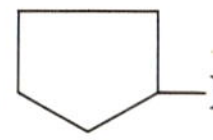$-Br + Na^+ \ ^-OCH_3 \longrightarrow$

(c) $CH_3CH_2CH_2I + K^+ \ ^-OC(CH_3)_3 \longrightarrow$

CHAPTER 15 ALCOHOLS, PHENOLS, AND ETHERS

OBJECTIVES

Naming the Alcohols *Name alcohols by the IUPAC system and as "alcohols."* * *Write structures for primary, secondary, and tertiary alcohols.*

Reactions of Alcohols and Phenols *Write equations for the reactions of primary and secondary alcohols with an oxidizing agent such as hot $KMnO_4$ solution.* * *Contrast the reactivities of phenols and alcohols toward NaOH.* * *Give equations for the preparation of salts of phenols and alcohols.* * *Describe some uses of phenolic compounds.* * *Write equations for the substitution reactions of alcohols with strong acids.* * *Write equations for the dehydration reactions of alcohols.* * *Write equations for the formation of esters from alcohols and carboxylic acids.* * *Write equations for the formation of inorganic esters from alcohols and H_2SO_4 and HNO_3.* * *Describe by equations the chemical properties of sulfate esters of alcohols.* * *Describe some of the uses of nitrate esters and phosphate esters of alcohols.*

Carbon, hydrogen, and oxygen—these are the "Big Three" elements in organic compounds in nature. Over 90 percent of the weight of the human body is made up of these three elements. Oxygen can form a variety of functional groups in combination with carbon. Oxygen forms covalent bonds by sharing *two* pairs of electrons. Thus, oxygen can form only two types of bonds to carbon: single bonds and double bonds.

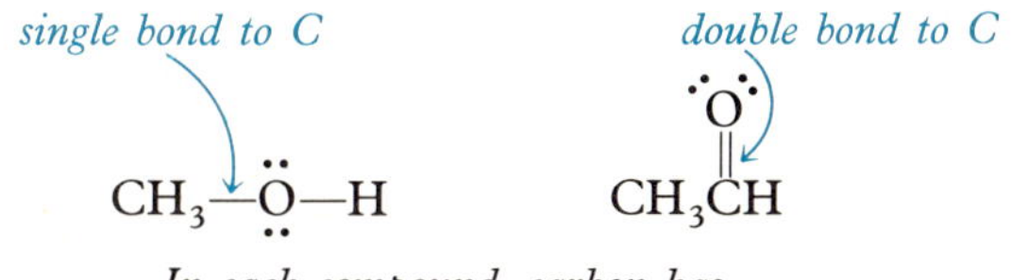

In each compound, carbon has four bonds and oxygen, two bonds.

Although there are only two ways that oxygen can bond to carbon, a variety of types of compounds can result. In this chapter, we will deal with compounds that contain the carbon-oxygen single bond. In Chapters 16 and 17, we will study compounds that contain the carbon-oxygen double bond.

There are three classes of compounds that contain oxygen with a single sigma bond to carbon: *alcohols, phenols,* and *ethers*. The alcohols and phenols both contain a hydroxyl group (—OH). In the alcohols, the hydroxyl group is bonded to an alkyl group, while in the phenols the hydroxyl group is attached to an aromatic ring. (Because of the interactions of the filled orbitals of oxygen with the aromatic pi cloud, the chemical properties of phenols are sometimes different from those of alcohols.)

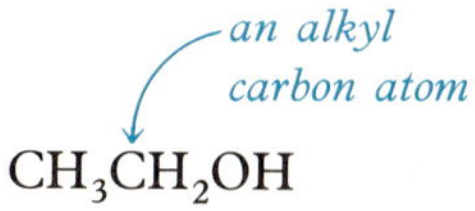

ethyl alcohol

an alcohol

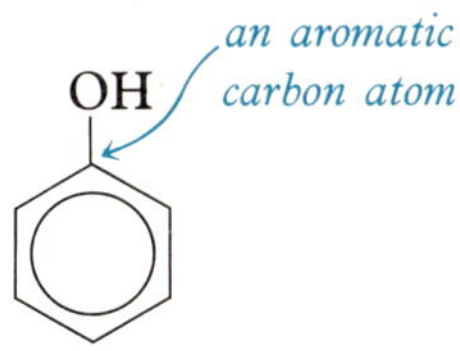

phenol

a phenol

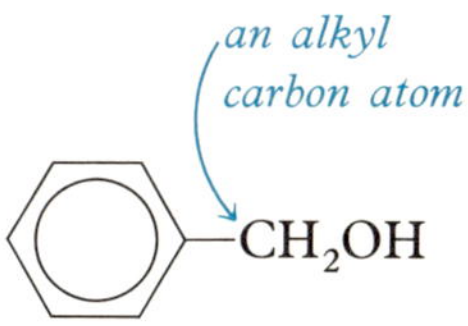

benzyl alcohol

an alcohol

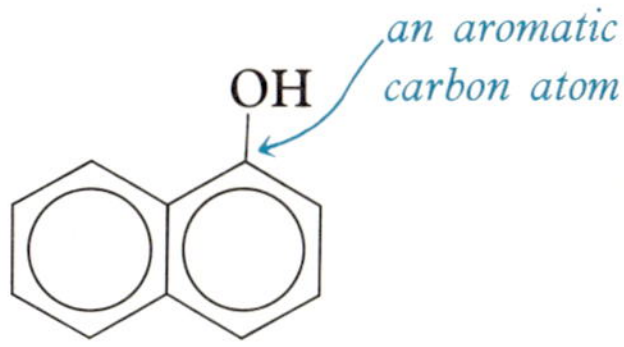

α-naphthol

a phenol

Like water, alcohols and phenols are capable of hydrogen bonding. An alcohol molecule may form a hydrogen bond with another alcohol molecule or with a water molecule. The lower-molecular weight alcohols are therefore water-soluble. The sugars (Chapter 22) are of high molecular weight, but contain several hydroxyl groups. These compounds, too, are generally water-soluble.

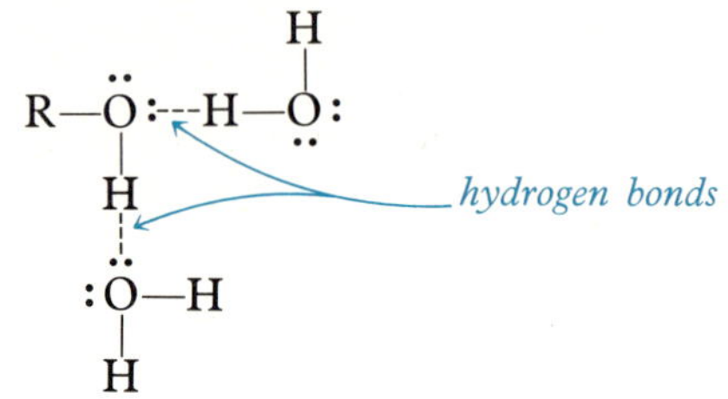

As a minor example of how we can use the hydrogen bonding between water and an alcohol to our advantage, consider automobile gas-line antifreeze. Gasoline will not freeze in the coldest of winter nights; however, traces of water in the gasoline may freeze and clog the carburetor and other engine parts. Methyl alcohol (CH_3OH) burns cleanly, dissolves in gasoline, and attracts water molecules. In cold winter areas, if a can of methyl alcohol is added to the gasoline tank of an automobile, the driver need not worry about a frozen gas line.

Ethers are the class of compound in which oxygen is bonded to *two* carbon atoms (either alkyl or aromatic). The most commonly encountered ether is diethyl ether, which has two ethyl groups attached to the oxygen. Diethyl ether is a volatile liquid (boiling point 35.4°C) that is used as a surgical anesthetic. This compound is also a widely used laboratory solvent.

$$CH_3CH_2OCH_2CH_3$$

diethyl ether
also called ethyl ether, or ether

In many respects, the chemistry of ethers is similar to that of the hydrocarbons. The ethers are not easily attacked by most mild laboratory reagents. However, ethers, like hydrocarbons, burn readily in the air if ignited. Diethyl ether, because of its high volatility, is especially dangerous in this regard.

$$\underset{\text{diethyl ether}}{CH_3CH_2OCH_2CH_3} + 6\,O_2 \longrightarrow 4\,CO_2 + 5\,H_2O + \text{Energy}$$

NAMING THE ALCOHOLS

Alcohols may be named by the IUPAC system or as alkyl alcohols. In the IUPAC system, the name of the hydrocarbon is used as the parent name. The final letter *-e* is dropped from the alkane name, and the ending *-ol* is added. Prefix numbers are used to show the position of the hydroxyl group.

$CH_3CH_2CH_2CH_3$

butane

$CH_3CH_2CH_2CH_2OH$

1-butanol

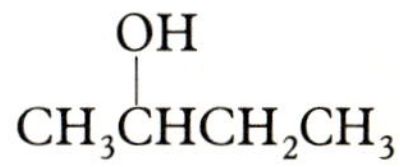

2-butanol

Simple alcohols are also named by using the name of the alkyl group that is attached to the oxygen of the hydroxyl group and adding the word *alcohol.*

CH_3OH

methyl alcohol (methanol)

a toxic alcohol, sometimes called wood alcohol

CH_3CH_2OH

ethyl alcohol (ethanol)

found in alcoholic beverages, sometimes called grain alcohol

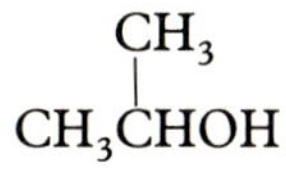

isopropyl alcohol (2-propanol)

used as rubbing alcohol

Alcohols may also be classified as *primary, secondary,* or *tertiary,* depending upon the number of carbon atoms attached directly to the "head" carbon, the carbon with the OH group. Alcohols of these three types exhibit somewhat different behavior in chemical reactions.

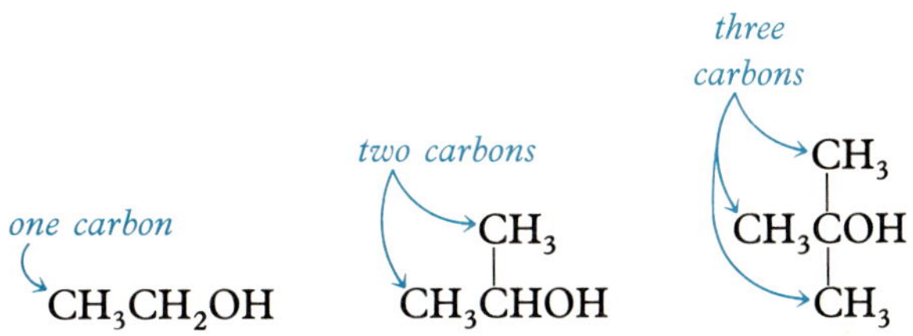

primary *secondary* *tertiary*

Sample Problem: Name the following alcohol by the IUPAC system, and classify the alcohol as primary, secondary, or tertiary:

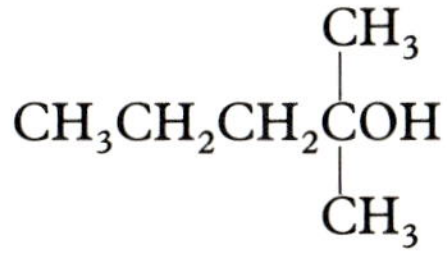

Solution: 2-methyl-2-pentanol, a tertiary alcohol.

REACTIONS OF ALCOHOLS AND PHENOLS

OXIDATION. Like most other organic compounds, alcohols and phenols burn when ignited. Flaming shish kabob and cherries jubilee depend upon the ignition of the vapors of ethyl alcohol from warmed brandy or another beverage with a high alcohol content.

$$\underset{\text{ethyl alcohol}}{CH_3CH_2OH} + 3\,O_2 \xrightarrow{\text{spark}} 2\,CO_2 + 3\,H_2O + \text{Energy}$$

The term *proof,* which refers to the alcohol content of a beverage, is related to the combustion of ethyl alcohol. In days of old, a sample of a shipment of rum (or other beverage) was poured on some gunpowder and ignited. If the gunpowder burned after the alcohol had burned away, it was *proof* that the rum was not watered down. (The number given as proof is twice the percent by volume of alcohol. For example, a whiskey that is 80 proof is 40 percent by volume alcohol.)

Because alcoholic beverages are taxed in most countries of the world, ethyl alcohol that is not taxed (that used for laboratory or industrial purposes) is often *denatured;* that is, a small quantity of toxic material is added to the alcohol so that it is unfit to drink. The denaturing agent used is dependent upon the use for which the alcohol is intended.

Alcohols may be partially oxidized by selective oxidizing agents. In general, we find that primary alcohols are oxidized by laboratory oxidizing agents to carboxylic acids. This reaction proceeds by way of an aldehyde (a compound with R and H attached to C═O), which cannot usually be isolated because of the ease of its further oxidation. Secondary alcohols form ketones (compounds with two alkyl groups attached to C═O). These reactions are similar to those of the oxidation of alkenes. (See Chapter 13.) Tertiary alcohols and phenols do not undergo these simple oxidations.

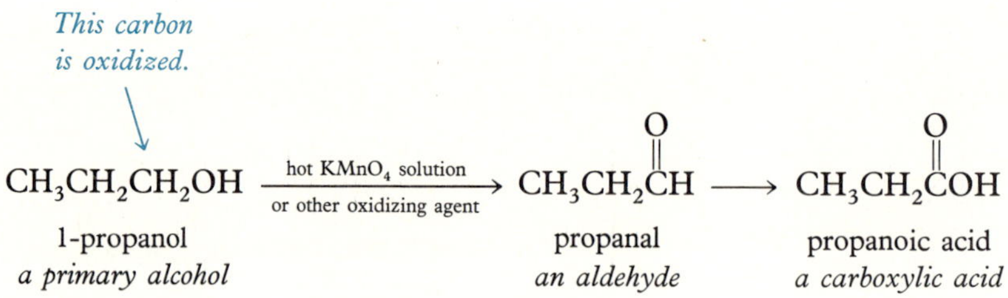

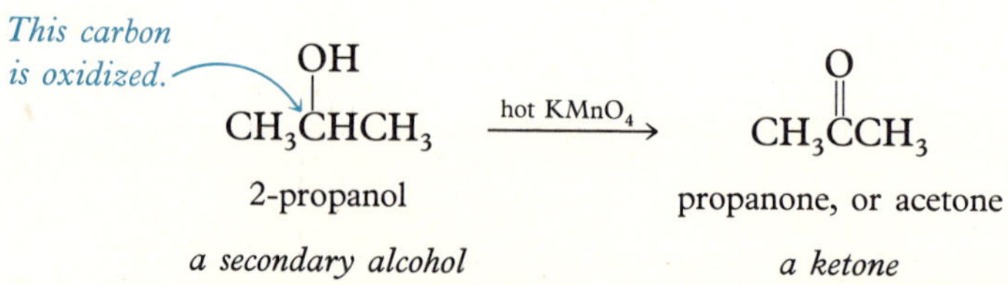

In living cells, oxidation of alcohols is catalyzed by enzymes. The following example shows the oxidation of a hydroxy acid to a keto acid, which in turn is normally cleaved into acetate ions. In diabetics, however, the conversion to acetate is incomplete, and some of the keto acids spill over into the urine. This topic will be discussed in Chapter 27.

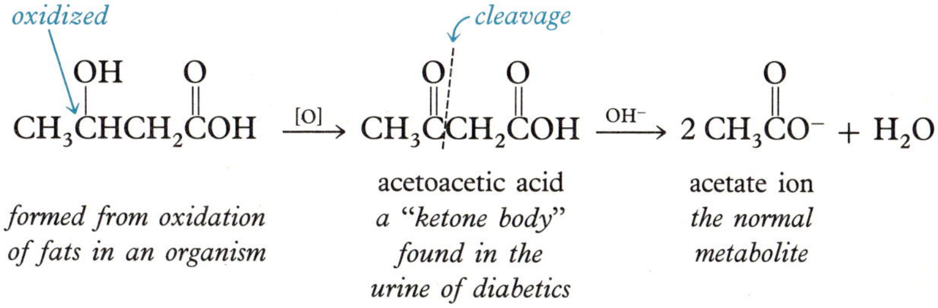

Ethyl alcohol does not occur naturally in the body. When it is ingested, however, ethyl alcohol is oxidized enzymatically to acetaldehyde, then to acetate ions, which are found in combination with coenzyme A. (See Chapter 17.) From here, the acetate ions can be converted to fats and other compounds or can be used for energy. (See Chapter 26.)

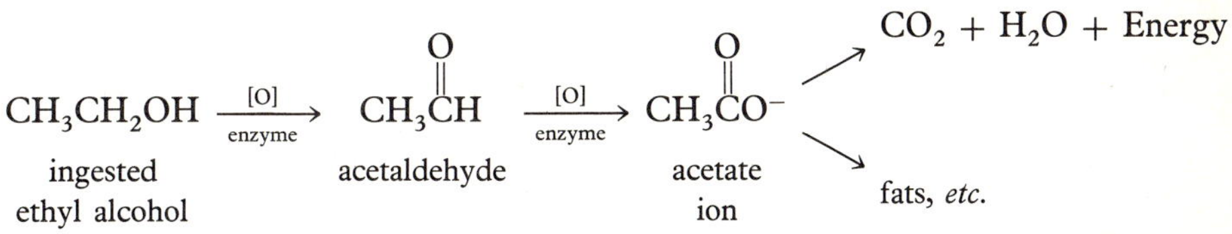

Methyl alcohol is poisonous because its oxidation results in the toxic compound formaldehyde.

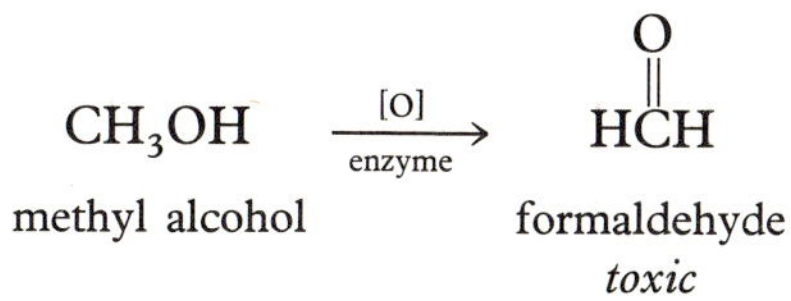

REACTION WITH BASES. The distinguishing feature between the alcohols and the phenols is that phenols are quite acidic, while alcohols are not. Although phenols are not as acidic as carboxylic acids, they react with dilute aqueous sodium hydroxide to yield sodium salts. (For comparison, the dissociation constants are given in the following examples.)

$$\underset{\text{acetic acid}}{CH_3CO_2H} + Na^+\ OH^- \longrightarrow \underset{\text{sodium acetate}}{CH_3CO_2^-\ Na^+} + H_2O$$

$K = 1.85 \times 10^{-5}$

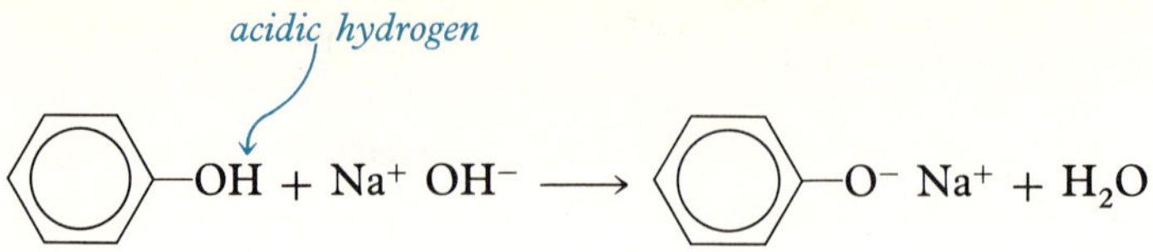

phenol　　　　　　　　　　sodium phenoxide

$K = 1.28 \times 10^{-10}$

$$CH_3CH_2OH + Na^+\ OH^- \longrightarrow \text{No reaction}$$

ethyl alcohol

$K \cong 10^{-18}$

Salts of alcohols may be formed by using extremely strong bases, such as $Na^+\ NH_2^-$, or with sodium metal. Both these reagents act as stronger bases than sodium hydroxide.

$$\underset{\text{ethyl alcohol}}{CH_3CH_2OH} + \underset{\substack{\text{sodamide}\\ \textit{an extremely}\\ \textit{strong base}}}{Na^+\ NH_2^-} \longrightarrow \underset{\text{sodium ethoxide}}{CH_3CH_2O^-\ Na^+} + NH_3$$

$$2\ CH_3CH_2OH + \underset{\substack{\text{sodium}\\ \text{metal}}}{Na} \longrightarrow 2\ CH_3CH_2O^-\ Na^+ + H_2\uparrow$$

Because of its acidity, phenol was originally called *carbolic acid*. The English surgeon Joseph Lister instigated its use as a hospital antiseptic in the late 1800's. Before this time antiseptics had not been used, and a surgical patient was very likely to succumb to infection, if not to unskilled surgery. Since then, more effective antiseptics that are less irritating to the skin have replaced phenol. Two such compounds are hexachlorophene and hexylresorcinol. (Hexachlorophene, once used in antiseptics and in deodorant soaps, is no longer routinely used this way because it can be absorbed through the skin.)

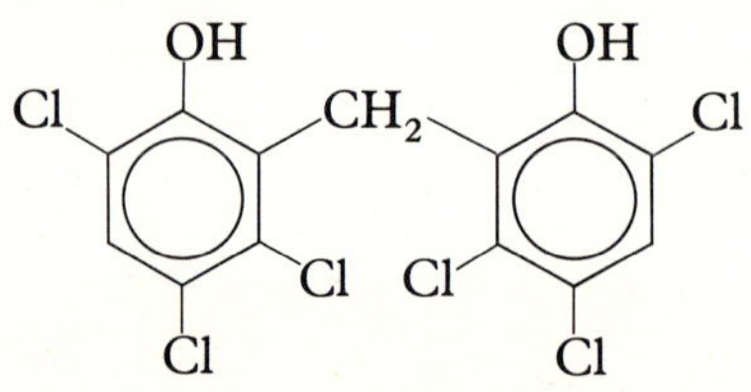

hexachlorophene

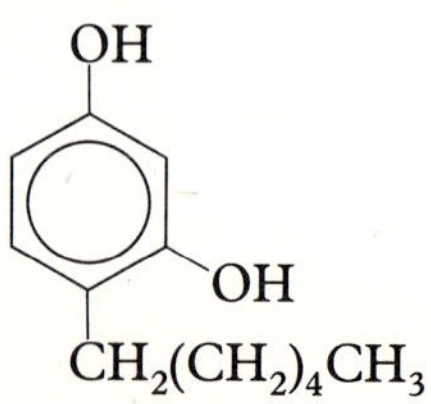

hexylresorcinol

Although phenol itself is no longer used as an antiseptic, the effectiveness of an antiseptic is still rated by a *phenol coefficient*. This is a number that compares antiseptic action with that of a 5 percent phenol solution.

Sample Problem: Complete the following equations. In (c), show only the organic product.

(a) 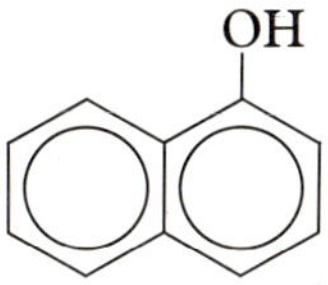 + NaOH $\longrightarrow$

(b) $CH_3OH + Na \longrightarrow$

(c) 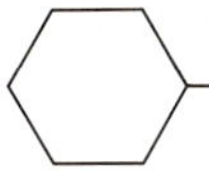 —OH + hot $KMnO_4 \longrightarrow$

Solution:

(a) 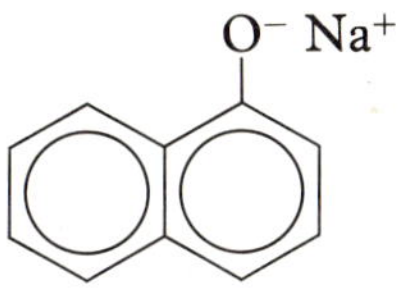 + H_2O

(b) $CH_3O^- Na^+ + H_2$

(c) 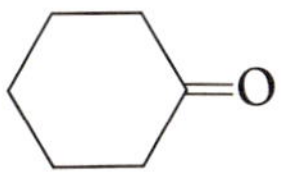

SUBSTITUTION REACTIONS OF ALCOHOLS. The hydroxyl group of an alcohol may be displaced by another group under acidic conditions. For example, the reaction of ethyl alcohol with hydrogen bromide yields ethyl bromide and water. In this reaction, the carbon-oxygen bond is broken and another group (in this example, bromide) becomes attached to the carbon.

$$CH_3CH_2{-}OH + H{-}Br \longrightarrow CH_3CH_2Br + H_2O$$

ethyl alcohol ethyl bromide

$$CH_3CH_2CH_2OH + HCl \xrightarrow{ZnCl_2} CH_3CH_2CH_2Cl + H_2O$$

propyl alcohol propyl chloride

Because HCl is a slightly weaker acid than HBr or HI, a catalyst (such as $ZnCl_2$) is often required for its reaction with alcohols. A catalyst is usually unnecessary for reaction with HBr or HI.

The reaction of alcohols with acids is similar to the reaction of alkyl halides with bases in that one group is substituted for another. The difference is in the reaction conditions involved—a base for alkyl halides, an acid for alcohols.

$$\underset{\text{an alkyl halide}}{RX} + \underset{\text{a base}}{OH^-} \longrightarrow ROH + X^-$$

$$\underset{\text{an alcohol}}{ROH} + \underset{\text{an acid}}{HX} \longrightarrow RX + H_2O$$

These displacement reactions are not observed with the phenols. The aromatic carbon-oxygen bond of the phenols is too strong to be broken by these acidic reagents.

DEHYDRATION OF ALCOHOLS. Besides resulting in the substitution of alcohols, acidic reagents can also cause an alcohol to undergo the loss of water and form an alkene. Because water is lost from the alcohol, this type of a reaction is called a *dehydration* reaction. Sulfuric acid, a good dehydrating agent, is usually the acid used in this reaction.

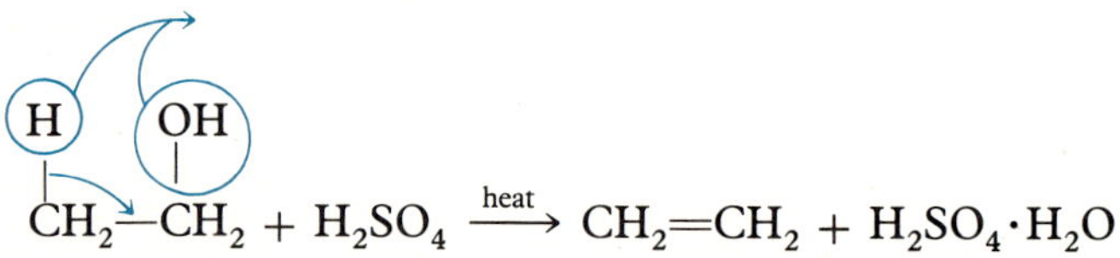

ethyl alcohol ethylene

Sample Problem: Write equations for a substitution reaction (with HBr) and an elimination reaction (with H_2SO_4) of cyclohexanol.

Solution: *Substitution:*

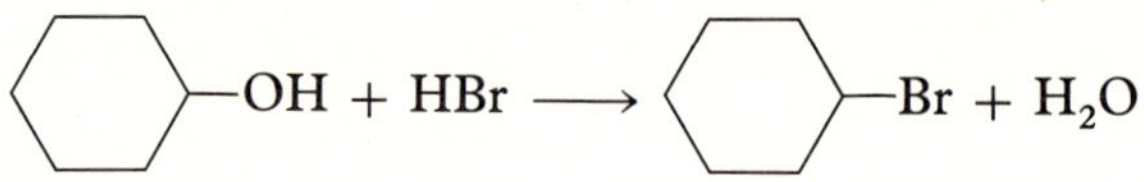

Elimination:

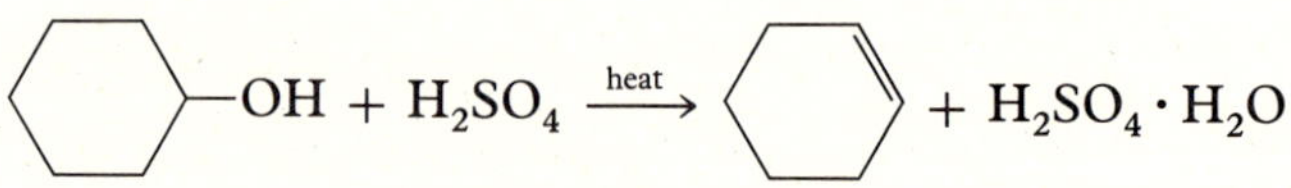

FORMATION OF ESTERS. An alcohol can react with an organic acid to form a product called an *ester*. A trace of mineral acid, such as H_2SO_4, is used to catalyze the reaction. The carboxylic acids and esters will be discussed at greater length in Chapter 17. (In the following equation, R′ is used to show that the group is not necessarily the same as R.)

$$CH_3\overset{\overset{\large O}{\|}}{C}{-}OH + H{-}OCH_3 \underset{}{\overset{H^+}{\rightleftharpoons}} CH_3\overset{\overset{\large O}{\|}}{C}OCH_3 + H_2O$$

acetic acid — methyl alcohol — methyl acetate, *an ester*

Generalized:

$$R\overset{\overset{\large O}{\|}}{C}OH + HOR' \overset{H^+}{\rightleftharpoons} R\overset{\overset{\large O}{\|}}{C}OR' + H_2O$$

a carboxylic acid — *an alcohol* — *an ester*

An alcohol can also form esters with some of the inorganic acids. These products are called *inorganic esters of alcohols. Sulfate esters* are formed by the reaction of sulfuric acid and an alcohol or phenol.

$$ROH + HO\overset{\overset{\large O}{\|}}{\underset{\underset{\large O}{\|}}{S}}OH \xrightarrow{\text{cold}} RO{-}\overset{\overset{\large O}{\|}}{\underset{\underset{\large O}{\|}}{S}}{-}OH + H_2O$$

an alkyl sulfate

The sulfates can undergo a variety of different reactions that are dependent upon the reaction conditions (temperature, concentration, and so forth). These reactions are often competitive. In any given reaction, a mixture of all these products can be expected.

Reactions of alkyl sulfates:

$$ROSO_3H + \text{excess } H_2O \longrightarrow ROH + H_2SO_4$$

an alcohol

$$ROSO_3H + \text{excess } ROH \longrightarrow ROR + H_2SO_4$$

an ether

$$RCH_2CH_2OSO_3H \xrightarrow{\text{heat}} RCH{=}CH_2 + H_2SO_4$$

an alkene

The sulfate esters of long hydrocarbon-chain alcohols are used as detergents. (See Chapter 6.) Sulfates are also formed in the body. For example, when phenols are detoxified by body processes, they are often eliminated in the urine as the sulfate.

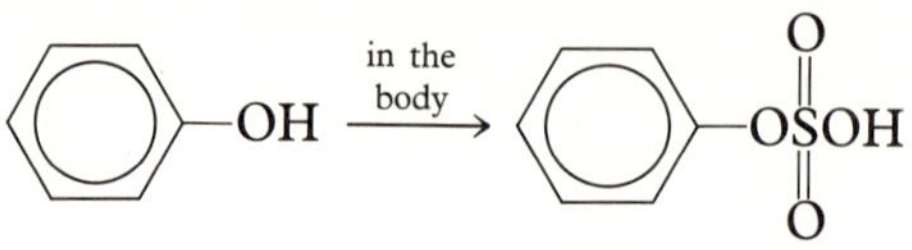

a urinary excretion product of phenol

Nitrate esters can be prepared from nitric acid and alcohols.

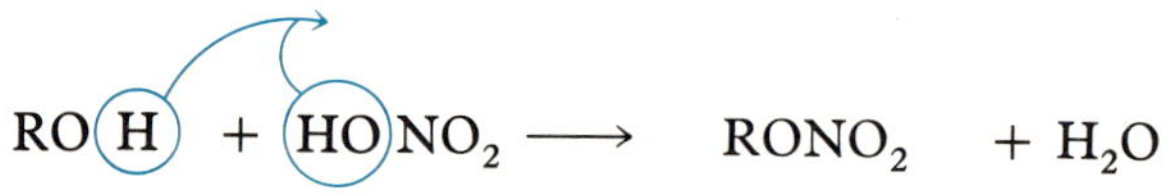

an alcohol nitric acid *an alkyl nitrate*

Because the combination of a strong oxidizing agent (the nitro group) and a reducing agent (the reduced carbons of an alcohol) within one molecule is an explosive situation, many of the nitrate esters are powerful explosives. Nitroglycerin, the trinitrate ester of glycerol, is just one example of this type of compound.

The nitrate esters are also strong smooth-muscle relaxants and vasodilators (agents that relax the stomach and heart muscles and dilate blood vessels). They find wide clinical use for lowering blood pressure and in treatment of the heart condition *angina pectoris*. Nitroglycerin and pentaerythritol tetranitrate are commonly used for these purposes.

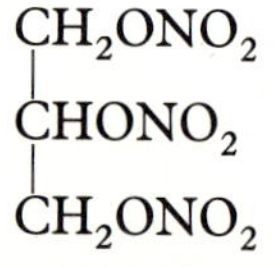

nitroglycerin

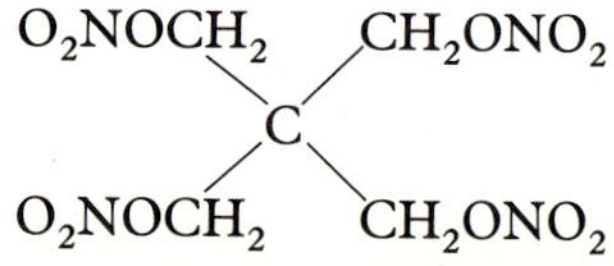

pentaerythritol tetranitrate

Phosphate esters are the inorganic esters of phosphoric acid and alcohols. In biological systems, phosphate esters and their anhydrides provide for both storage and transfer of chemical energy for living systems. (See Chapter 25.) Phosphate groups are also found in nucleic acids. (See Chapter 28.)

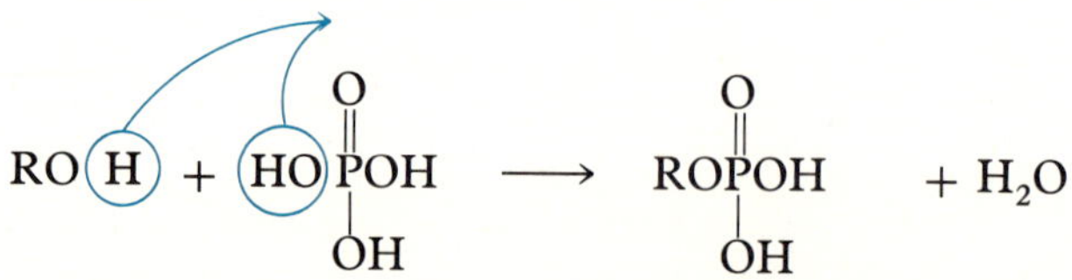

an alkyl phosphate

Synthetic organic phosphates are often potent poisons, and many of this class of compounds are used as insecticides.

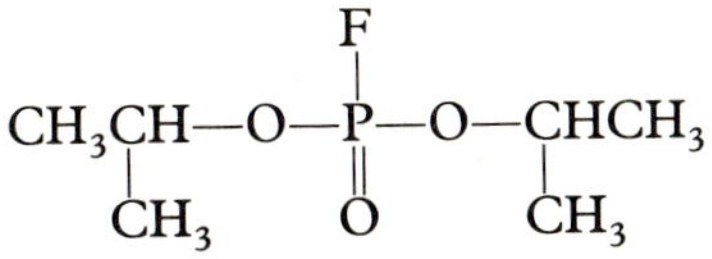

DFP

a potent nerve gas

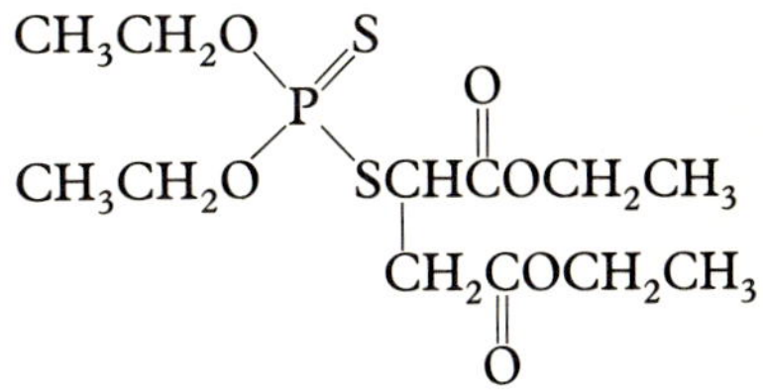

malathion

an insecticide

STUDY QUESTIONS

15-1. Identify each of the following compounds as an ether, an alcohol, or a phenol:

(a) 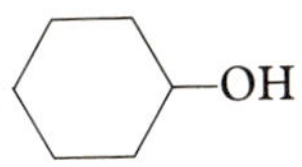 (b) 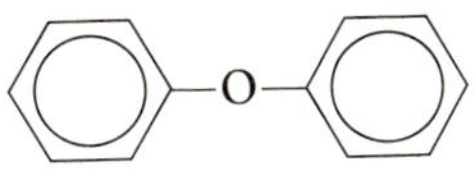

(c) 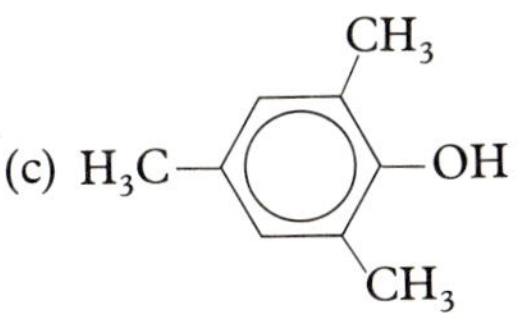 (d) 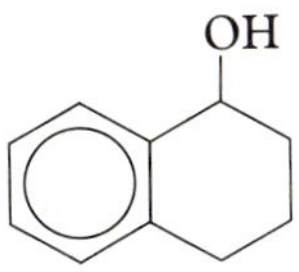

15-2. Show how hydrogen bonds form between the following:
(a) 1-propanol and 1-propanol
(b) cyclohexanol and water

15-3. Name the following by (1) the IUPAC system and (2) the functional group method (that is, as alkyl alcohols).

(a) $(CH_3)_2CHOH$ (b) $CH_3CH_2CH_2CH_2OH$

(c) 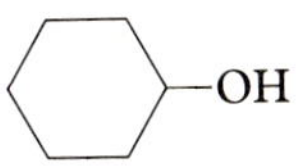 (d) $(CH_3)_3COH$

15-4. Classify the compounds in the preceding problem as primary, secondary, or tertiary.

15-5. Which of the following compounds could be used as denaturing agents for ethyl alcohol?

(a) sugar (b) CH_3CHO (with C=O)
(c) $(CH_3)_2CHOH$ (d) H_2O

15-6. Write an equation for the partial oxidation of cyclohexanol. What would be the products of the complete oxidation of cyclohexanol?

15-7. Is the reaction of sodium metal with ethyl alcohol an acid-base reaction or an oxidation-reduction reaction? Explain.

15-8. Which of the following compounds (if any) would you expect to form a salt when treated with NaOH?

(a) CH_3CH_2OH (b) 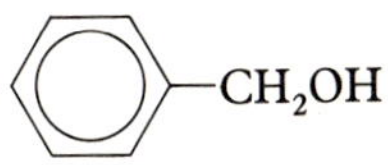

(c) 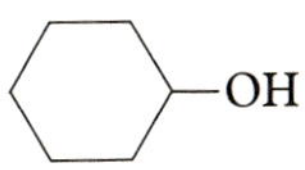 (d) 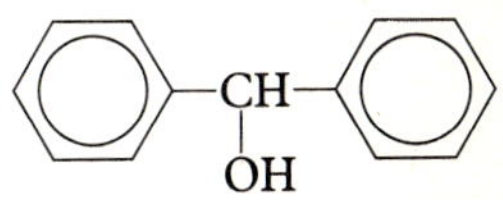

15-9. Tell what reagent you would use to prepare the following from an alcohol or a phenol:
(a) potassium phenoxide (the potassium salt of phenol)
(b) sodium propoxide (the sodium salt of 1-propanol).

15-10. Complete the following substitution reactions. (If no reaction is expected to occur, write *No reaction.*)

(a) 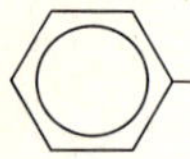 $—OH + HCl \longrightarrow$

(b) 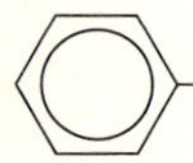 $—CH_2OH + HBr \longrightarrow$

(c) $CH_3CH_2CH_2CH_2OH + HI \longrightarrow$

(d) $(CH_3)_2CHOH + HBr \longrightarrow$

15-11. How many alkenes would you expect from the dehydration of each of the following alcohols?

(a) $CH_3CH_2CH_2\overset{CH_3}{\overset{|}{C}}HCH_2OH$

(b) 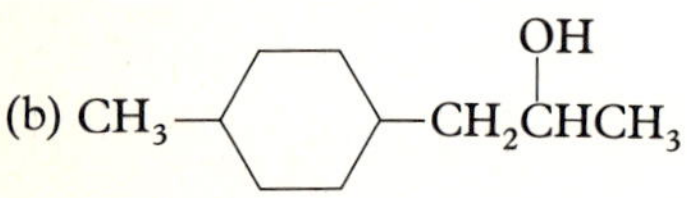

(c) 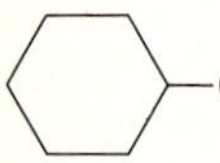 $—OH$

(d) 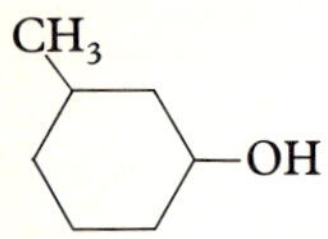

15-12. Show both the elimination and the substitution products that would be formed when the following alcohol is treated with HI:

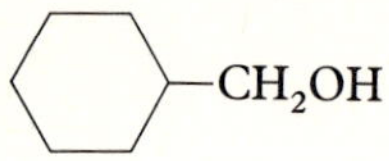

15-13. Write the equation for the formation of an ester from benzoic acid ($C_6H_5CO_2H$) and *sec*-butyl alcohol.

15-14. Which of the following compounds are inorganic esters of alcohols?

(a) $CH_3O\underset{\underset{O}{\|}}{\overset{\overset{O}{\|}}{S}}OH$

(b) 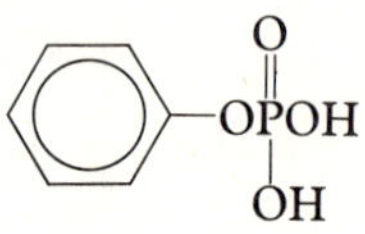

(c) 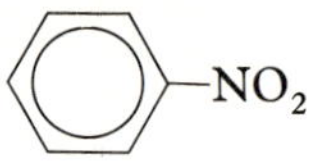

(d) $CH_3\underset{\underset{O}{\|}}{\overset{\overset{O}{\|}}{S}}OCH_3$

15-15. Complete the following equations:

(a) $CH_3CH_2\overset{CH_3}{\overset{|}{C}}HOH + H_2SO_4 \xrightarrow{\text{heat}}$

(b) $CH_3CH_2CO_2H +$ 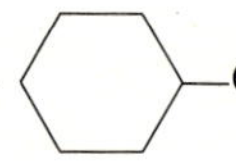 $—OH \overset{H^+}{\rightleftharpoons}$

(c) $CH_3CH_2CH_2CH_2OH + H_2SO_4 \xrightarrow{\text{warm}}$

(d) $CH_3OSO_3H + H_2O \longrightarrow$

(e) $CH_3OSO_3H + CH_3CH_2OH \longrightarrow$

(f) $CH_3CH_2CH_2OSO_3H \xrightarrow{\text{heat}}$

(g) 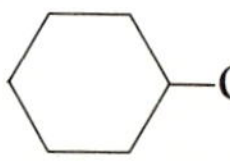 $—OH + HNO_3 \longrightarrow$

CHAPTER 16 ALDEHYDES AND KETONES

OBJECTIVES

The Carbonyl Group *Describe the structure and polarity of the carbonyl group.*

Classes of Carbonyl Compounds *Define aldehyde and ketone and show examples of each.*

Naming Aldehydes and Ketones *Give the structures for formaldehyde, acetaldehyde, and acetone. * Name aldehydes and ketones by the IUPAC system.*

Important Aldehydes and Ketones *Describe some of the uses of formaldehyde and acetone.*

Addition of Alcohols to the Carbonyl Group *Write equations for the reaction of alcohols with aldehydes. * Define hemiacetal and acetal. Write a structure for each.*

Reactions of Aldehydes and Ketones with Nitrogen Compounds *Write equations for the reactions of NH_3, RNH_2, and NH_2NH_2 with aldehydes and ketones.*

Reduction of Aldehydes and Ketones *Write equations for the reduction of aldehydes and ketones.*

Oxidation of Aldehydes *Write equations for the reactions of aldehydes with mild oxidizing agents.*

The Effect of the Carbonyl Group on Adjacent Groups *Write an equation that shows the ionization of a compound that contains an α-hydrogen.* * *Write an equation to illustrate tautomerism.*

THE CARBONYL GROUP

The group that contains the carbon-oxygen double bond is called the *carbonyl group*. The carbonyl group is found as part of a large number of functional groups (such as —CO_2H) and is found in proteins, fats, carbohydrates, hormones, and other biological compounds.

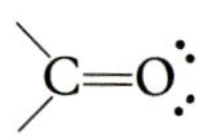

the carbonyl group

The chemistry of carbonyl compounds results partly from the structure of the carbonyl group. The carbon atom and the oxygen atom are joined by a *sigma bond*, formed from an sp^2 orbital of the carbon and an orbital of the oxygen, plus a *pi bond*, formed from a p orbital of the carbon and an orbital of the oxygen. In addition, the oxygen atom has *two filled orbitals* that contain unshared electrons.

Unlike a carbon-carbon double bond, the carbon-oxygen double bond is *polar*. The carbonyl group contains loosely held pi electrons that are readily drawn toward the more electronegative oxygen atom. Thus, the carbon atom of the group carries a partial positive charge, while the oxygen atom carries a partial negative charge.

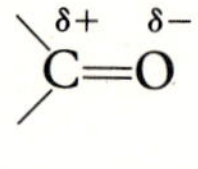

polarity of the carbonyl group

Because of its polarity, a variety of reactants are able to attack the carbonyl group. A positively charged reagent can attack the negative oxygen, while a negatively charged reagent can attack the positive carbon. When reagents add across the carbonyl group, the positive portion of the reagent always becomes attached to the oxygen of the carbonyl. The negative portion of the reagent always becomes attached to the carbon of the carbonyl group.

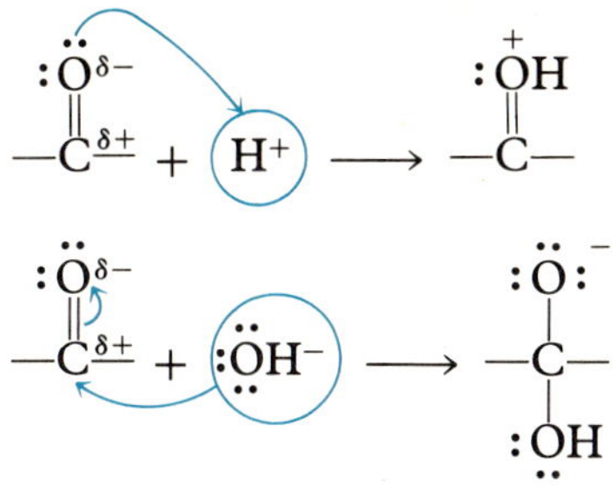

Carbonyl compounds undergo hydrogen bonding with compounds that contain NH or OH bonds. It is just such hydrogen bonds that hold protein molecules in their shapes. We shall see in Chapter 23 that the shapes of protein molecules are one of their distinguishing features. For example, muscle fibers have a different shape from that of hemoglobin.

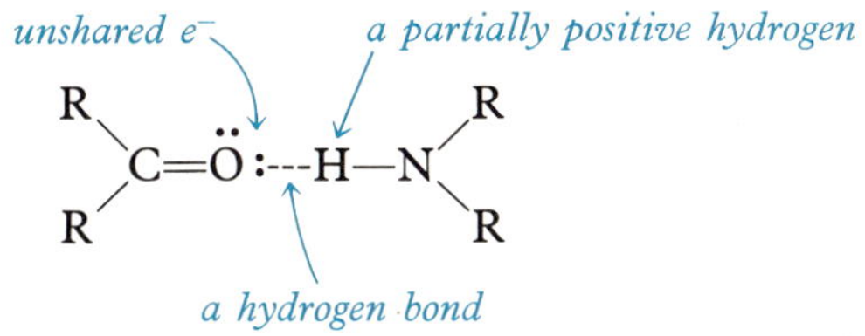

CLASSES OF CARBONYL COMPOUNDS

The carbonyl group is found in a variety of compounds. These are summarized in Table 16-1. In this chapter, we will look at the *aldehydes* (compounds in which the

Table 16-1. Some Classes of Carbonyl Compounds

Name of Class	Functional Group	Example
Aldehyde	—C(=O)—H	$CH_3C(=O)H$
Ketone	—C(=O)—	$CH_3C(=O)CH_3$
Carboxylic acid	—C(=O)—OH	$CH_3C(=O)OH$
Ester of a carboxylic acid	—C(=O)—O—	$CH_3C(=O)OCH_2CH_3$
Amide	—C(=O)—N<	$CH_3C(=O)NH_2$

carbonyl group has a hydrogen attached) and *ketones* (compounds in which the carbonyl group is attached to two carbon atoms).

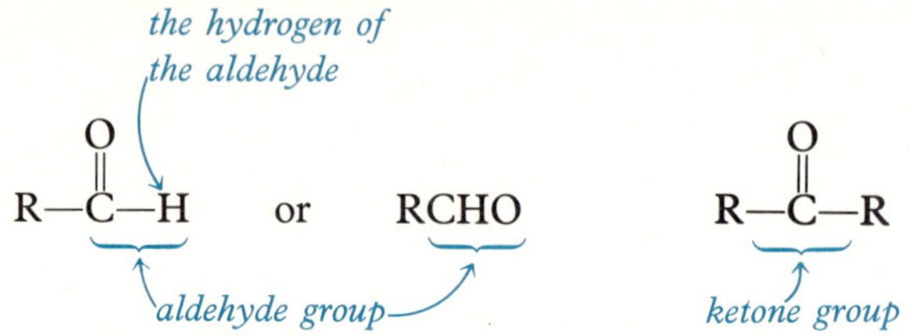

Sample Problem: Circle and name the functional groups in the following compounds:

(a) 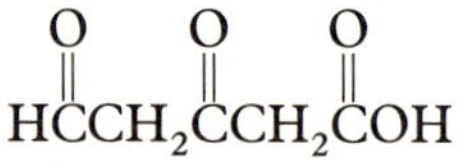 (b) 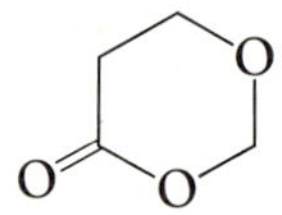

Solution:

(a) 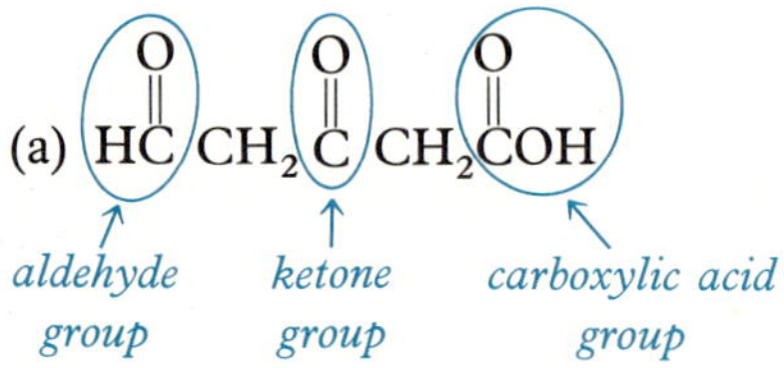

(b) 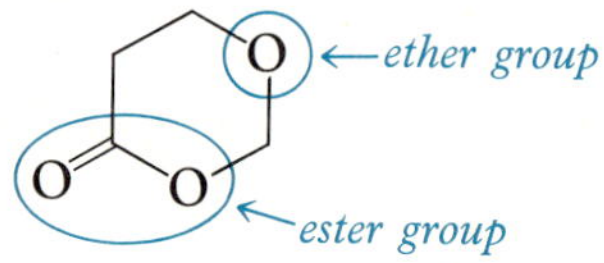

NAMING ALDEHYDES AND KETONES

In the IUPAC system, aldehydes are named by dropping the final *e* from the name of the parent alkane and adding *al.*

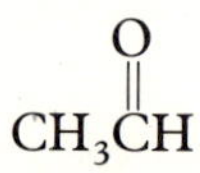

ethanal

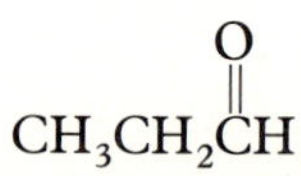

propanal

Notice that the carbonyl carbon in an aldehyde must be at the end of a carbon chain because it has a hydrogen atom attached. In numbering the parent chain, the carbonyl carbon atom of an aldehyde is considered the first carbon. (However, we do not use the number in the name; it is understood that the aldehyde group is at position 1.)

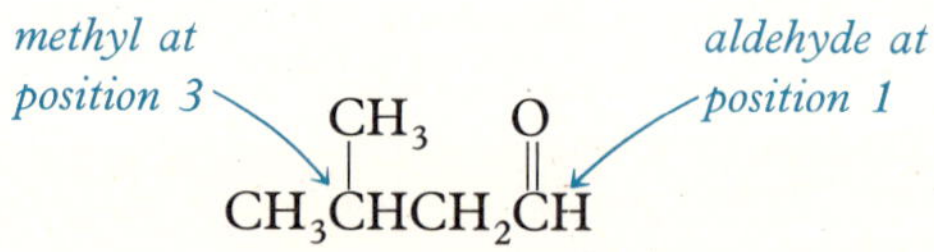

3-methylbutanal

Ketones are named in the IUPAC system with the ending *-one* on the parent name. If there is more than one possible place for a ketone group to appear in a molecule, a prefix number is used to designate the position.

$$CH_3\overset{\overset{\displaystyle O}{\|}}{C}CH_2CH_2CH_3 \qquad\qquad CH_3CH_2\overset{\overset{\displaystyle O}{\|}}{C}CH_2CH_3$$

2-pentanone 3-pentanone

Small aldehydes and ketones are generally called by their trivial names. Some of these are shown in Table 16–2.

Table 16–2. Names of Some Aldehydes and Ketones

Formula	IUPAC Name	Trivial Name
Aldehydes:		
HCHO	Methanal	Formaldehyde
CH_3CHO	Ethanal	Acetaldehyde
CH_3CH_2CHO	Propanal	Propionaldehyde
$CH_3CH_2CH_2CHO$	Butanal	Butyraldehyde
Ketones:		
$CH_3\overset{\overset{O}{\|}}{C}CH_3$	Propanone	Acetone
$CH_3\overset{\overset{O}{\|}}{C}CH_2CH_3$	Butanone	Methyl ethyl ketone

Sample Problem: Write the structure for 3-propyl-1-cyclopentanone.

Solution: *Step 1:* Draw and number the parent skeleton.

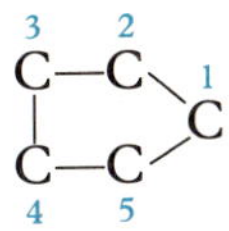

Step 2: Add functional groups and branches: ketone at position 1, propyl at position 3.

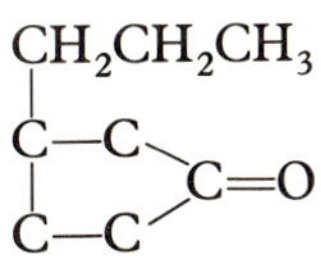

Step 3: Add hydrogens.

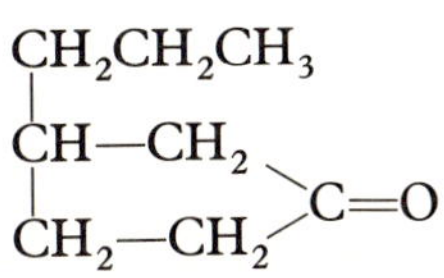

IMPORTANT ALDEHYDES AND KETONES

The gas *formaldehyde* (HCHO) is an irritant formed in photochemical smog. This compound is also encountered in biology classes where, in a water solution called *formalin*, it is used as a preservative for specimens.

Formaldehyde forms polymers readily, both with itself and with other compounds. The formaldehyde polymer, called *paraformaldehyde*, is a solid that regenerates gaseous formaldehyde when heated. This polymer provides a convenient method for shipping and storing formaldehyde as a solid, rather than as a gas. Paraformaldehyde is also a sedative when taken internally and is occasionally used for that purpose in medical practice.

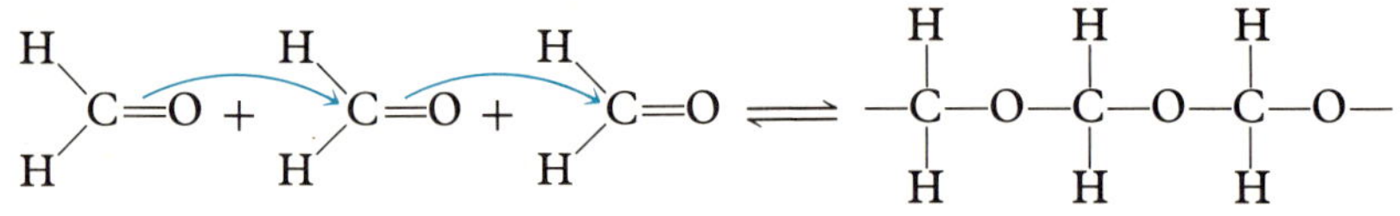

formaldehyde — *portion of a paraformaldehyde molecule*

Industrially, formaldehyde is polymerized with phenol to give Bakelite, a hard plastic used as an insulator and as handles for pots and pans.

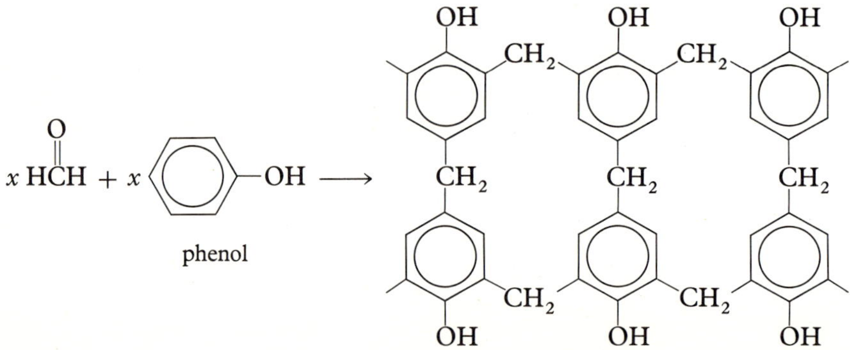

phenol

portion of a Bakelite molecule

Formaldehyde is also used to synthesize such widely sold polymeric products as Formica counter tops, plastic radio cabinets, adhesives, translucent light panels, buttons, and plastic dinnerware (Melmac).

Acetone (CH_3COCH_3) is a ketone that is used as a solvent in industry and the laboratory. It is useful because it is relatively inexpensive and is miscible with water as well as with organic solvents.

Examples of biologically important aldehydes and ketones are the sugar glucose (p. 297) and the hormone estrone (p. 398).

ADDITION OF ALCOHOLS TO THE CARBONYL GROUP

We have already mentioned that, because the carbonyl group is polar, reagents may add to the carbonyl carbon or oxygen. Let us take a look now at some of the addition reactions of these compounds.

The addition reactions of *alcohols* to aldehydes and ketones are catalyzed by a trace of acid and are reversible reactions. In the equilibrium, the aldehyde or ketone is usually favored over the addition product. Therefore, in an equilibrium mixture, we would expect to find mainly the carbonyl compound plus the alcohol and a small amount of product.

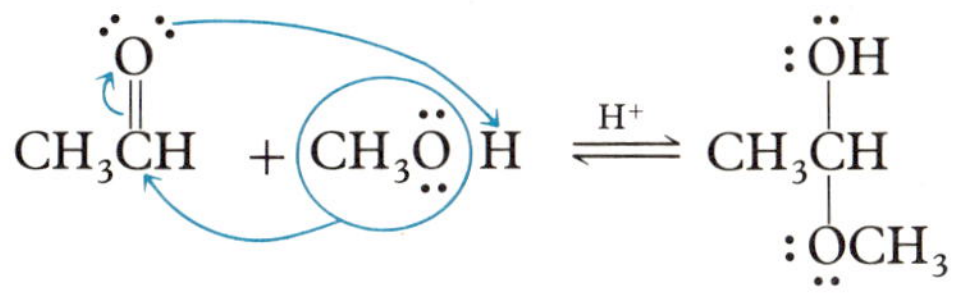

an aldehyde *an alcohol* *a hemiacetal*

The product of the addition of an alcohol to an aldehyde is called a *hemiacetal.* This class of compounds contains an —OH group and an —OR group on the same carbon atom.

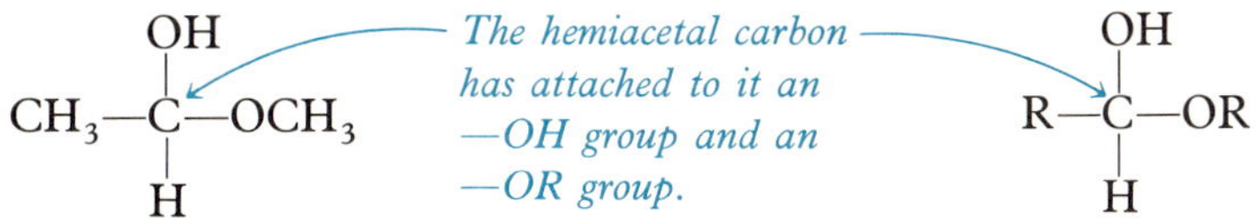

Of utmost importance in sugar chemistry is that a second molecule of an alcohol can undergo reaction with the hemiacetal to give a product with *two* —OR groups on the same carbon. This product is called an *acetal.*

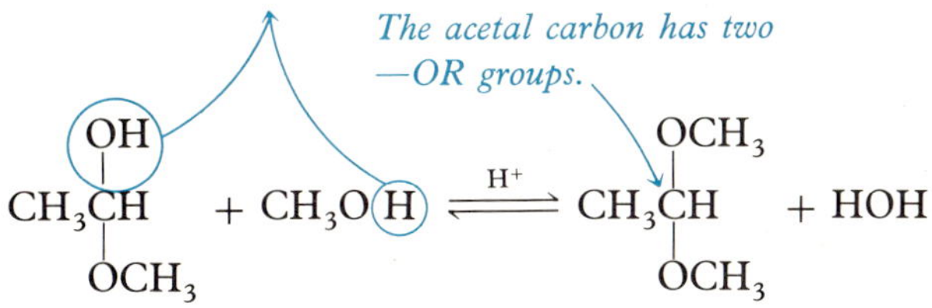

a hemiacetal *an alcohol* *an acetal*

This reaction, too, is reversible in acidic solution. Thus, in an acidic solution of an aldehyde and an alcohol, we will also find the hemiacetal and acetal.

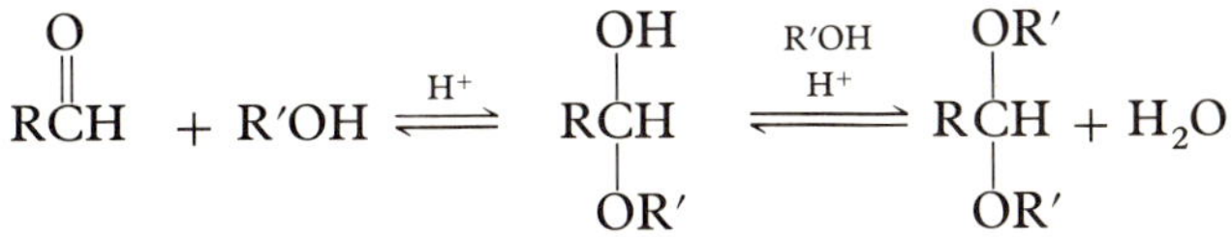

aldehyde *alcohol* *hemiacetal* *acetal*

The importance of hemiacetals and acetals in sugar chemistry is twofold. 1. Five- and six-membered cyclic hemiacetals are quite stable, and sugars are structurally arranged to form these cyclic products.

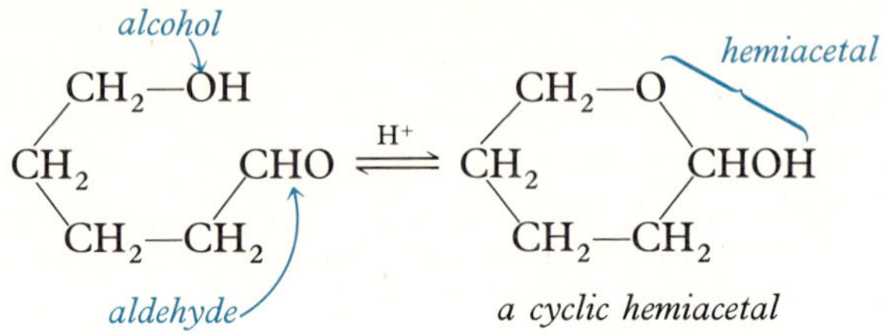

2. Acetals are stable in alkaline and neutral solutions. Thus, in the absence of an acidic catalyst, acetals are not in equilibrium with the aldehyde and hemiacetal. Starch and cellulose are polymers formed from sugar units linked together by acetal bonds.

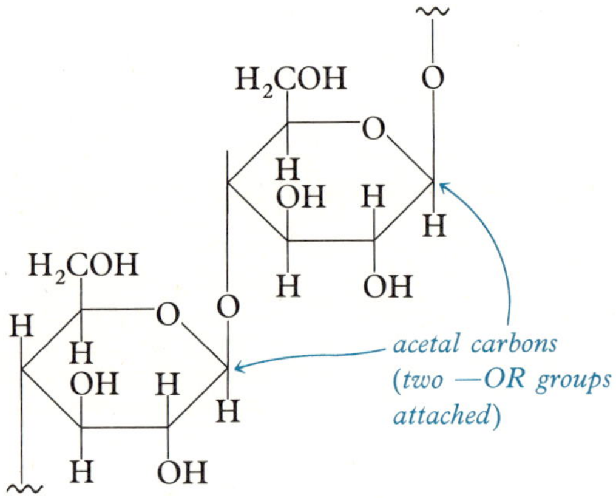

portion of a molecule of cellulose

Sample Problem: Write the structures of the hemiacetal and the acetal from formaldehyde and methyl alcohol.

Solution:

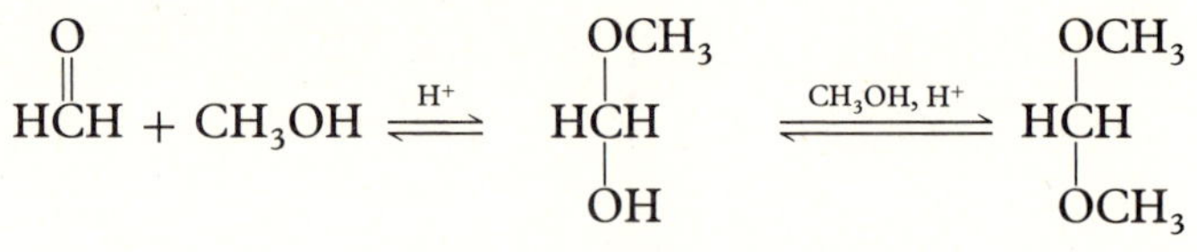

the hemiacetal *the acetal*

REACTIONS OF ALDEHYDES AND KETONES WITH NITROGEN COMPOUNDS

Aldehydes and ketones react with ammonia and other nitrogen compounds. The product of the reaction with NH_3 or RNH_2 is called an *imine*.

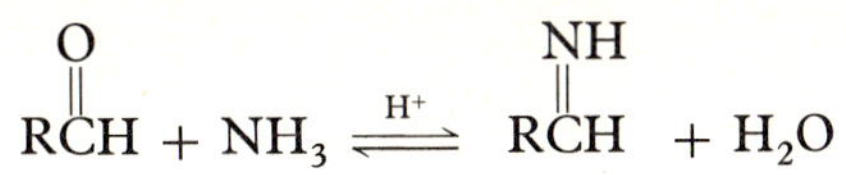

Imines are of greater concern to the biochemist than to the organic chemist. For example, if an organism needs a particular amino acid to synthesize a protein, the amino acid may sometimes be synthesized enzymatically from another amino acid by transfer of an amino group ($—NH_2$). The intermediates in this complex reaction sequence are imines. (The following sequence is somewhat simplified.)

A transamination sequence:

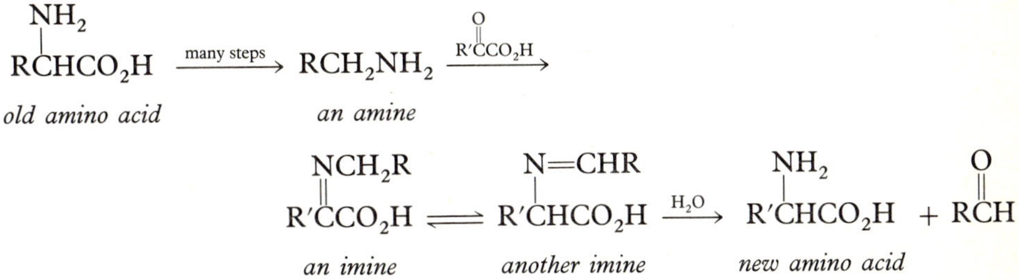

Other nitrogen-containing reagents react with aldehydes and ketones in a manner analogous to the reaction of ammonia. Typical of these reagents is *hydrazine,* NH_2NH_2. The product of hydrazine's reaction with either an aldehyde or ketone is called a *hydrazone.*

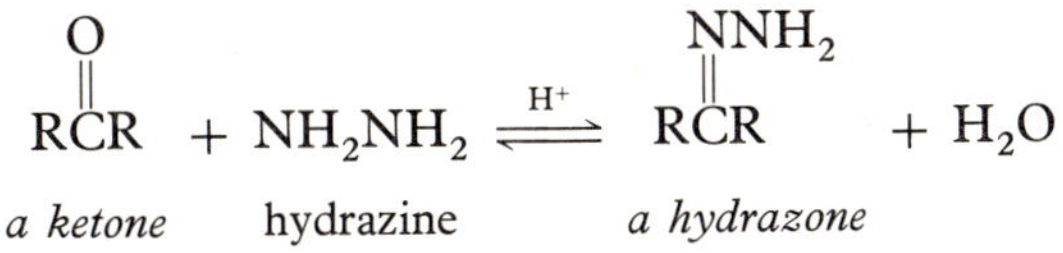

Sample Problem: Predict the organic products:

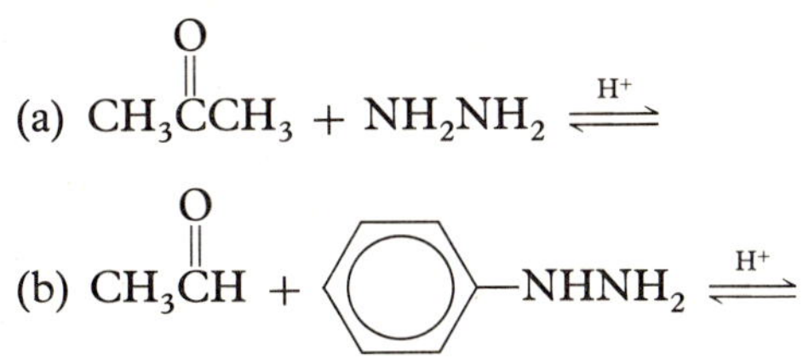

Solution:

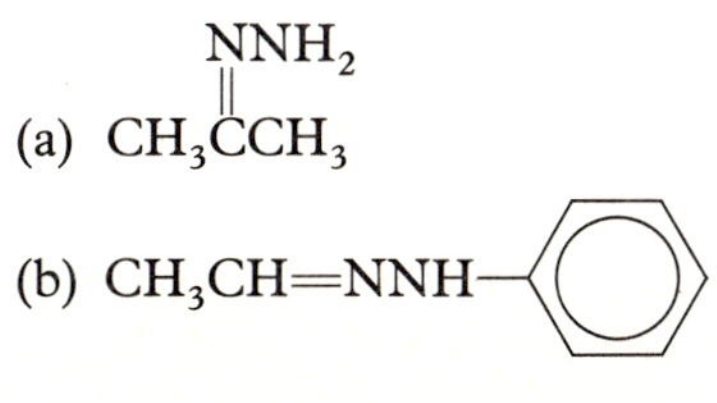

REDUCTION OF ALDEHYDES AND KETONES

In the presence of hydrogen gas and a suitable catalyst, such as finely divided nickel, the double bond of a carbonyl group may be reduced. The hydrogen gas adds across the carbonyl pi bond in a reaction that is similar to the hydrogenation of alkenes.

$$CH_2{=}CH_2 + H_2 \xrightarrow{\text{catalyst}} CH_3{-}CH_3$$

an alkene *an alkane*

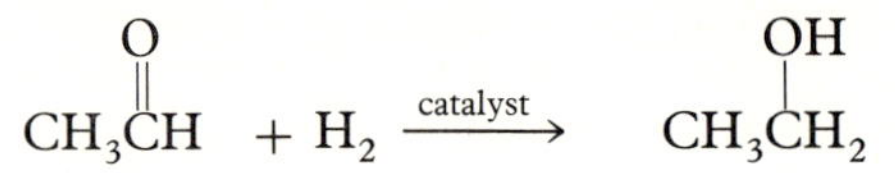

an aldehyde *a primary alcohol*

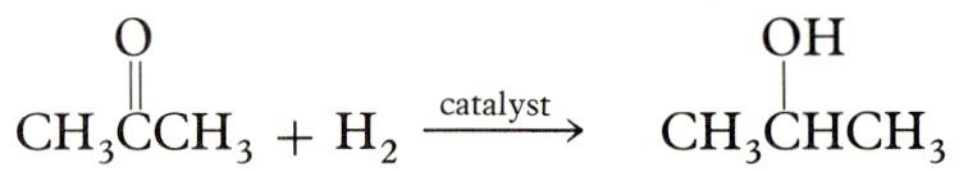

a ketone *a secondary alcohol*

As may be seen in the examples, the reduction of an aldehyde gives a primary alcohol, while that of a ketone gives a secondary alcohol.

Because of their convenience compared to that of hydrogen gas, the solid metal hydrides are often used in laboratory work to carry out the reduction of carbonyl groups.

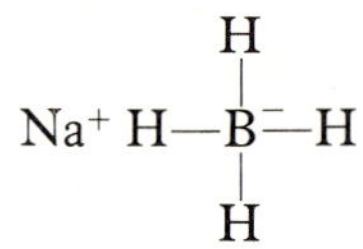

sodium borohydride ($NaBH_4$)
a metal hydride

Reduction by a metal hydride leads to the salt of the alcohol. The alcohol itself may be obtained after treatment of this salt with aqueous acid.

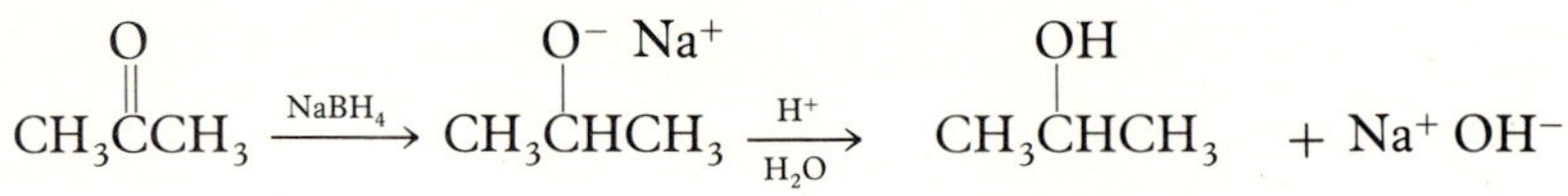

a ketone *a salt* *a secondary alcohol*

OXIDATION OF ALDEHYDES AND KETONES

Under mild oxidizing conditions, aldehydes are easily oxidized to carboxylic acids. Ketones resist oxidation.

$$\underset{\text{acetaldehyde}}{CH_3\overset{\overset{\displaystyle O}{\|}}{C}H} \xrightarrow[\text{oxidizing agent}]{\text{mild}} \underset{\text{acetic acid}}{CH_3\overset{\overset{\displaystyle O}{\|}}{C}OH}$$

$$\underset{\text{acetone}}{CH_3\overset{\overset{\displaystyle O}{\|}}{C}CH_3} \xrightarrow[\text{oxidizing agent}]{\text{mild}} \text{No reaction}$$

Generalized:

$$\underset{\textit{an aldehyde}}{RCHO} \xrightarrow[\text{oxidizing agent}]{\text{mild}} \underset{\textit{a carboxylic acid}}{RCO_2H}$$

Many of the chemical tests used to distinguish aldehydes from other functional groups are based upon the easy oxidation of this functional group. *Tollens reagent* is an alkaline solution of the silver-ammonia complex ion. An aldehyde (but not a simple ketone) reacts with the silver ion (Ag^+) of Tollens reagent to form the anion of the carboxylic acid and silver metal. The silver plates out on the container, forming a silver mirror. A similar reaction is used in the manufacture of silver mirrors.

$$\underset{\textit{an aldehyde}}{CH_3\overset{\overset{\displaystyle O}{\|}}{C}H} + \underset{\substack{\text{Tollens reagent} \\ \textit{(a mild oxidizing agent)}}}{2\,Ag(NH_3)_2^+} + 2\,OH^- \longrightarrow \underset{\substack{\textit{the anion} \\ \textit{of a carboxylic} \\ \textit{acid}}}{CH_3\overset{\overset{\displaystyle O}{\|}}{C}O^-} + \underset{\substack{\textit{silver} \\ \textit{metal}}}{2\,Ag\downarrow} + H_2O + NH_4^+ + 3\,NH_3$$

Many sugars contain an aldehyde group and therefore react with Tollens reagent to yield a silver mirror. In the past, this reaction has been used as a clinical test for glucose in the urine, which is one symptom of diabetes. Currently, an enzyme test that is specific for glucose is used in clinical testing.

THE EFFECT OF THE CARBONYL GROUP ON ADJACENT GROUPS

The carbonyl group affects the chemistry of groups adjacent to it in a molecule. These adjacent groups are said to be *alpha* (α) to the carbonyl group. The next group is called a *beta* (β) group, and so on, according to the Greek alphabet. It is the hydrogen atoms on the alpha-carbon, the α-hydrogens, that we shall discuss now.

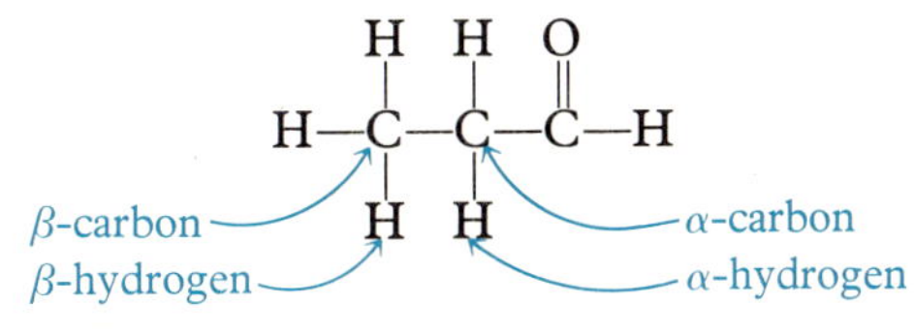

The carbon-hydrogen bond is usually quite stable. We do not expect a hydrogen atom attached to a carbon to be acidic. However, α-hydrogens are different. An α-hydrogen of a ketone, an aldehyde, or an ester is slightly more acidic than hydrogens attached to other carbons in an organic molecule.

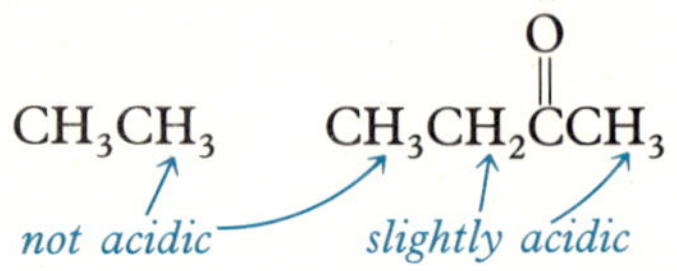

Part of the reason for the acidity of the α-hydrogen is that the partially positive carbon atom in the carbonyl group draws nearby electrons close to itself. This means that the α-carbon is also partially positive and draws electrons away from the α-hydrogen. (These shifts in electron density are indicated by arrowheads on the bonds in the following structure.) The result of the electrons being drawn away from the α-hydrogen is that this hydrogen carries a partial positive charge and can be lost to a base.

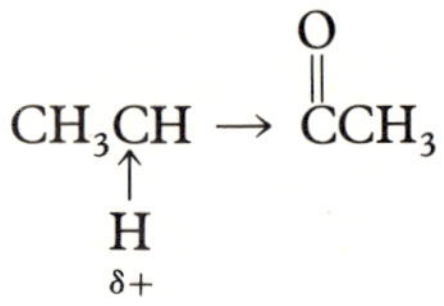

A hydrogen atom that is alpha to two carbonyl groups is even more acidic than a hydrogen atom alpha to only one carbonyl group. A compound with such an α-hydrogen is slightly more acidic than an alcohol (but not nearly as acidic as a carboxylic acid).

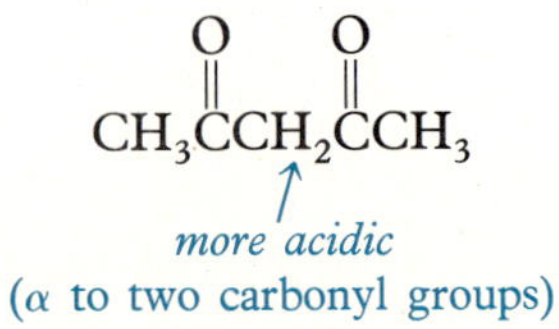

The acidity of a hydrogen atom alpha to two carbonyl groups means that we can treat a compound containing such a hydrogen with a strong base and prepare a salt of the carbonyl compound. This reaction is an acid-base reaction and is reversible.

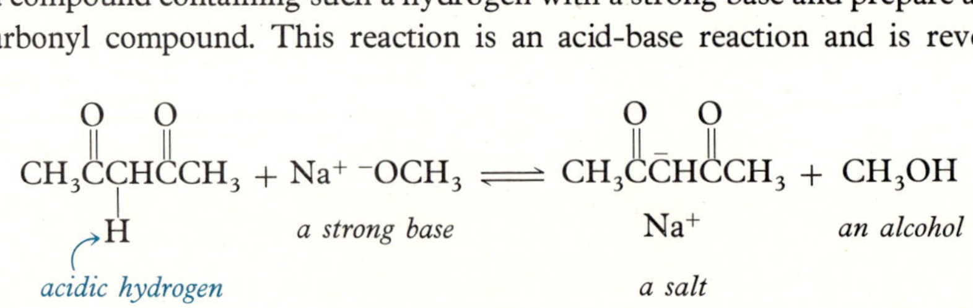

TAUTOMERISM. Because an α-hydrogen is acidic and the carbonyl group is polar, an interesting equilibrium exists in solution.

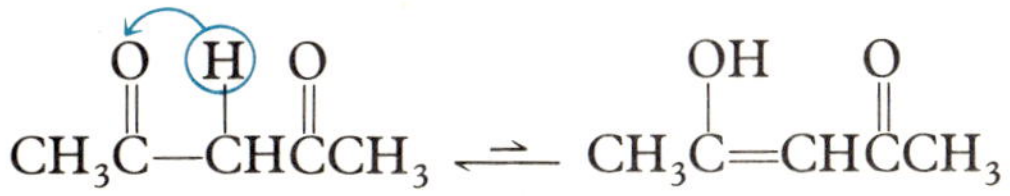

The reaction involves the shift of the acidic hydrogen atom to the negative oxygen of the carbonyl group. Electrons from the carbonyl pi bond are used to form a sigma bond between the oxygen and the hydrogen. The electrons from the old carbon-hydrogen bond form a new carbon-carbon pi bond.

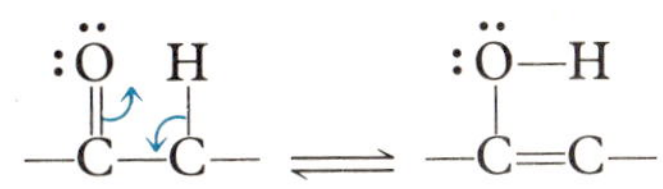

partial structures showing shift of H

A reaction that involves the shift of a hydrogen atom alpha to a carbonyl group is called a *tautomerism reaction,* and the two forms of the carbonyl compound are called *tautomers*. The carbonyl tautomer is called the *keto form* of the compound, while the tautomer with a hydroxyl group and a carbon-carbon double bond is called the *enol* form (from alk*ene* and alcoh*ol*).

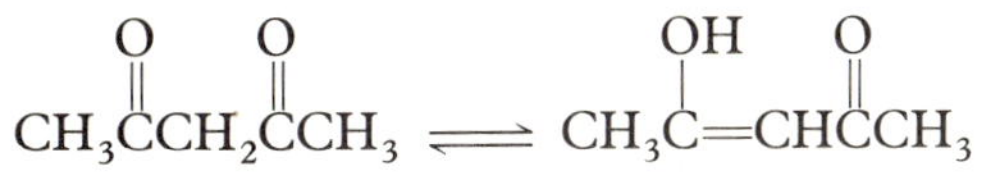

keto form *enol form*

tautomers

One result of tautomerism is that the sugar fructose reduces Tollens reagent even though fructose does not contain an aldehyde group. The reason is that fructose is converted to the aldehyde sugar glucose by way of an *enediol,* an intermediate with a double bond and two OH groups.

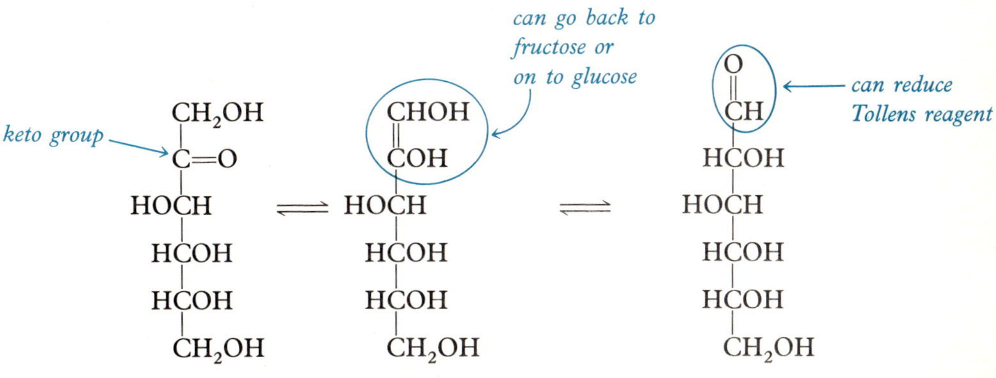

fructose *an enediol* glucose

Sample Problem: Draw the enediol form of glyceraldehyde and dihydroxyacetone. Would you expect these two compounds to be interconvertible?

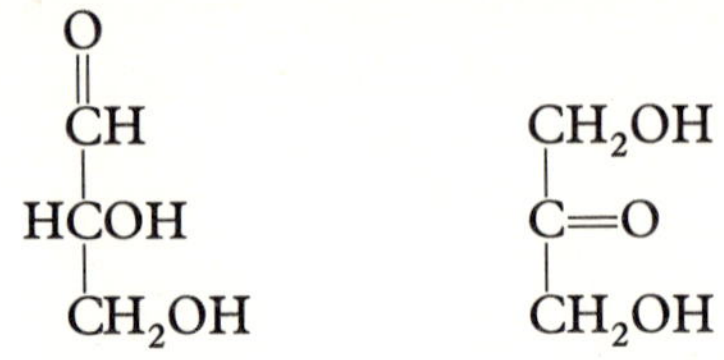

glyceraldehyde dihydroxyacetone

Solution: Both have the same enediol form and therefore can be interconverted:

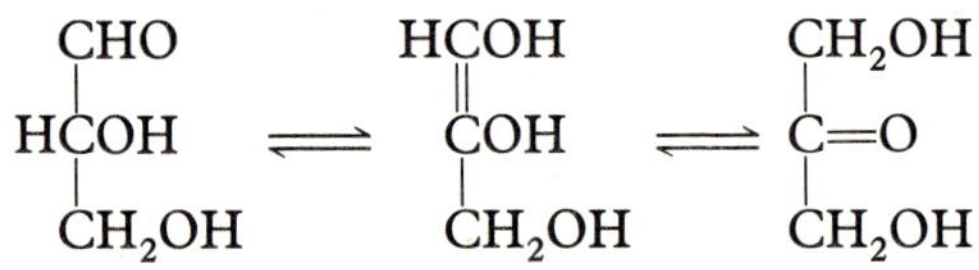

the enediol

STUDY QUESTIONS

16-1. Which of the following statements are true?

(a) The carbonyl carbon is in the sp^3-hybrid state.

(b) The oxygen of the carbonyl group is positive in comparison to the carbon.

(c) A positive reagent attacks the oxygen of a carbonyl group.

(d) The carbonyl group has a pi bond and a sigma bond.

16-2. Label each of the following compounds as to its class (Table 16-1):

(a) 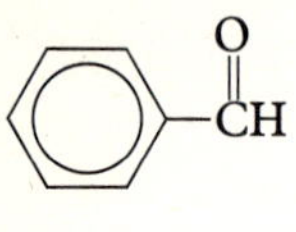

(b) 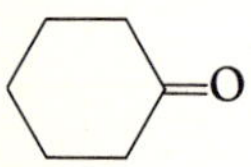

(c) 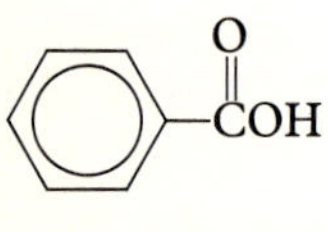

(d) 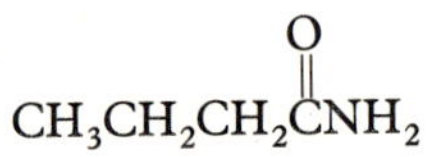

(e) 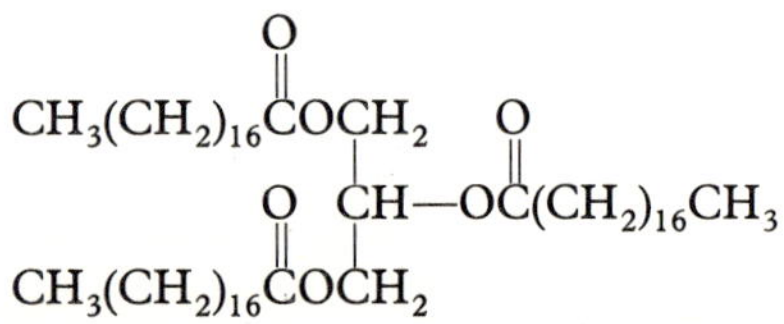

16-3. Name the following compounds using the IUPAC system:

(a) 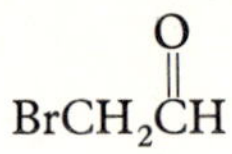

(b) 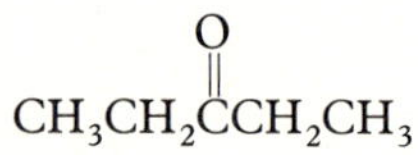

(c) 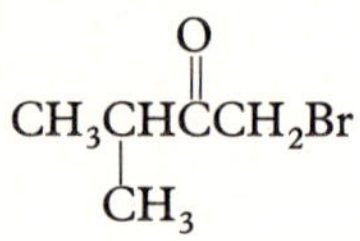

(d) 

16-4. Why is "1-pentanone" an incorrect name?

16-5. Write structures for the following compounds:

(a) 2,4-dimethyl-3-pentanone

(b) diisopropyl ketone

(c) 4-chloro-1-cyclohexanone
(d) 3-ethyl-3-methylhexanal

16-6. Which of the following structures are hemiacetals?

(a)  (b) 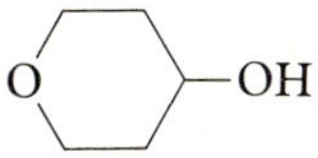

(c) 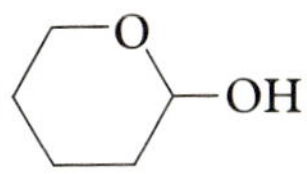 (d) $CH_3CH_2\overset{OH}{\overset{|}{C}}HOCH_2CH_3$

16-7. Which of the following structures are acetals?

(a) $CH_3CH_2\overset{OCH_2CH_3}{\overset{|}{C}}HOCH_3$

(b) 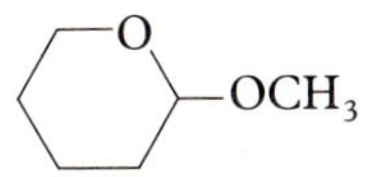

(c) 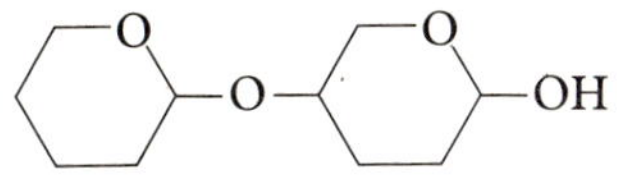

(d) 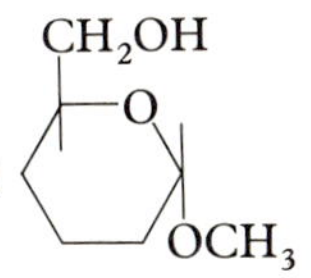

16-8. Using ethanal and methanol, give the equation that shows the aldehyde in equilibrium with the acetal.

16-9. Can a tertiary alcohol be prepared by the reduction of an aldehyde or ketone? Explain.

16-10. Complete the following equations:

(a) 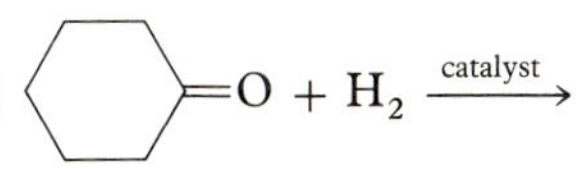

(b) $CH_3CH_2CHO + \begin{matrix}(1)\ NaBH_4\\(2)\ H_2O, H^+\end{matrix} \longrightarrow$

(c) 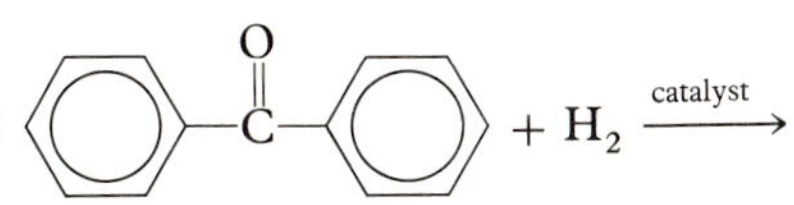

(d) $CH_3CH_2CH_2CHO + \begin{matrix}(1)\ NaBH_4\\(2)\ H_2O, H^+\end{matrix} \longrightarrow$

16-11. A compound with the molecular formula C_3H_6O is treated with Tollens reagent, and a silver mirror is observed. What is the structure of the compound?

16-12. How many (1) α-hydrogens and (2) β-hydrogens do each of the following contain?

(a) $CH_3\underset{CH_3}{\underset{|}{C}}HCHO$ (b) 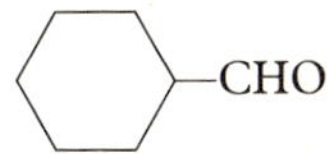

(c) 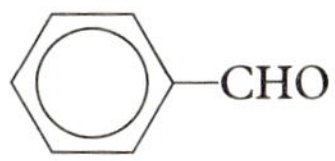 (d) $CH_3CH_2\overset{O}{\overset{\|}{C}}CH_2CH_3$

16-13. Circle the α-hydrogens in each of the following structures:

(a) $CH_3CH_2CH_2CHO$

(b) $CH_3CH_2O\overset{O}{\overset{\|}{C}}CH_2CO_2CH_2CH_3$

(c) $CH_3\overset{O}{\overset{\|}{C}}CH_3$

(d) 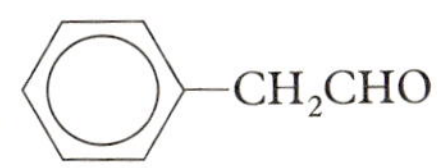

16-14. Explain why acetaldehyde is the product of the reaction between acetylene and water in the presence of acid.

$$\underset{\text{acetylene}}{CH{\equiv}CH} + H_2O \xrightarrow{H^+} \underset{\text{acetaldehyde}}{CH_3\overset{O}{\overset{\|}{C}}H}$$

16-15. Draw the enol tautomers for the following carbonyl compounds:

(a) CH_3CH_2CHO (b) 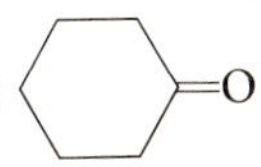

CHAPTER 17 CARBOXYLIC ACIDS AND THEIR DERIVATIVES

OBJECTIVES

Carboxylic Acids *Write the structure of the carboxyl group.* ∗ *Give the structures and trivial names for the carboxylic acids containing one, two, three, and four carbon atoms.* ∗ *Derive the IUPAC names for carboxylic acids.* ∗ *Write equations for the reactions of carboxylic acids with NaOH and* $NaHCO_3$. ∗ *Write equations for esterification of carboxylic acids.*

Esters of Carboxylic Acids *Derive the names of esters of carboxylic acids.* ∗ *Write equations for the hydrolysis and the saponification of esters.* ∗ *Show by an equation how a soap may be prepared.* ∗ *Describe the structure of a wax.* ∗ *Write the equation for the conversion of HSCoA to* $CH_3\overset{\overset{\large O}{\|}}{C}SCoA$. ∗ *Write equations for the ester condensations of ethyl acetate and of acetylcoenzyme A.*

Amides *Write the structure of the amide group.* ∗ *Name an amide.* ∗ *Give equations for the reactions of amides with* H_2O *in acid and in base.*

As we have mentioned, a variety of compounds besides aldehydes and ketones contain the carbonyl group. These other compounds are the *carboxylic acids* and their *derivatives* (compounds that are *derived* from the carboxylic acids). These compounds are different from the aldehydes and ketones because the carbonyl group is affected by an adjoining electronegative group.

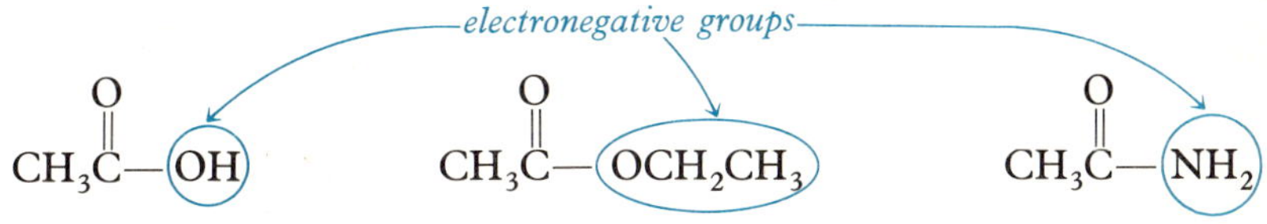

a carboxylic acid *an ester* *an amide*

CARBOXYLIC ACIDS

A carboxylic acid is a compound that contains a hydroxyl group connected to the carbon of a carbonyl group. The functional group of the carboxylic acids, called a *carboxyl* group, may be represented in a number of ways:

$$-\overset{\overset{\displaystyle O}{\|}}{C}OH \quad \text{or} \quad -COOH \quad \text{or} \quad -CO_2H$$

the carboxyl group

We have encountered acetic acid (CH_3CO_2H) many times already. Commercially, acetic acid is made by the oxidation of ethyl alcohol or of acetaldehyde. Dilute solutions (vinegar) may be made by a bacterial oxidative fermentation of ethyl alcohol, which, in turn, is produced from sugar in a fermentation process with yeast. Acetic acid inhibits microbial growth, and vinegar has been used since ancient times as a preservative for meat, fish, and vegetables.

Such diverse compounds as the analgesic aspirin; the weed killer 2,4-D; and lysergic acid, the precursor of LSD (p. 482), contain carboxyl groups. Many other compounds containing this functional group will be encountered in the biochemical chapters of this book.

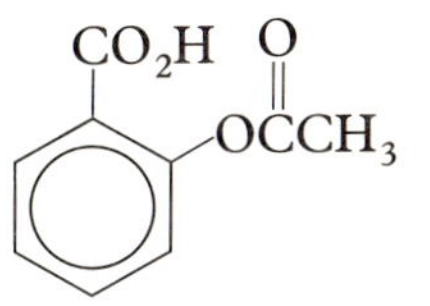

acetylsalicylic acid
(aspirin)

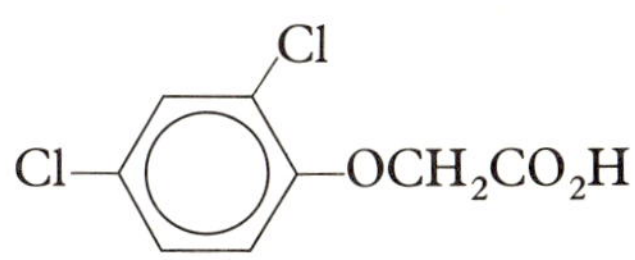

2,4-dichlorophenoxyacetic acid
(2,4-D)

Carboxylic acids of up to four carbons are water soluble and ionize to a limited extent in water. Carboxylic acids of more than four carbons are insoluble in water.

$$CH_3\overset{\overset{\displaystyle O}{\|}}{C}OH \xrightleftharpoons{H_2O} CH_3\overset{\overset{\displaystyle O}{\|}}{C}O^- + H^+$$

acetic acid
a water-soluble carboxylic acid

acetate ion

All carboxylic acids react with bases such as NaOH or $NaHCO_3$ to form *carboxylate salts.* Most sodium and potassium carboxylates are soluble in water (even when the carboxylic acid itself is not). High molecular weight sodium carboxylates, used as soaps (page 132), are only sparingly soluble.

$$CH_3\overset{\overset{\displaystyle O}{\|}}{C}OH + Na^+\,OH^- \longrightarrow CH_3\overset{\overset{\displaystyle O}{\|}}{C}O^-\,Na^+ + H_2O$$

acetic acid

sodium acetate
a salt

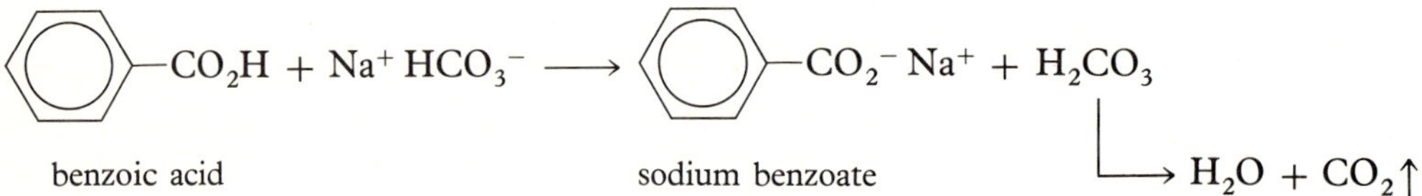

benzoic acid

sodium benzoate
a salt

In general, liquid carboxylic acids boil at higher temperatures than would be expected from their molecular weights. This higher boiling point is attributed to hydrogen bonding. Instead of random hydrogen bonding, as is found in substances like water, carboxylic acids form *dimers* in which two molecules are joined together by a pair of hydrogen bonds.

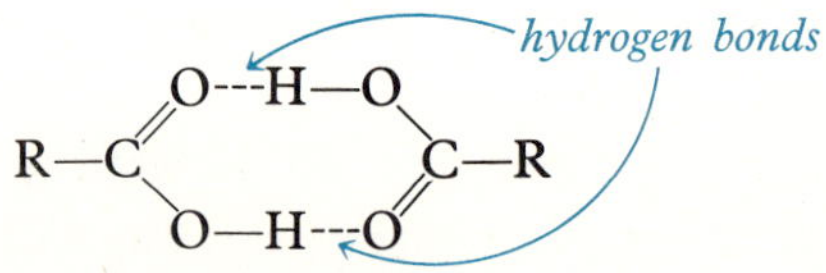

carboxylic acid dimer

NAMING CARBOXYLIC ACIDS. The carboxylic acids of low enough molecular weight to be volatile have strong, and often very unpleasant, odors. The smell of vinegar is due to acetic acid. Some of the common names of the carboxylic acids are based upon their odors; for example, the carboxylic acid with four carbon atoms is called butyric acid because it is the odorous component of rancid butter. Six-, eight-, and ten-carbon acids, all found in skin secretions of goats (Latin, *caper*), have unforgettable permeating odors. Formic acid was named by the alchemists, who obtained this compound by the distillation of red ants (*formica*).

The IUPAC names of the carboxylic acids are derived similarly to those of the aldehydes. The number of carbon atoms in the longest chain containing the carboxyl group determines the parent. The carboxyl group is always found at an end of a chain; the carbonyl carbon is designated as carbon 1. (Again, we do not use the prefix number 1 in the name; the number is understood.) The ending of the IUPAC name for a carboxylic acid is *-oic acid*. Table 17-1 lists the IUPAC and trivial names of some carboxylic acids.

Table 17-1. Names of Some Carboxylic Acids

Structure	Trivial Name	Latin Origin	IUPAC Name
$H\overset{\overset{O}{\|}}{C}OH$	Formic acid	*Formica*, ants	Methanoic acid
$CH_3\overset{\overset{O}{\|}}{C}OH$	Acetic acid	*Acetum*, vinegar	Ethanoic acid
$CH_3CH_2CH_2\overset{\overset{O}{\|}}{C}OH$	Butyric acid	*Butyrum*, butter	Butanoic acid
$CH_3(CH_2)_4\overset{\overset{O}{\|}}{C}OH$	Caproic acid	*Caper*, goat	Hexanoic acid
$CH_3(CH_2)_6\overset{\overset{O}{\|}}{C}OH$	Caprylic acid	*Caper*, goat	Octanoic acid
$CH_3(CH_2)_8\overset{\overset{O}{\|}}{C}OH$	Capric acid	*Caper*, goat	Decanoic acid

Sample Problem: Give the IUPAC name for the following:

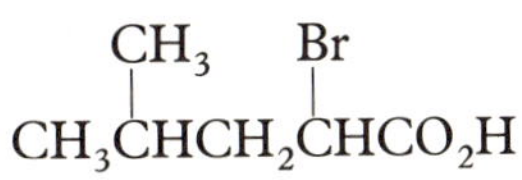

Solution: *Step 1:* Number the chain.

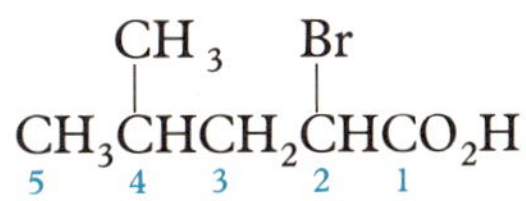

Step 2: Name the parent.

pentanoic acid

Step 3: Add the branch and the functional group.

2-bromo-4-methylpentanoic acid

REACTION OF CARBOXYLIC ACIDS WITH ALCOHOLS. Carboxylic acids undergo reaction with alcohols in the presence of an acidic catalyst to yield *esters*. These reactions leading to esters are called *esterification reactions*.

An esterification reaction:

$$CH_3\overset{\overset{\displaystyle O}{\|}}{C}OH + HOCH_2CH_3 \xrightleftharpoons{H^+} CH_3\overset{\overset{\displaystyle O}{\|}}{C}OCH_2CH_3 + H_2O$$

acetic acid — ethyl alcohol — ethyl acetate, *an ester*

Esterification in acidic solution is a reversible reaction. In the laboratory, an excess of the alcohol is usually used to drive the equilibrium to the right. The reverse of the esterification reaction is the hydrolysis reaction used to prepare a carboxylic acid from its ester. (*Hydrolysis* means cleavage with water.) In this case, an excess of water would be used to drive the equilibrium to the carboxylic acid side.

Esterification:

$$RCO_2H + \underset{excess}{R'OH} \xrightleftharpoons{H^+} RCO_2R' + H_2O$$

Hydrolysis (*the reverse reaction*):

$$RCO_2R' + \underset{excess}{H_2O} \xrightleftharpoons{H^+} RCO_2H + R'OH$$

Sample Problem: What are the esterification products of benzoic acid (page 302) and methyl alcohol?

Solution:

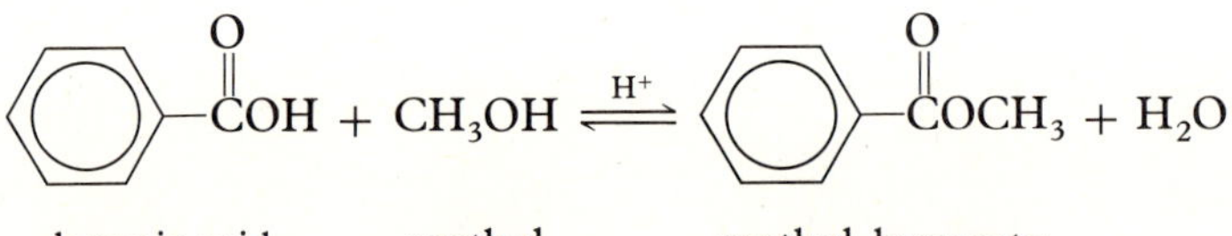

benzoic acid — methyl alcohol — methyl benzoate, *an ester*

ESTERS OF CARBOXYLIC ACIDS

Esterification reactions lead us to that exceedingly important class of compounds, the esters of carboxylic acids. An ester, as we have seen, is a compound that contains a —CO_2R grouping.

$$\overset{\displaystyle O}{\overset{\|}{—C}}OR \quad \text{or} \quad —COOR \quad \text{or} \quad —CO_2R$$

ester group

Unlike carboxylic acids, volatile esters often have sweet, pleasant, often fruity, odors.

$$CH_3CH_2CH_2\overset{\displaystyle O}{\overset{\|}{C}}OCH_2CH_2CH_2CH_3$$

butyl butyrate
odor of pineapple

$$CH_3CH_2\overset{\displaystyle O}{\overset{\|}{C}}OCH_2\overset{\displaystyle CH_3}{\overset{|}{C}}HCH_3$$

isobutyl propionate
odor of rum

NAMING ESTERS. Esters of carboxylic acids are named similarly to salts of carboxylic acids. The alkyl or aryl group that has replaced the proton of the carboxylic acid precedes the name as a separate word. The ending *-ic acid* is then changed to *-ate.*

$$CH_3CH_2\overset{\displaystyle O}{\overset{\|}{C}}OH$$

propanoic acid
(propionic acid)
a carboxylic acid

$$CH_3CH_2\overset{\displaystyle O}{\overset{\|}{C}}O^-\ Na^+$$

sodium propanoate
(sodium propionate)
a salt

$$CH_3CH_2\overset{\displaystyle O}{\overset{\|}{C}}OCH_3$$

methyl propanoate
(methyl propionate)
an ester

Note that the first portion of the ester name is the name of the alkyl or aryl group that is attached to the oxygen. The latter part of the ester name refers to the part of the molecule that contains the carbonyl group.

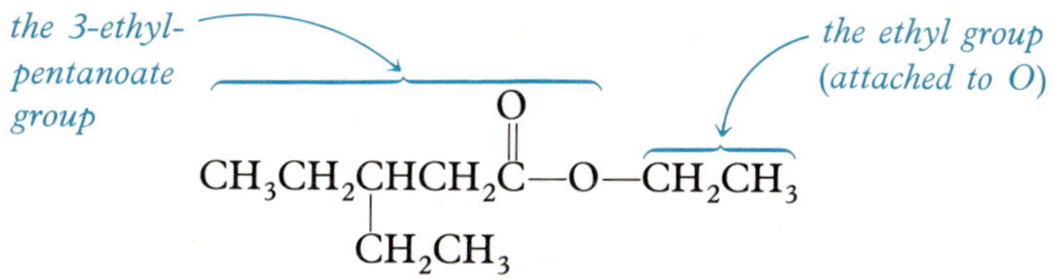

ethyl 3-ethylpentanoate

Sample Problem: Name the following ester by the IUPAC system:

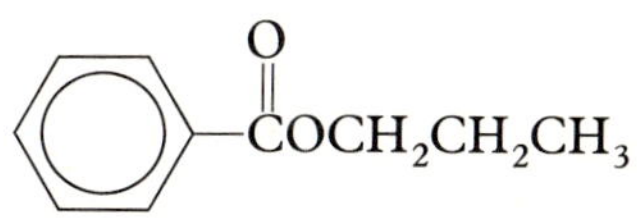

Solution: *Step 1:* Name the parent acid.

benzoic acid

Step 2: Change the ending to *-ate* and add the alkyl group.

propyl benzoate

HYDROLYSIS OF ESTERS. We have already mentioned that hydrolysis of esters in acidic solution is a reversible reaction and gives carboxylic acids and alcohols as the products. Esters may also be hydrolyzed by a water solution of a *strong base* to the salt of a carboxylic acid. This reaction, called a *saponification reaction,* is not reversible.

Saponification in alkaline solution:

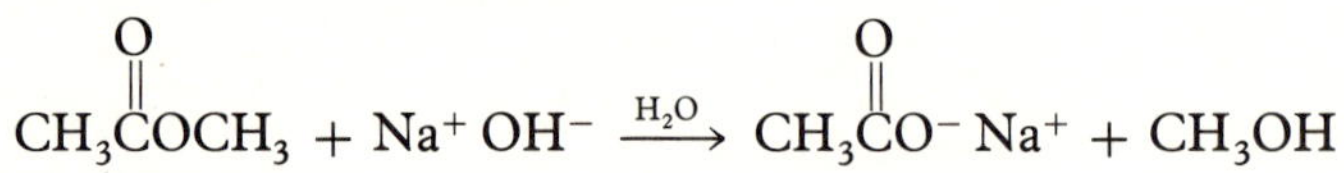

methyl acetate *an ester* — sodium acetate *a salt*

The carboxylic acid may be formed by acidification of the solution following the saponification.

Step 1 (*saponification*): $RCO_2R' + OH^- \longrightarrow RCO_2^- + R'OH$

an ester — *the anion of a carboxylic acid*

Step 2 (*acidification*): $RCO_2^- + H^+ \longrightarrow RCO_2H$

the anion — *a carboxylic acid*

The term "saponification" comes from the word *soap*. A soap is the sodium salt of a long-chain carboxylic acid, called a *fatty acid*. In nature, these acids are found in fats and oils as esters of the triol glycerol (also called glycerin). Tristearin, a fat found in cattle, is the ester from one molecule of glycerol and three molecules of the fatty acid stearic acid. Saponification of tristearin with sodium hydroxide yields a mixture of sodium stearate (a soap) and glycerol. The properties that allow the salts of fatty acids to act as soaps were discussed in Chapter 6. A detailed discussion of the properties of fats and oils will follow in Chapter 24.

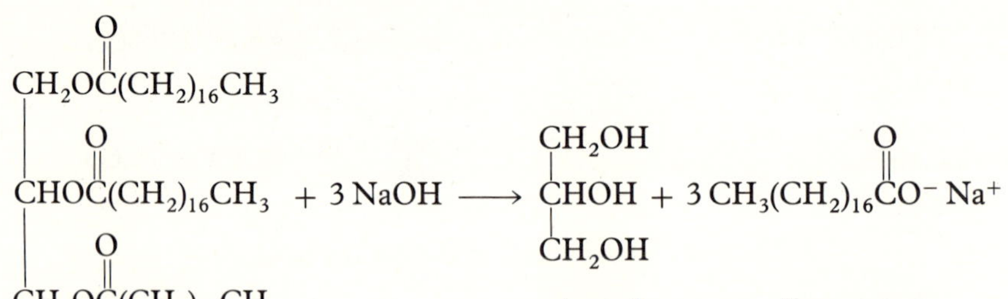

tristearin *a fat* — glycerol *a triol* — sodium stearate *a soap*

Sample Problem: Write equations for (a) the hydrolysis and (b) the saponification of propyl benzoate.

Solution:

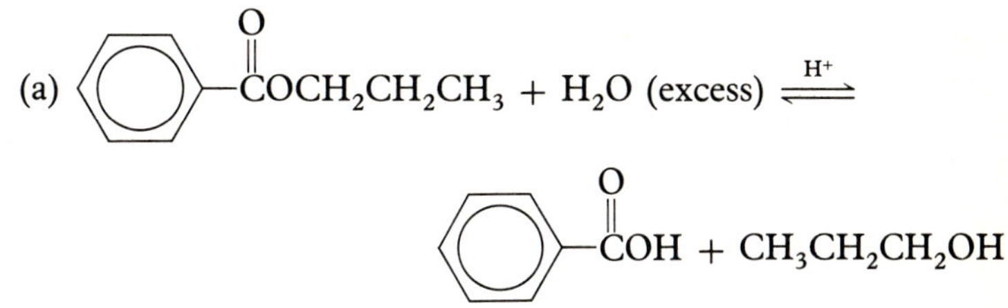

benzoic acid 1-propanol

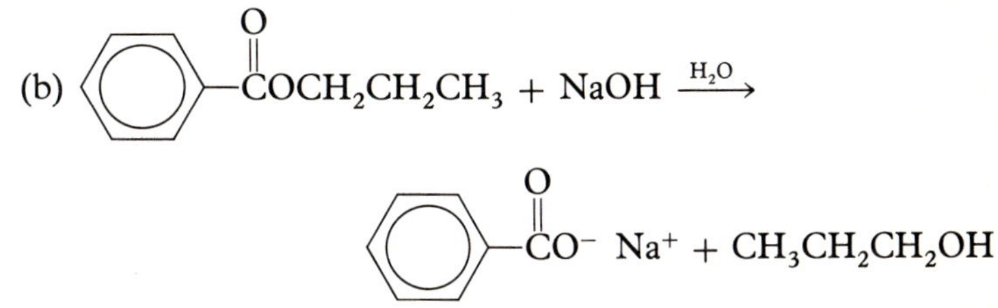

sodium benzoate 1-propanol

SOME IMPORTANT ESTERS. *Waxes* are esters of long-chain fatty acids and long-chain alcohols. For example, beeswax, the material from which bees build their honeycombs, is a mixture of esters of C_{26} and C_{28} carboxylic acids, and C_{30} and C_{32} alcohols. Carnauba wax, which occurs as a coating on the leaves of a Brazilian palm tree and is used in candles and for polishing floors and cars, is a mixture of esters composed mainly of C_{24} and C_{28} acids, and of C_{32} and C_{34} alcohols.

$$\underbrace{CH_3(CH_2)_{24}\overset{\overset{\Large O}{\|}}{C}}_{\text{the } C_{26} \text{ acid}}\underbrace{-O-CH_2(CH_2)_{28}CH_3}_{\text{the } C_{30} \text{ alcohol}}$$

one of the esters in beeswax

Polyesters are polymers containing units joined together by ester linkages. Dacron is a polyester used as a textile fiber.

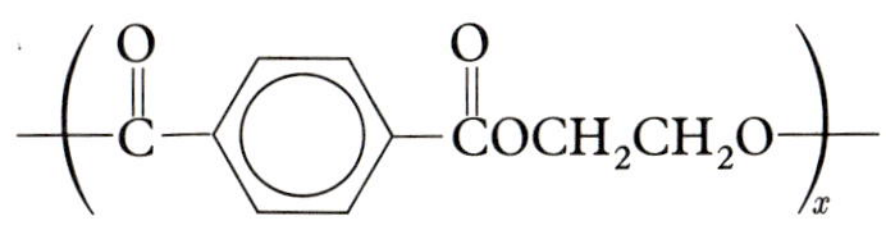

Dacron
a polyester

A *thioester* is an ester that has a sulfur atom in place of the —OR oxygen.

an ester *a thioester*

Acetylcoenzyme A is a biologically important thioester. It is formed by the enzymatic esterification of coenzyme A, a *thiol,* with acetate ions. (A thiol, RSH, is the sulfur analog of an alcohol, ROH.)

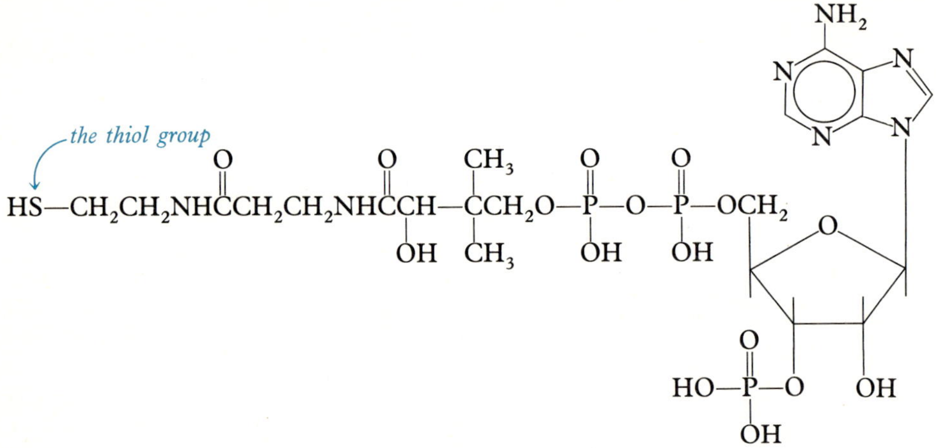

coenzyme A
(abbreviated CoA or HSCoA)

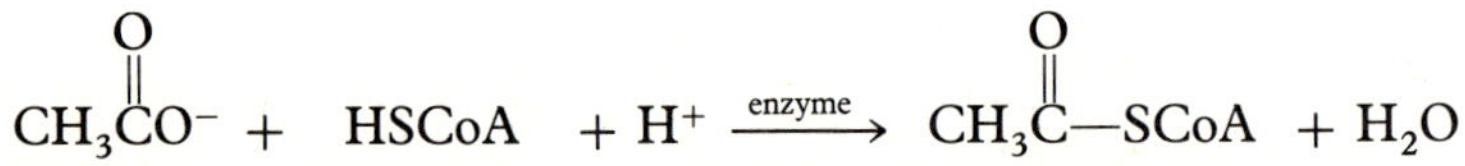

acetate ion coenzyme A *a thiol* acetylcoenzyme A *a thioester*

In biological systems, acetate ions are usually found, not free, but as the acetyl group (CH_3CO—) in acetylcoenzyme A. We will find in later chapters that these acetyl groups can be converted to CO_2, H_2O, and energy. These acetyl groups can also be converted to fats, to some hormones, to carbohydrates, and to other compounds.

ESTER CONDENSATIONS. Recall from Chapter 16 that the α-hydrogen of an aldehyde or a ketone is slightly acidic. The α-hydrogen of an ester is also acidic.

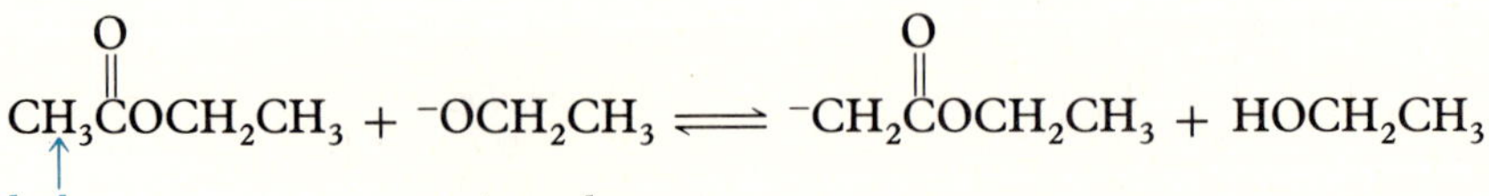

a strong base

The negative carbon atom in the ester anion can attack the partially positive carbon atom of a carbonyl group of another ester.

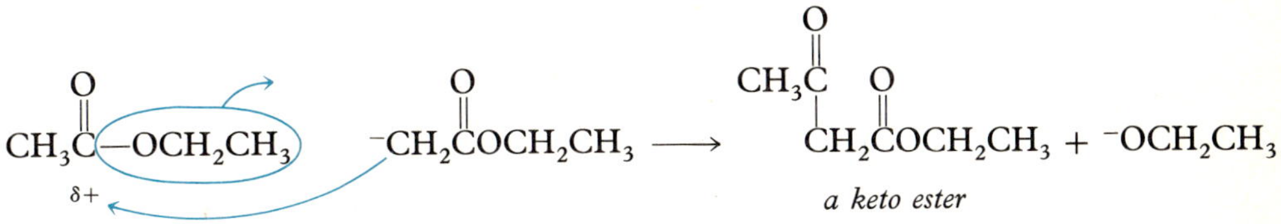

a keto ester

A reaction in which two small molecules combine to form a larger molecule is called a *condensation reaction.* Condensation reactions are very useful for the laboratory syntheses of complex structures. Of more concern to us now is that condensation reactions are one way that an organism can build the molecules it needs. For example, a mammal can synthesize long-chain compounds from the acetyl groups in acetyl-coenzyme A by a series of condensation reactions. In the following equation, we show how a four-carbon chain is formed from two acetyl groups.

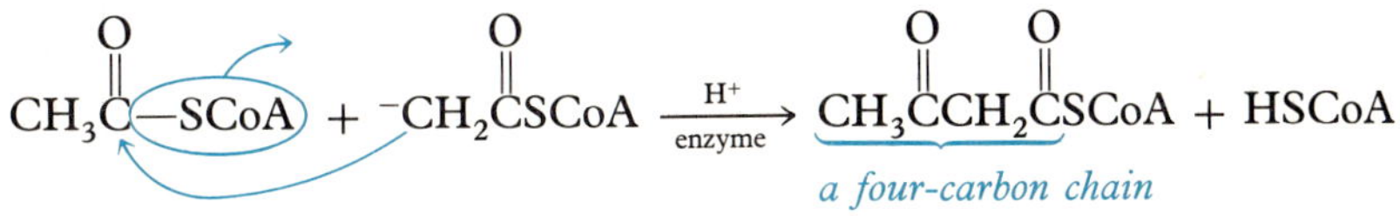

a four-carbon chain

AMIDES

An *amide* is a compound that contains a nitrogen atom attached to a carbonyl group:

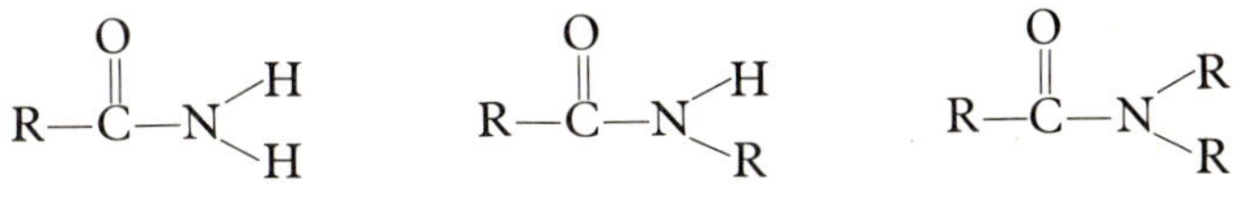

An amide is named by dropping the *-ic acid* or *-oic acid* ending of the carboxylic acid name and adding the ending *-amide.*

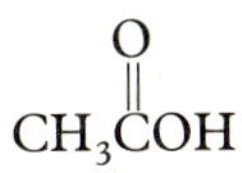

acetic acid

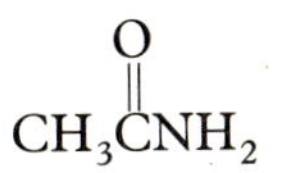

acetamide

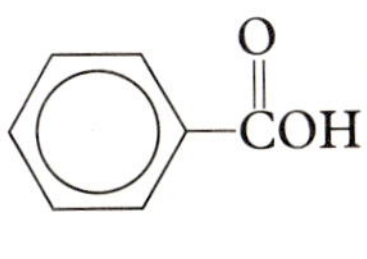

benzoic acid

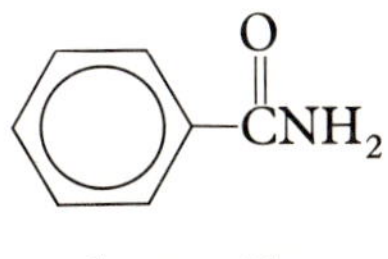

benzamide

If an alkyl or aryl group is attached to the nitrogen, the group is designated in the name by N- (for nitrogen) followed by the name of the group.

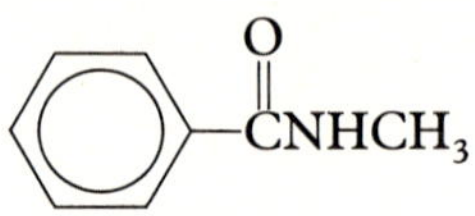

N-methylbenzamide

Proteins are natural polyamides, while nylon is a synthetic polyamide. These two types of polymer differ in the hydrocarbon portions of the monomers and also in the order of attachment of the amide links.

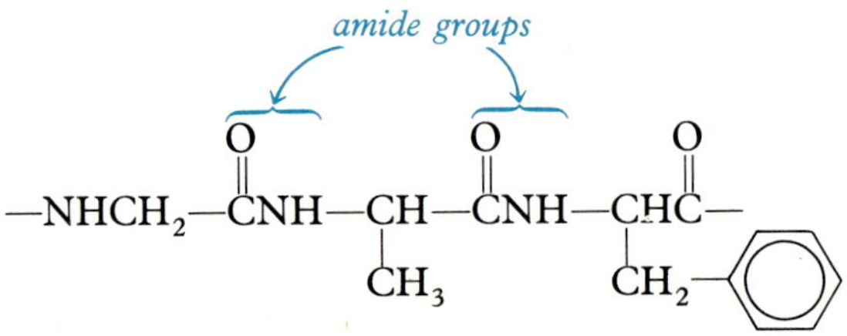

portion of a protein molecule

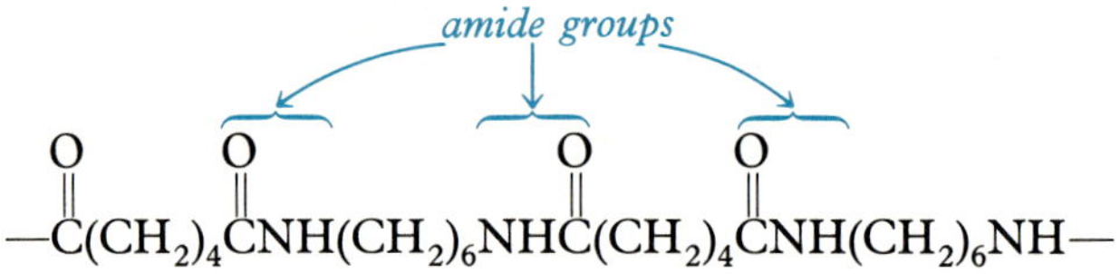

portion of a nylon molecule

Amides, like esters, can be hydrolyzed in acidic or basic solution to the carboxylic acid or its anion, respectively.

Hydrolysis in acid:

$$CH_3\overset{\overset{\Large O}{\|}}{C}{-}NH_2 + H_2O + H^+ \longrightarrow CH_3\overset{\overset{\Large O}{\|}}{C}OH + NH_4^+$$

Hydrolysis in base:

$$CH_3\overset{\overset{\Large O}{\|}}{C}{-}NH_2 + OH^- \longrightarrow CH_3\overset{\overset{\Large O}{\|}}{C}O^- + NH_3$$

Sample Problem: The protein molecule shown above is subjected to acid hydrolysis. What are the structures of the organic products from this portion of the protein molecule?

Solution:

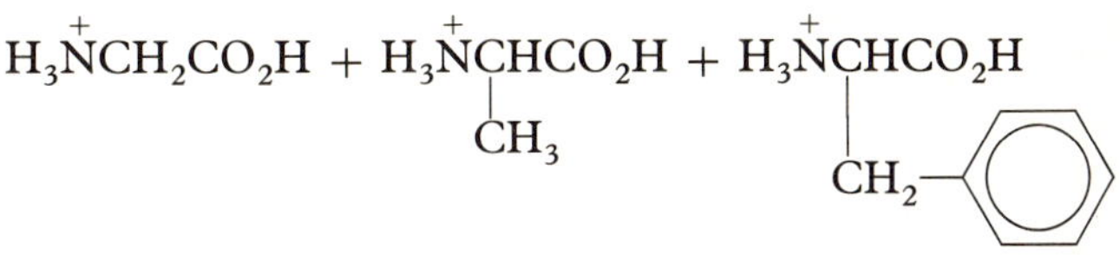

STUDY QUESTIONS

17-1. Complete the following equations:

(a) 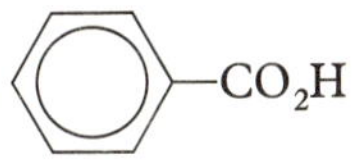 $+ LiOH \longrightarrow$

(b) $HO_2C{-}CO_2H + 2\,NaOH \longrightarrow$

(c) $CH_3CH_2CH_2CO_2H + KOH \longrightarrow$

(d) 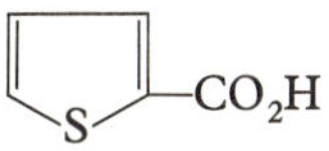 $+ NaOH \longrightarrow$

17-2. Draw the hydrogen-bonded carboxylic acid dimer for each of the following acids:

(a) $(CH_3)_2CHCO_2H$ (b) 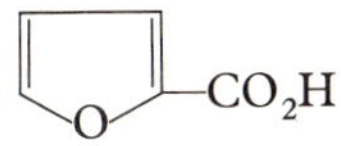

17-3. Name the following carboxylic acids using the IUPAC system:

(a) $CH_3CH{-}CHCO_2H$ (with Br on C-3 and CH_3 on C-2)

(b) $CH_3CH_2CO_2H$

(c) HCO_2H

17-4. Give the structure of each of the following:

(a) formic acid

(b) 3,3-dimethylbutanoic acid

(c) 5-chloropentanoic acid

17-5. What are the two classes of organic compounds that are readily oxidized to carboxylic acids? (*Hint:* See Chapters 15 and 16.)

17-6. Complete the following equations:

(a) 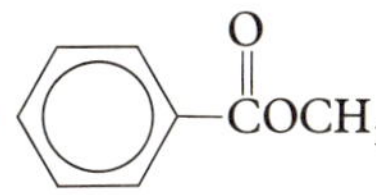 $+ H_2O \xrightleftharpoons{H^+}$

(b) 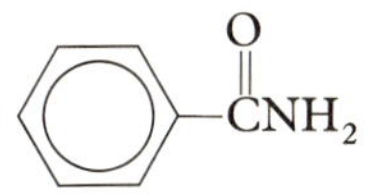 $+ H_2O + H^+ \longrightarrow$

17-7. Using benzoic acid, write esterification reactions for the following alcohols:

(a) CH_3CHCH_2OH (with CH_3 on C-2) (b) $(CH_3)_2CHOH$

(c) 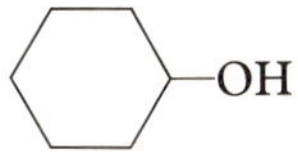 (d) 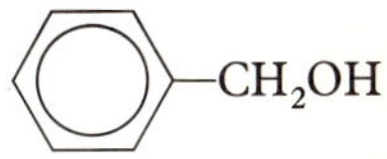

17-8. What is the difference in laboratory procedure between ester hydrolysis and esterification of a carboxylic acid?

17-9. Name the following esters using the IUPAC system:

(a) $CH_3CH_2CH_2CH_2CO_2CH_3$

(b) $(CH_3)_2CHCH_2CO_2CH_2CH_2CH_3$

(c) $CH_3CH_2CHCH_2CCH_2CO_2CH_2CH_3$ (with CH_3 on C-3, and two CH_3 on C-5)

17-10. Write the structures of the saponification products of the esters in Problem 17-9.

17-11. Describe in words how you could obtain butanoic acid from butyl butanoate by a saponification reaction.

17-12. Complete the following equations for ester condensations:

(a) $2\,CH_3\overset{\overset{O}{\|}}{C}OCH_3 \xrightarrow{\text{base}}$

(b) $2\,CH_3\overset{\overset{O}{\|}}{C}SCH_3 \xrightarrow{\text{base}}$

17-13. Name the following amides:

(a) $CH_3CH_2\overset{\overset{O}{\|}}{C}NH_2$ (b) 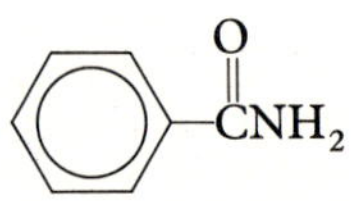

(c) $H\overset{\overset{O}{\|}}{C}NH_2$ (d) $CH_3CH_2\underset{\underset{CH_3}{|}}{C}H\overset{\overset{O}{\|}}{C}NH_2$

17-14. Write the equations for the acid hydrolysis reactions of the amides in Problem 17-13.

17-15. An amide of unknown structure is treated with an aqueous acid. Benzoic acid and methylamine (CH_3NH_2) are finally isolated. What was the structure of the amide?

17-16. Circle the amide groups in *Nembutal,* a barbiturate used as a sedative.

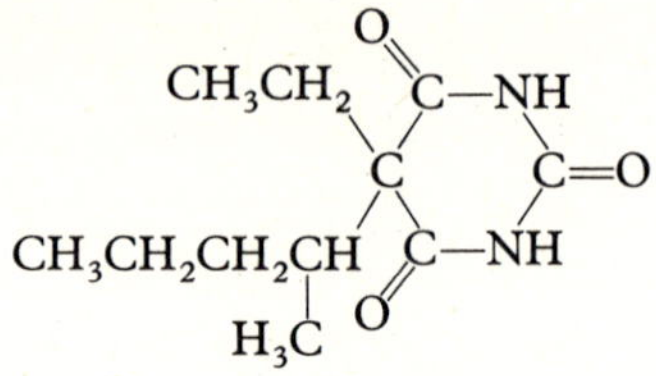

Nembutal

17-17. The product of acid hydrolysis of *nicotinamide,* one of the B vitamins, is *nicotinic acid.* What is the structure of nicotinic acid?

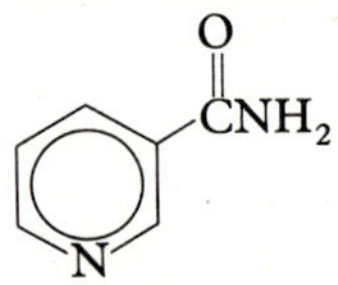

nicotinamide
(niacinamide)

17-18. *Xylocaine* is used as a local anesthetic. What are the organic products of its hydrolysis in basic solution?

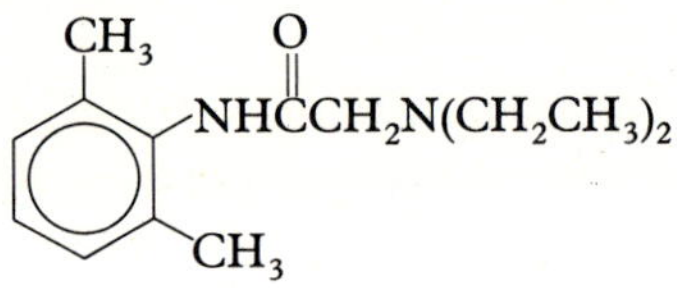

xylocaine
(lidocaine)

CHAPTER 18 AMINES

OBJECTIVES

Structure and Classification of Amines *Draw the structure of an amine, showing the orbital containing unshared electrons.* * *Write structures for a primary, a secondary, and a tertiary amine.* * *Define and give an example of a heterocyclic compound.*

Naming Amines *Name various amines.*

Properties of Amines *Describe (with formulas) the hydrogen bonding of amines.*

Basicity of Amines *Write an equation for the reaction of an amine with an acid and of an amine salt with a base.*

After carbon, hydrogen, and oxygen, nitrogen is the most important element in organic compounds. Life as we know it depends upon nitrogen. Some less hospitable planets of our solar system have atmospheres composed primarily of methane and ammonia. Speculation about whether amino acids, the components of living proteins, could arise from these simple compounds has led to research along these lines. Indeed, amino acids can be formed when a discharge of electricity is passed through such an atmosphere. In addition, nonterrestrial amino acids, along with other organic compounds, have been isolated from carbonaceous meteorites that have fallen on the earth.

Nitrogen forms a variety of types of compounds with carbon. Among these are the *amines,* compounds in which nitrogen is attached to alkyl or aryl groups by single bonds; *amides,* in which nitrogen is attached to a carbonyl group; and *aromatic heterocyclic compounds,* in which nitrogen is contained in an aromatic ring. The amides have already been discussed in Chapter 17, and the aromatic nitrogen compounds will be discussed in Chapter 19. In this chapter, we will discuss the amines.

Examples of amines:

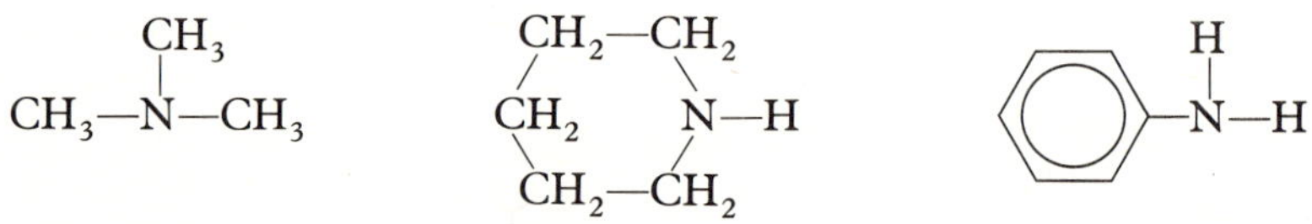

trimethylamine piperidine aniline

STRUCTURE AND CLASSIFICATION OF AMINES

Amines are an important group of compounds that are related to ammonia in their structures and properties. The nitrogen atom in an amine is in the sp^3-hybrid state, just as it is in ammonia, and is connected by three sigma bonds to alkyl groups, aromatic rings, or hydrogen atoms. A filled sp^3 orbital remains on the amine nitrogen, just as in ammonia. (See Figure 18-1.) We will see that much of the chemistry of amines depends primarily upon this filled orbital.

Figure 18-1. Bonding in ammonia and some amines.

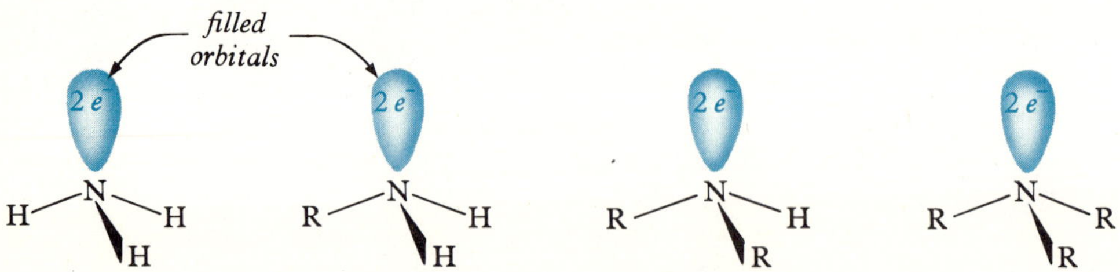

ammonia an alkylamine a dialkylamine a trialkylamine

Amines may be classified by the number of organic groups attached to the nitrogen. An amine with one organic group on the nitrogen is a *primary* amine; with two groups, a *secondary* amine; and with three groups, a *tertiary* amine. (These designations are slightly different from those for alcohols. Amine classification depends on the number of carbon atoms bonded directly to the *nitrogen* atom. Alcohol classification depends on the number of carbon atoms bonded directly to the *carbon* atom with the —OH group.)

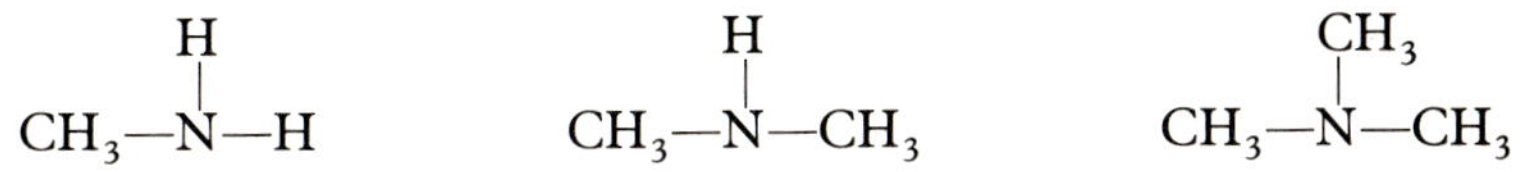

methylamine *a primary amine* · dimethylamine *a secondary amine* · trimethylamine *a tertiary amine*

The amino group may be part of a ring system. This type of compound is called a *heterocyclic compound* (from *hetero-*, different). Piperidine, shown on page 314, is an example of a heterocyclic compound. (Piperidine is also a secondary amine.) Nicotine contains two types of heterocyclic rings.

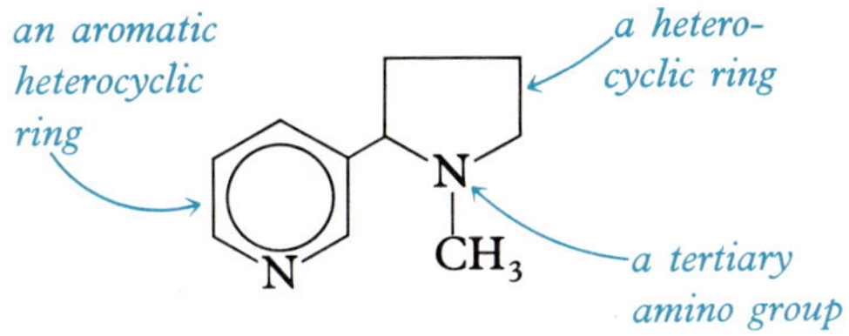

nicotine

NAMING AMINES

The most common way of naming a simple amine is by listing the alkyl or aryl groups that are attached to the nitrogen and adding the ending *-amine*. Methylamine, dimethylamine, and trimethylamine are examples. (Note that amine names are run together as one word.)

If necessary, a number is used to designate the position of the amino group.

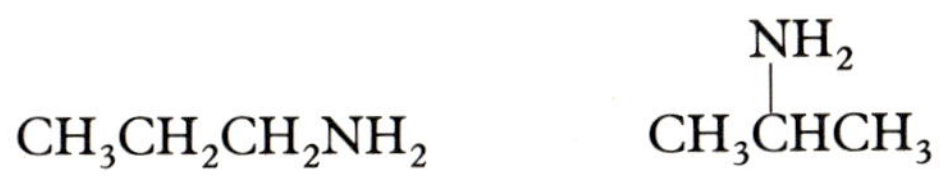

1-propylamine or *n*-propylamine · 2-propylamine or isopropylamine

In more complex amines, the longest chain is the parent. Other substituents on the nitrogen are indicated with the prefix N, as the following examples show.

two methyls on N

$CH_3CH_2N(CH_3)_2$ — N,N-dimethylethylamine

$CH_3CH_2CH_2N(CH_3)_2$ — 1-(N,N-dimethyl)propylamine

Arylamines, such as aniline (page 314), and the heterocyclic amines, such as piperidine (page 314), have their own individual names. A few other important examples follow:

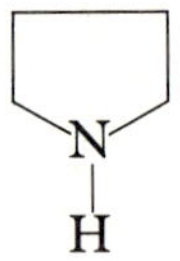

pyrrolidine

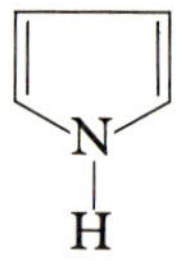

pyrrole

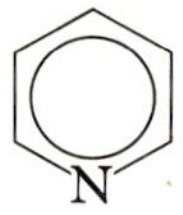

pyridine

Sample Problem: Name the following amines:

(a) $(CH_3CH_2)_3N$ (b) CH_3CHCH_3 with $N(CH_2CH_3)_2$ on the middle carbon

Solution: (a) triethylamine (b) 2-(N,N-diethyl)propylamine

PROPERTIES OF AMINES

Amines that have low molecular weights are gases or low-boiling liquids and often have objectionable odors. For example, trimethylamine smells like dead salt-water salmon and piperidine smells like dead fresh-water fish. Two diamines, putrescine and cadaverine, help give rise to the odor of decaying flesh. These compounds come from the bacterial degradation of the two amino acids lysine and ornithine. (See Chapter 27.)

$H_2NCH_2CH_2CH_2CH_2CH_2NH_2$ — cadaverine (1,5-pentanediamine)

$H_2NCH_2CH_2CH_2CH_2NH_2$ — putrescine (1,4-butanediamine)

Like water, amines are capable of forming hydrogen bonds. Because of this ability, low-molecular weight amines are water soluble. We will also find, in Chapter 28, that the hydrogen bonding of nitrogen compounds is part of the molecular basis of the genetic code and of the synthesis of proteins within an organism.

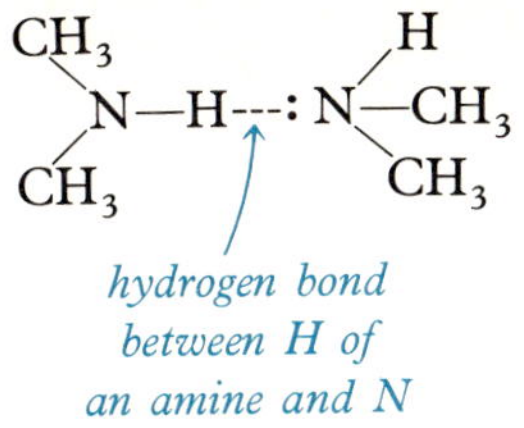

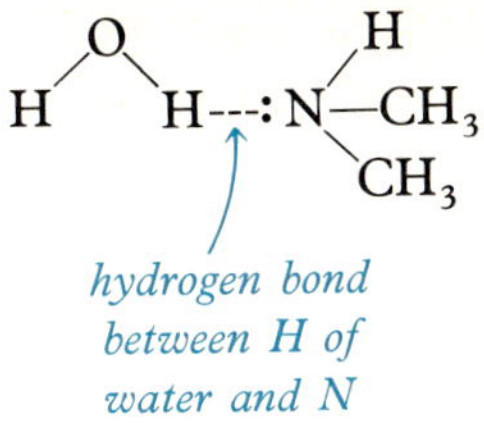

BASICITY OF AMINES

Amines, like ammonia, are weak bases in water solution and undergo reaction with acids to form salts. (The equilibrium constants for the ionization of a few amines are shown in Table 18-1.)

Partial ionization in water:

$$\ddot{N}H_3 + H_2O \rightleftharpoons NH_4^+ + OH^-$$

ammonia
a weak base

$$CH_3\ddot{N}H_2 + H_2O \rightleftharpoons CH_3NH_3^+ + OH^-$$

methylamine
a weak base

Salt formation:

$$\ddot{N}H_3 + HCl \longrightarrow NH_4^+\ Cl^-$$

ammonium chloride
a salt

$$CH_3\ddot{N}H_2 + HCl \longrightarrow CH_3NH_3^+\ Cl^-$$

methylammonium chloride
(methylamine hydrochloride)
a salt

Table 18-1. Equilibrium Constants for the Ionization of Some Amines in Water

Amine	Formula	K
Ammonia	NH_3	1.8×10^{-5}
Methylamine	CH_3NH_2	4.4×10^{-4}
Trimethylamine	$(CH_3)_3N$	6×10^{-5}
Aniline	$C_6H_5NH_2$	4.2×10^{-10}
Pyridine	C_5H_5N	2.3×10^{-9}

Amines and ammonia are bases for the same reason. In either, the unshared pair of electrons on the nitrogen may be donated to form a sigma bond with a hydrogen ion.

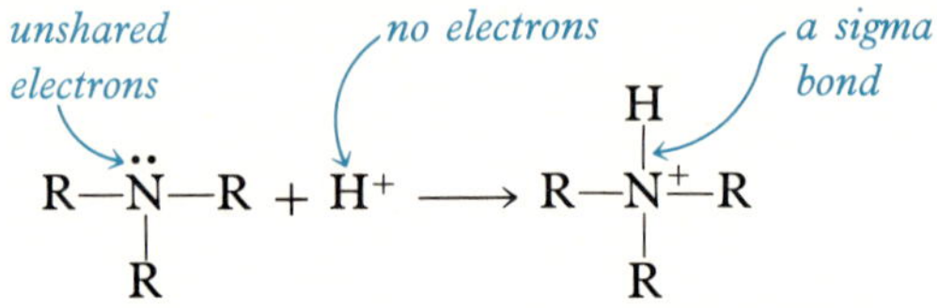

Salts formed from amines and acids are called *amine salts* and are usually named as derivatives of ammonium salts or as amine hydrohalides, as may be seen in the examples. Amine salts are high-melting, water-soluble, colorless compounds, similar in physical properties to inorganic salts. Unlike some of their parent amines, amine salts are odorless.

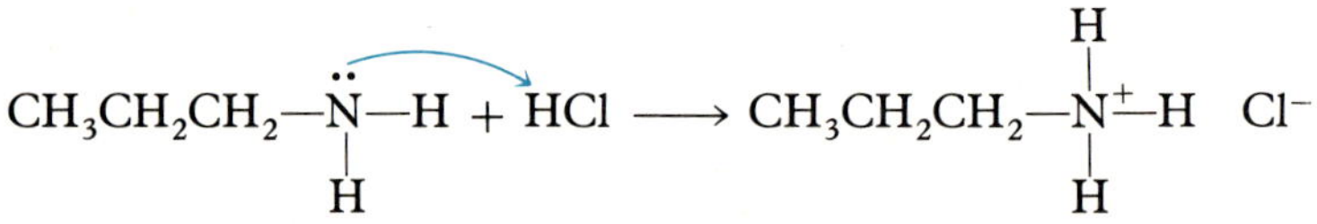

propylamine

propylammonium chloride (propylamine hydrochloride) *an amine salt*

Generalized:

$$R_3N + HX \longrightarrow R_3NH^+ X^-$$

an alkylammonium salt

Many naturally occurring amines are biologically active, and some are used as drugs. These amines, usually of high molecular weight and water insoluble, may be made soluble and extracted from plant material by an acidic solution. Because they react with acids, these compounds are called *alkaloids* ("like a base"). The salts of amine-containing drugs are also useful in medicine. By treatment with an acid, an insoluble amine-containing drug can be made soluble for easier injection and absorption.

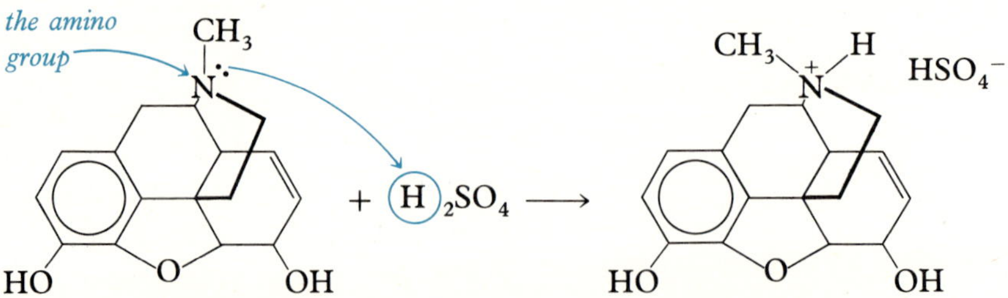

morphine *water insoluble*

morphine sulfate *water soluble*

Sample Problem: Write the equation for the reaction of acetic acid and methylamine. Name the product.

Solution:

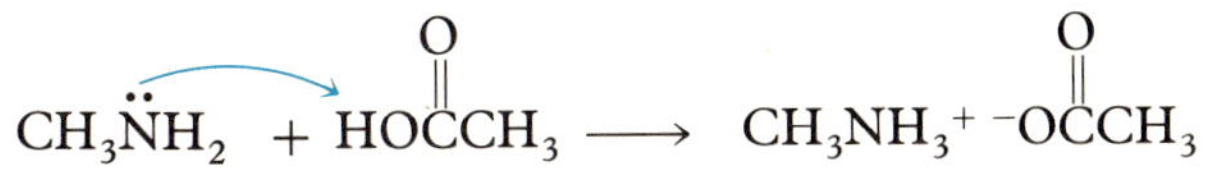

methylamine acetic acid methylammonium acetate

If an amine salt is treated with a base, the free amine is generated.

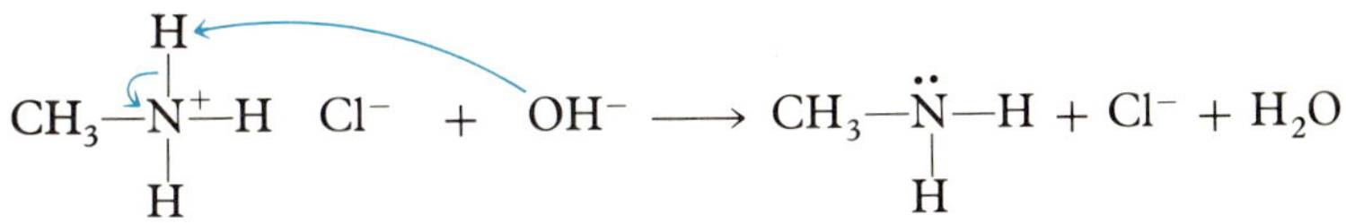

methylammonium chloride *an amine salt* methylamine *an amine*

Generalized:

$$R_3NH^+ X^- + OH^- \longrightarrow R_3N + H_2O + X^-$$

an amine salt *a base* *an amine*

STUDY QUESTIONS

18-1. Draw the structures of ammonia and methylamine; show the filled sp^3 orbitals.

18-2. Classify each of the following structures as a primary, secondary, or tertiary alcohol or as a primary, secondary, or tertiary amine.

(a) CH_3CH_2OH (b) $CH_3CH_2NH_2$
(c) $(CH_3)_2CHNH_2$ (d) $CH_3CH_2NHCH_2CH_3$
(e) $(CH_3)_3CNH_2$ (f) $(CH_3)_3COH$
(g) $(CH_3)_3CNHCH_3$ (h) $(CH_3)_3CN(CH_3)_2$

(i) 

18-3. Name the compounds in Problem 18-2.

18-4. Give the structures of the following amines:
(a) 2-butylamine
(b) N,N-dimethylcyclohexylamine
(c) N-ethylaniline
(d) 2-(N-methyl)pentylamine

18-5. Show the hydrogen bonding between molecules of cyclohexylamine.

18-6. Show all the different types of hydrogen bonds that could be formed in an aqueous methylamine solution.

18-7. Which of the following liquids undergoes hydrogen bonding? Explain your answer.
(a) pure triethylamine
(b) aqueous triethylamine

18-8. Complete the following equations:

(a) $CH_3CH_2NH_2 + H_2SO_4 \longrightarrow$

(b) 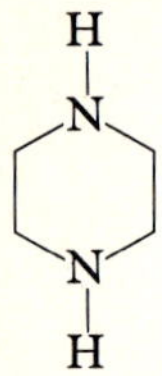 $+ \ 2\ CH_3CO_2H \longrightarrow$

(c) $(CH_3)_3CNH_2 + HCl \longrightarrow$

(d) 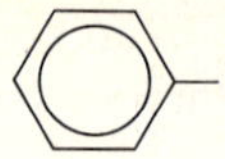 $-NH_2 + H_3PO_4 \longrightarrow$

18-9. What is the pH of a 0.100 *M* aqueous solution of trimethylamine? (See Table 18-1.)

18-10. When an aqueous solution of aniline hydrochloride is treated with NaOH, an oil separates from the solution. Explain this observation.

CHAPTER 19 AROMATIC COMPOUNDS

OBJECTIVES

The Aromatic Pi Cloud in Benzene *Draw the p orbital structure of benzene.* * *Describe the characteristic chemistry of benzene in reactions with most laboratory reagents.* * *Write the equation for the reaction of an alkylbenzene with* $KMnO_4$.

Naming Aromatic Compounds *Give the structures and names for some common monosubstituted benzene compounds, disubstituted benzene compounds, and benzene compounds with more than two substituents.*

Aromatic Substitution Reactions *Give equations for the aromatic bromination and nitration of benzene.* * *List some common o-,p-directors and m-directors.*

Aromatic Heterocyclic Compounds *Draw the orbital structure of pyridine.* * *Give equations for the reactions of aromatic nitrogen heterocycles with acids.*

An *aromatic compound* is a compound that contains one or more rings and has *p* electrons forming an aromatic pi cloud. Aromatic compounds are used in practically every phase of organic technology and are found in food, drugs, dyes, plastics, and explosives.

Benzene is one aromatic compound that we have already encountered many times. Naphthalene is another aromatic compound; its ring system is formed from two rings that share two carbon atoms, called *fused* rings. Pyridine is an aromatic compound that contains a nitrogen atom in the ring.

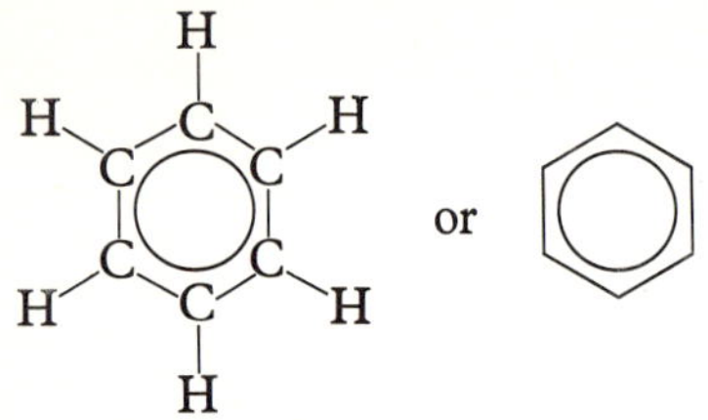

benzene

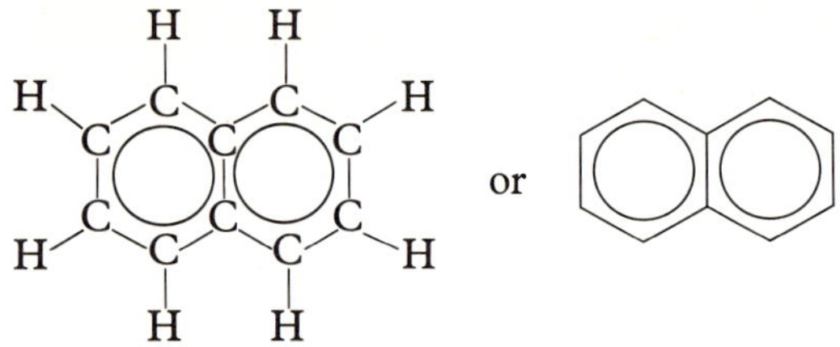

naphthalene

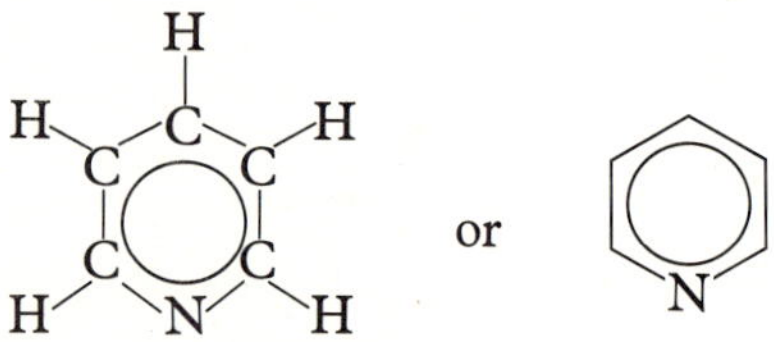

pyridine

Aromatic compounds are obtained from coal tar, the distillate from the destructive distillation (distillation in the absence of air) of coal. Coal tar is rich in alkylbenzenes, naphthalene, and other aromatic compounds, including some carcinogens (cancer-causing agents) and phenols.

Some aromatic hydrocarbons found in coal tar:

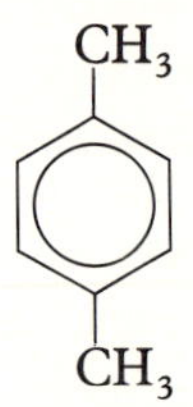

p-xylene

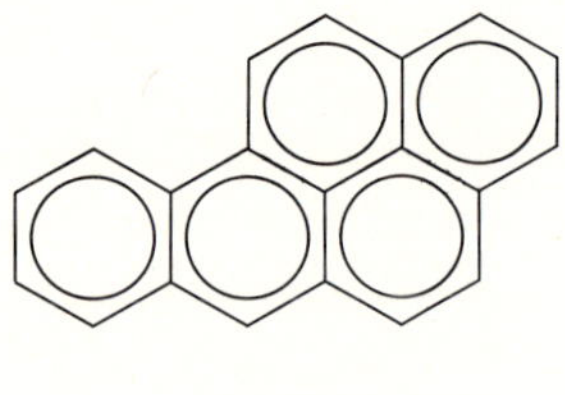

benzopyrene
a carcinogen

THE AROMATIC PI CLOUD IN BENZENE

Before we consider the chemistry of the aromatic compounds, let us quickly review some of the structural and electronic features of benzene, the simplest aromatic compound. (Some of these features have been discussed in Chapter 4.) The six carbon atoms in the benzene ring are joined to each other by sigma bonds formed from sp^2-hybrid orbitals. The sp^2 bonds are planar, and therefore the benzene ring is flat. Each carbon in the ring is also bonded to one hydrogen atom by an sp^2-hybrid orbital. The bonds to these hydrogens are in the plane of the ring. Each carbon atom also has one electron in a *p* orbital, which is perpendicular to the plane of the ring. Each *p* orbital completely overlaps with both its neighbors to form the aromatic pi cloud. (See Figure 19-1.)

Figure 19-1. The bonding in benzene.

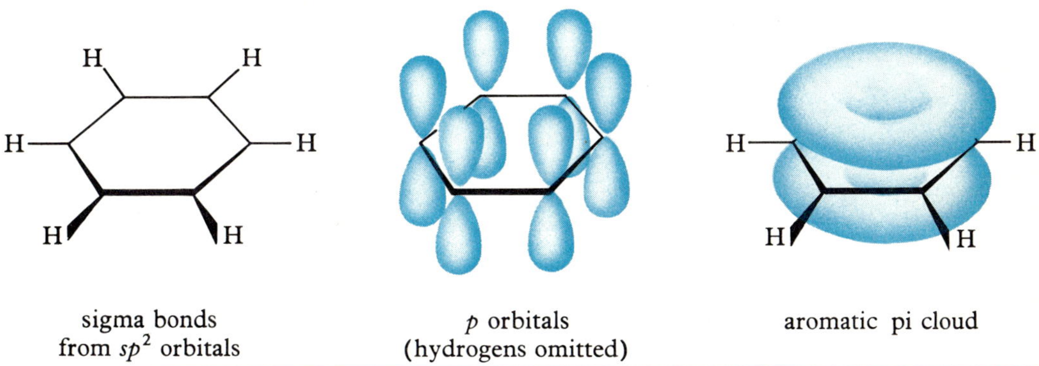

The sharing of the *p* electrons by all the ring carbon atoms *stabilizes* the benzene ring. In general, we can say that aromatic compounds are less reactive than other compounds containing *p* electrons. Without a catalyst or other harsh conditions, benzene (for example) does not react with most laboratory reagents, such as dilute acids or bases, oxidizing agents ($KMnO_4$), or reducing agents (Na).

When an alkylbenzene is treated with an oxidizing agent, the side chain is oxidized, but not the aromatic ring. Regardless of the length of the side chain, the oxidation of an alkylbenzene results in benzoic acid.

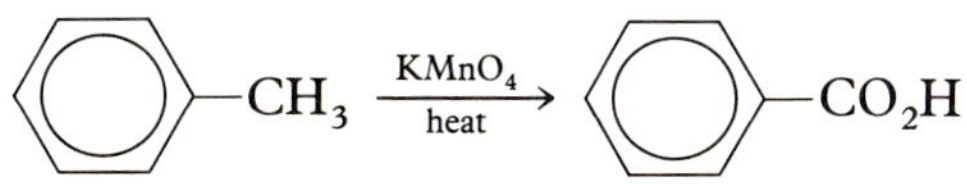

benzoic acid

Like all compounds containing pi systems, benzene can be hydrogenated. However, high temperature and pressure are required. Benzene is not hydrogenated as readily as an alkene.

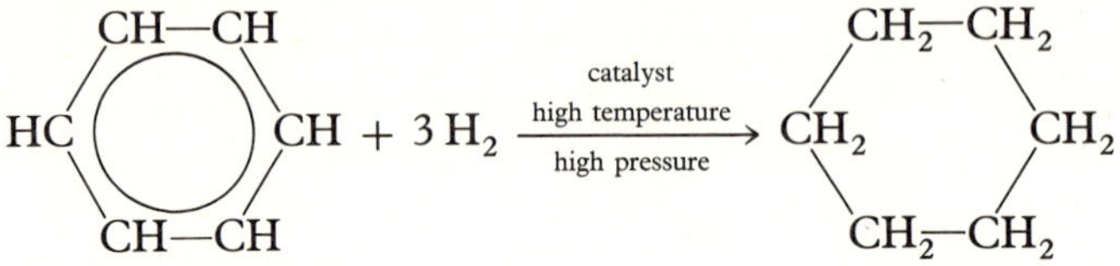

benzene cyclohexane

NAMING AROMATIC COMPOUNDS

Aromatic compounds are named somewhat differently from the aliphatic compounds and the cycloalkanes. Several of the substituted benzenes have specific names.

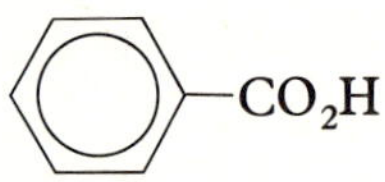

benzoic acid

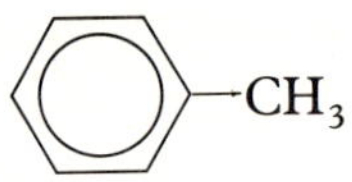

toluene

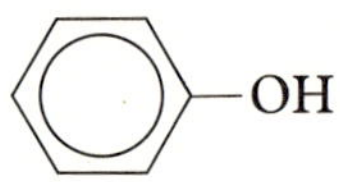

phenol

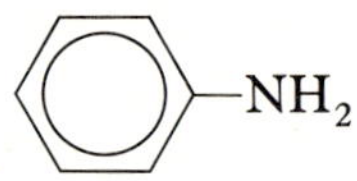

aniline

When benzene is named as a substituent or radical, its prefix name is *phenyl-*:

phenyl group

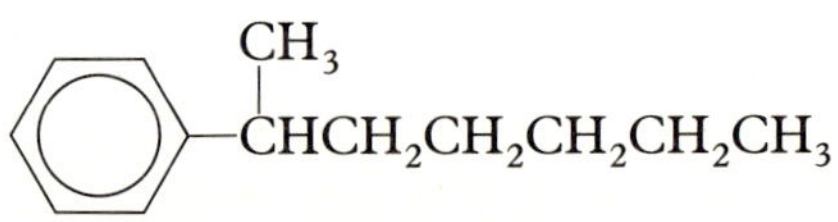

2-phenylheptane

Some functional groups and alkyl groups attached to the benzene ring are designated by prefixes:

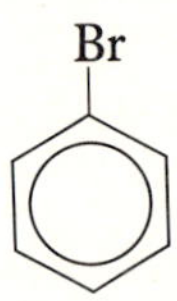

bromobenzene

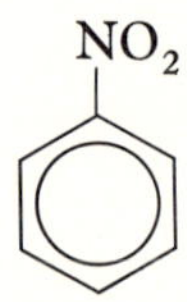

nitrobenzene

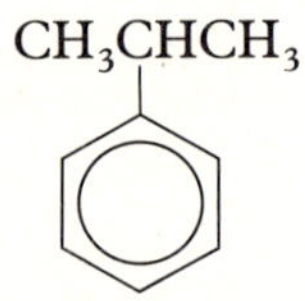

isopropylbenzene

When a benzene ring has two substituents, there is the possibility of three structural isomers:

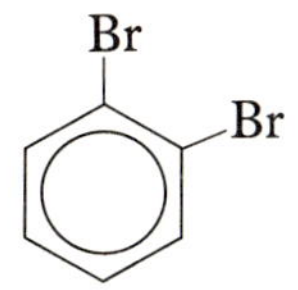

o-dibromobenzene

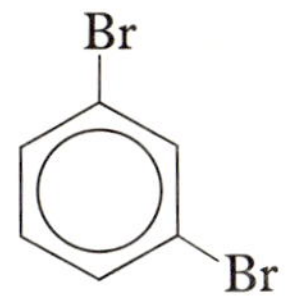

m-dibromobenzene

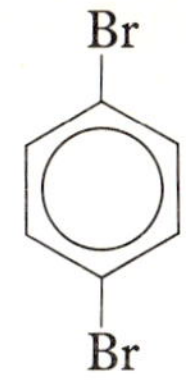

p-dibromobenzene

These three isomers are distinguished from each other by the prefixes *ortho, meta,* and *para,* abbreviated *o*-, *m*-, and *p*-. *Ortho* means that the two groups are on adjacent carbon atoms on the benzene ring, or in the 1,2-positions. *Meta* means the substituted carbon atoms are separated by one carbon atom, or that the groups are in the 1,3-positions. *Para* means that the two groups are in the 1,4-positions. (The *o, m,* and *p* designations are reserved solely for disubstituted benzene compounds. The substituted cyclohexanes are named using numbers, never as *o, m,* or *p*.) Using these prefixes then, we can designate the positions of two substituents on a benzene ring as we have done for the dibromobenzenes and for the following examples.

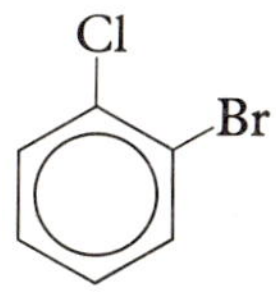

o-bromochlorobenzene

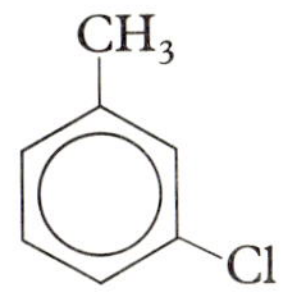

m-chlorotoluene

When there are *more than two* substituents on a benzene ring, their positions are designated by numbers. The ring is numbered so that the prefix numbers are as small as possible. More than one of the same group are grouped together in the prefix, for example, as *tribromo*- or *trimethyl*-. In numbering a substituted benzene ring that has a specific name, such as a substituted toluene, that key group is considered at position 1. In other words, if the parent name toluene is to be used, as in 2,4,6-trinitrotoluene, the methyl is understood to be at position 1.

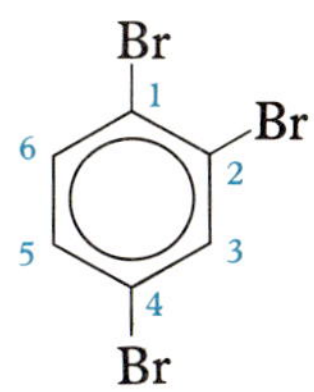

1,2,4-tribromobenzene

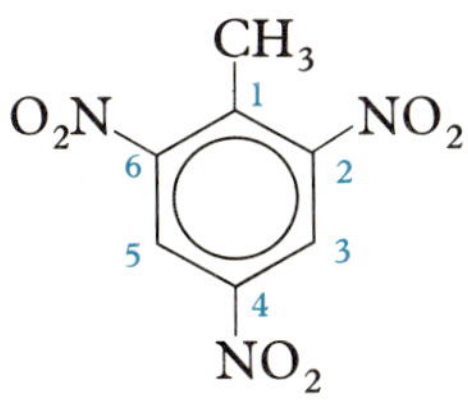

2,4,6-trinitrotoluene

(TNT)

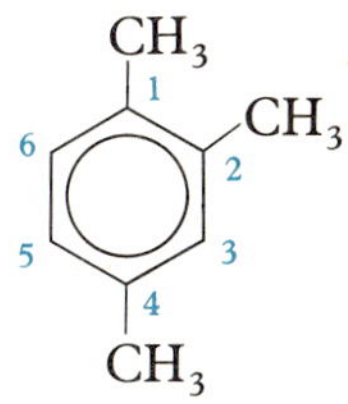

1,2,4-trimethylbenzene

(*not 2,4-dimethyltoluene, which is inconsistent in the grouping of the methyls*)

Sample Problem: Name the following compound:

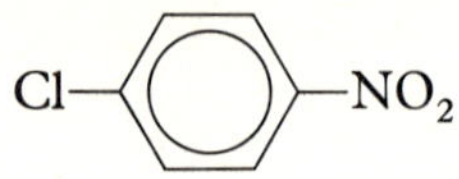

Solution: Since there are two groups attached to the benzene ring, the *o*, *m*, *p* system is used. This structure is named *p*-chloronitrobenzene.

Sample Problem: Give the structure of 2,4,6-tribromoaniline.

Solution: *Step 1:* The parent is aniline.

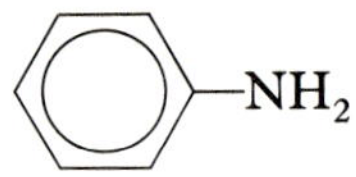

Step 2: Number the ring with the $—NH_2$ group at position 1, and add the bromo substituents.

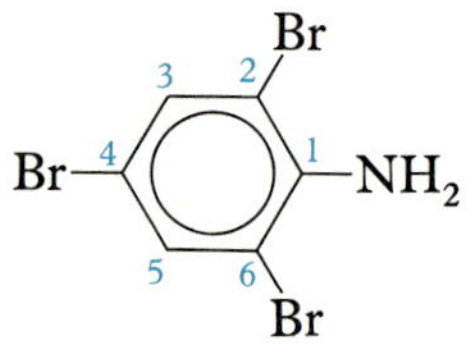

AROMATIC SUBSTITUTION REACTIONS

Although aromatic rings tend to be unreactive, there is one important class of reactions that aromatic rings undergo. These reactions, called *aromatic substitution reactions,* involve substitution of a hydrogen atom attached to the ring by another atom or group. The aromatic pi cloud is not lost in these reactions, but is retained in the product.

One example of an aromatic substitution reaction is *bromination* of benzene. When benzene is treated with bromine in the presence of an iron(III) bromide catalyst, a bromine atom displaces a hydrogen atom on the benzene ring. The products of this bromination reaction are bromobenzene and hydrogen bromide. *Nitration,* the replacement of a hydrogen atom by a nitro group ($—NO_2$), is another example of aromatic substitution.

Bromination:

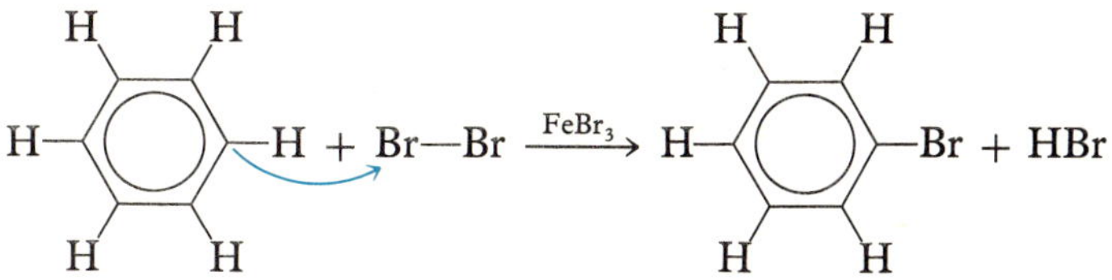

bromobenzene

Nitration:

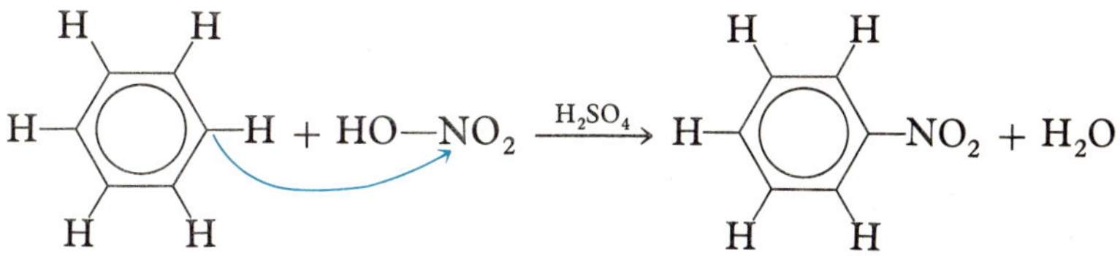

nitric acid   nitrobenzene

SUBSTITUTION OF A SECOND GROUP ON BENZENE. A substituted benzene ring can undergo further aromatic substitution reactions. The presence of the first group on the benzene ring determines the position at which another group will attack. For example, bromobenzene undergoes attack at the *ortho-* and *para*-positions, regardless of the attacking reagent. Very little attack occurs at the *meta*-position.

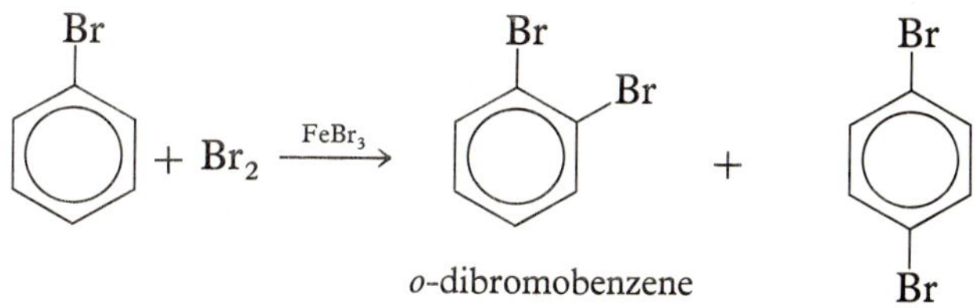

o-dibromobenzene   *p*-dibromobenzene

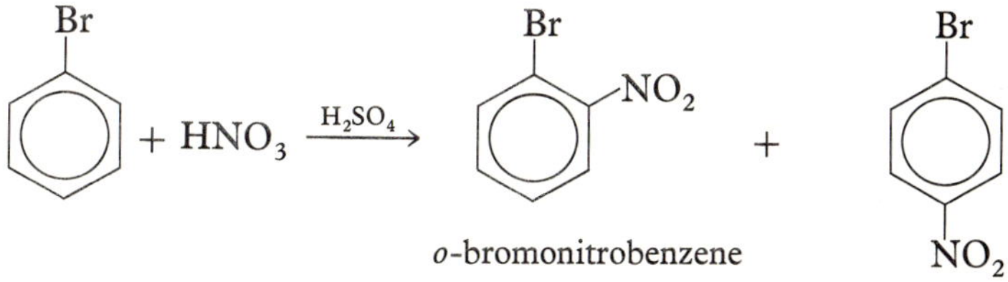

o-bromonitrobenzene   *p*-bromonitrobenzene

Because it directs incoming groups to either the *ortho*-position or the *para*-position, bromine is called an *ortho-,para*-director. Other *ortho-,para*-directors are shown in Table 19-1. Notice that all of the *ortho-,para*-directors except R have

Table 19–1. Common *o-,p-* and *m*-Directing Groups

ortho-,para- Directing Groups	*meta*-Directing Groups
$—\ddot{N}H_2$, $—\ddot{N}R_2$	$—C(=O)—R$
$—\ddot{\underset{..}{O}}H$, $—\ddot{\underset{..}{O}}R$	$—C\equiv N$
$—R$	$—S(=O)_2OH$
$—\ddot{\underset{..}{Cl}}:$, $—\ddot{\underset{..}{Br}}:$	$—N^+(=O)—O^-$
$—\ddot{N}HC(=O)R$	$—C(=O)OH$

unshared electrons on the atom attached to the benzene ring. This is a good rule of thumb for predicting whether or not a group is an *ortho-,para*-director.

In contrast to bromobenzene, if nitrobenzene is brominated or nitrated, we observe the *meta*-isomer as the product, and not the *ortho-* or *para*-isomer. In other words, the nitro group is a *meta*-director.

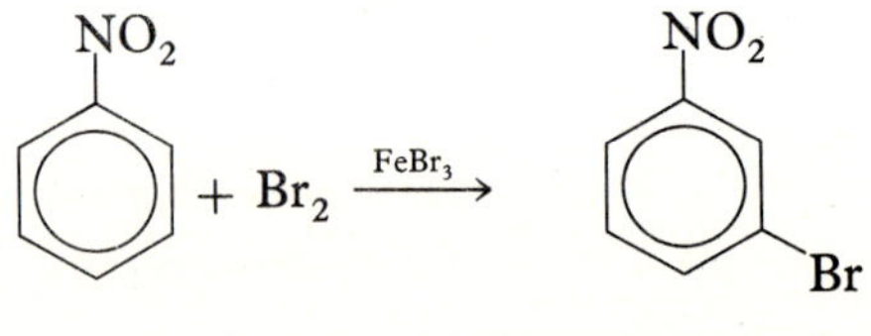

m-bromonitrobenzene

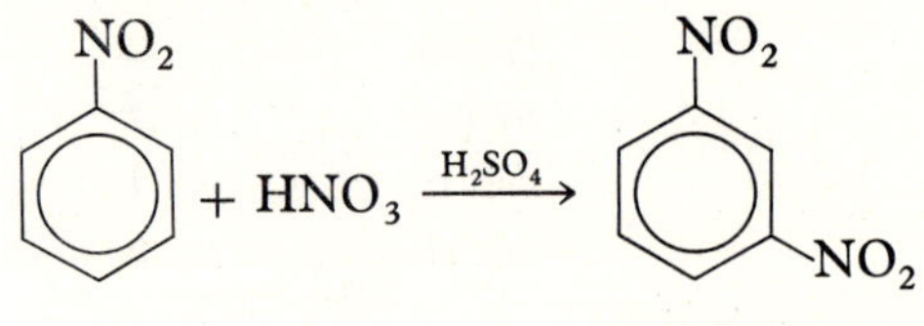

m-dinitrobenzene

Other *meta*-directors besides the nitro group are shown in Table 19–1. Note that none of the *meta*-directing groups have unshared electrons on the atom attached to the ring.

Sample Problem: Complete the following equations:

(a) 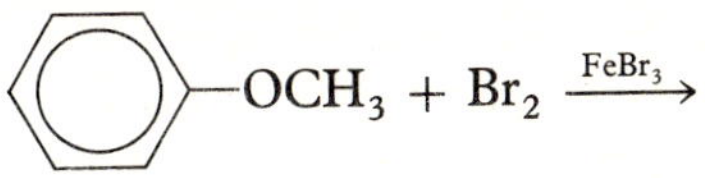

(b) 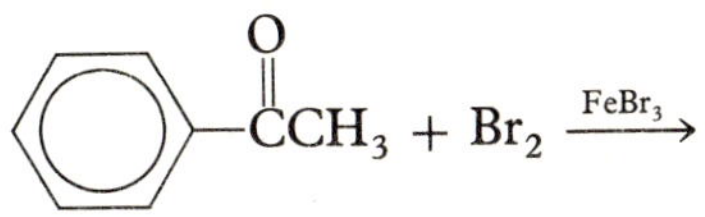

Solution:

(a) 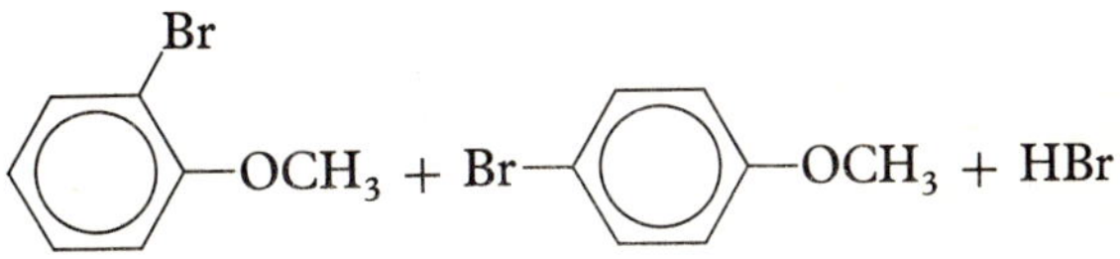

(b) 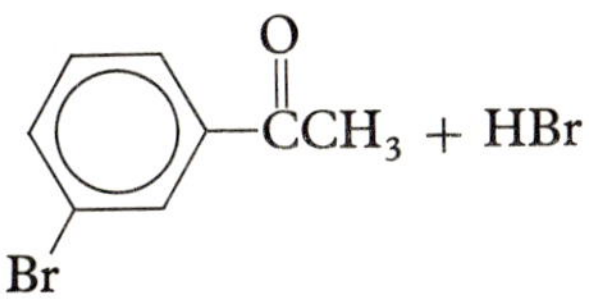

Aromatic substitution reactions involving a second substitution are the basis of many industrial procedures. Salicylic acid, used to make aspirin, may be prepared from phenol by an aromatic substitution reaction.

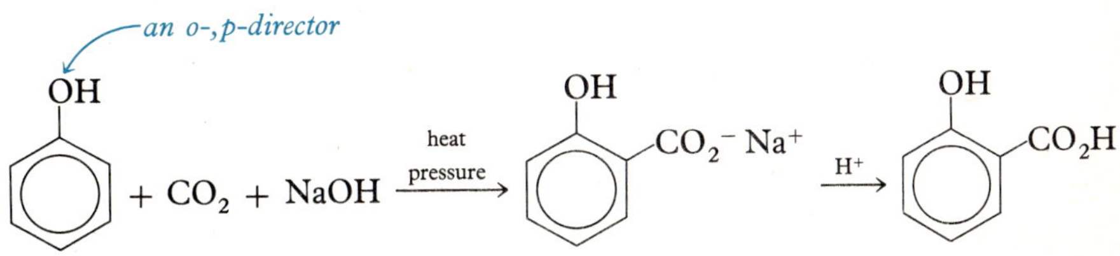

phenol — sodium salicylate — salicylic acid

Some aromatic substitution reactions can be accomplished enzymatically in animal organisms. The amino acid phenylalanine from proteins is normally converted to tyrosine before it is broken down to acetyl groups. This substitution reaction is catalyzed by the enzyme phenylalanine hydroxylase.

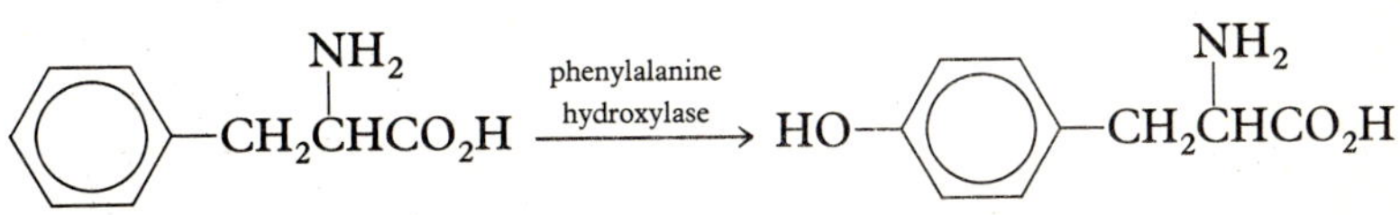

phenylalanine — tyrosine

One out of every ten thousand humans lacks the enzyme that catalyzes this conversion to tyrosine. In these individuals, phenylalanine is converted to phenylpyruvic acid by a transamination reaction.

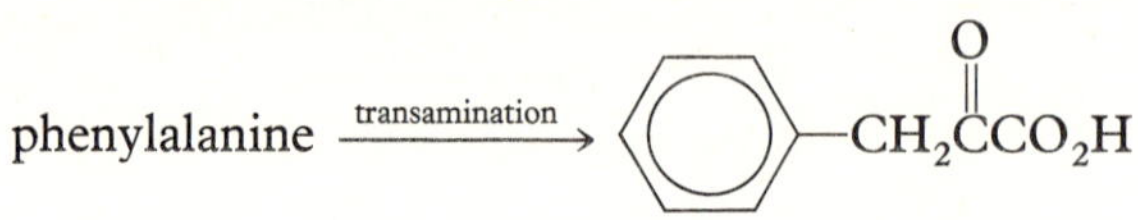

phenylpyruvic acid

The phenylpyruvic acid is not broken down further; instead, it builds up in the bloodstream and spills over into the urine. This condition is known as *phenylketonuria* (meaning "a phenyl ketone in the urine"), often abbreviated PKU. The presence of phenylpyruvic acid in the blood prevents normal development of the brain, resulting in mental retardation. PKU is controlled by limiting the phenylalanine in the diet.

AROMATIC HETEROCYCLIC COMPOUNDS

As we have mentioned, an aromatic ring may contain atoms other than carbon atoms. Nitrogen can form sp^2 bonds, just as carbon can. Nitrogen, too, can contribute a p electron toward an aromatic pi cloud. Pyridine and pyrimidine are six-membered aromatic rings that contain one and two nitrogen atoms, respectively.

pyridine pyrimidine

Pyridine is a pungent-smelling liquid used as a laboratory reagent. It is especially useful as a scavenger for hydrogen ions because the filled sp^2 orbital on the nitrogen (see Figure 19-2) can form a sigma bond with the hydrogen ion just as a filled sp^3 orbital of ammonia or an amine can.

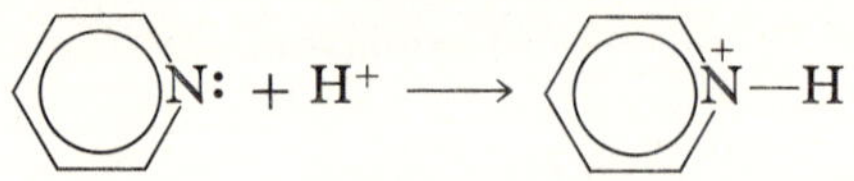

A pyridine ring is part of the structure of the nicotinamide molecule. The filled orbital of the nitrogen in nicotinamide is used to form a sigma bond in the coenzyme *nicotinamide adenine dinucleotide* (NAD^+), which is used in biological oxidation and reduction reactions. (See Chapter 25.)

Figure 19-2. Orbital picture of pyridine. (The hydrogen atoms on the ring carbons are not shown.)

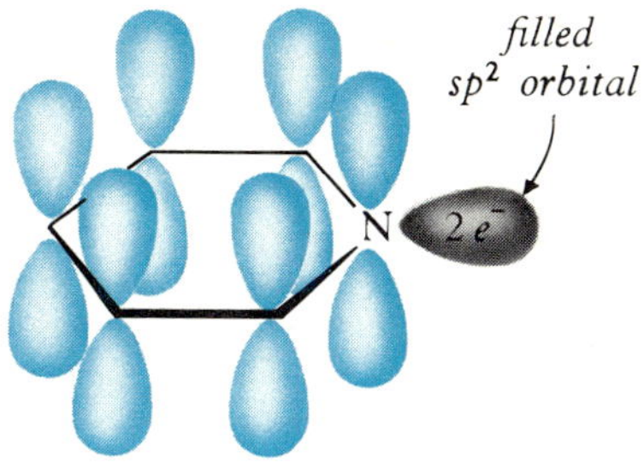

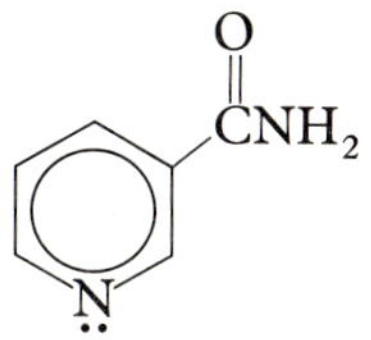

nicotinamide

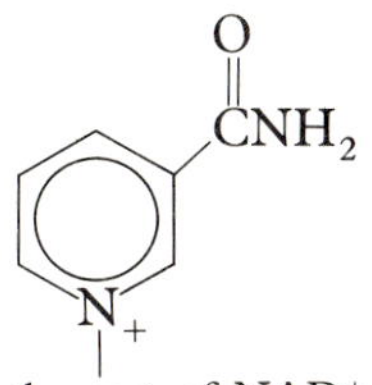

the rest of NAD⁺

nicotinamide in NAD⁺

The nitrogen of a pyrimidine ring also can donate electrons to form a sigma bond. Derivatives of pyrimidine bond with sugar molecules to form an important structural feature of the nucleic acids. (See Chapter 28.)

Reaction with hydrogen ion:

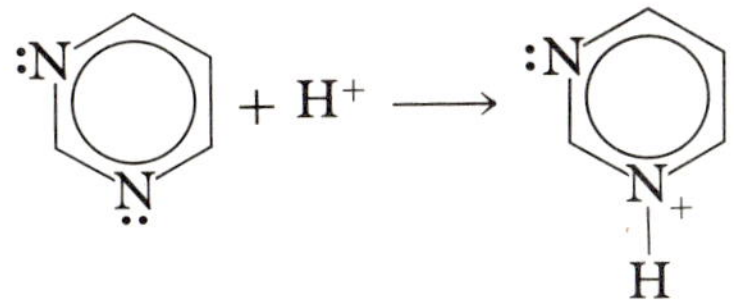

Other important aromatic heterocyclic compounds are quinoline, whose ring system is found in the antimalarials quinine and chloroquine; indole, which is found in proteins; and purine, another heterocycle found as part of nucleic acids.

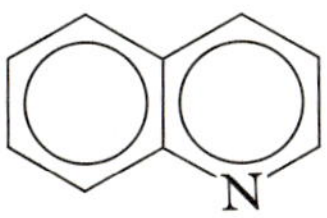

quinoline

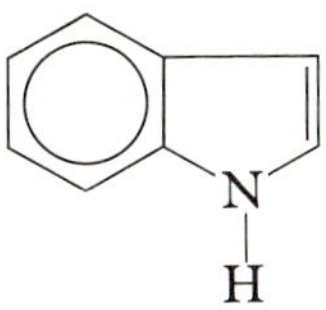

indole

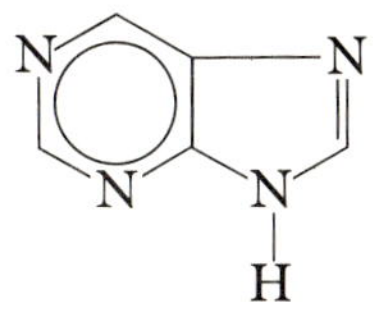

purine

Compounds containing the quinoline ring system:

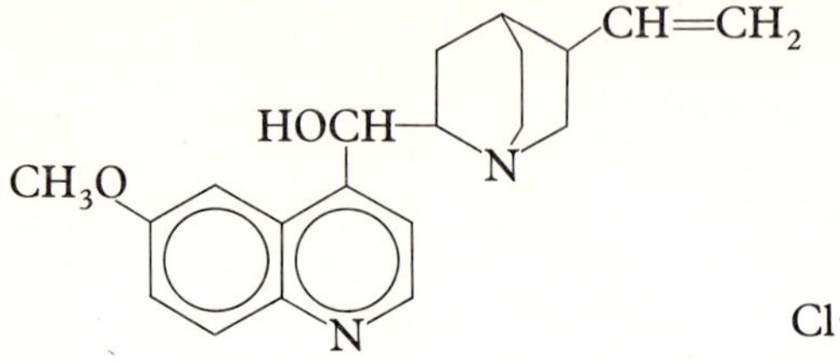

quinine
a natural antimalarial

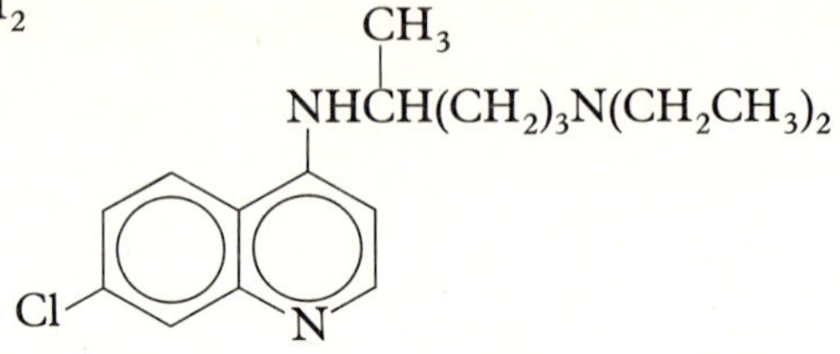

chloroquine
a synthetic antimalarial

A compound containing the indole ring system:

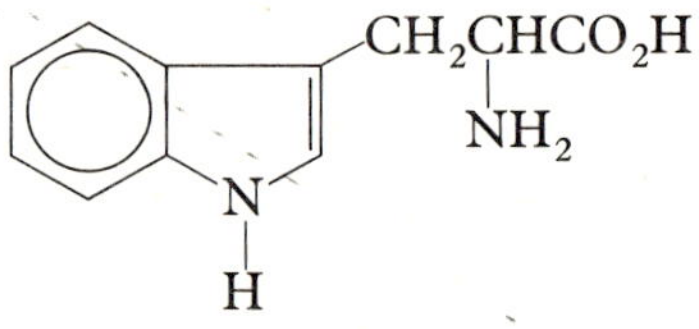

tryptophan
from proteins

Compounds containing the purine ring system:

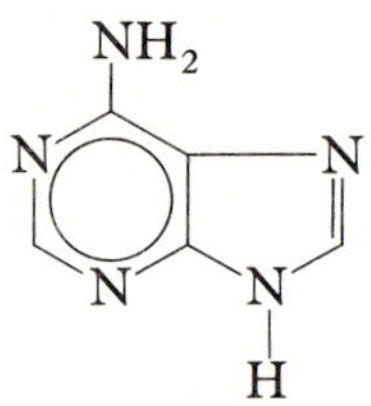

adenine
found in nucleic acids and in NAD+

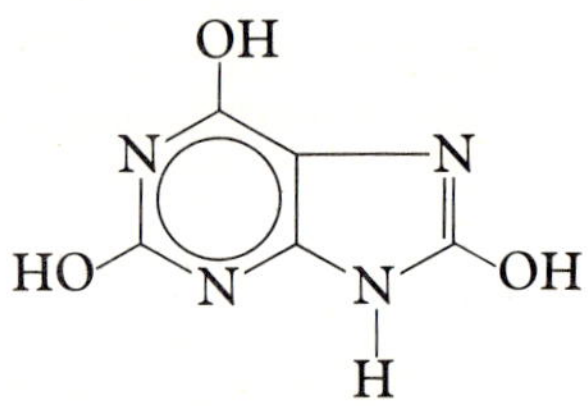

uric acid
associated with gout

STUDY QUESTIONS

19-1. Which of the following are aromatic compounds?

(a) 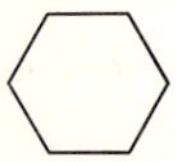

(b) 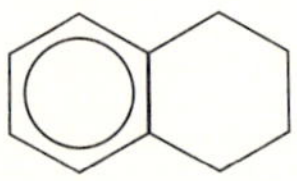

(c) 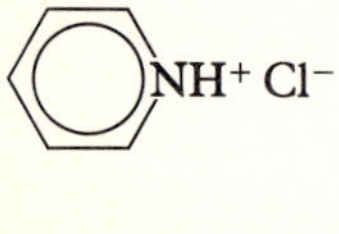

(d) 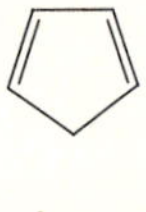

19-2. Draw the *p* orbital picture of naphthalene; show clearly the hydrogen atoms, the sigma-bond structure, and the *p* orbitals that form the aromatic pi cloud.

19-3. How many electrons are present in the aromatic pi cloud of naphthalene?

19-4. How many *p* orbitals are used in the formation of the aromatic pi cloud in each of the following compounds?

(a) 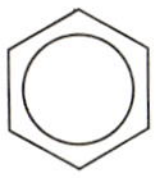

(b) 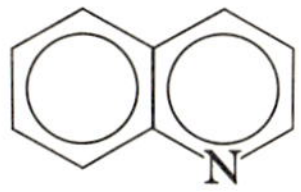

(c) 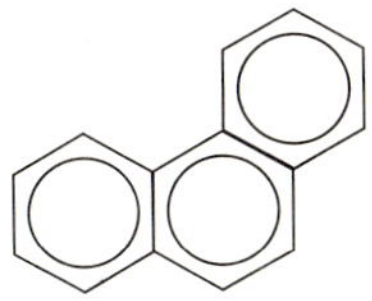

(d) 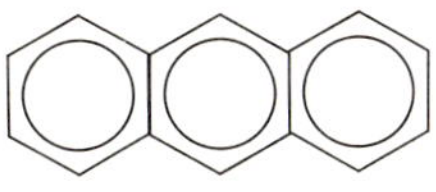

19-5. How many moles of hydrogen gas are required to hydrogenate completely one mole of each of the compounds in Problem 19-4?

19-6. Draw the structures of the products of complete hydrogenation of the compounds in Problem 19-4.

19-7. Complete the following equations:

(a) 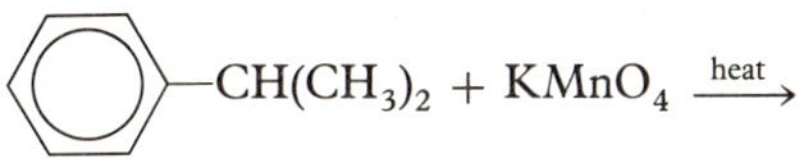

(b) 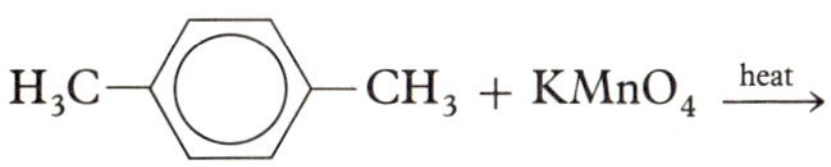

(c) 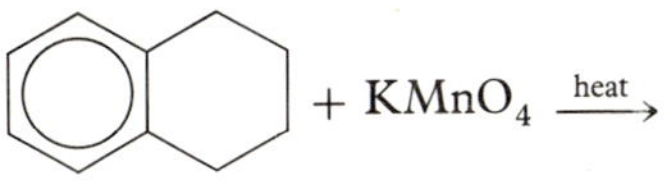

19-8. Give the structure for each of the following:
(a) benzoic acid
(b) phenol
(c) aniline
(d) 4-phenyl-1-cyclohexanol

19-9. Name the following compounds:

(a) 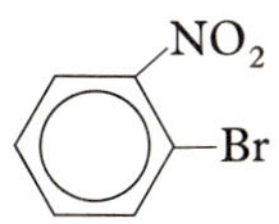

(b) 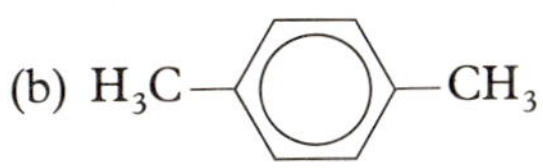

(c) 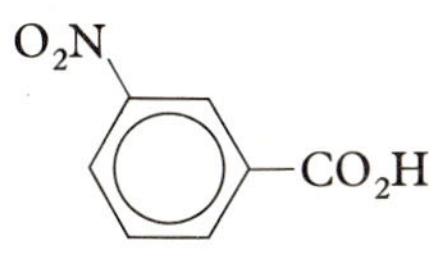

(d) 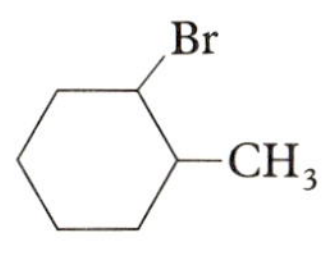

19-10. Give the structure for each of the following:
(a) *o*-dichlorobenzene
(b) *p*-*t*-butylphenol
(c) *m*-nitroaniline
(d) 2,4,6-trichlorobenzoic acid

19-11. Complete the equations for the following aromatic monosubstitution reactions:

(a) 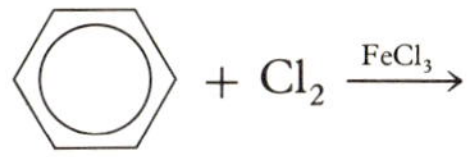

(b) 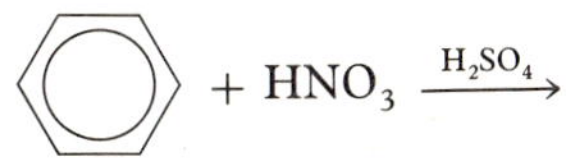

19-12. Write the equations for the monobromination of each of the following compounds:
(a) phenol
(b) benzoic acid
(c) benzonitrile (C_6H_5CN)

19-13. Write the equation for the nitration of each of the compounds in the preceding problem.

19-14. Which of the following compounds would be difficult to obtain from benzene by a direct aromatic substitution reaction of a substituted benzene?
(a) *o*-dichlorobenzene
(b) *p*-dinitrobenzene
(c) *m*-bromoaniline
(d) *m*-bromonitrobenzene

19-15. Which of the following compounds contains an aromatic heterocyclic ring?
(a) α-D-glucose (p. 364)
(b) FAD (p. 407)
(c) estrone (p. 398)

19-16. Give the products from the reactions of each of the following with dilute HCl:

(a) 

(b) 

(c) 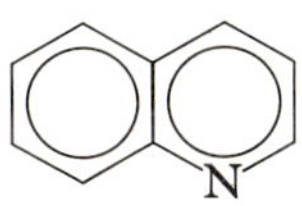

CHAPTER 20 MOLECULES IN THREE DIMENSIONS

OBJECTIVES

Optical Isomerism *Describe a plane of symmetry.* * *Describe the relationship between* (*1*) *a symmetrical object and its mirror image and* (*2*) *an unsymmetrical object and its mirror image.* * *Define chirality.*

Chiral and Achiral Carbon Atoms *Describe* (*with structures*) *the relationship between* (*1*) *an achiral molecule and its mirror image and* (*2*) *a chiral molecule and its mirror image.* * *Describe how to determine if a carbon atom is chiral or not.* * *Define enantiomer and give examples.*

Fischer Projections *Describe the features of and write an example of a Fischer projection.* * *Differentiate a* D-*form from an* L-*form for a compound with one chiral carbon atom.*

More Than One Chiral Carbon Atom *Differentiate a* D-*form from an* L-*form for a compound with more than one chiral carbon atom.*

Rotation of Plane-Polarized Light *Define plane-polarized light.* * *Describe the rotation of the plane of vibration of plane-polarized light by a pair of separated enantiomers.* * *Define racemic mixture, and describe the effect of a racemic mixture upon plane-polarized light.* * *Define dextrorotatory, levorotatory,* (+), *and* (−).

Importance of Optical Activity *Describe why optical isomerism is important to the student of biochemistry or medicine.*

Cyclic Compounds in Three Dimensions *Give the reasons that cyclohexane rings are puckered and exist primarily in the chair form.* * *Define eclipsed and staggered.*

OPTICAL ISOMERISM

Besides structural and geometric isomerism, a molecule is capable of an additional type of isomerism that arises from the *right-* or *left-handed arrangement of atoms and groups within a molecule.* This type of isomerism is called *optical isomerism.* Optical isomerism is a special type of *stereoisomerism,* which is isomerism due to the arrangement of the atoms of a molecule in space rather than to the order of attachment of the atoms. Geometric isomerism is also a type of stereoisomerism.

Optical isomerism is a somewhat more subtle concept than other types of isomerism that we have discussed. However, it is so important in the chemistry of living systems that we must take a fairly close look at this type of stereoisomerism.

First, let us consider some properties of symmetrical and unsymmetrical objects. Most common symmetrical objects, such as a ball or a chair, have a *plane of symmetry.* This plane of symmetry is a hypothetical plane that can be passed through the object to divide it into two opposite but otherwise identical halves. In other words, a symmetrical object can be divided into a pair of mirror-reflective halves. Some symmetrical objects, a ball for example, can have more than one plane of symmetry. Figure 20-1 shows three symmetrical objects divided by planes of symmetry.

Let us take a symmetrical object, such as a coffee cup, and remove the hypothetical plane. Now we will hold the coffee cup up to a mirror. With a little imagination, we find we can superimpose the symmetrical coffee cup upon its mirror image. (See Figure 20-2.) The shape of a symmetrical object is identical with the shape of its mirror image, and the shapes of the two can be merged together. *A symmetrical object can be superimposed upon its mirror image.*

The human hand has no plane of symmetry; there is no way you can pass a plane through your hand and get two mirror-reflective halves. Your hand is an unsymmetrical object. At first glance, your two hands may appear identical but, in actuality, they are mirror images of each other. Hold your left hand up to a mirror and you will see a reflection that looks like a right hand. (See Figure 20-3.)

Figure 20–1. A symmetrical object may be divided into a pair of mirror-reflective halves by a plane of symmetry.

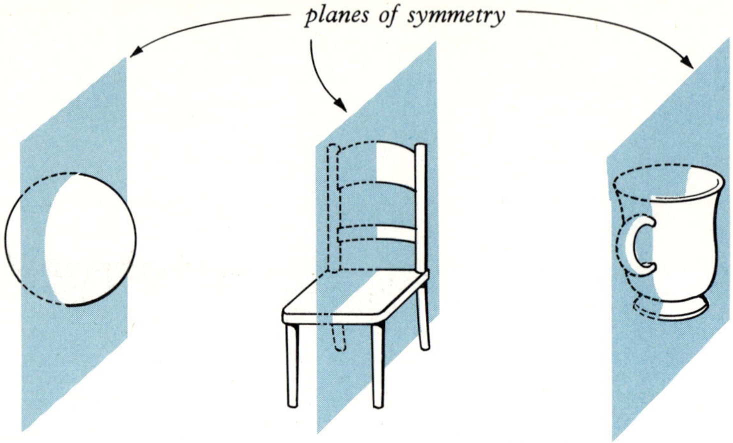

Figure 20–2. A symmetrical object can be superimposed upon its mirror image.

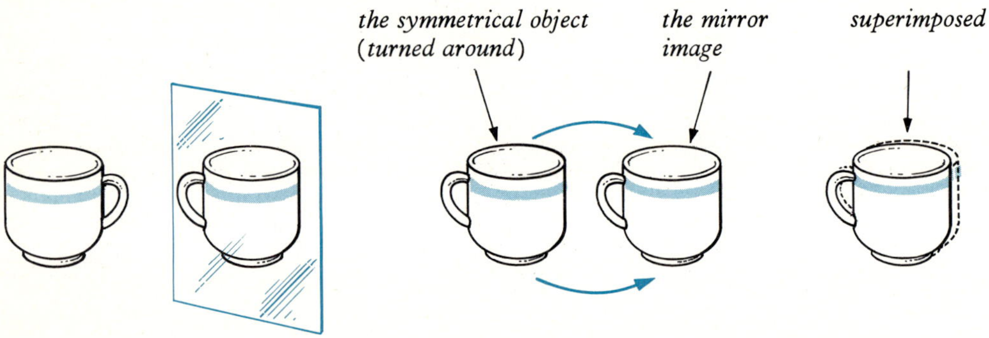

Figure 20–3. Your two hands are mirror images.

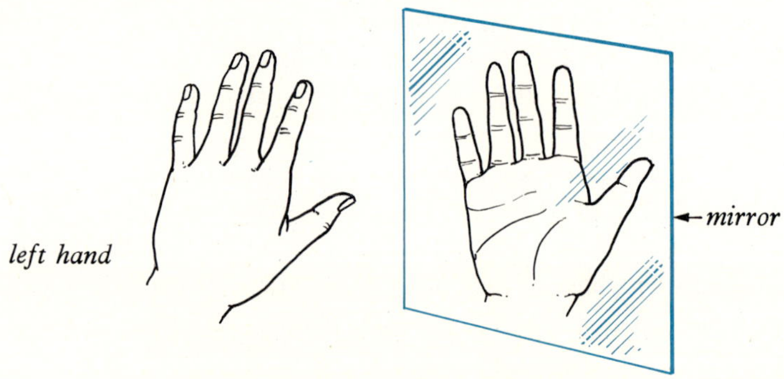

Unless you happen to have two left hands, you cannot superimpose one of your hands upon the other or upon its mirror image. If you keep the backs of your hands up and try to superimpose them, your thumbs will be on opposite sides. *An unsymmetrical object cannot be superimposed upon its mirror image.* (See Figure 20–4.) Any object that cannot be superimposed on its mirror image is said to be *chiral* (from the Greek *cheir,* handedness). Conversely, an object that is symmetrical and thus superimposable on its mirror image is *achiral* (not chiral).

Figure 20–4. An unsymmetrical object cannot be superimposed on its mirror image.

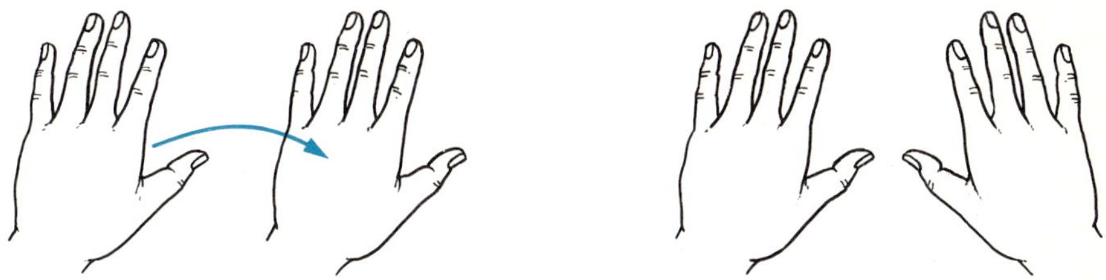

Two left hands being superimposed.

A left hand cannot be superimposed on a right hand.

CHIRAL AND ACHIRAL CARBON ATOMS

Molecules can also be chiral or achiral. The structure of the molecule as a whole determines chirality (or lack of it); however, we can obtain a clue as to whether or not a molecule is chiral by inspecting the attachments to each carbon in the structure.

When a carbon atom is in the sp^3-hybrid state, its four bonds point toward the corners of a tetrahedron. When two or more of the groups attached to the carbon atom are the same, a plane of symmetry can be drawn through the molecule. (When picturing a plane of symmetry passing through a molecule, it is perfectly acceptable that the plane passes through bonds or the centers of atoms.) By appropriate rotating and flipping upside down, a molecule with a plane of symmetry can be superimposed on its mirror image. The molecule and its mirror image are identical if they can be superimposed. They are not isomers. This molecule is therefore achiral. (See Figures 20–5 and 20–6.)

It is difficult to visualize symmetrical molecules and the superimposition of mirror images on paper. We strongly urge you to use an inexpensive molecular model kit and verify the statements made in this chapter for yourself.

Now let us take a different case, that of a carbon atom with no two attached groups alike. When all four groups attached to the tetrahedral sp^3-hybridized carbon atom are different, *no plane of symmetry can be drawn through the molecule.* In this

Figure 20–5. Because each carbon has two or more identical attachments (H), this molecule is symmetrical and can be superimposed upon its mirror image. The molecule and its mirror image are identical.

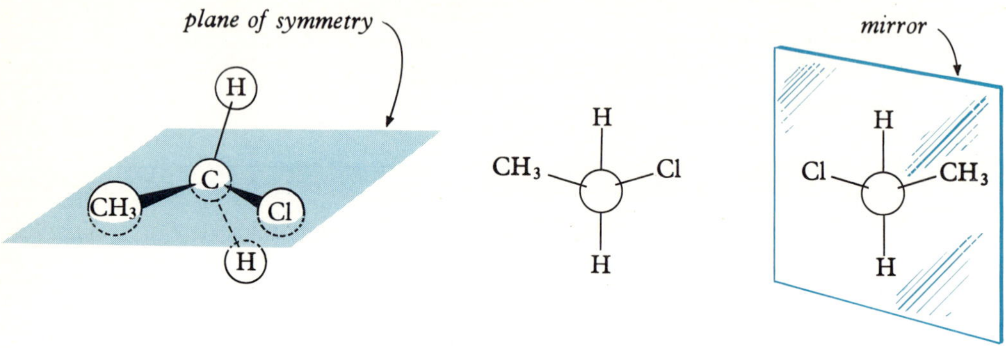

Figure 20–6. To be superimposed on its mirror image, the molecule must be turned backwards, then flipped upside-down.

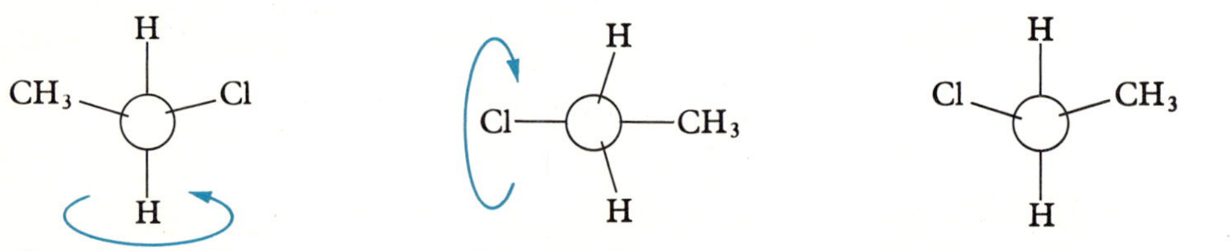

case, no amount of turning, twisting, and flipping will allow the molecule to be superimposed on its mirror image. Since they cannot be superimposed, the two molecules are chiral and the two mirror images are different compounds—they are isomers. (Again, in order to see this for yourself, construct a molecular model with the four atoms attached to the carbon having four different colors. Construct the mirror image. Now you will be able to see that the two cannot be superimposed.)

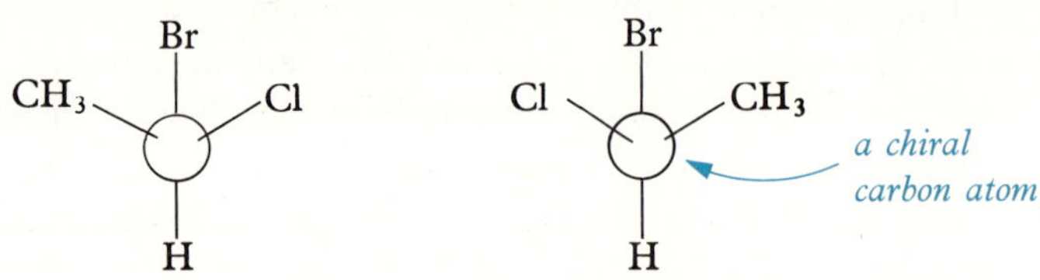

I *no plane of symmetry*

II *mirror image of* **I**

I *and* ***II*** *cannot be superimposed.*

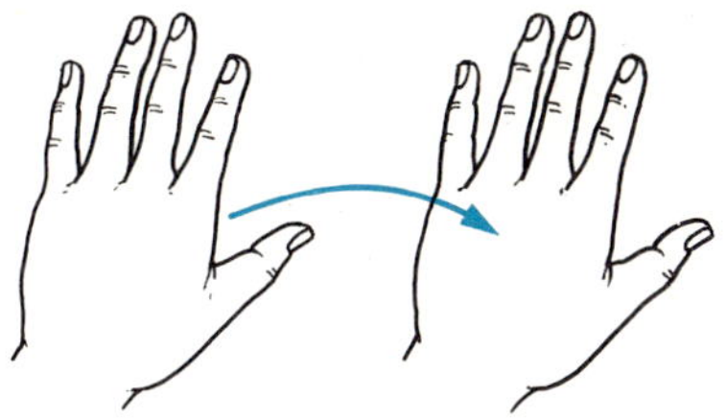

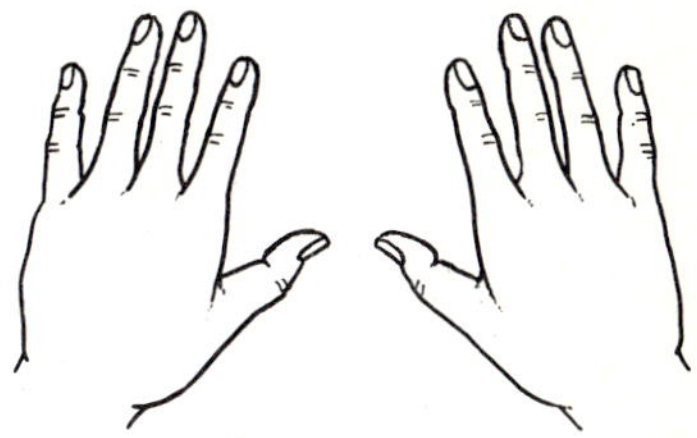

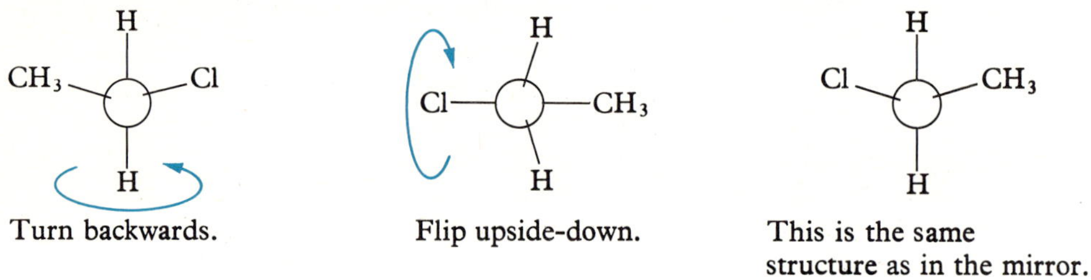

Figure 20–7. Inexpensive molecular models made from gumdrops and toothpicks may be used to illustrate chiral and achiral carbons.

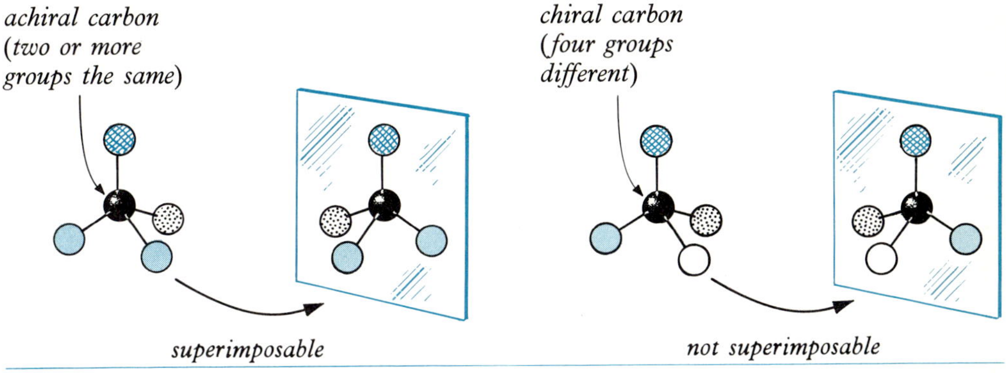

Any carbon atom that has four different groups attached is called a *chiral carbon atom.* A molecule with a chiral carbon atom can generally exist in two isomeric forms called *enantiomers,* or nonsuperimposable mirror images of each other.

In practice, chemists do not always construct models, try to draw planes of symmetry, or attempt superimposing mirror images in order to see if a carbon atom is chiral or achiral. The quickest way to determine whether or not a carbon is chiral is by inspecting the attached groups. If there are four different groups attached, the carbon atom is chiral.

Examples of chiral carbon atoms, each with four different attached groups:

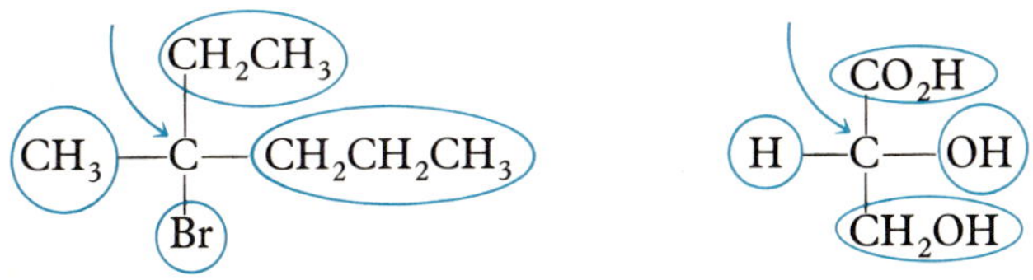

Many naturally occurring compounds, such as sugars and steroids, contain chiral carbon atoms as part of a ring system. To determine symmetry or asymmetry in such a case, go around the ring to see if one side is like the other side or not.

Not a chiral carbon. Sides (1) and (2) are identical attachments.

A chiral carbon. Sides (1) and (2) are different; thus, the carbon has four different attached groups.

Sample Problem: Which of the following compounds have chiral carbon atoms?

(a) 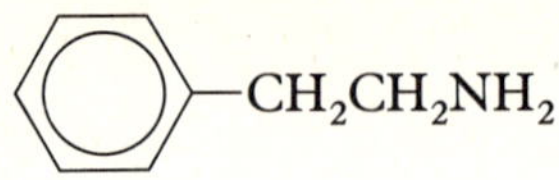 (b) 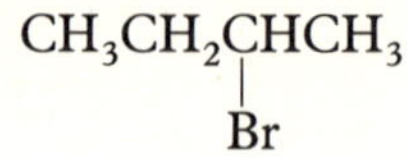

(c) 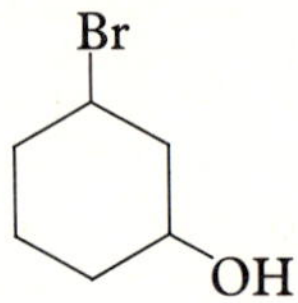

Solution: (b) and (c). The chiral carbon atoms are starred in the following formulas:

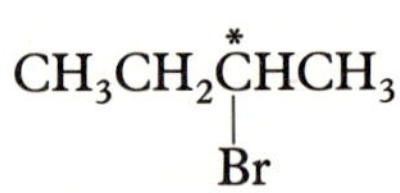 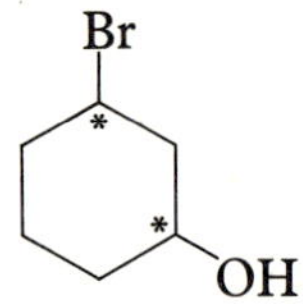

FISCHER PROJECTIONS

Since it is inconvenient to write three-dimensional structures on paper, certain conventions have been adopted for the representing of compounds with chiral carbon atoms.

Glyceraldehyde has been chosen as the standard for the purpose of comparing the configurations (order of attachment around a chiral atom) of these compounds; therefore, we will use this compound as the initial example. Glyceraldehyde has a chiral center at carbon atom number 2 and thus has a pair of enantiomeric forms, called the D-form and the L-form. To represent the three-dimensional structures of the enantiomers of glyceraldehyde, we can use a two-dimensional projection. One projection commonly used is called the *Fischer projection.* In this text, we use a modified Fischer projection, with wedges representing bonds angled toward the viewer.

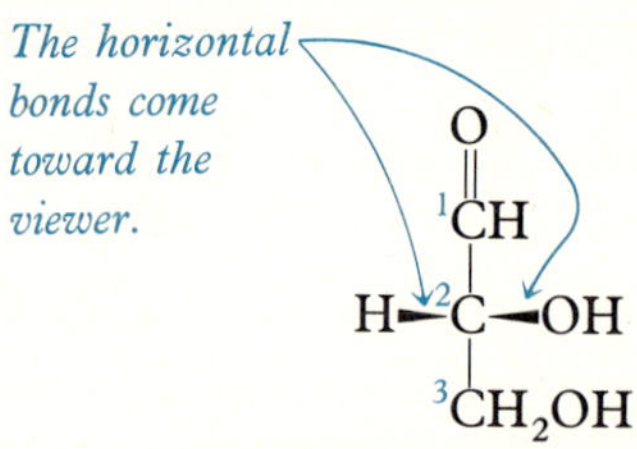

Fischer projection of D-glyceraldehyde

means

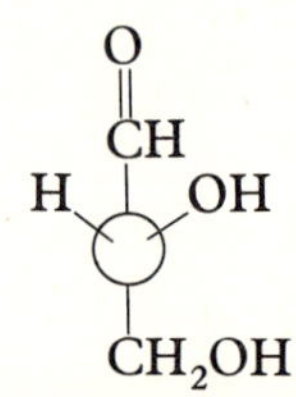

three-dimensional structure of D-glyceraldehyde

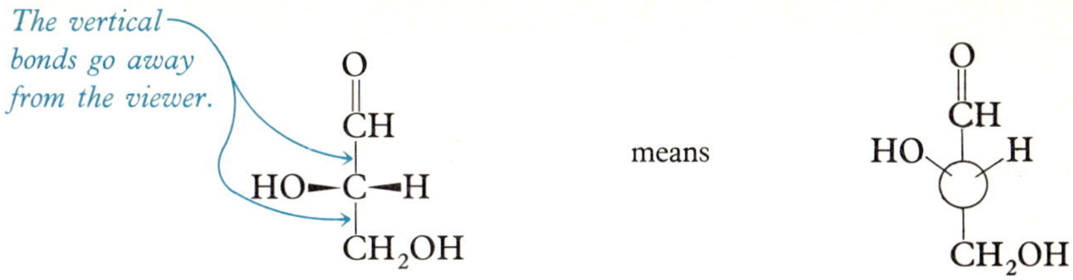

Fischer projection of L-*glyceraldehyde*

three-dimensional structure of L-*glyceraldehyde*

In the Fischer projection, glyceraldehyde has the aldehyde group (carbon 1) positioned at the top. Any horizontal bonds in the Fischer projection are defined as coming out toward the viewer. The vertical bonds go away from the viewer.

In comparing Fischer projections, certain rules must be followed. The projected structures may be slid around on the paper, but they must not be lifted out of the plane of the paper and flipped over. (If a Fischer projection is flipped over, the horizontal bonds no longer come toward the viewer and the structure is no longer a Fischer projection.)

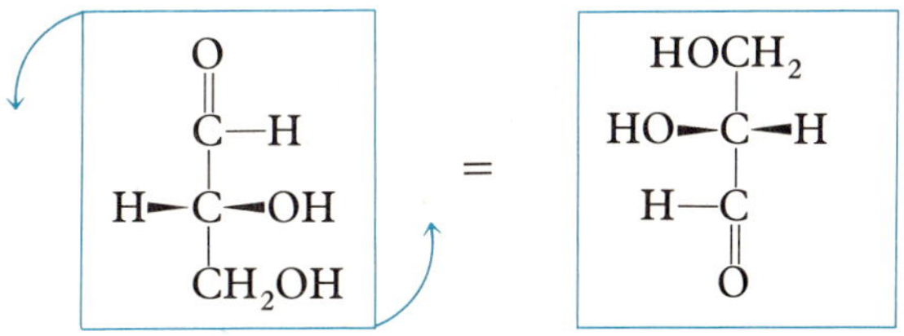

In the name D-glyceraldehyde, the prefix letter D means that the hydroxyl group on carbon 2 is to the *right* in the Fischer projection when the aldehyde group is at the top. The prefix letter L in the name L-glyceraldehyde means that the hydroxyl group is projected to the *left*. The prefixes D and L are used to tell us the absolute configuration, or exact relationship, of the groups around the chiral carbon atom.

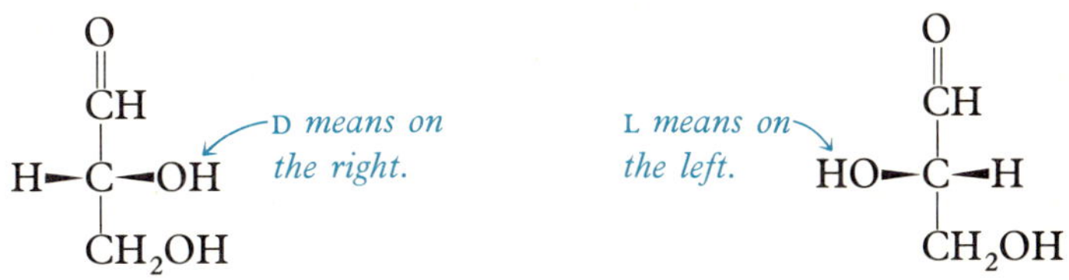

D-glyceraldehyde

L-glyceraldehyde

Glyceric acid has a structure similar to that of glyceraldehyde. Instead of an aldehyde group at carbon 1, glyceric acid has the closely related carboxylic-acid group. If glyceric acid is written in a Fischer projection, the hydroxyl group at carbon

2 must be either to the right or to the left. Just as in D-glyceraldehyde, the enantiomer in which this hydroxyl is projected to the right is called D-glyceric acid. The enantiomer in which it is to the left is called L-glyceric acid.

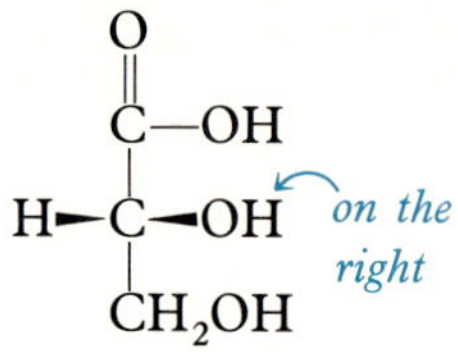

D-glyceric acid

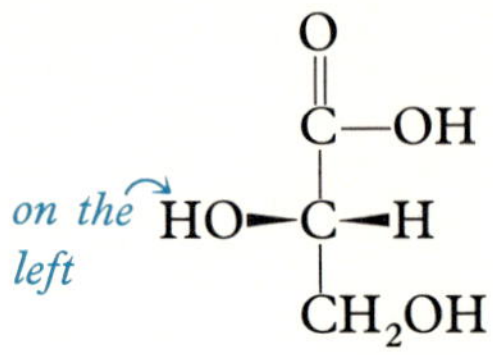

L-glyceric acid

Sample Problem: Redraw the following structure in a Fischer projection:

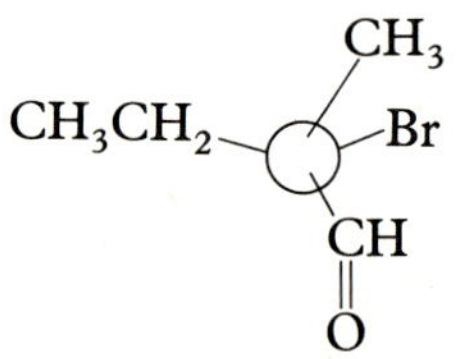

Solution: *Step 1:* Place the aldehyde group at the top.

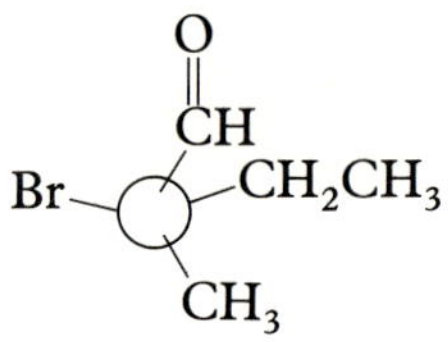

Step 2: Flip over so the horizontal bonds come out toward the viewer.

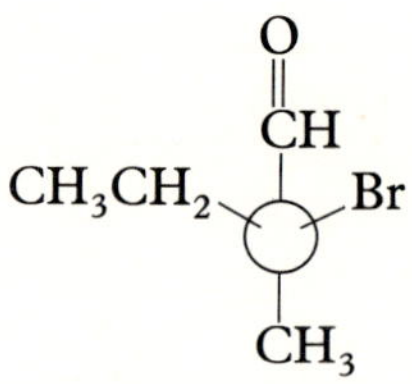

Step 3: Redraw the formula.

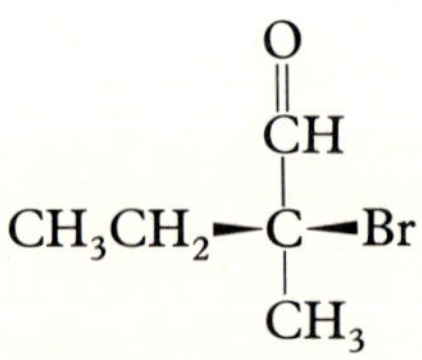

Sample Problem: Which of the following pairs of Fischer projections represent enantiomers?

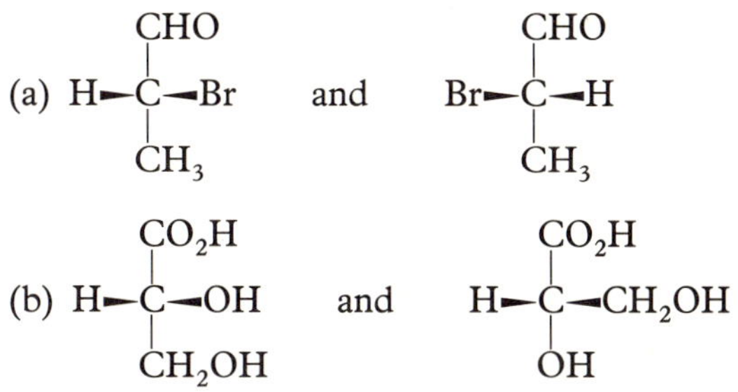

Solution: Both pairs represent enantiomeric pairs. (Construct models and verify this for yourself.) Note that the interchange of *any two groups* around the chiral carbon atom results in the enantiomer of the original compound. This statement is true for any structure with a single chiral carbon atom.

MORE THAN ONE CHIRAL CARBON ATOM

A molecule can have more than one chiral carbon atom. The following sugar is named erythrose. Carbon atoms 2 and 3 each has four different groups attached; they are both chiral carbon atoms.

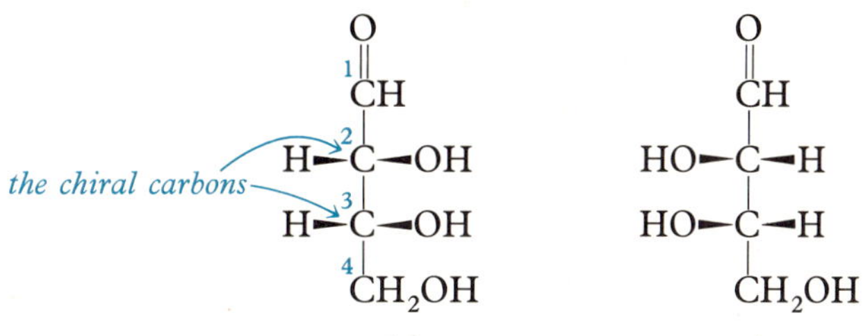

D-erythrose L-erythrose

D-Erythrose and L-erythrose are enantiomers. Note that enantiomers are quickly recognized when we use Fischer projections. The groups attached to the chiral carbons are simply reversed left to right (or right to left) in a pair of enantiomers.

When there is more than one chiral carbon in a molecule, how do we know if a compound is D or L? By definition, if the hydroxyl group on the chiral carbon farthest from carbon 1 projects to the *right,* the compound is a member of the D-series.

If the hydroxyl group on the farthest chiral carbon projects to the *left,* the compound is a member of the L-series.

The hydroxyl groups that place these compounds in the D-series are circled:

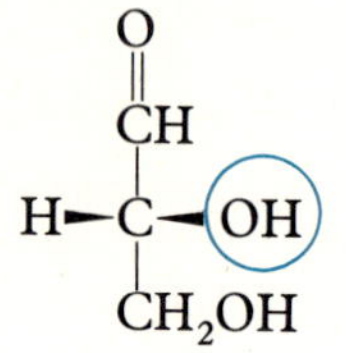

D-glyceraldehyde

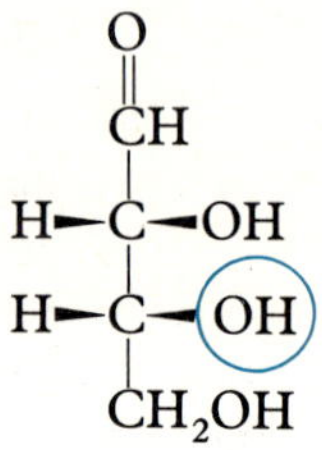

D-erythrose

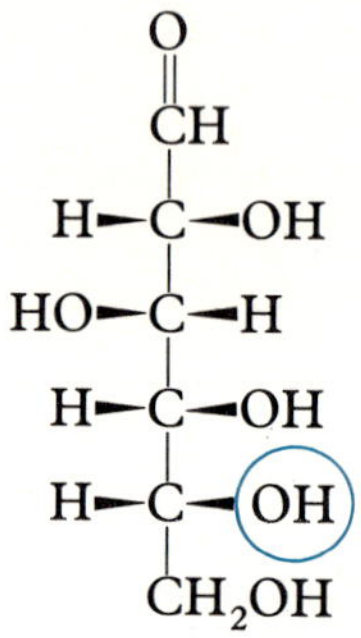

D-glucose

ROTATION OF PLANE-POLARIZED LIGHT

A beam of light passed through a filter that blocks out all the light waves except for those vibrating in one plane is called *plane-polarized light.* (See Figure 20-8.) A popular brand of sunglasses uses this principle in filtering out a portion of the sun's radiation.

Figure 20-8. The plane polarization of light.

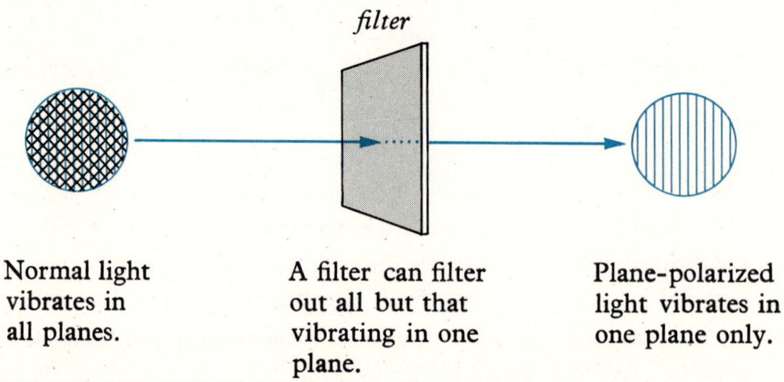

A pair of enantiomers, such as D- and L-glyceraldehyde, are exceedingly similar to each other in structure. As you might guess, most of the chemical and physical properties of a pair of enantiomers are the same. The notable exception is the ability to *rotate the plane of vibration of plane-polarized light*.

When plane-polarized light is passed through a solution containing one member of a pair of enantiomers, the light is rotated and emerges vibrating in a different plane. The enantiomer is said to be *optically active*.

The rotation of plane-polarized light by an optically active compound may be measured with a *polarimeter*, an instrument designed to polarize light, pass it through a solution, and measure the rotation of the plane of vibration of the light. With any single pair of enantiomers, it has been found that one enantiomer (**I** in Figure 20–9) rotates the plane-polarized light in one direction; the other enantiomer (**II**) rotates it to the *same degree* but in the *opposite direction*.

Plane-polarized light passing through a mixture of equal parts of **I** and **II** is not rotated. The effects of the two enantiomers on the light cancel each other. A mixture of equal amounts of a pair of enantiomers is called a *racemic mixture*.

If a particular optically active compound rotates plane-polarized light to the right, it is said to be *dextrorotatory* (Latin *dexter*, right). This may be indicated in the

Figure 20–9. The rotation of the plane of vibration of plane-polarized light by optically active compounds. (1) When plane-polarized light passes through **I** in a polarimeter, it is rotated so that it vibrates in a different plane. (2) When plane-polarized light passes through **II** (the enantiomer of **I**), it is rotated in the opposite direction to the same degree. (3) When plane-polarized light passes through an equal mixture of **I** and **II**, the effect is cancelled. No rotation is observed.

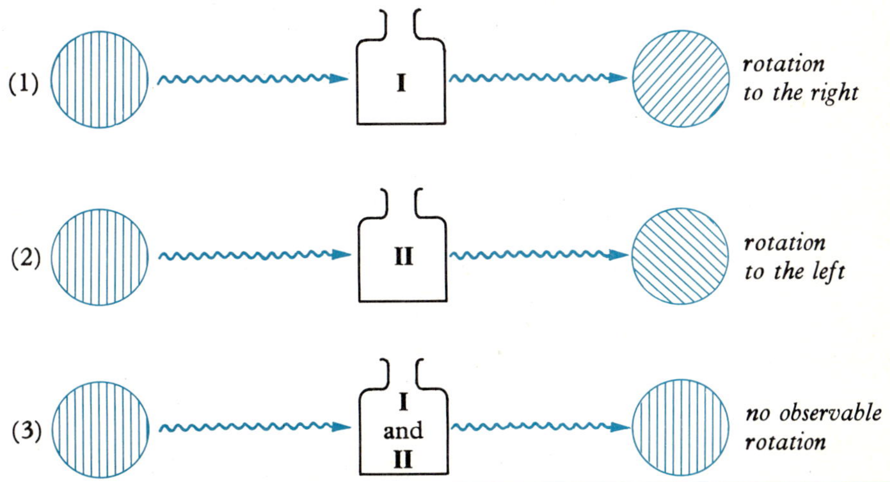

name of the compound by the prefix sign (+); for example, D-(+)-glyceraldehyde. If the compound rotates the light to the left, the compound is said to be *levorotatory* (Latin *laevus*, left). The prefix sign (−) is used to designate a levorotatory compound.

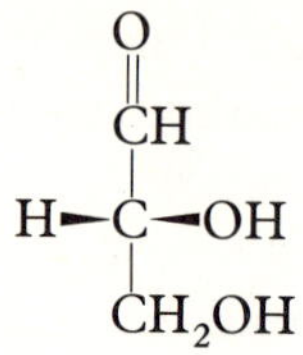

D-(+)-*Glyceraldehyde rotates plane-polarized light to the right.*

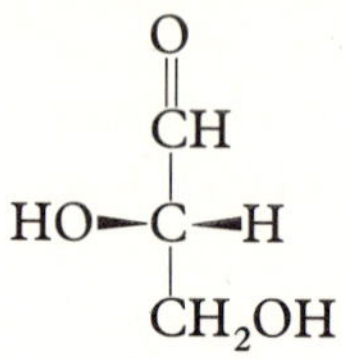

L-(−)-*Glyceraldehyde rotates plane-polarized light to the left.*

Capital D and L refer to the configuration around the chiral carbon atom. The symbols (+) and (−) refer to the direction of rotation of plane-polarized light, which is a measured physical constant. The two are not necessarily related. A compound in the D-series may rotate plane-polarized light to the right or it may rotate it to the left. For example, both D-glyceraldehyde and D-glyceric acid have the hydroxyl group at carbon 2 projected to the right. Yet D-glyceraldehyde is dextrorotatory and D-glyceric acid is levorotatory.

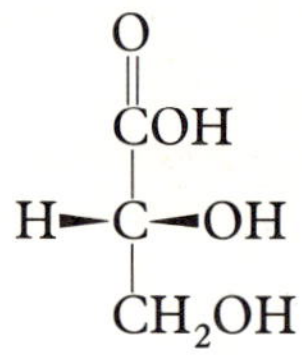

D-(−)-*Glyceric acid rotates plane-polarized light to the left.*

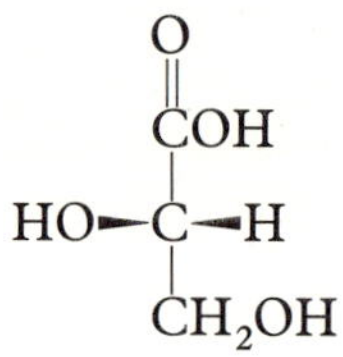

L-(+)-*Glyceric acid rotates plane-polarized light to the right.*

IMPORTANCE OF OPTICAL ACTIVITY

If a racemic mixture of certain amino acids is fed to a living system, the organism uses the L-forms to build proteins and rejects the D-forms. Almost all naturally occurring amino acids belong to the L-series. On the other hand, most sugars found in nature belong to the D-series. Living systems, both plant and animal, are storehouses of optically active compounds. Because of this, in the field of medicine it is frequently the case that one enantiomer of a compound is physiologically active and the other is not.

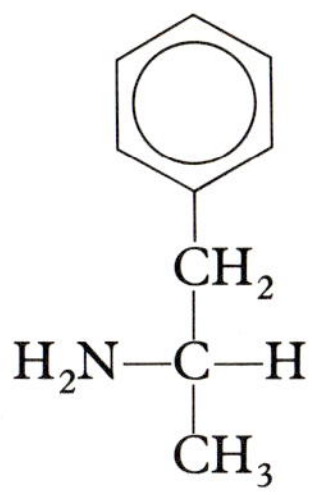

amphetamine
(benzedrine)

The dextrorotatory enantiomer (dextroamphetamine or Dexedrine) is much more active than the levorotatory enantiomer or the racemic mixture.

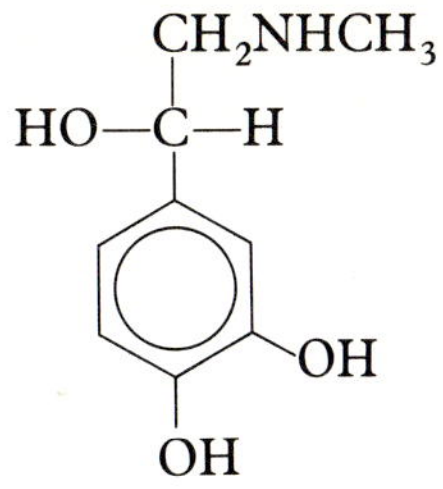

epinephrine
(adrenaline)

The levorotatory enantiomer is far more active than the dextrorotatory enantiomer.

CYCLIC COMPOUNDS IN THREE DIMENSIONS

One other aspect of stereochemistry should be mentioned before we can go on to topics in biochemistry. This is the conformational structure of cyclic compounds. We have, for the sake of simplicity, represented cyclic compounds as planar polygons. For benzene, the carbon atoms of which are in the sp^2 state, the representation as a flat hexagon with bond angles of 120° is reasonably correct. But what of cyclohexane, the carbon atoms of which are in the sp^3-hybrid state? Recall that atoms in the sp^3 state favor bond angles of 109° or close to that value. To force the bonds to assume other angles is to create a strain in the molecule. However, a cyclohexane molecule is strain-free because the cyclohexane ring is *puckered,* not flat. A six-membered ring can pucker in a number of ways, but the most common conformation (shape) is the *chair form,* in which all bond angles are approximately 109°.

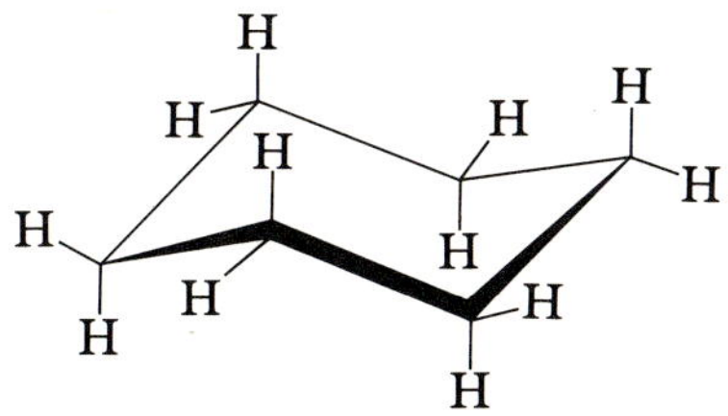

chair form of cyclohexane

Another aspect of the chair form of cyclohexane is that the hydrogen atoms on adjacent carbon atoms are *staggered* (as far apart as possible). If a cyclohexane molecule were flat, all the hydrogen atoms would be *eclipsed* (as close together as possible). Because nonbonded atoms repel each other, the eclipsed-hydrogen structure, as in the flat ring, is not favored. Instead, the chair form, with the hydrogen atoms farther apart, is the favored conformation.

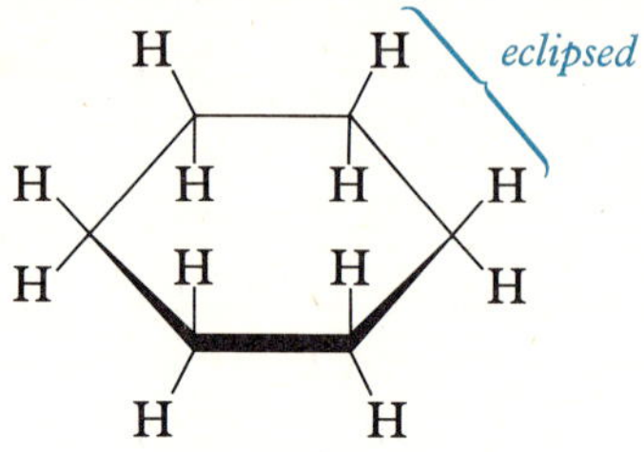

not favored

STUDY QUESTIONS

20-1. Define and give examples to illustrate the following terms:
(a) isomerism
(b) structural isomerism
(c) geometric isomerism
(d) optical isomerism

20-2. Sketch (a) a bowl, (b) a rectangular box, and (c) a cup with a handle. Draw a plane of symmetry through each.

20-3. How many different planes of symmetry can be drawn through (a) the bowl, (b) the rectangular box, and (c) the cup with a handle?

20-4. Using expanded formulas with correct bond angles, show the planes of symmetry through:
(a) methyl chloride
(b) ethyl chloride
(c) propyl chloride

20-5. Star the chiral carbon atoms (if any) in each of the following:
(a) $(CH_3)_2CHBr$

(b) 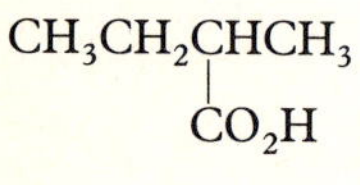

(c) 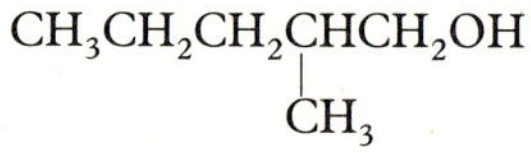

(d) 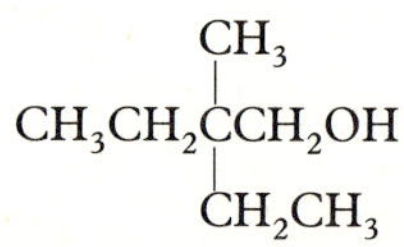

20-6. Which of the following statements are true for D-glyceric acid in the Fischer projection?
(a) The hydroxyl group at carbon 2 is in front of the paper, toward the viewer.
(b) The hydrogen atom on carbon 2 is behind the paper and away from the viewer.
(c) Carbon atom 1 is projected away from the viewer.

20-7. Match each structure on the left with its Fischer projection on the right:

(a) 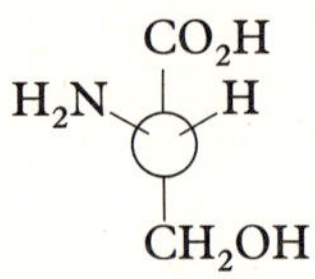

(1) 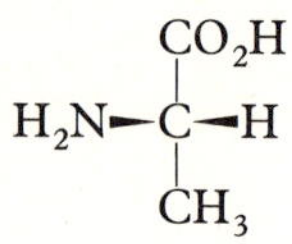

(b) 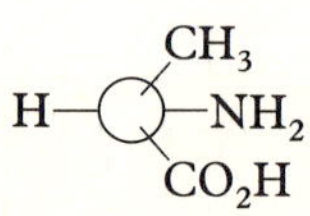

(2) 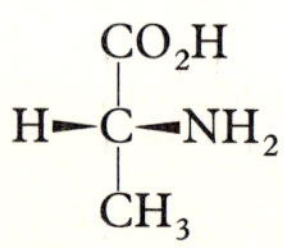

(c) 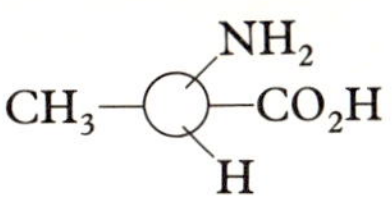

(3) 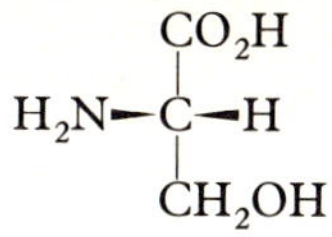

(d) 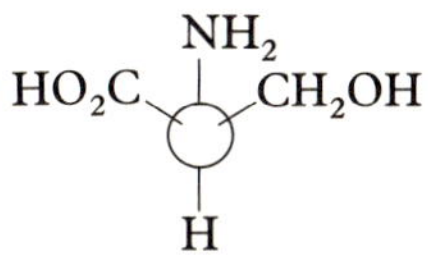

(4) 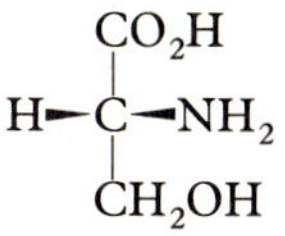

20-8. Rewrite each of the following structures using a Fischer projection. Remember to place the aldehyde carbon or carboxyl carbon (carbon 1) on top.

(a) 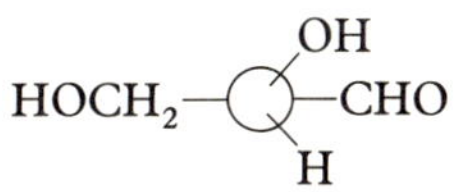

(b) 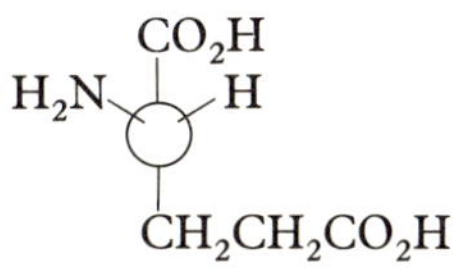

(c) 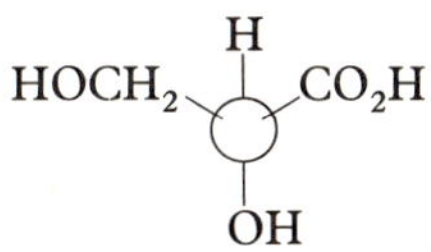

20-9. Which of the Fischer projections in Problem 20-7 represents a compound that belongs to the L-series?

20-10. Circle the four different groups for each chiral carbon in D-erythrose, page 344.

20-11. Star the chiral carbons:

(a) 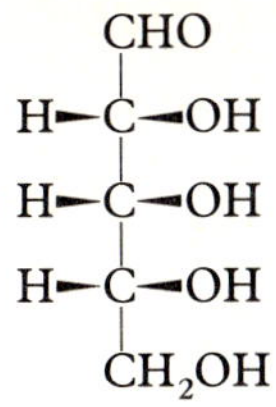

(b) 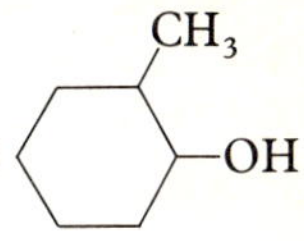

(c) 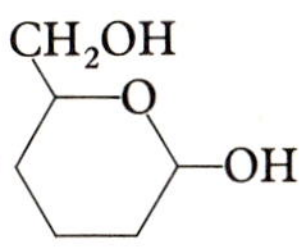

(d) 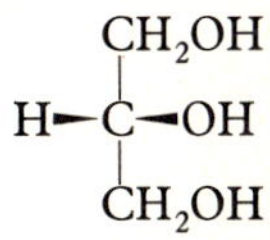

20-12. Draw the four stereoisomers of the following compound. Tell which formulas are enantiomers of each other.

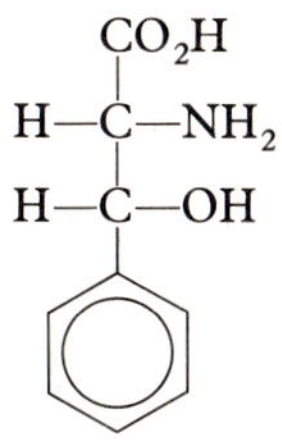

20-13. Which of the following statements are true?

(a) A compound of the D-series always rotates plane-polarized light to the right.

(b) The prefix (+) means that a compound always rotates plane-polarized light to the right.

(c) The term to denote rotation of plane-polarized light to the left is dextrorotatory.

20-14. Draw the chair forms of (a) *cis*-1,4-dimethylcyclohexane and (b) *trans*-1,4-dimethylcyclohexane.

CHAPTER 21 INTRODUCTION TO BIOCHEMISTRY

OBJECTIVES

What Is Biochemistry? *Define biochemistry. * List the four elements that account for almost all the mass of a living system. * Write the generalized equation for the polymerization of amino acids to proteins.*

What About the Vital Force? *Describe briefly and generally the way in which biochemical reactions occur. * Describe the general function of enzymes.*

The Study of Living Systems *Define (in a general fashion) the following: carbohydrates, proteins, and fats and oils.*

A Look at the Cell *Diagram the structure of a living cell, and list the functions of the more important organelles.*

WHAT IS BIOCHEMISTRY?

A living system is unique—it grows, it sustains itself, and it reproduces itself. The most amazing thing about a living system is that it is composed of nonliving atoms and molecules. What is the magic that imparts life to a cell? Is there some mysterious vital force as was once believed? Or is a cell just one big set of chemical reactions? Ah, sweet mystery of life. Is it any wonder that man has taken up the challenge of trying to understand the what's, the how's, and the why's of living systems?

The pursuit of knowledge of what goes on chemically within a living system is, of course, known as *biochemistry*. In our present era, biochemists are probing deeper and deeper into the questions of how the molecules of a living system are organized, how they react, and how they interact.

Biochemistry is not entirely mystery and romance. Knowledge gleaned from the study of biochemistry has immense practical application in medical science, veterinary science, agriculture, and related fields. All of the biological sciences feel the impact of biochemical knowledge.

How does a biochemist approach the complexities of living systems? One approach is the *analytical* approach. Of what chemical elements are living systems composed? From this standpoint, we find that living systems are rather simple. There are only 16 chemical elements that are found in all living species. (See Tables 21-1 and 21-2.) Of these elements, only four—carbon, hydrogen, oxygen, and nitrogen—account for more than 95 percent of the mass of a living system.

In a living system, the chemical elements are found in combination with each other as compounds. Since it would be impractical to begin a study of all the compounds found in a living system such as those that are found in an elephant, biochemists usually work with less complex (and less massive) systems. A favorite laboratory subject is *Escherichia coli* (or *E. coli*, as it is more commonly called), an intestinal bacterium that is estimated to contain only 5000 different kinds of organic molecules. Five thousand types of molecules does not sound like an inordinately large number. Again, from this standpoint, a living system sounds rather simple.

Another example of simplicity in living systems, at least at first glance, is found

Table 21-1. Elements that are Found in All Living Systems

Oxygen	Phosphorus	Magnesium	Iron
Carbon	Sulfur	Calcium	Cobalt
Nitrogen	Sodium	Chlorine	Copper
Hydrogen	Potassium	Manganese	Zinc

Table 21-2. Elements that are Found in the Human Body[a]

Element	Approximate Percent
Oxygen	65
Carbon	20
Hydrogen	10
Nitrogen	3
Calcium	1.5
Phosphorus	1.0
Potassium	Less than 1.0
Sulfur	Less than 1.0
Chlorine	Less than 1.0
Sodium	Less than 1.0
Magnesium	Less than 1.0

[a]Other elements are also found, but only in trace amounts.

in proteins. In any living system, all proteins are polymers built from amino acids, and there are only about 20 different amino acids that are used to make up different proteins.

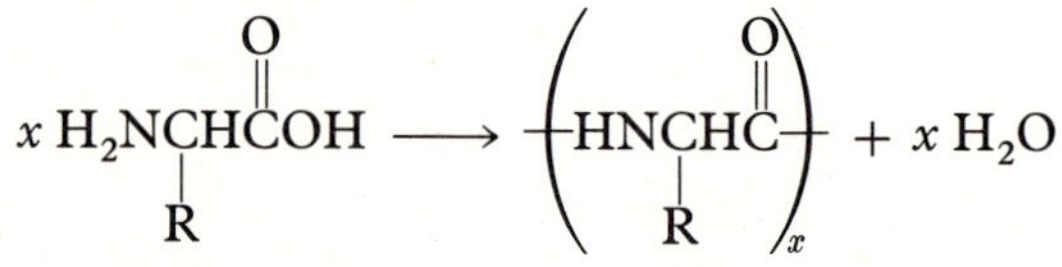

an amino acid: the monomer *a protein: the polymer*

Again, proteins do not sound terribly complex. Or are they? These few amino acids can be put together in different ways to make somewhere around 10^{12} different kinds of protein! These different proteins form skin, hair, muscles, antibodies, and many other tissues and necessary units of an organism.

We will also find that only a few different monomers go into the making of nucleic acids, but the results of how these units are put together are incredible. The genetic code of each living system is absolutely dependent on the construction of its nucleic acids from these few monomeric units. We must quickly arrive at the conclusion that a living system is at once simple and yet very, very complex.

WHAT ABOUT THE VITAL FORCE?

Chemists no longer believe that there is a unique and vital force associated with living systems. Rather, molecules within a living system follow the same physical and chemical laws as molecules that compose crystals in a flask or clouds in the sky. In a living system, there are no unique chemical bonds and no unique chemical reactions. Sigma bonds are still sigma bonds and equilibrium constants are still equilibrium constants.

Let us look at a living system from this rather prosaic viewpoint: A living system is a chemical machine in which chemical reactions occur.

In the laboratory, we can often force reactions to occur by heating the reactants. A living system also uses energy to make reactions occur. The energy needed by the system is taken from the environment, either directly in the form of sunlight or indirectly in the form of matter (that is, food). Through a series of elaborate chemical transformations, part of the energy absorbed by a living system is trapped and stored in "high-energy" compounds. In this form, the stored energy may be tapped and drawn upon when the system needs the energy. The part of the energy that is not trapped is released, usually as heat.

What then does a living system do with this stored-up energy that it has stolen from the environment? Of course, the energy is used to cause chemical reactions. In a living system, reactions must occur within a narrow range of pH and temperature. A cell cannot boil a mixture of reactants to force a reaction to proceed. Instead, reactions in a living system are catalyzed by compounds called *enzymes*. With the help of these biological catalysts, reactions occur under mild conditions—reactions that otherwise might require quite vigorous conditions, such as boiling with a strong acid.

The result of biochemical reactions is a conversion of randomly organized molecules to a highly organized network of specific organic compounds. Herein lies the secret of a living system: its ability to transform and organize organic molecules. A living system can organize organic molecules into cells that perform certain prescribed tasks, such as food absorption or transporting oxygen to other cells. Within a cell, molecules are organized so that the cell can reproduce itself for growth and maintenance as well as for reproduction of the entire organism.

THE STUDY OF LIVING SYSTEMS

Studying the chemical machine that is the living system might be compared to a study of a radio. We know that if we plug the radio into an electrical socket, the radio works. If we take the radio apart to see which component makes the radio work, we have a problem. We find that we have in front of us a lot of radio parts, but that no one part is wholly responsible for the functioning of the radio. Thus, we conclude that the *organization* of the parts within the radio is the characteristic that allows the radio to take electrical energy and convert it into, say, rock music.

On the other hand, if we wish to study the radio, we might well begin with the individual parts. We might try to deduce how each part functions and what its characteristics are. Then, we might progress to the point of trying to interrelate the parts. Such is the case with a study of a living system. We will first look at some of the individual components of the system; then we will look at some of the organization of these components within the system.

If we dismantle a living system into its molecules, we find that the molecules may be fitted very nicely into a few classes. One such class of compounds is known as the *carbohydrates*.

A carbohydrate is a compound made up from polyhydroxyaldehydes and ketones. A carbohydrate may be a simple sugar, such as glucose, or it may be a polymer. The carbohydrates starch and cellulose, for example, are polymers of glucose. About 75 percent of plant material is composed of carbohydrates.

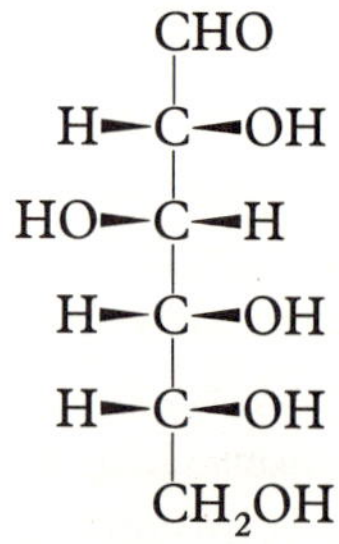

glucose

a carbohydrate

Carbohydrates play an important role in the interrelationship between plants and animals. Plants produce carbohydrates. Animals eat the plants, then use the carbohydrates for energy and as raw materials for the biosyntheses of other compounds.

Proteins are polyamides made up from amino acid units. Although proteins form diverse parts of a living system, all proteins are polyamides. An animal ingests plant protein and protein from other animals. The animal then breaks the protein polymers into individual amino acids and reforms new protein as required. Protein foods may also be converted to energy or to other types of compounds within a living system.

Another class of biological compounds is that of the *lipids*. The lipids are a diverse group of compounds that include the fats and oils and the steroids, such as

Table 21–3. Approximate Composition of Some Animal and Plant Materials

	Carbohydrate	Protein	Lipid	Water
Lean muscle meat	1%	20%	5%	75%
Codfish	0	15	5	80
Dry soybeans	35	35	18	5

cholesterol (page 397). Fats and oils are esters of glycerol that are found in both plants and animals. Along with carbohydrates and proteins, the fats and oils comprise the third major category of foodstuffs for mammals, including man. Fats and oils, like carbohydrates and proteins, are used both for energy and as starting materials for other types of compound. Table 21-3 shows the approximate percents of carbohydrates, proteins, and lipids in some biological substances.

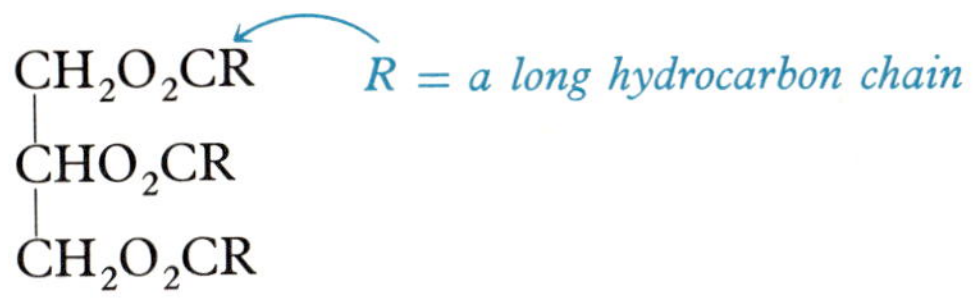

a fat or oil, one type of lipid

A living system takes in food and uses some of it for energy; however, a living system is also capable of rearranging matter. If an animal system takes in more carbohydrates than it can use, it can convert the carbohydrate to fat. In this way, the animal system can build a reserve food supply. If needed, the fat reservoir can be tapped and used for energy at a later time. The interrelations between carbohydrates, fats, proteins, and steroids are outlined in Figure 21-1. Many of the features of these interrelationships will be looked at in succeeding chapters.

Figure 21-1. General relationships between the major metabolism pathways.

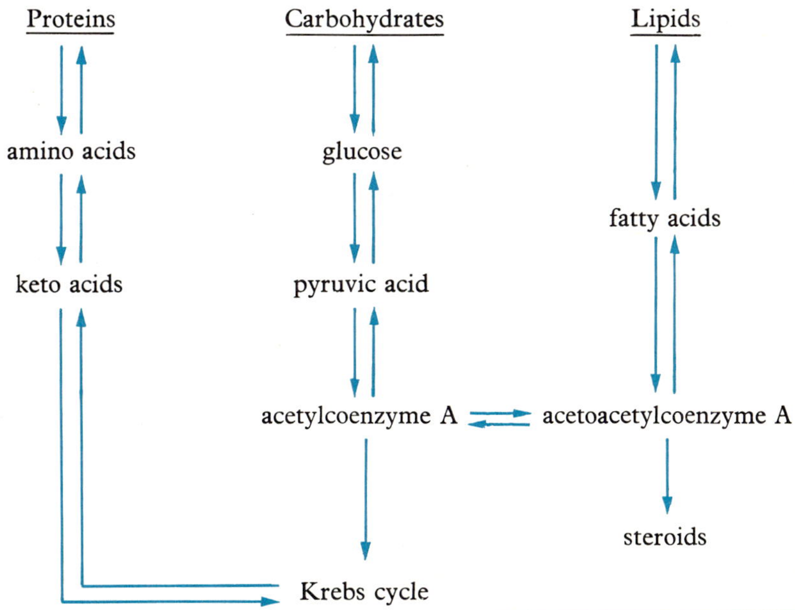

A LOOK AT THE CELL

Where does the molecular organization in living systems exist? The answer, of course, is in the cell. A typical animal cell is not just a jumble of molecules. Rather, the cell has a highly organized architecture. This architecture permits chemical reactions to be coupled to one another (or separated from one another) in a way that would not be possible if reactants and enzymes were scattered randomly throughout the cell.

The door to the cell is its outer membrane, called the *plasma membrane.* (See Figure 21-2.) The plasma membrane keeps the extracellular fluids out, yet lets food and waste pass. The plasma membrane is not a passive tissue; it is an active component of the cell, organized in such a way that it controls the electrolyte concentration of the interior of the cell. The plasma membrane does not simply *allow* nutrients to pass, but forces them into the cell against a concentration gradient—that is, from lower concentration to higher. For example, although the extracellular fluid is rich in sodium ions and low in potassium ions, the interior of the cell is rich in potassium ions but low in sodium ions. *How* substances are transported across the plasma membrane is one of the subjects of current research.

extracellular inorganic ions: Na^+, Cl^-, HCO_3^-

intracellular inorganic ions: K^+, Mg^{2+}, SO_4^{2-}, HPO_4^{2-}, $H_2PO_4^-$

The plasma membrane is composed of two opposing layers of *phospholipids,* which are fat molecules that have an ionic phosphate "head," studded with proteins. The fatty hydrocarbon chains of the phospholipid molecules point toward one another, while their ionic, water-soluble heads point outward. (See Figure 21-3.)

Figure 21-2. Diagram of a typical animal cell.

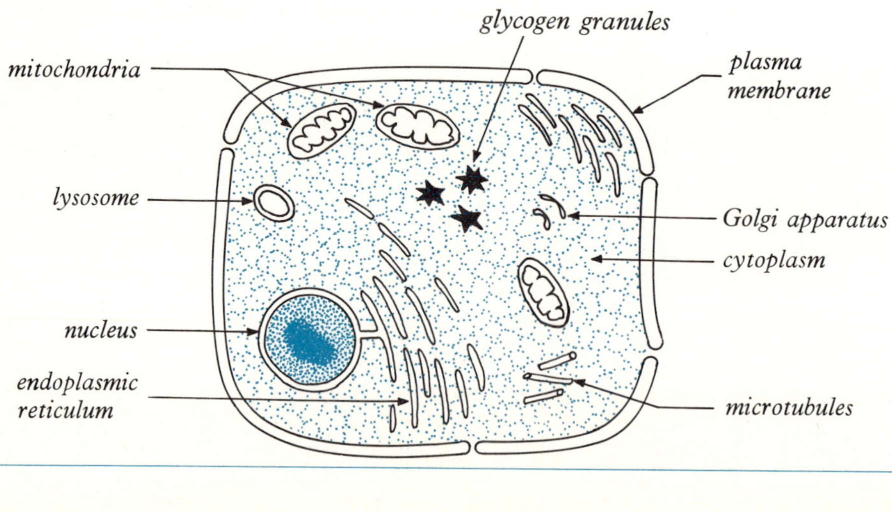

Figure 21–3. The plasma membrane is composed of a phospholipid bilayer imbedded with protein molecules.

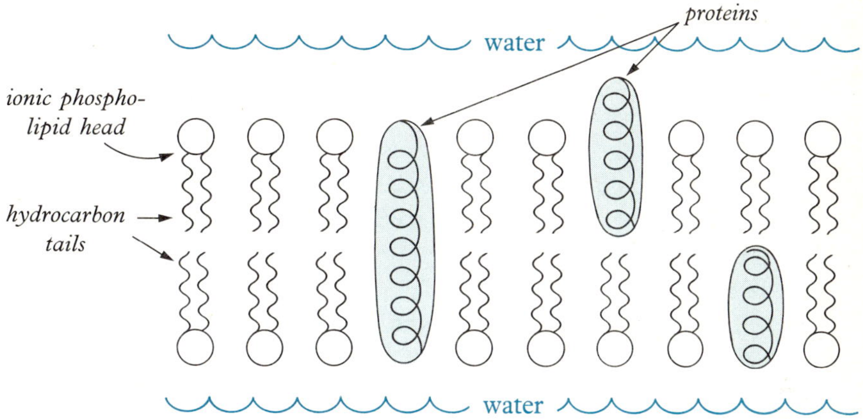

Surrounding the plasma membrane is a coating of carbohydrates and proteins. This coating is responsible for the immunological properties of the cell and also acts as "glue" between the cells of a multicelled animal.

A bacterium plasma membrane is surrounded by a rigid cell wall, which is not present in the cells of higher animals. This cell wall, called a *murein* (Latin *murus,* wall) is composed of long carbohydrate molecules cemented together by proteins. Penicillin blocks the synthesis of the cell wall in growing bacteria by preventing the protein cross-linking that occurs in the final step of cell-wall synthesis. Penicillin has no effect upon the cells of higher animals because animal cells do not have cell walls.

Inside the plasma membrane of a typical animal cell, we encounter the *cytoplasm,* a pool of reactants along with hundreds of enzymes that direct the synthesis of cell constituents. Embedded within the cytoplasm are numerous *organelles,* smaller structures within the cell that have distinct membranous boundaries of their own, along with filamentous structures (*microfilaments*) and channels (*microtubules*). Many of the organelles are "packages" of enzymes that catalyze particular reactions. For example, the *lysosomes* contain hydrolysis enzymes and are the sites of many hydrolysis reactions. Both microtubules and the microfilaments help maintain the cell shape and play a role in cell movement. The microtubules also provide pathways for intracellular transport, a sort of intracity rapid-transit system.

Of the various organelles, the *nucleus* is the largest. The nucleus is the site of genetic control and cellular duplication. It is composed of dense aggregates of nucleic acid-protein complexes, called *chromatin granules.* The nucleus is surrounded by a double membrane, the inner one being the true nuclear membrane. In many cells, the space between these two membranes is connected to the *endoplasmic reticulum,* a network of membrane-bound channels extending from the nuclear membrane throughout the cytoplasm and occasionally to the exterior of the cell. The endoplasmic reticulum provides the sites for the syntheses of proteins and steroids and also acts as a precursor for other cell membranes.

The *mitochondria*, often called the "power packs" of the cell, are seen as spheres, rods, or filaments under the electron microscope. Within the mitochondria are concentrated the oxidative and respiratory enzymatic systems; that is, the mitochondria play a role in conversion of nutrients to biological energy. The *glycogen granules* are composed of stored carbohydrates and the enzymes used in their formation and degradation. Liver cells are especially rich in these organelles. The *Golgi apparatus* is a package of substances secreted by the cell and is also instrumental in the formation of cell membranes.

Our look at the cell has been brief, but sufficient to give an overall picture. Our purpose for presenting this quick look at the cell is that the breakdown of glucose to CO_2 and the role of nucleic acids in cellular duplication, both topics that will be covered, occur not just on the pages in a book but in living cells.

STUDY QUESTIONS

21-1. How many different elements are found in all living systems? Which are the four most abundant?

21-2. To what does the term *E. coli* refer?

21-3. Draw the general structure for a protein polymer. What is the principal functional group found in the protein polymer?

21-4. If a protein were completely hydrolyzed in acidic solution, what would be the products?

21-5. What is an enzyme?

21-6. Match the following. (More than one answer may be correct.)

(a) 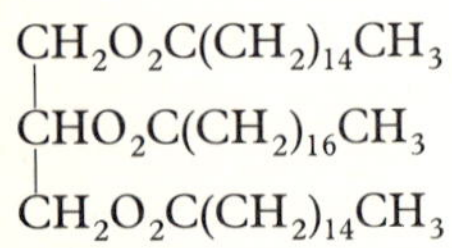

(1) a lipid
(2) a carbohydrate
(3) an amino acid
(4) a sugar
(5) a fat

(b) 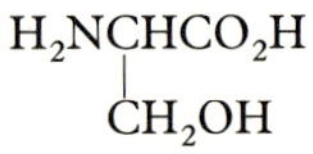

(c) 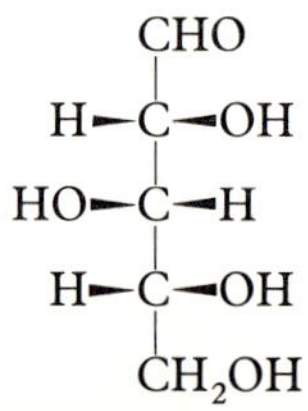

21-7. Match the following:

(a) mitochondria	(1) site of genetic control
(b) plasma membrane	(2) site of protein synthesis
(c) endoplasmic reticulum	(3) cell covering
(d) nucleus	(4) cell "power packs"

21-8. In what respect do the outer coverings of bacteria and higher animals differ?

CHAPTER 22 SUGARS AND OTHER CARBOHYDRATES

OBJECTIVES

Monosaccharides *Give the Fischer projections for the aldehyde forms of* D-*glucose,* D-*fructose,* D-*ribose, and 2-deoxy-*D-*ribose.* * *Define aldose, ketose, and hexose. Give examples.* * *Give equations for the cyclization of glucose.*

Glycosides *Write the structure of a typical glycoside.*

Disaccharides *Describe and write an equation for the formation of a disaccharide.* * *Write the structure of maltose.*

Polysaccharides *Differentiate between the structures of amylose and cellulose.*

Carbohydrates serve biological systems in two principal capacities, as foodstuffs and as structural components. Some carbohydrates also serve special functions; for example, *heparin* is a carbohydrate derivative that prevents the coagulation of blood.

Carbohydrates are synthesized by plants from carbon dioxide and water in a process known as *photosynthesis.*

$$\textit{in green plants:}\ \ CO_2 + H_2O \xrightarrow{\text{sunlight}} \text{carbohydrates} + O_2$$

Typical carbohydrates used by animals as food are *sucrose* (table sugar), which we obtain from sugar cane and sugar beets, and *starch,* which is found in fruits, vegetables, and grains. *Cellulose,* which forms woody fibers in plants, is an example of a carbohydrate used as a structural component. Both starch and cellulose are polymers of the simple sugar *glucose* (also called *dextrose,* because the naturally occurring form is dextrorotatory; *blood sugar,* because it is found in the blood; and *grape sugar,* because it is found in grapes).

By definition, a *sugar* is a carbohydrate such as glucose or sucrose that is sweet-tasting and water-soluble. However, sweetness is not a characteristic unique to sugars. Saccharin and calcium cyclamate are examples of sweet-tasting compounds that are not sugars.

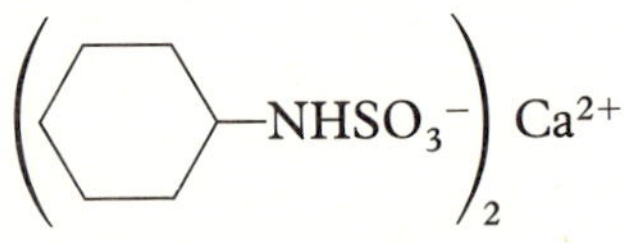

calcium cyclamate
500 times sweeter than sucrose

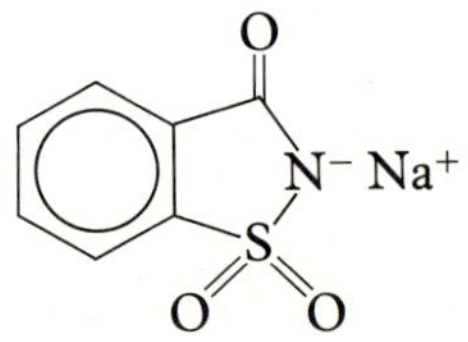

saccharin
30 times sweeter than sucrose

In this chapter, we will look at the structures of a few carbohydrates. We will begin with the sugars and then consider how glucose is bonded together to form the polymers, starch and cellulose. In Chapter 26, we will discuss the metabolism of the carbohydrates—how these compounds serve as a foodstuff and yield their energy to a living organism.

MONOSACCHARIDES

The name "carbohydrates" was derived from the fact that the elemental analyses of compounds in this class often have molecular formulas that are multiples of CH_2O. For example, the formula for glucose is $C_6H_{12}O_6$, or $(CH_2O)_6$. At one time, the carbohydrates were erroneously thought to be hydrates of carbon. We know now that the carbohydrates are made up of polyhydroxyaldehydes and polyhydroxyketones.

The simplest carbohydrates, which contain only one sugar unit, are called the *monosaccharides*. Although a large number of monosaccharides are found in nature, we will confine our discussion here to only four of them: glucose, fructose, ribose, and 2-deoxyribose.

Glucose is the monomer for many of the larger carbohydrate structures. Alone or in combination, glucose is probably the most abundant organic compound on the earth. Fructose (also called *levulose* because the naturally occurring form is levorotatory) is a sugar that is found with glucose in honey and fruit juices. It is the sweetest of all the sugars. Ribose is found in nature as part of the structures of vitamin B_{12} and the ribonucleic acids (RNA). 2-Deoxyribose is an important constituent of the deoxyribonucleic acids (DNA), which form our chromosomes. (The prefix *deoxy-* means minus an oxygen.) The naturally occurring forms of these monosaccharides belong to the D series. Because of the large number of hydrogen-bonding hydroxyl groups in their structures, all of these monosaccharides are water soluble.

Fischer projections for four monosaccharides:

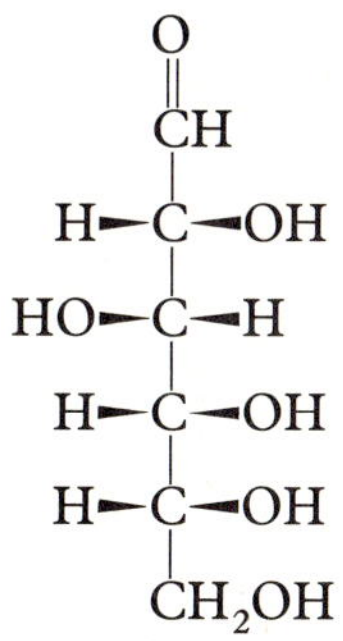

D-glucose

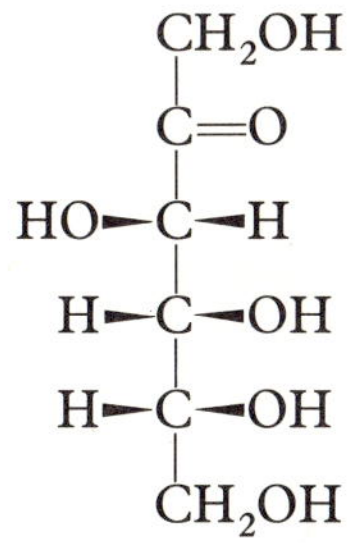

D-fructose

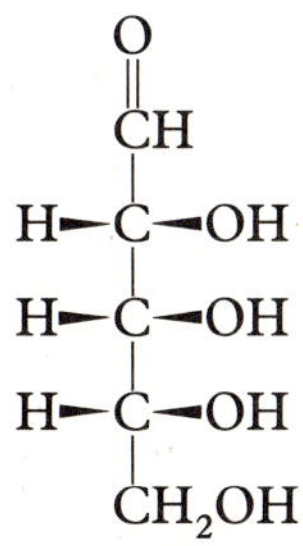

D-ribose

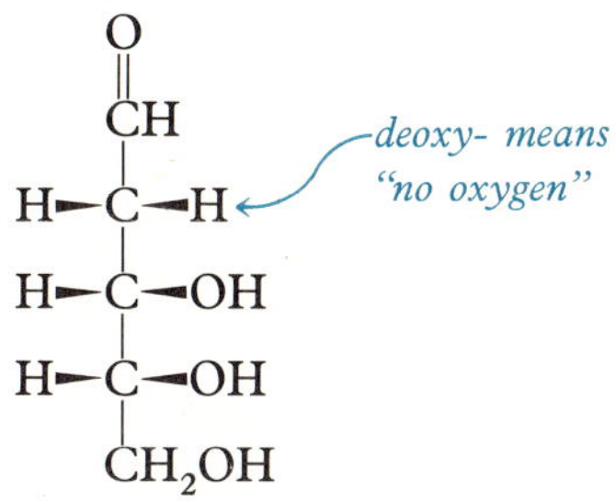

2-deoxy-D-ribose

Monosaccharides may be classified in a number of ways. For example, a monosaccharide with an aldehyde group is called an *aldose* (*ald-* for aldehyde; *-ose* meaning sugar), while one with a keto group is called a *ketose* (*ket-* for ketone). A monosaccharide may also be classified according to the number of carbon atoms it contains:

Number of carbon atoms:	*Class of monosaccharide:*
Three	Triose
Four	Tetrose
Five	Pentose
Six	Hexose

Often the terms are combined. Glucose contains six carbon atoms and an aldehyde group. It is an *aldohexose.*

Sample Problem: Classify each of the following monosaccharides as an aldose or a ketose and also as to the number of carbon atoms each contains.

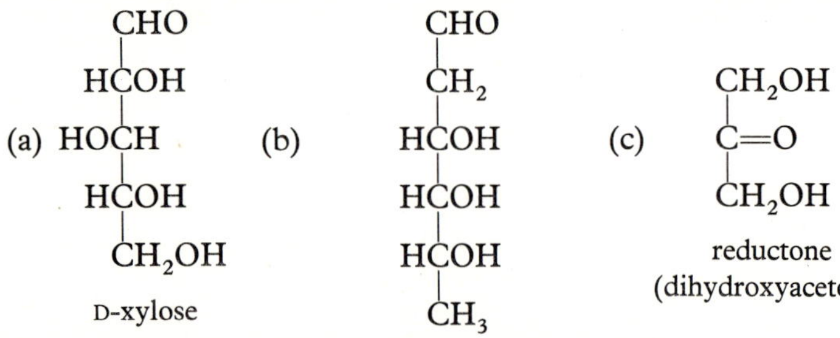

(a) D-xylose (b) 2,6-dideoxy-D-allose (c) reductone (dihydroxyacetone)

Solution: (a) is an aldose and a pentose (or an aldopentose).
(b) is an aldose and a hexose (or an aldohexose).
(c) is a ketose and a triose.

CYCLIZATION OF MONOSACCHARIDES. Aldehydes react with alcohols to form hemiacetals (Chapter 16). If the aldehyde group and the hydroxyl group are in the same molecule and can react to form a five- or six-membered ring, a cyclic hemiacetal is formed.

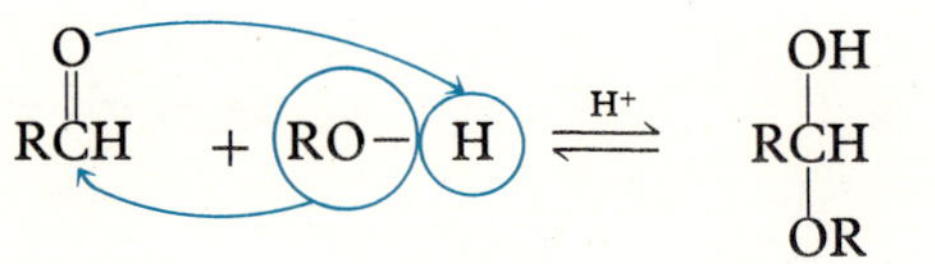

an aldehyde *an alcohol* *a hemiacetal*

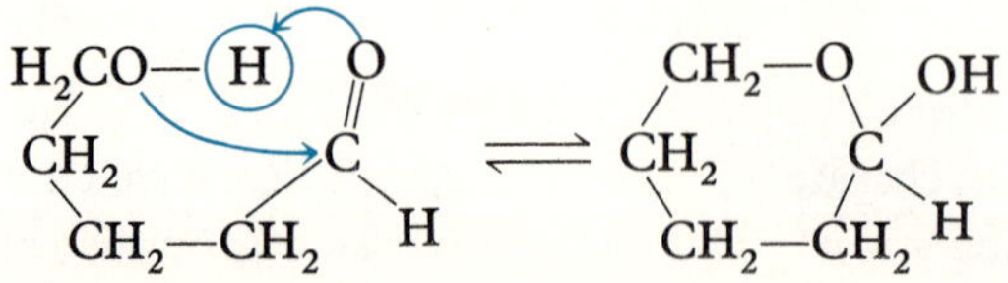

an aldehyde and an alcohol in the same molecule *a cyclic hemiacetal*

Glucose contains an aldehyde group and hydroxyl groups that are correctly positioned in the molecule for the formation of either a five- or six-membered cyclic hemiacetal. Glucose preferentially forms the six-membered ring. The cyclic hemiacetal of glucose has two isomeric forms called the α and β forms. They differ only by the configuration of the hydroxyl group at carbon 1 (the aldehyde carbon before cyclization). This aspect of sugar chemistry will be discussed shortly.

The cyclic hemiacetals of the monosaccharides are quite stable and, in water solution, sugars cyclize readily. The equilibrium between the open-chain form and the cyclic hemiacetal lies toward the cyclic hemiacetal side of the equation. In water, 99.8 percent of glucose is in the cyclic form and only 0.2 percent in the open-chain aldehyde form.

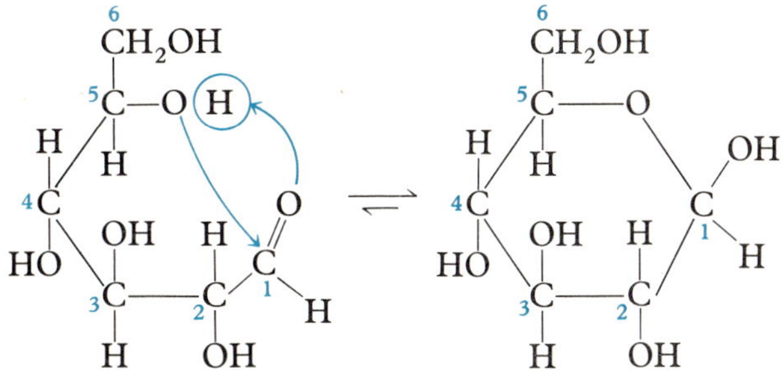

D-glucose

β-D-glucose
a cyclic hemiacetal

Many types of formula and projection have been used to represent sugar molecules. The Fischer projections, although handy for open-chain sugars, are awkward for cyclic structures. The Haworth formulas, which are cyclic formulas, are easy to work with, and are the representations we will use. In actuality, cyclic sugar molecules form puckered rings, just as cyclohexane does. (Note that the symbols for the hydrogen atoms on the rings that follow are omitted; the omission is for the sake of simplicity.)

Different representations for β-D-glucose:

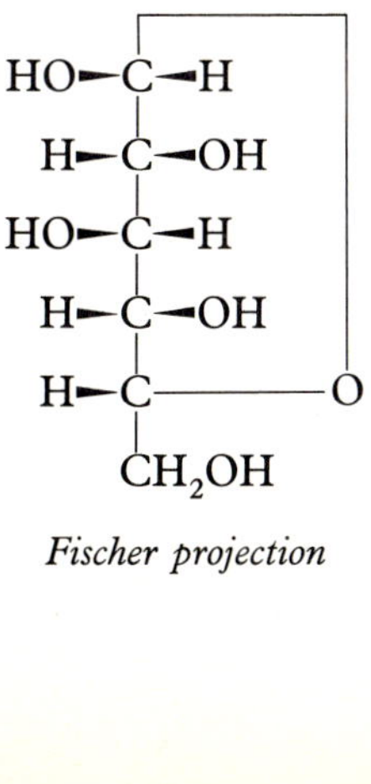

Fischer projection

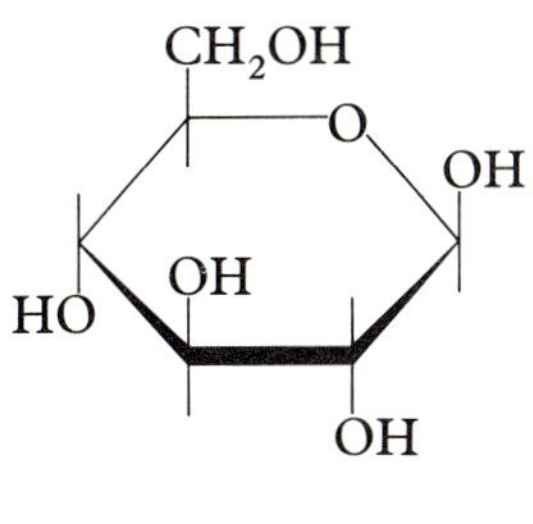

Haworth formula

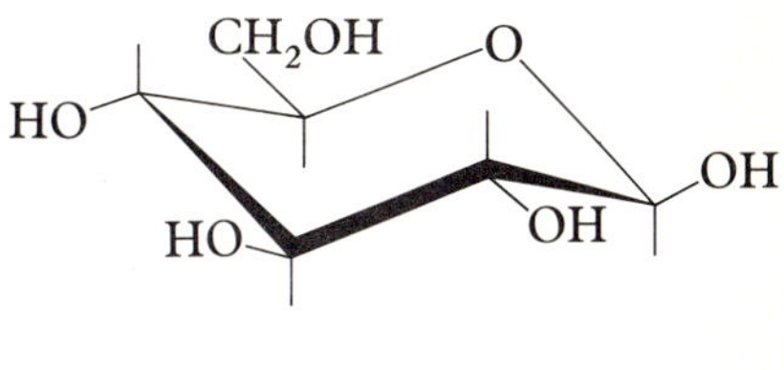

chair form

Sample Problem: Write the formula for a cyclization product of D-mannose:

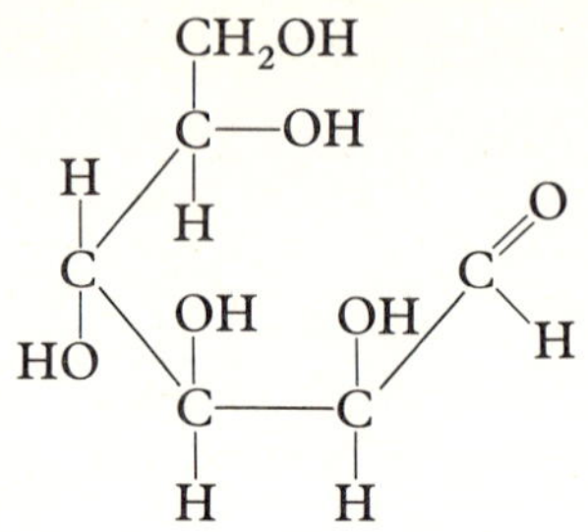

Solution:

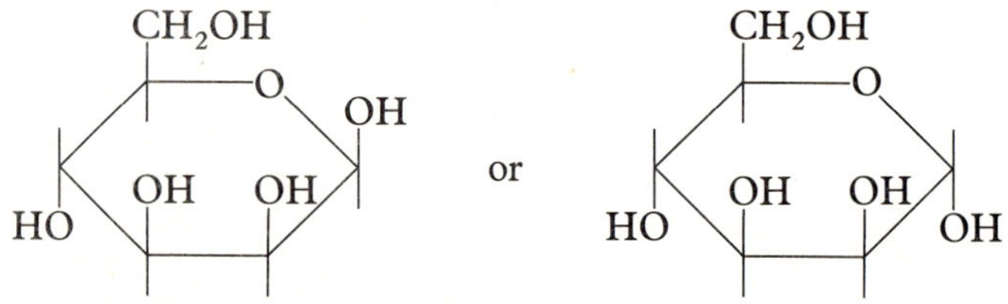

THE CHIRAL CENTER FORMED BY CYCLIZATION. In its open-chain form, glucose has four chiral carbons: carbons 2, 3, 4, and 5. When a sugar cyclizes to the hemiacetal, a new chiral center is formed at carbon 1. Two compounds result from the formation of this chiral center because the hydroxyl group at carbon 5 can attack the carbonyl group of the aldehyde (carbon 1) from either side of that carbon. Therefore, the new hydroxyl group of the hemiacetal can be either above or below the plane of the ring. The pair of cyclic compounds that are formed are not mirror images of each other because the other chiral centers in the sugar are not changed. The two compounds are stereoisomers that differ only in the configuration of the hydroxyl group at carbon 1.

In the Haworth formula of a D-sugar, if the hydroxyl group at carbon 1 is below the plane of the ring and *trans* to the —CH_2OH group, the hydroxyl group is said to be *alpha* (α). If it is above the plane of the ring, *cis* to the —CH_2OH group, it is *beta* (β).

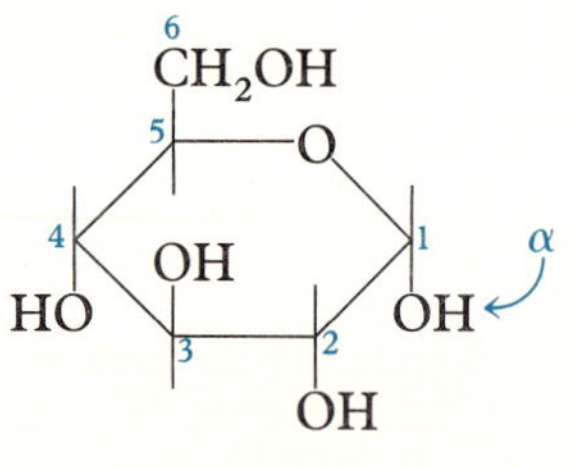

α-D-glucose

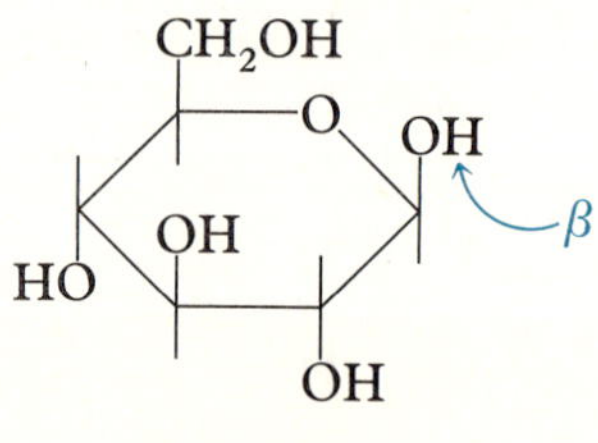

β-D-glucose

The α and β forms of D-glucose can be separated from each other, and are stable in the solid state. In water solution, however, α- and β-D-glucose are not stable. Each form is in equilibrium with the open-chain aldehyde. Hence, in water the α and β forms are in equilibrium with each other as well.

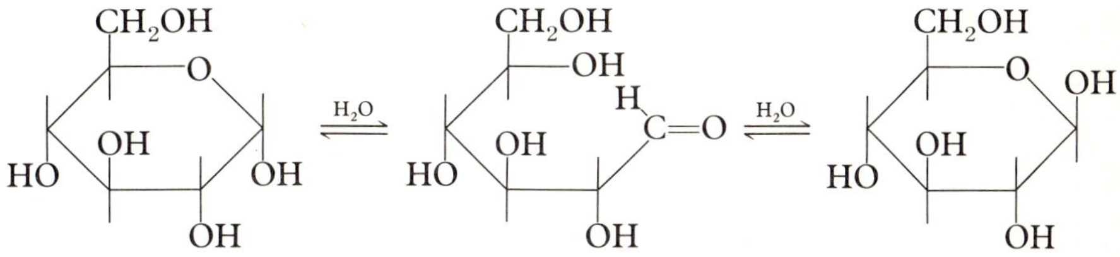

α-D-glucose D-glucose *aldehyde form* β-D-glucose

Sample Problem: Star the chiral carbons in β-D-glucose.

Solution:

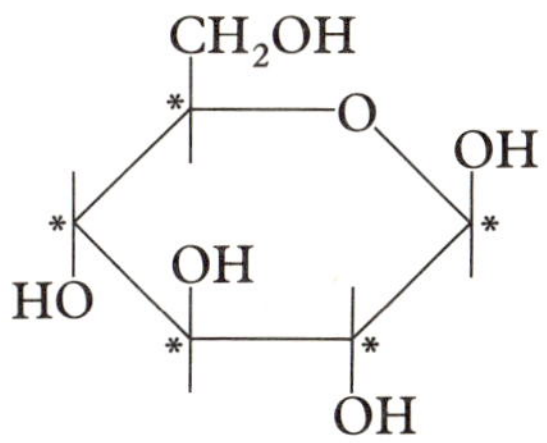

GLYCOSIDES

A hemiacetal can react with an alcohol in acidic solution to form an acetal. The hemiacetal-hydroxyl group of a monosaccharide can similarly react with an alcohol to form an acetal, which, in carbohydrate chemistry, is called a *glycoside*. Like other acetals, glycosides are stable in neutral or alkaline solution; only in acidic solution are they in equilibrium with their hemiacetals or aldehydes.

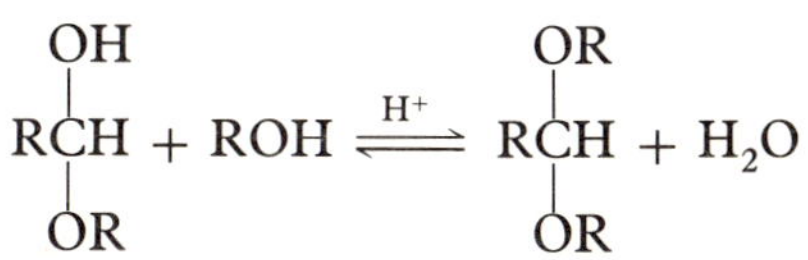

a hemiacetal *an acetal*

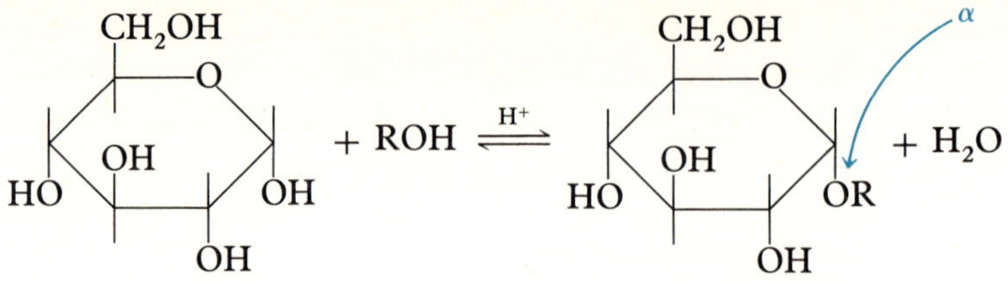

α-D-glucose

an alkyl α-D-glucoside
a glycoside

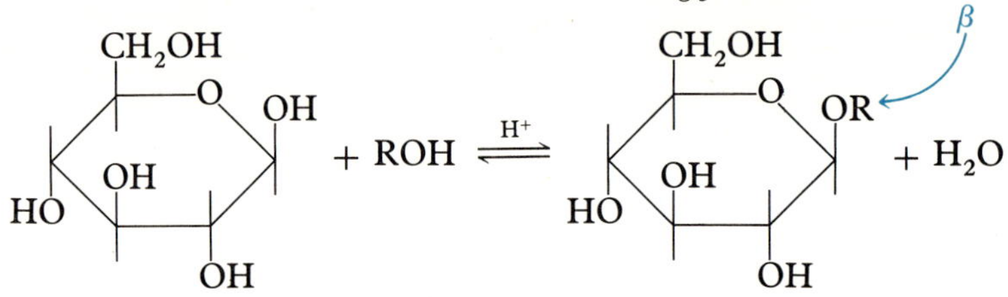

β-D-glucose

an alkyl β-D-glucoside
a glycoside

DISACCHARIDES

Sugars form glycosides, not only with the hydroxyl groups of alcohols such as methanol, but also with the hydroxyl groups of other sugar molecules. A compound composed of two monosaccharide units joined to each other by a glycoside linkage is called a *disaccharide*. Maltose, a sugar used in baby food and malted milk, is a disaccharide of glucose. The acetal link joins carbon 1 of one glucose unit to carbon 4 of the other glucose unit.

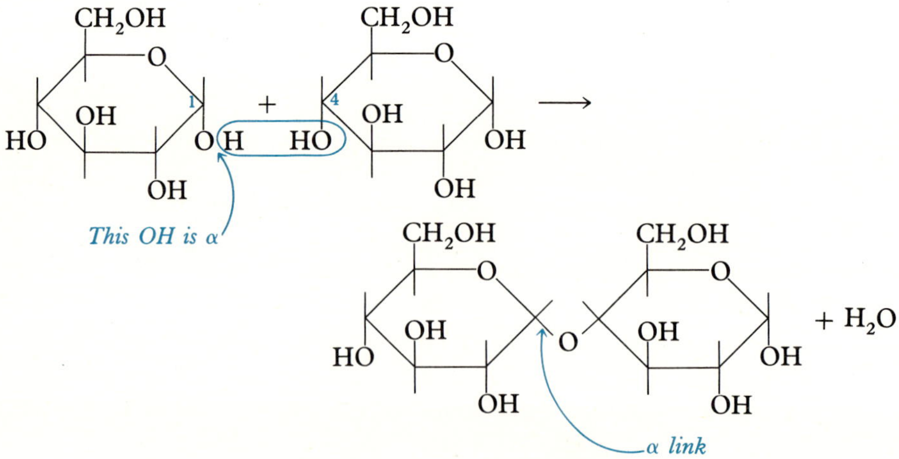

α-maltose
a disaccharide

An important structural feature in disaccharide formation is the configuration of the linkage that joins the two sugar molecules together. The hydroxyl group at carbon 1 of a glucose molecule can be either α or β. Therefore, glucose can form a glycoside with either an α link or a β link. In maltose, the glycoside link is an α link from carbon 1 of the left-hand unit, as we have shown the structure.

An example of a disaccharide with a β link is lactose, or milk sugar, found in milk and milk products. Lactose contains two different monosaccharides, D-galactose and D-glucose, that are joined together by a β link from carbon 1 of the galactose unit to carbon 4 of the glucose unit.

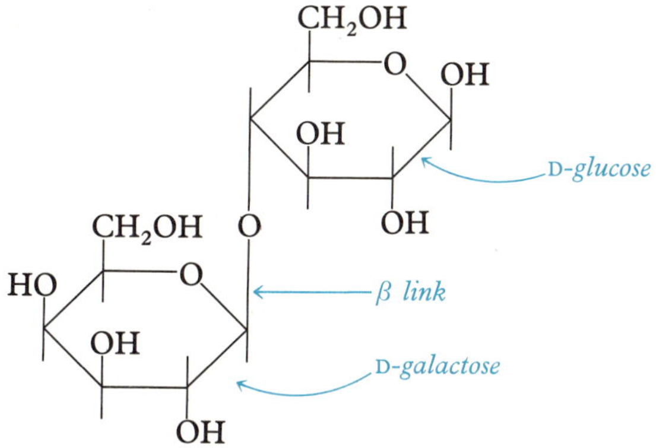

lactose

Sucrose is yet another example of a disaccharide, this one formed by an acetal link joining the first carbons of both glucose and fructose. Sucrose is found in all plants and in large amounts of sugar cane and sugar beets. Our commercial supply comes from these two plants. (The processed sugars from both sources are chemically identical.) Sucrose is commercially processed in greater quantity than any other pure organic chemical. The average person in the United States consumes over one hundred pounds of sucrose per year!

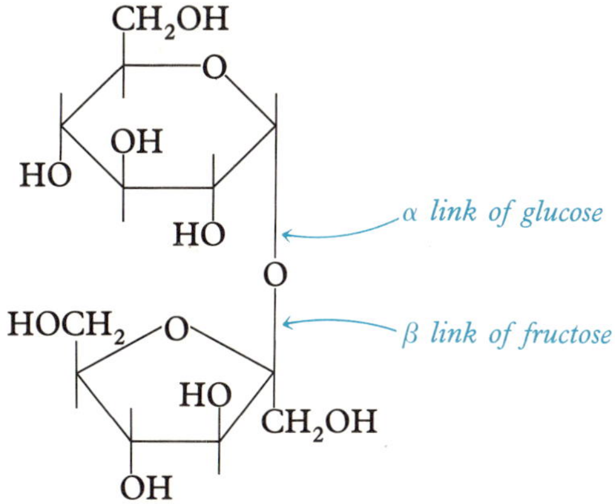

sucrose

a disaccharide formed from α-D-glucose and β-D-fructose

Invert sugar is the mixture of acid-hydrolysis products from sucrose; that is, a mixture of glucose and fructose. (The term "invert" derives from the fact that invert sugar has an opposite sign of rotation of the plane of vibration of plane-polarized light from that of sucrose.) Bees contain an enzyme that catalyzes the hydrolysis of sucrose; honey contains invert sugar. Because fructose is sweeter than sucrose, invert sugar is also sweeter than sucrose.

POLYSACCHARIDES

Starch and cellulose are *polysaccharides* composed of repeating units of glucose. Because of the very size of starch and cellulose molecules, these substances are not

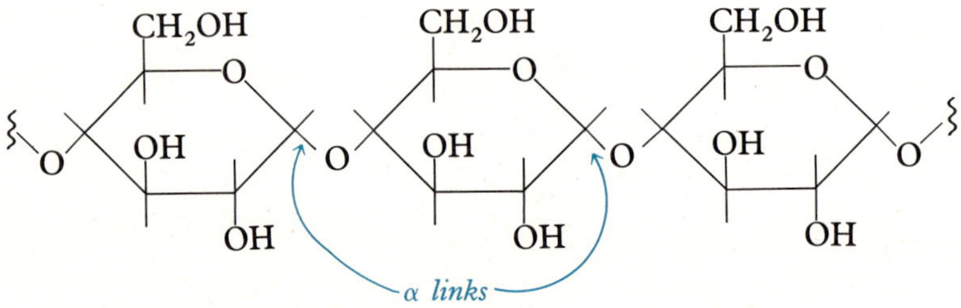

amylose

a polysaccharide in starch formed from units of α-D-glucose

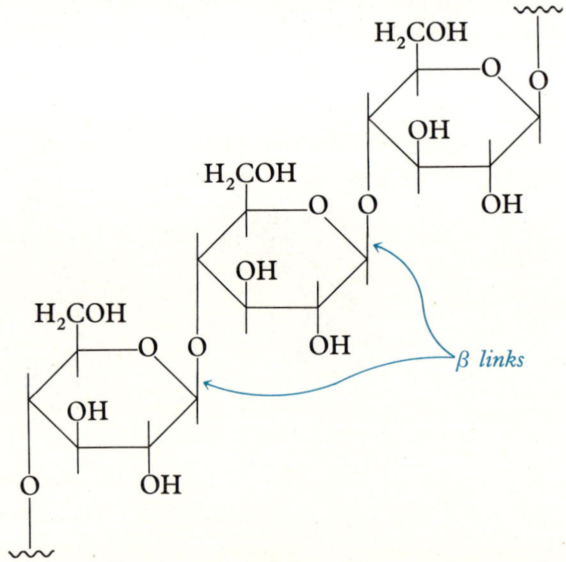

cellulose

a polysaccharide formed from units of β-D-glucose

soluble in water as are the mono- and disaccharides. Starch is an α-linked polysaccharide, while cellulose is β-linked.

Cellulose is one of the main components of wood and other plant materials. Cotton is about 98 percent pure cellulose. It is estimated that cellulose consists of chains of up to 14,000 units of β-D-glucose.

Enzymes produced by certain bacteria and protozoa are capable of hydrolyzing the β links in cellulose. The higher animals, including vertebrates, do not have the enzymes that can break β-linked polymeric glucose into monomeric glucose. Humans cannot utilize cellulose as food. Animals, such as deer, cows, or termites, that use cellulose for food must have colonies of the proper bacteria and protozoa in their digestive tracts to break the cellulose down into glucose.

Starch is a mixture of two types of polysaccharide: amylose and amylopectin. Amylose is a polymer composed of D-glucose units that are joined by an α link from carbon 1 of one glucose to carbon 4 of the next glucose. Amylopectin also contains units of 1,4-α-linked D-glucose. In addition, amylopectin has cross-linking at the 6 position of some of the glucose units. The cross-linking joins chains together into branched structures.

Amylose forms a deep blue complex with free iodine (I_2). If tincture of iodine (an alcoholic solution of I_2) is dropped on the cut surface of a raw potato, a deep blue coloration is observed. Starch is commonly used as a test for the presence of iodine and, conversely, iodine is used to detect the presence of starch.

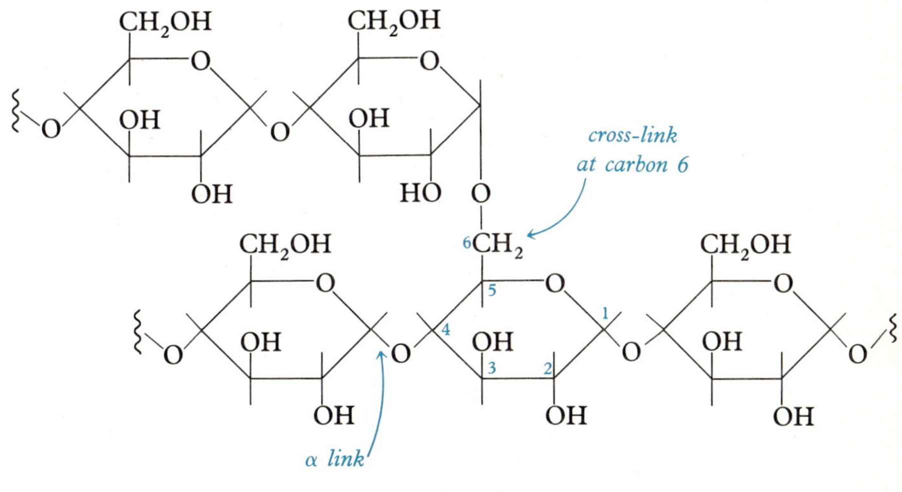

amylopectin
a cross-linked polysaccharide in starch

The human body, which does not have enzymes for breaking down the β-linked glucose in cellulose, does have enzymes that cleave the α links between glucose units in starch. Starch digestion begins in the mouth. When a starchy food such as corn or bread is chewed, the enzyme amylase, which is found in saliva, begins to break down the large starch molecules into maltose, which has a sweet taste. Further breakdown to the monosaccharide glucose occurs in the intestines. The glucose is then absorbed into the bloodstream and carried to cells throughout the body for use as food.

STUDY QUESTIONS

22-1. Match the following:

(a) glucose	(1) carbohydrate
(b) cellulose	(2) polyhydroxyaldehyde
(c) fructose	(3) monosaccharide
(d) ribose	(4) polysaccharide

22-2. (a) In its aldehyde form, ribose has ______ chiral carbon atoms. (b) In its hemiacetal form, ribose has ______ chiral carbon atoms.

22-3. From memory, give the Fischer projections for D-glucose, D-fructose, D-ribose, and 2-deoxy-D-ribose.

22-4. Why does the presence of a large number of hydroxyl groups impart water solubility to a compound?

22-5. Classify each of the following monosaccharides as an aldose or a ketose:

(a) 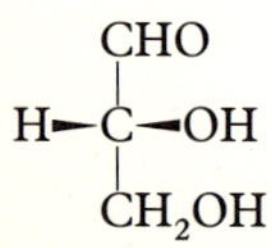

(b) 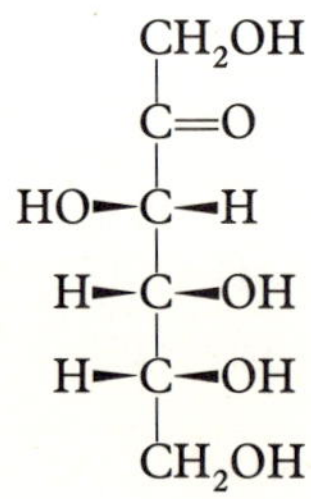

(c) 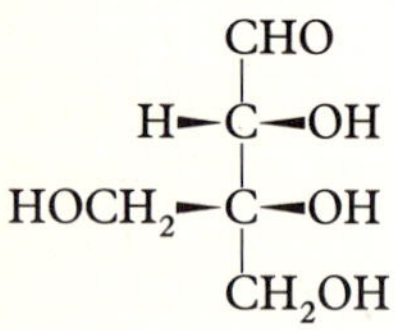

(d) 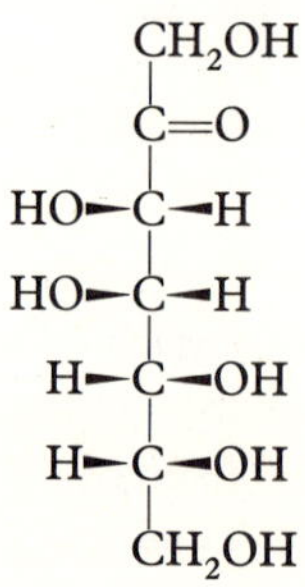

22-6. Classify each of the monosaccharides in the preceding problem as a triose, a tetrose, a pentose, a hexose, or a heptose.

22-7. There are eight aldohexoses with the D-configuration. Draw the Fischer projection of each one.

22-8. Which of the following structures are hemiacetals?

(a) 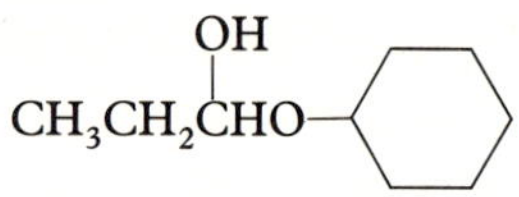

(b) 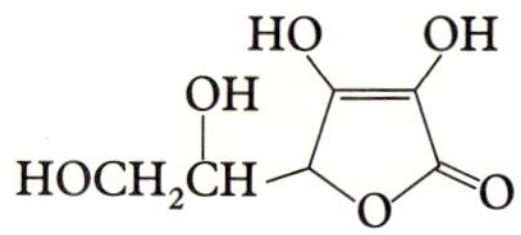

(c) 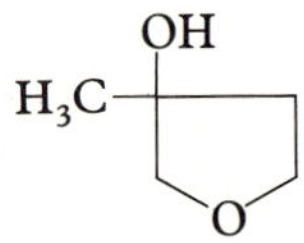

(d) 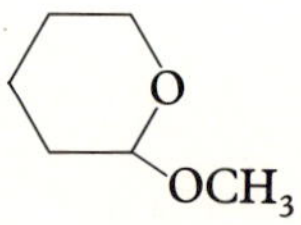

22-9. Complete the following equations:

(a) $CH_3CHO + CH_3OH \xrightleftharpoons{H^+}$

(b) $HOCH_2CH_2CH_2CH_2CHO \xrightleftharpoons{H^+}$

(c) $HOCH_2CH_2CH_2CHO \xrightleftharpoons{H^+}$

(d) $HOCH_2CH_2\underset{\underset{OH}{|}}{C}HCH_2CHO \xrightleftharpoons{H^+}$

22-10. Draw β-D-glucose in the Fischer projection, the Haworth projection, and the chair form. Number the carbon atoms in each, starting with the hemiacetal carbon.

22-11. Are α- and β-D-glucose enantiomers? Explain.

22-12. In water solution, what structures would you expect to be in equilibrium with the following?

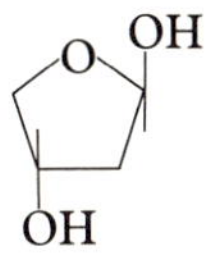

22-13. The sugar D-fructose can form a pair of five-membered-ring hemiketals as well as a pair of six-membered-ring hemiketals. Draw the Fischer projections for the five-membered-ring hemiketals. (A hemiketal is analogous to a hemiacetal; the only difference is that a hemiketal is formed from a ketone.)

22-14. A glycoside is
(a) a hemiacetal.
(b) an acetal.
(c) stable in a dilute solution of HCl.
(d) stable in a dilute solution of NaOH.

22-15. Which of the following structures is an acetal?

(a) 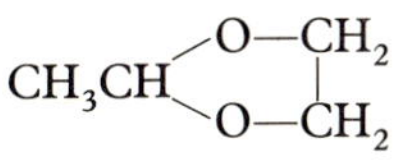

(b) 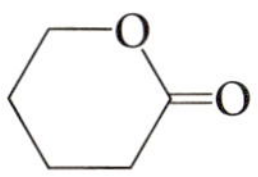

(c) 

(d) 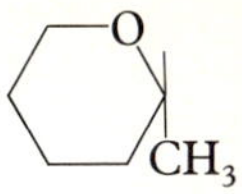

22-16. Write an equation that shows the formation of the five-membered ring of methyl β-D-riboside from D-ribose. (Use Haworth formulas.)

22-17. Write the equation for the formation of invert sugar from sucrose.

22-18. Which of the following statements are true for sucrose?
(a) It is a disaccharide.
(b) The glucose unit is β-linked to the fructose unit.
(c) It is obtained from amylose by acid hydrolysis.
(d) It contains a glycosidic linkage between both sugar units.

22-19. Draw α-maltose, then circle and label the following features:
(a) the α link
(b) the glycosidic bond
(c) carbon 1 of both sugar units
(d) carbon 4 of both sugar units

22-20. Draw the structure of β-maltose. (*Hint:* The glycoside link in β-maltose is still an α link. Ask yourself what link is the β link.)

22-21. Cellobiose is a disaccharide formed in the hydrolysis of cellulose. Draw its structure.

CHAPTER 23 AMINO ACIDS AND PROTEINS

OBJECTIVES

Amino Acids *Write the structures and names of some common amino acids. * Define and give an example of a dipolar ion. * Differentiate between polar and nonpolar amino acids. * Write the equation for the equilibrium between cystine and cysteine. * Define essential amino acid. * Describe the ninhydrin test for amino acids.*

Peptides and Proteins *Differentiate between a peptide and a protein. * Write a peptide structure. Point out the peptide bonds and the N-terminal and C-terminal amino acids. * Describe the higher structures of keratin and collagen. * Describe the properties and function of hemoglobin. * Describe denaturation.*

Enzymes *Describe the action of enzymes in a system. * Deduce the function of an enzyme from its name. * Describe the induced fit theory of enzyme activity. * Differentiate between competitive and noncompetitive enzyme inhibition. * Describe enzyme regulation and list some regulatory factors.*

About 50 percent of the dry weight of an animal cell is protein. Proteins are the end result of the translation of the genetic code of a cell into organic structures, and they perform a wide range of functions within an organism.

Protein molecules are polymers. They are found in different sizes and shapes with different physical and chemical properties, according to their function. One type of protein, the *fibrous* proteins, are long and threadlike, tough and water-insoluble; they form muscles, tendons, skin, hair, fur, fingernails, and feathers. Another type of protein is the class called *globular* proteins, so named because they are somewhat spherical in shape. Globular proteins are more soluble than fibrous proteins and carry on much of the work of a living system. Insulin, hemoglobin, albumins, enzymes, antibodies, and many hormones are globular proteins.

All proteins are polyamides. The hydrolysis of an ordinary amide yields a carboxylic acid and an amine. The hydrolysis of a protein yields amino acids, in which the carboxyl group and the amino group are in the same molecule. A living organism can hydrolyze a protein, then use the amino acids as building blocks for its own proteins. Before we can discuss protein structure, we must first take a look at these amino acids.

Hydrolysis of an amide:

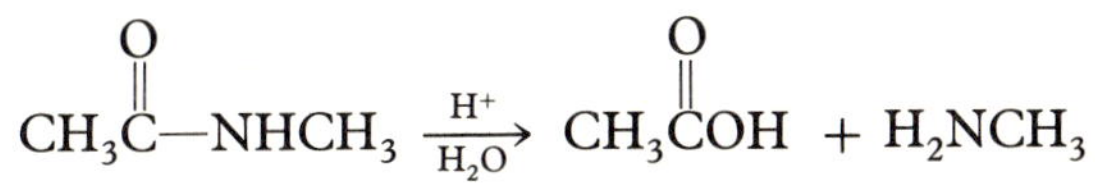

an amide *a carboxylic acid* *an amine*

Hydrolysis of a polyamide:

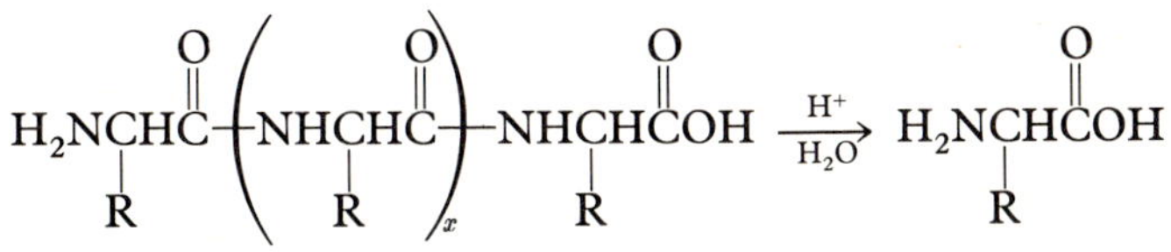

a polyamide as is found in a protein molecule *amino acids*

AMINO ACIDS

The term *amino acid* refers to an organic compound that contains an amino group and a carboxylic-acid group in the same molecule. About 25 amino acids have been identified from the hydrolysis of proteins. Of these, 20 are widespread and found in most proteins. Each of these is an α-amino acid; that is, the amino group is on the carbon atom adjacent, or alpha, to the carboxyl group. The variations in the structures of the amino acids occur in the side chain (R) on the α-carbon.

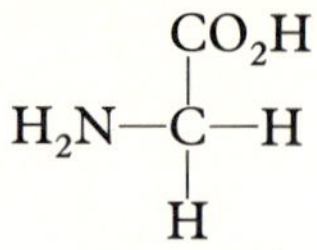

glycine
(*abbreviated* gly)

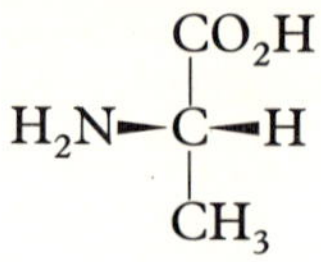

alanine
(*abbreviated* ala)

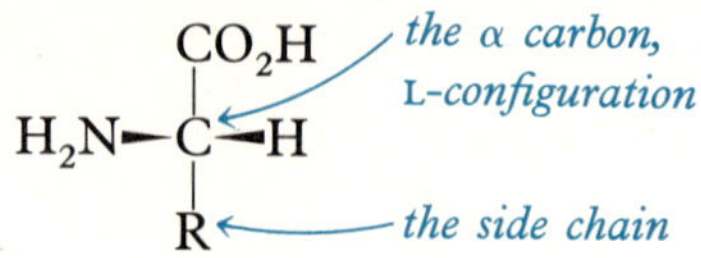

an α-amino acid

Glycine (R = H) and alanine (R = CH_3) are shown as examples of the two simplest amino acids. Although the amino acids could be named by the IUPAC system, the common names are in general use. The names for amino acids found in proteins are frequently abbreviated. For example, *gly* and *ala* are abbreviations for glycine and alanine.

Glycine does not have a chiral carbon atom and thus does not have any optical isomers. Alanine and all the other amino acids from proteins have a chiral carbon at position 2, the α-carbon atom. Both D- and L-amino acids are found in nature; however, most proteins contain only the L-amino acids.

Sample Problem: Star the chiral carbons in valine and threonine (Table 23–1).

Solution:

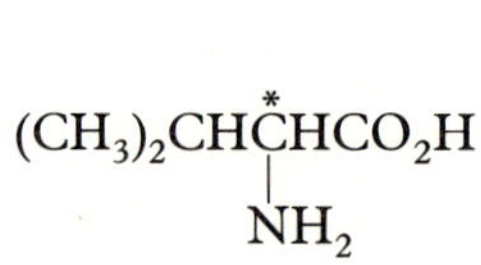

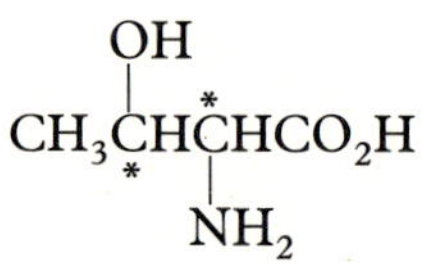

DIPOLAR IONS. Amino acids are crystalline, high-melting solids that are more soluble in water than in organic solvents. They act more like inorganic salts than do most organic compounds. The reason for this anomalous behavior is that an amino acid contains a carboxylic-acid group, which can *lose* a proton, and a basic amino group, which can *gain* a proton, in the same molecule.

In water, an internal acid-base reaction leads to a structure called a *dipolar ion,* which has both a positive and a negative ionic charge. The saltlike properties of amino acids indicate that the equilibrium lies on the dipolar-ion side of the equation. Since the atoms in a dipolar ion are held together by covalent bonds, the ionic charges cannot be separated in water solution as can the ions of sodium chloride. Rather, the ionic charges must remain together in the same molecule. The dipolar ion has a net ionic charge of zero and thus is neutral.

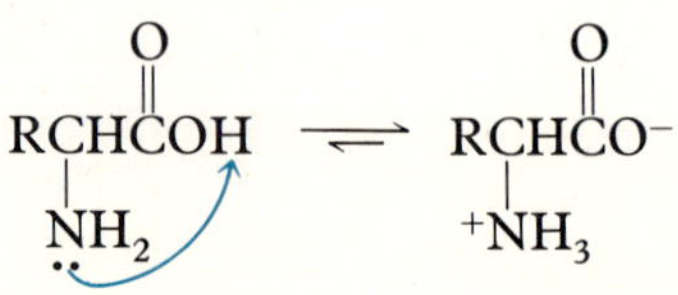

an amino acid *a dipolar ion*

Sample Problem: Write the equations for the formation of the dipolar ions of phenylalanine and proline (Table 23-1).

Solution:

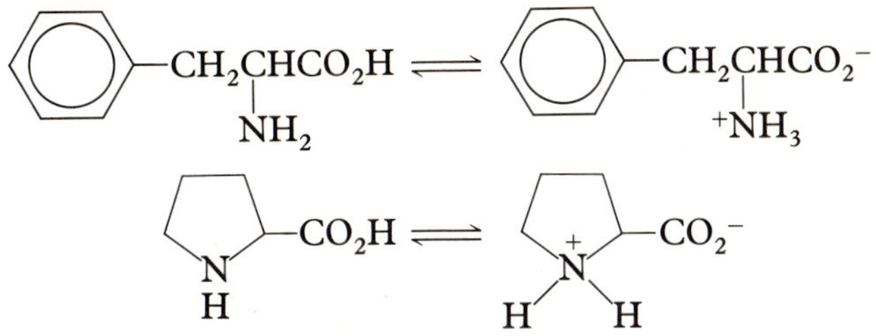

Table 23-1. The Twenty Common Amino Acids Found in Proteins

Name	Structure	Name	Structure
Alanine (ala)	CH_3CHCO_2H NH_2	Leucine (leu)[a]	$(CH_3)_2CHCH_2CHCO_2H$ NH_2
Arginine (arg)[a]	$H_2NCNHCH_2CH_2CH_2CHCO_2H$ NH NH_2	Lysine (lys)[a]	$H_2NCH_2CH_2CH_2CH_2CHCO_2H$ NH_2
Asparagine (asp-NH_2)	O $H_2NCCH_2CHCO_2H$ NH_2	Methionine (met)[a]	$CH_3SCH_2CH_2CHCO_2H$ NH_2
Aspartic acid (asp)	$HO_2CCH_2CHCO_2H$ NH_2	Phenylalanine (phe)[a]	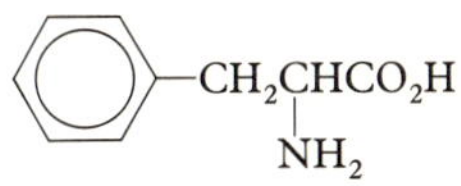
Cysteine (cys)	$HSCH_2CHCO_2H$ NH_2	Proline (pro)	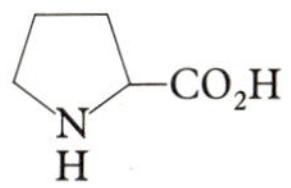
Glutamine (glu-NH_2)	O $H_2NCCH_2CH_2CHCO_2H$ NH_2	Serine (ser)	$HOCH_2CHCO_2H$ NH_2
Glutamic acid (glu)	$HO_2CCH_2CH_2CHCO_2H$ NH_2	Threonine (thr)[a]	OH $CH_3CHCHCO_2H$ NH_2
Glycine (gly)	CH_2CO_2H NH_2	Tryptophan (try)[a]	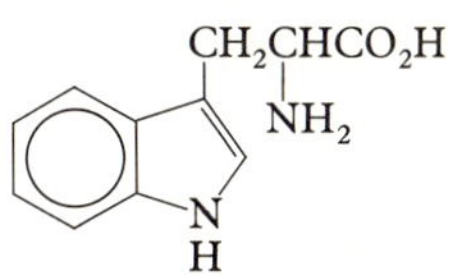
Histidine (his)[a]	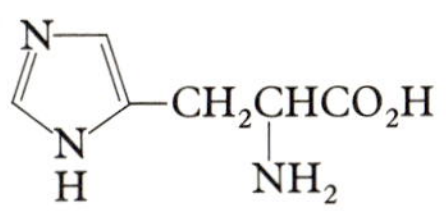	Tyrosine (tyr)	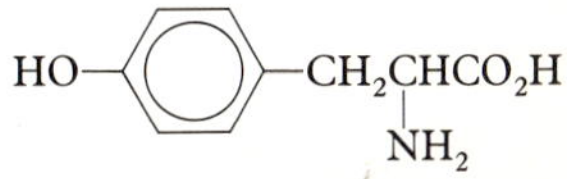
Isoleucine (ileu)[a]	CH_3 $CH_3CH_2CHCHCO_2H$ NH_2	Valine (val)[a]	$(CH_3)_2CHCHCO_2H$ NH_2

[a]Commonly referred to as an essential amino acid.

VARIATIONS IN THE STRUCTURES OF AMINO ACIDS. Table 23–1 lists the twenty amino acids commonly found in proteins. These amino acids may be grouped into two broad categories based upon the nature of their side chains: *nonpolar amino acids* and *polar amino acids*.

The amino acids that contain only alkyl or aryl groups in their side chains, such as alanine or phenylalanine, are considered nonpolar. Proteins that contain nonpolar amino acids tend to be nonreactive and insoluble in water. The proteins of silk, wool, and skin are rich in these nonpolar amino acids—for example, proteins in silk are composed of about 45 percent glycine and 30 percent alanine.

Typical nonpolar amino acids:

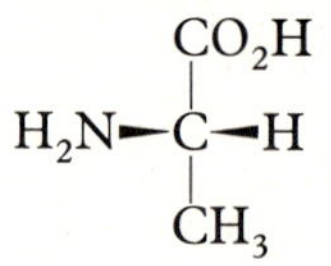

alanine (ala)

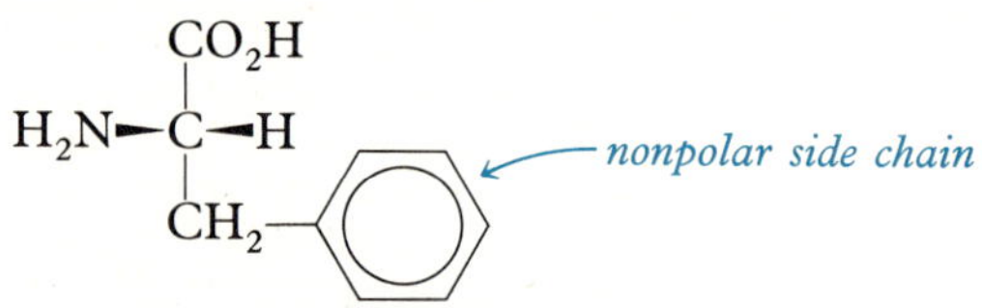

phenylalanine (phe)

Amino acids may contain $—CO_2H$, $—NH_2$, —OH or other polar groups in their side chains. These amino acids, called polar amino acids, are found in hormones and other water-soluble proteins.

Typical polar amino acids:

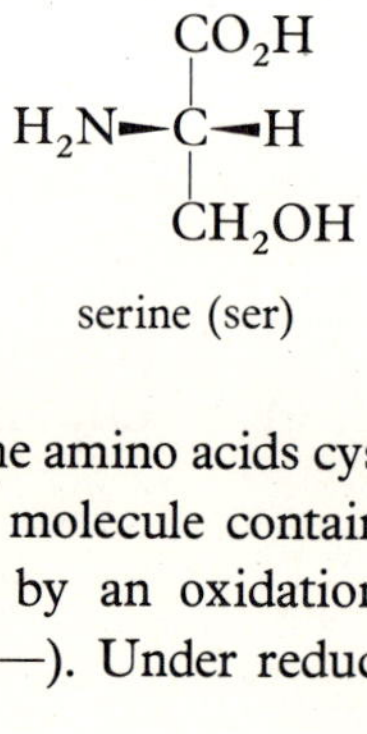

serine (ser)

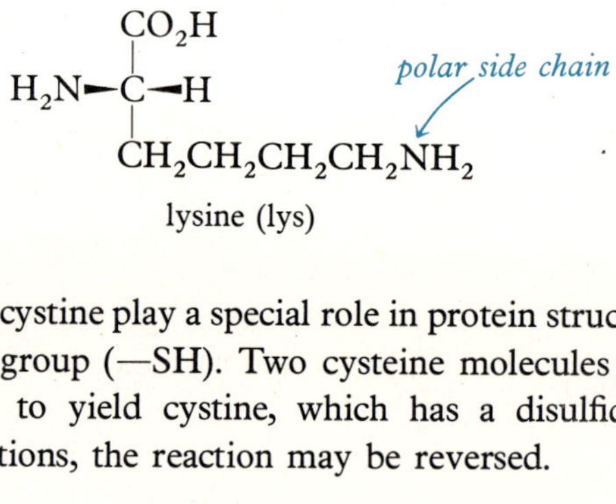

lysine (lys)

The amino acids cysteine and cystine play a special role in protein structures. A cysteine molecule contains a thiol group (—SH). Two cysteine molecules may be coupled by an oxidation reaction to yield cystine, which has a disulfide bond (—S—S—). Under reducing conditions, the reaction may be reversed.

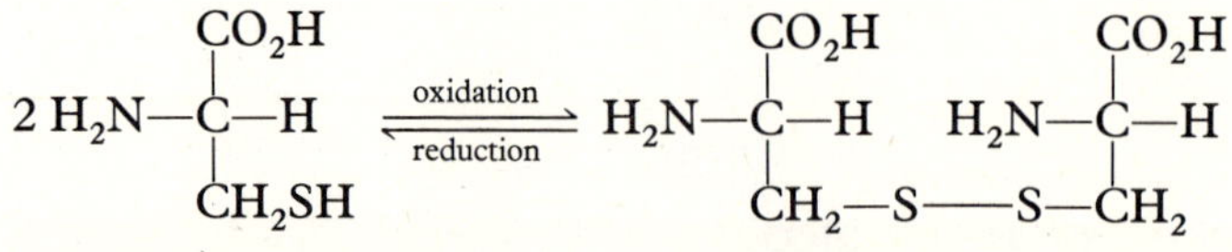

cysteine (cys) cystine (cys-cys or cyS—cyS)

The cystine disulfide bonds may join two protein chains together; thus, cystine provides a means of cross-linking protein chains. This cross-linking is an important feature in the structure of insulin and many other proteins.

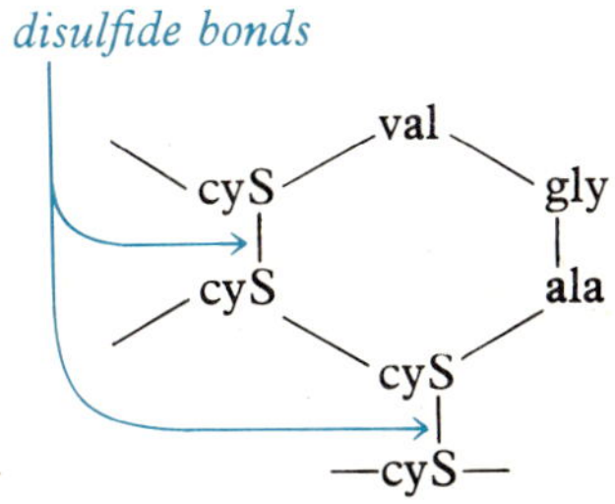

a fragment of the structure of sheep insulin, which is used in the treatment of diabetes

Hair is about 14 percent cystine. Curling or straightening hair by means of home permanents involves breaking the disulfide bond of the cystine units in the hair protein with a mild reducing agent. The hair is then curled and set in the desired fashion and treated with a mild oxidizing agent to reform the disulfide bonds from different cysteine units. The shape of the protein fibers in the hair is changed by this technique. Overdoing permanent waves results in weakened protein fibers and frizzled hair because of the broken disulfide bonds.

The decomposition of sulfur-containing proteins gives rise to volatile and often foul-smelling organic sulfur compounds and hydrogen sulfide. The odor of rotten eggs comes from these sulfur-containing decomposition products.

Sample Problem: Can methionine (Table 23–1) be coupled in the same manner as cysteine?

Solution: No. Methionine does not have a thiol group, a necessary requirement of the oxidative coupling reaction.

ESSENTIAL AMINO ACIDS. The body can synthesize many of the amino acids it needs from other amino acids or from ammonium salts and keto acids, which arise from carbohydrate metabolism. The amino acids that the body cannot synthesize must be supplied in the diet and are called the *essential amino acids.* The need for these amino acids varies not only from species to species, but also from individual to individual. What may be essential for a rat may not be essential for a human; what may be essential for one person may not be essential for another. Ten amino acids commonly referred to as essential are marked in the list of amino acids in Table 23–1.

Of the essential amino acids, the body can convert a few (tryptophan, phenylalanine, methionine, and histidine) from the D-form to the necessary L-form. Therefore, a mixture of the D- and L-stereoisomers of these amino acids is as nutritious as the pure L-form. Of the other amino acids, only the L-form can be utilized; if the D-form is fed, it is not used by the body to synthesize new protein molecules.

NINHYDRIN TEST FOR AMINO ACIDS. Amino acids undergo reaction with *ninhydrin* to yield a blue-violet product called *Ruhemann's purple*. This reaction is widely used to test for the presence of amino acids. The intensity of the blue-violet color provides an estimate of the amount of amino acids present. Under controlled conditions, the reaction can be used for quantitative analysis of amino acids.

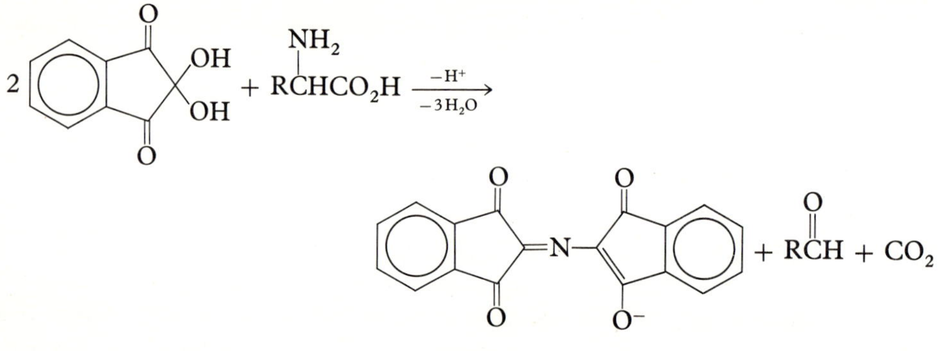

Ruhemann's purple

PEPTIDES AND PROTEINS

Peptides and proteins are formed from amino acids joined together by amide bonds. A peptide is differentiated from a protein by the number of amino acid units, or *residues,* that would be formed by hydrolysis. Generally, a polyamide with fewer than fifty amino acid residues is classified as a peptide, and one larger as a protein. Thus, a peptide is really just a small protein.

Peptides are further classified by the number of amino acid units they contain. A peptide composed of two amino-acid units is a *dipeptide;* three amino-acid units, a *tripeptide;* and so on. A large peptide is referred to as a *polypeptide.*

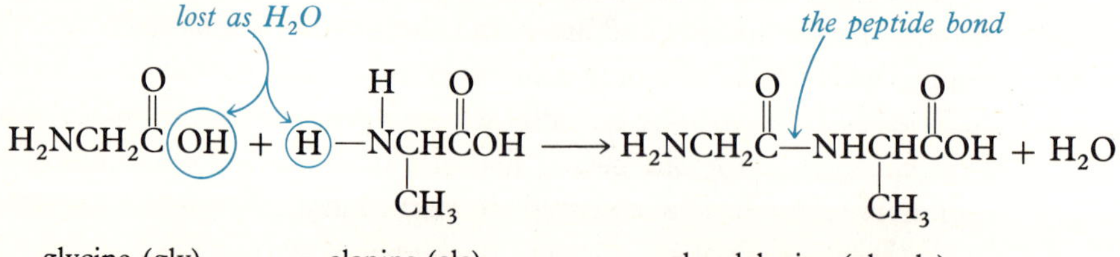

glycine (gly) alanine (ala) glycylalanine (gly-ala) *a dipeptide*

The linkage between the α-amino group of one amino acid and the carbonyl group of the other is frequently called a *peptide bond.* This term refers specifically to the amide linkage between two α-amino acids. A peptide has a free amino group at one end of the chain. The amino acid at that end is called the *N-terminal amino acid.* The other end of the chain has a free carboxylic acid, and the amino acid at that end is called the *C-terminal amino acid.*

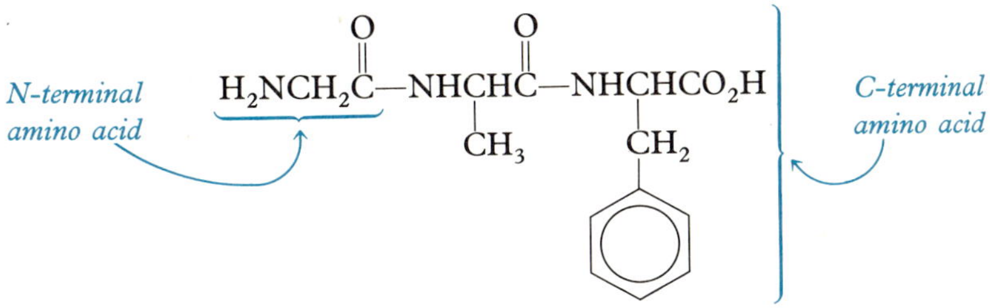

glycylalanylphenylalanine (gly-ala-phe)

a tripeptide

There are two ways of joining two amino acids to form a dipeptide. Glycine and alanine may be joined into the dipeptide gly-ala, shown on page 378, or into a different dipeptide, ala-gly. In the first, glycine is the N terminal and alanine is the C terminal. In the second, alanine is the N terminal and glycine the C terminal. (In using the abbreviated names for peptides, it is understood that the N terminal is on the left, the C terminal on the right.)

Sample Problem: Write the structure of ala-gly.

Solution:

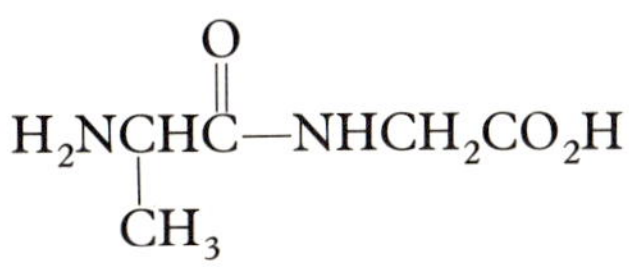

As the size of a peptide increases, the possibilities for variation in structure skyrocket. The different ways of combining the amino acids give rise to the fantastic number of variations possible in protein structure. A decapeptide contains ten amino-acid units and has a molecular weight of something over 1000. It is far smaller than a protein. Even so, the number of different decapeptides that can be constructed (using each amino acid no more than one time in each structure) is over 4,000,000,000,000! Most proteins have molecular weights between 12,000 and 1 million. Proteins may contain the same amino acid more than once in a molecule and may also form cross-linked structures—for example, by cystine bonds. The number of variations is beyond comprehension.

Three amino acids may be joined in six different ways to form six tripeptides:

gly-ala-phe	ala-gly-phe	ala-phe-gly
phe-ala-gly	phe-gly-ala	gly-phe-ala

A slight difference in the structure of a peptide may cause a difference in how it behaves in an organism. *Oxytocin* and *vasopressin* are two small peptides that are released by the pituitary gland. From the standpoint of their amino acid sequences, these two peptides are very similar. They differ by only two amino acids, circled in the following formulas. This small difference in structure is sufficient to result in different physiological effects. Although both peptides stimulate uterine contractions, oxytocin lowers the blood pressure while vasopressin raises it.

oxytocin:

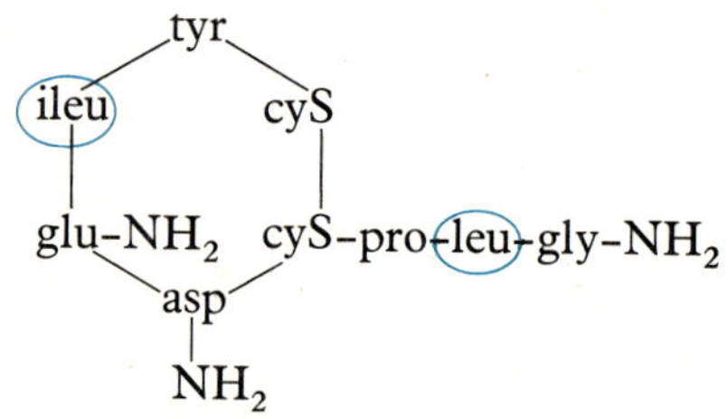

vasopressin:

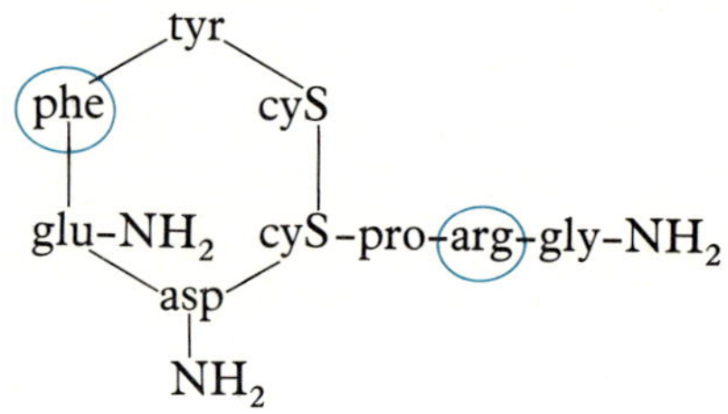

THE HIGHER STRUCTURES OF PROTEINS. The amino-acid sequence is called the *primary structure* of a protein. If the sequence of amino acids were all that there was to protein structures, life as we know it would not exist. Proteins obtain their individual properties partly by the ordered shape into which the backbones of the molecules arrange themselves. This shape is referred to as the *secondary structure* of a protein.

Figure 23-1. Higher structures of fibrous and globular proteins.

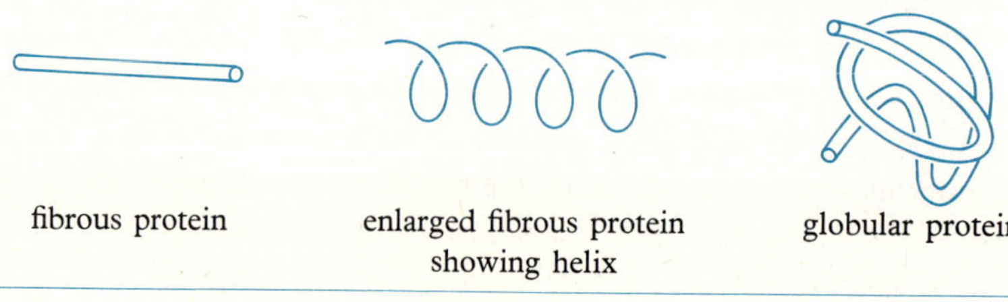

We will soon see that an orderly arranged protein molecule may interact further (by folding or bending, for example) into a *tertiary structure*.

Let us take a look at some common secondary structures. It has been discovered by *x*-ray analysis (which gives a three-dimensional map of electron density) that most fibrous protein molecules are helical, or spiral, in shape. This shape is maintained by hydrogen bonding between the carbonyl group from one amino acid unit and the hydrogen attached to the nitrogen of another unit in the helix. The helical secondary structure gives rise to a rather flexible and elastic structural material. *Keratin* (a protein in hair, skin, fur, and feathers) is an example of a fibrous protein with this type of helical structure.

Figure 23–2. Hydrogen bonds hold a protein molecule in a helix. (The nonparticipating groups attached to carbon atoms have been omitted.)

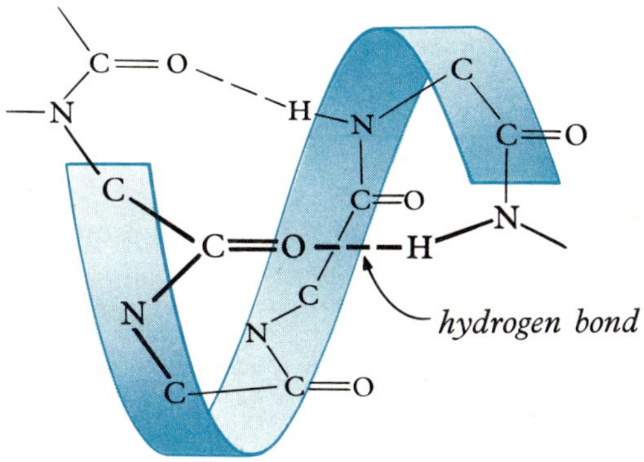

The fundamental structure of collagen (which forms cartilage, bone, tendons, ligaments, and skin) appears to be that of a *tropocollagen* molecule, which is composed of superhelices from the intertwining of *three* helical polypeptides in a single chain. The three-strand helix is also stabilized by hydrogen bonding. The intertwining of the three strands gives extra strength where it is needed in the structure of an organism.

Figure 23–3. A tropocollagen molecule consists of three helical protein chains.

The third common type of secondary structure is the *pleated sheet*. The term is derived from the general appearance of this type of protein, which resembles corrugated metal roofing. (See Figure 23–4.) Silk fibroin (a fibrous protein found in silk) is

Figure 23–4. In silk fibroin, the protein molecules are lined up and held together by hydrogen bonding, giving the general impression of corrugated roofing.

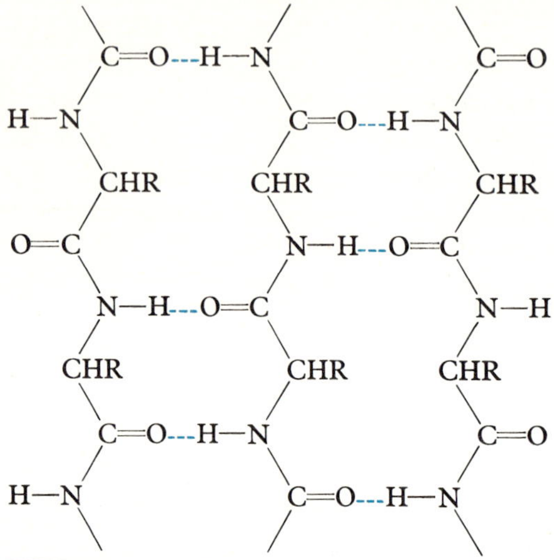

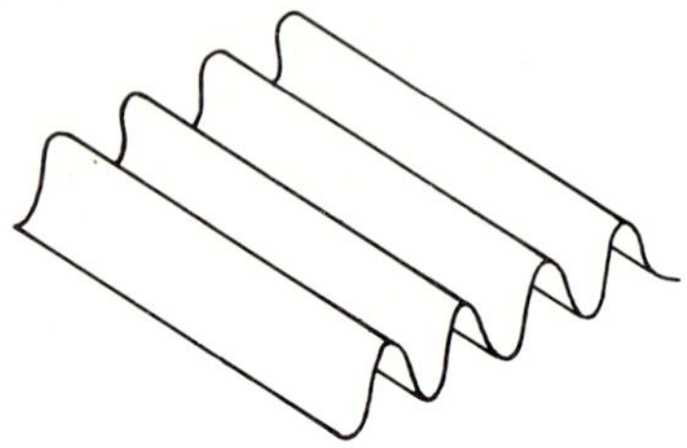

an example of a pleated sheet protein. The protein chains in silk fibroin are lined up side by side and held together by hydrogen bonds.

Globular proteins, such as enzymes and hemoglobin, are completely dependent on their tertiary structures in order to carry out their roles in an organism. Interactions between polar side chains allow these proteins to be held together in intricate folded shapes. The final folded molecule presents a unique surface, usually with polar side chains appearing on the outside and nonpolar groups on the inner surfaces. The nonpolar inner surface allows an enzyme to catalyze organic reactions in a nonaqueous medium, while the polar outer surface assures solubility. The positioning of functional groups is crucial for the proper functioning of a globular protein. A properly folded molecule with its unique surface is capable of "recognizing" other specific organic molecules. This recognition means that a globular protein can be used to catalyze specific reactions of specific compounds.

To illustrate the structure and function of a globular protein, let us consider hemoglobin, the protein in red blood cells responsible for transporting oxygen from the lungs to the cells of the body. A unit of hemoglobin consists of four protein chains called *globins*. Each globin is folded around a pocket that holds a molecule of *heme*, which is a porphyrin ring system holding an encaged, or chelated, iron(II) ion (p. 188).

The surface of each globin has functional groups that hold the heme in position. The surface of each globin is also structured so that four of these units fit together perfectly to make up the hemoglobin. Each iron(II) ion in a heme unit can carry an oxygen molecule. Therefore, each hemoglobin structure can carry four oxygen molecules from the lungs by way of the bloodstream to the cells of the body. It is interesting that hemoglobin picks up the fourth oxygen with greater ease than the first three, so hemoglobin does not leave the lungs without a full load.

In *carbon monoxide poisoning,* carbon monoxide molecules replace oxygen molecules in the hemoglobin. The affinity of hemoglobin for carbon monoxide is more than 200 times greater than its affinity for oxygen! The hemoglobin with carbon monoxide cannot carry as much oxygen as hemoglobin without carbon monoxide, and hemoglobin carrying both does not release its oxygen to the cells. The result is that the hemoglobin cannot function as an oxygen transporter if it is carrying carbon monoxide.

Sickle cell anemia is a hereditary disease caused by a defect in hemoglobin and is characterized by abnormally shaped erythrocytes (red blood cells). Severe cases, in which the sufferer has inherited the disease from both parents, are usually fatal. This disorder will serve to illustrate the consequences of a slight error in protein biosynthesis.

The difference between normal hemoglobin and the hemoglobin of a person affected with sickle cell anemia is simply that one glutamic acid unit has been replaced by a valine unit. Where normal hemoglobin has an acidic side chain, the hemoglobin of a victim of sickle cell anemia has a nonpolar side chain. Because the abnormal hemoglobin is slightly less soluble than normal hemoglobin, it may precipitate and alter the shape of the red blood cell from round to sickle-shaped. It has been shown (by the use of glycine labelled with nitrogen-15) that the abnormal cells have a very short lifetime compared to the usual 100-day lifespan of a normal erythrocyte. What are the consequences of the lower solubility and short life of the abnormal cells? The oxygen-carrying capacity of the blood is decreased. The affected red blood cells may clump together and clog capillaries. The work load on the heart is increased. Yet the only difference between normal and abnormal hemoglobin is that one amino acid in a chain of 146 has been changed.

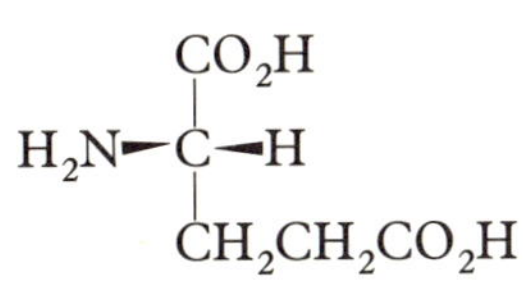

glutamic acid

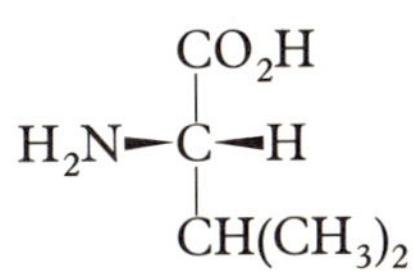

valine

DENATURATION AND RENATURATION. One of the great difficulties in the study of the structure of proteins is that if the normal environment of a living protein molecule is changed even slightly, such as by a change in pH, in temperature, or in ionic concentration, the hydrogen bonds are disturbed and broken. When attractions between and within protein molecules are destroyed, the chains separate from each other, globules unfold, and helices uncoil. We say that the protein has been *denatured.*

Figure 23–5. Denaturation causes loss of the higher structure of the protein.

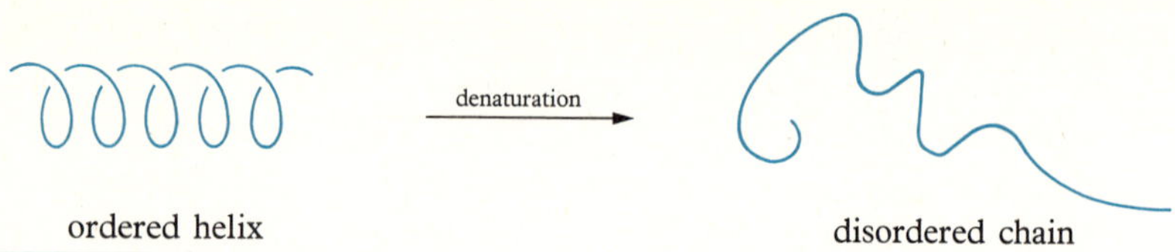

Denaturation might be compared to stretching a child's Slinky toy permanently out of shape, causing the once neat coils of the spring to become a tangled mass of metal. In the case of a denatured protein, the chemist is left with a long-chain polymer that has lost its important characteristics.

Denaturation is seen in our daily life in many forms. The curdling of milk is caused by bacteria in the milk that produce lactic acid. The change in pH caused by the lactic acid causes denaturation, coagulation, and precipitation of the milk proteins. Similarly, the cooking of an egg causes precipitation of the albumin proteins in the egg white. A denatured protein is easier to digest than a nondenatured protein; the cooking of food increases its digestibility by denaturation. Some proteins (such as those in skin, fingernails, and the stomach lining) are extremely resistant to denaturation.

Depending on the denaturation conditions, some denatured proteins can spontaneously refold themselves back into their ordered state when placed into a solution identical to their natural environment. The process, called *renaturation*, is usually very slow compared to denaturation, which is fast.

Sample Problem: Which of the following denaturation conditions might yield a denatured protein that could be renatured?

(a) A change in pH from 7.6 to 8.4.
(b) The addition of 0.05 M aqueous $(NH_4)_2SO_4$.
(c) A change in temperature from 37°C to 95°C.

Solution: Conditions (a) and (b) are mild changes in environment; one would expect that the denatured protein could revert to its natural form when placed in its natural environment. Condition (c) would be expected to do permanent damage to the protein structure.

ENZYMES

One of the highly important classes of protein in living systems is the class called *enzymes*. Enzymes are catalysts. Virtually every chemical reaction that occurs in a living system is catalyzed by an enzyme. Every known enzyme is a globular protein.

Figure 23-6. Energy diagram for the hydrolysis of a peptide.

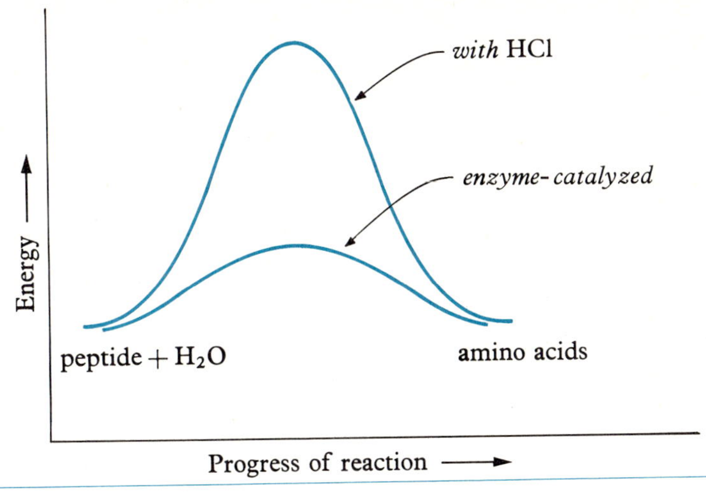

Just like chemical catalysts, enzymes are needed in only small quantities. Similar to the action of chemical catalysts, enzymes lower the energy barrier that reactants must pass over to go to products. For example, a *peptidase* enzyme induces hydrolysis of the amide links in peptides. A reaction that requires boiling hydrochloric acid in the laboratory occurs at nearly neutral pH and at moderate temperatures when catalyzed by an enzyme. (See Figure 23-6.)

$$\text{Peptide} + H_2O \xrightarrow{\text{peptidase}} \text{Amino acids}$$

Enzymes are selective catalysts. An enzyme may be specific for only one chemical change in one particular compound. For example, the enzyme *aspartase* catalyzes the reversible addition of ammonia to fumaric acid but is inactive toward its geometric isomer, maleic acid.

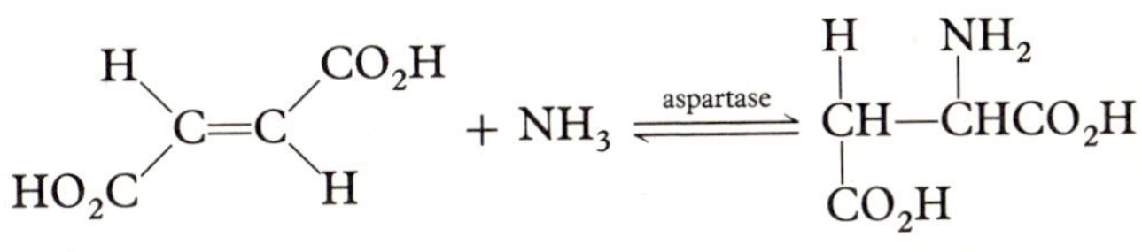

fumaric acid (*trans*) — aspartic acid

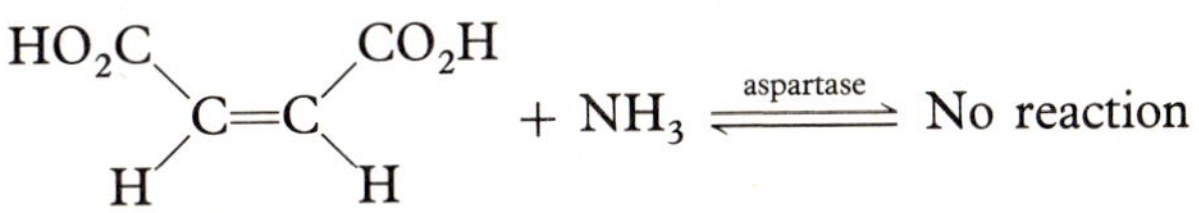

maleic acid (*cis*)

Other enzymes are more general and catalyze reactions for whole classes of compounds. Typical of this group are the digestive enzymes trypsin, chymotrypsin, and pepsin, all of which catalyze the hydrolysis of peptide bonds.

Enzymes are classified as to the general type of reaction they catalyze. A few of these broad classes are given in Table 23-2. Note that the enzyme class names are combinations of the names of the reaction types plus *-ase*. Specific enzymes are usually named after the compound upon which they act plus the suffix *-ase*. For example, *maltase* is the enzyme that catalyzes the hydrolysis of maltose to glucose. The names of many enzymes, however, are trivial and bear no relationship to the functions they perform. Trypsin, chymotrypsin, and pepsin are examples of trivial enzyme names.

Sample Problem: What would be the function of each of the following enzymes?

(a) phenylalanine hydroxylase
(b) acetyltransferase

Solution: (a) Oxidation of phenylalanine by insertion of a hydroxyl group into the molecule.
(b) Transfer of an acetyl group between molecules.

Exactly how enzymes function is only partially understood. An enzyme is generally a larger molecule than the molecules of the *substrate,* the compound on which the enzyme acts. In many cases, an enzyme requires the assistance of metallic ions or of other organic molecules (called *cofactors* or *coenzymes*) for activity. Many vitamins serve as part of cofactor structures—for example, nicotinamide in NAD^+ and riboflavin in FAD. (See Chapters 25 and 30.)

It is believed that an enzyme, along with its cofactor, if any, fits itself around the substrate molecule to form an enzyme-substrate complex. In some cases, the enzyme

Table 23-2. Some Enzyme Classifications

Class Name	Function
Transferase	Transfers a portion of one molecule to another
Lyase	Catalyzes addition to double bonds
Isomerase	Catalyzes isomerization, such as a *cis,trans* interconversion
Hydrolase	Catalyzes a hydrolysis reaction, such as a peptide or ester cleavage
Oxido-reductase	Catalyzes an oxidation-reduction reaction

undergoes a change in its conformation or shape to accommodate the substrate. In other words, the enzyme is induced to fit the substrate. This view of enzyme activity is therefore referred to as the *induced fit theory*.

$$\underset{\text{enzyme}}{E} + \underset{\text{substrate}}{S} \rightleftharpoons \underset{\textit{enzyme-substrate complex}}{E\text{-}S}$$

The attractions that hold the enzyme and substrate together may be due to unique and specific electronic attractions or may be actual covalent bonds. Regardless of how the complex is formed, the shape and polarity of the substrate determine whether or not a complex can be formed with a particular enzyme.

Generally, only a portion of the substrate is acted upon at one time. The portion of the enzyme that actually catalyzes reaction is called the *active site*. (Often this active site is the cofactor.) The rest of the enzyme molecule serves as a "holder" for the rest of the substrate molecule. After reaction occurs, the shapes of the enzyme and substrate may change so that the enzyme can no longer fit around the substrate. Then the enzyme releases the substrate and is ready to work again. Figure 23-7 depicts how these steps may occur.

$$\underset{\textit{enzyme-substrate complex}}{E\text{-}S} \rightleftharpoons \underset{\textit{enzyme-product complex}}{E\text{-}P} \rightleftharpoons \underset{\text{enzyme}}{E} + \underset{\text{product}}{P}$$

Figure 23-7. The induced fit theory of enzyme activity.

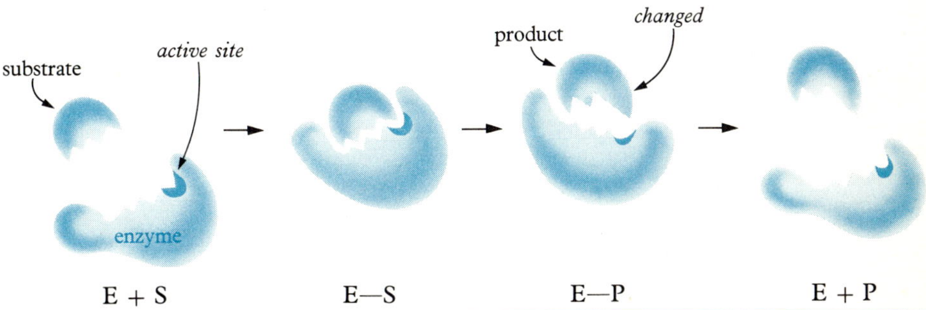

The induced fit theory of enzyme activity explains why enzymes are so selective toward substrates. If an enzyme cannot be induced to fit a substrate and form a complex, the reaction occurs slowly, if at all. This theory explains why one enantiomer may be biologically active and the other not. (See Figure 23-8.)

Figure 23-8. Only one optical isomer will fit onto an enzyme surface.

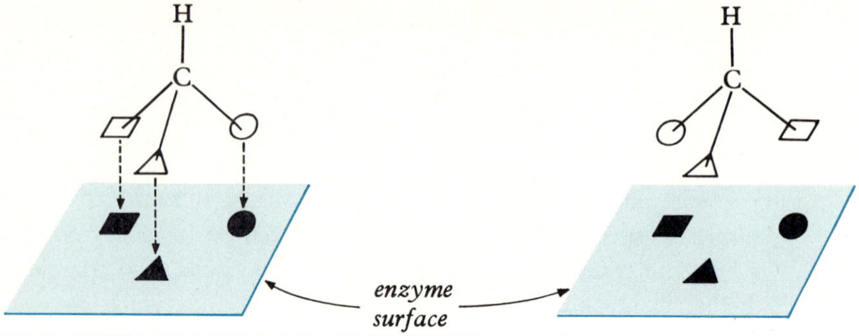

ENZYME INHIBITION. Certain organic compounds can inhibit the normal action of an enzyme. For example, the normal action of succinic acid dehydrogenase is to catalyze the loss of two hydrogen atoms from succinic acid. Malonic acid inhibits this reaction.

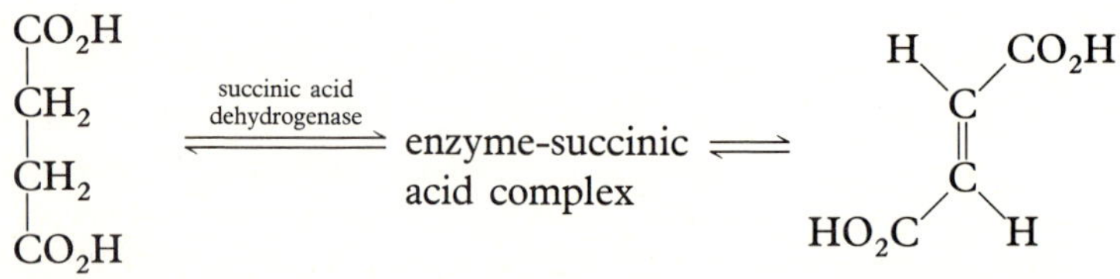

succinic acid fumaric acid

Malonic acid has a structure very similar to that of succinic acid. Apparently, the active site of succinic acid dehydrogenase can also fit around malonic acid. Malonic acid cannot lose two hydrogen atoms and form a double bond; therefore, the enzyme-malonic ester complex simply does nothing. In this fashion, the malonic acid ties up the enzyme and prevents it from performing its normal role.

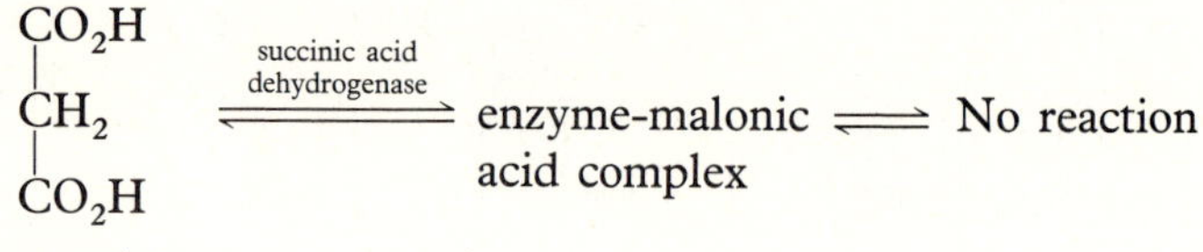

malonic acid

This example of enzyme inhibition is caused by two different molecules (malonic acid and succinic acid) competing for the active site on the enzyme and is therefore called *competitive inhibition.* Competitive inhibition is a common mode of action of many poisons and drugs. Because competitive inhibition is a reversible process, the extent of inhibition depends upon the relative concentrations of inhibitor and substrate.

Sample Problem: Which of the following compounds might be a competitive inhibitor of succinic acid dehydrogenase?

(a) HO_2CCO_2H
(b) $HO_2CCH_2CH_2CH_2CO_2H$
(c) $HOCH_2CH_2CH_2CH_2OH$

Solution: (a) and (b), which are both dicarboxylic acids similar to succinic acid.

Sulfa drugs act by competitive inhibition. In many microorganisms infectious to man, *p*-aminobenzoic acid (PABA) is required for the synthesis of the coenzyme folic acid. The sulfa drugs can be biologically hydrolyzed to *sulfanilamide*, which is similar in shape and structure to PABA. In the microorganism, the sulfanilamide competes with PABA for one of the enzymes necessary to the synthesis of folic acid. The synthesis is thus blocked, and the microorganism cannot grow or reproduce. Because sulfanilamide is a competitive inhibitor, a high and constant concentration is required in the blood of an animal for effective action.

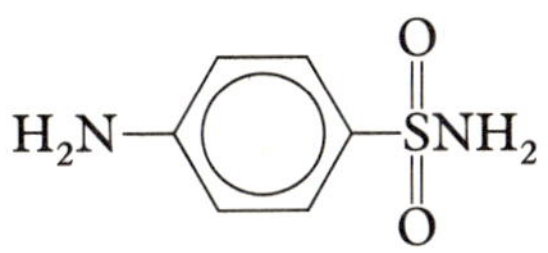
sulfanilamide

p-aminobenzoic acid (PABA)

Competitive inhibition requires that one molecule mimic the normal substrate. However, one substance can block the action of an enzyme by other means, as a *noncompetitive inhibitor*. A noncompetitive inhibitor is one that reacts with the enzyme either at the active site or elsewhere. In this case, either the active site becomes filled or the enzyme conformation becomes sufficiently altered that it cannot be induced to fit the substrate. In either case, the enzyme cannot form a complex with the substrate and consequently is inactive.

Many heavy metal ions, such as Hg^{2+} or Pb^{2+}, are poisonous because they act as noncompetitive inhibitors for numerous enzymes. One way these ions deactivate an enzyme is by forming insoluble salts with the enzyme's —SH groups.

$$\underset{\textit{active enzyme}}{2\,RSH} + Hg^{2+} \longrightarrow \underset{\textit{inactive enzyme}}{(RS)_2Hg} + 2\,H^+$$

ENZYME REGULATION. The different reaction sequences in an organism must be regulated; otherwise the system would be completely out of control. One common method of regulation depends upon the fact that almost all biochemical reactions are *coupled reactions,* in which the product of one reaction is the starting material for the next reaction. In this type of reaction sequence, *end-product inhibition,* or *feedback inhibition,* is an effective regulatory mechanism.

In end-product inhibition, the end product of a reaction pathway is a competitive inhibitor for the first enzyme in the pathway. As more of the end product is formed, it progressively inhibits its own formation. Conversely, as the end product is consumed by some biological process, the inhibition is less and the product is formed at a greater rate. The net result is that the concentration of the end product is maintained at its optimum value.

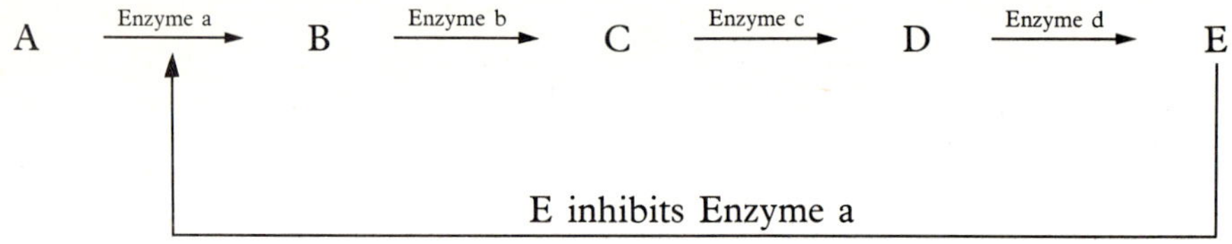

Another way an organism can control reaction pathways is by the synthesis or nonsynthesis of enzymes. The molecules within an organism, including enzymes, are in a constant state of destruction and resynthesis. If an organism needs a certain reaction sequence, hormones can signal the synthesis of the proper enzymes. When the reaction sequence is to be suppressed, hormones can signal "Stop synthesizing these enzymes." Then, as enzymes are consumed and not replaced, the reaction sequence is inhibited. The synthesis of proteins and the regulation of the synthetic sequences will be discussed at greater length in Chapter 28.

STUDY QUESTIONS

23-1. Write the equations for the acid hydrolysis of (a) N,N-dimethylacetamide and (b) the following peptide:

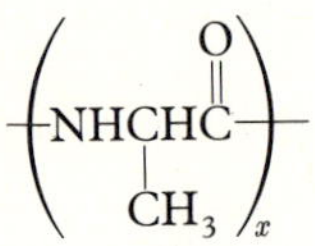

23-2. Which of the following structures are α-amino acids?

(a) 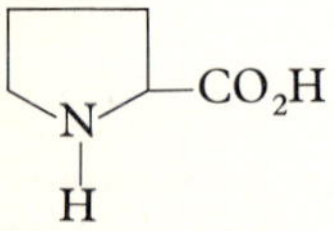

(b) $NH_2CH_2CH_2CO_2H$

(c) 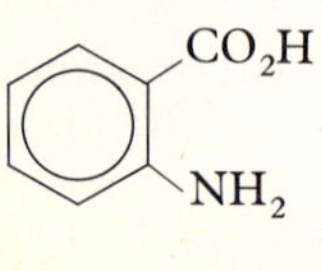

(d) 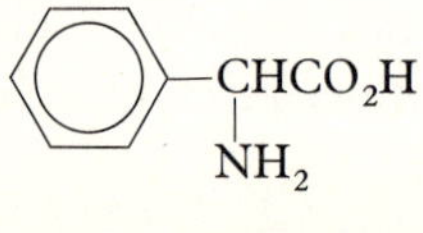

23-3. Give the name and structure for each of the following amino acids:

(a) ala (b) phe (c) tyr (d) ser

23-4. Show by an equation how each of the following amino acids becomes a dipolar ion: (a) alanine (b) methionine

23-5. Complete the following equations:

(a) $CH_3\underset{\substack{|\\NH_3^+}}{C}HCO_2^- + H^+ \rightleftharpoons$

(b) $CH_3\underset{\substack{|\\NH_3^+}}{C}HCO_2^- + OH^- \rightleftharpoons$

23-6. Circle the peptide bonds in the partial protein structure in Figure 23-4 (page 382).

23-7. (a) Write the equation for the reaction that leads to phe-leu from its amino acids; circle the OH and H that are lost as H_2O. (b) Do the same for the formation of leu-phe from its amino acids.

23-8. The C-terminal amino acid in a peptide contains a (free carboxyl group/free amino group) while the N-terminal amino acid contains a (free carboxyl group/free amino group).

23-9. Partial hydrolysis of an unknown tetrapeptide yielded the following fragments: ala-ser-val and ser-val-glu. What was the structure of the tetrapeptide?

23-10. Show all the different ways that his, glu, and asp may be joined together to form a tripeptide. (Use the abbreviated names.)

23-11. Define the following as related to proteins:
(a) primary structure
(b) secondary structure
(c) tertiary structure

23-12. What occurs when a protein is denatured?

23-13. Match the following enzyme names with the functions:

(a) DNA polymerase	(1) catalyzes hydrolysis of proteins
(b) transaminase	(2) catalyzes hydrolysis of esters
(c) proteinase	(3) catalyzes polymerization of DNA units
(d) esterase	(4) catalyzes the transfer of an amino group

23-14. Write the structural formulas for the following peptides:
(a) leu-gly-cys (b) lys-lys-ser-try

23-15. The digestive enzyme chymotrypsin catalyzes hydrolysis of the peptide bond of (a) phe-ala, (b) phe-gly, and (c) phe-val. Write the structural formulas of the hydrolysis products.

23-16. Chymotrypsin also catalyzes the hydrolysis of the following compounds. What are the hydrolysis products?

(a) 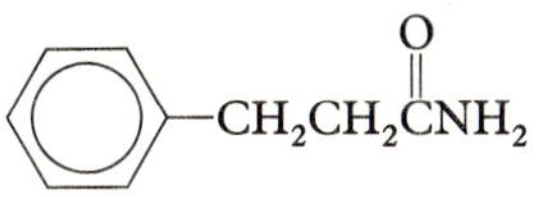

(b) 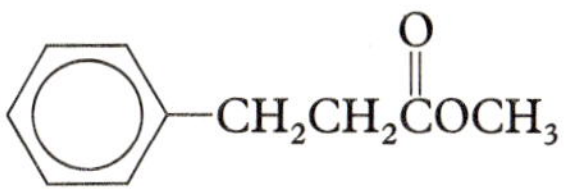

23-17. Which of the following compounds might be hydrolyzed with chymotrypsin as a catalyst?

(a) phe-ileu (b) ala-phe-ileu

(c) 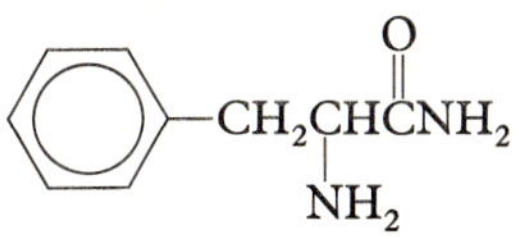

(d) 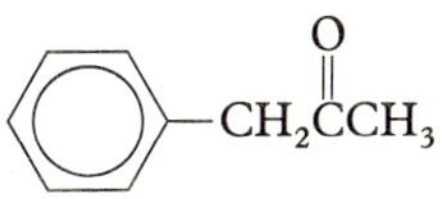

(e) 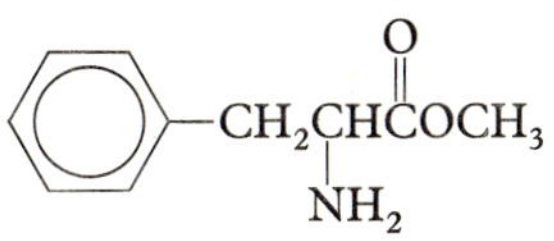

23-18. Write an equation that shows how silver nitrate ($AgNO_3$) might cause inhibition of an enzyme that contains cysteine.

CHAPTER 24 FATS, OILS, AND STEROIDS

OBJECTIVES

Fats and Oils *Write structures for a typical fatty acid and a typical triglyceride.* * *Write an equation to illustrate formation of a triglyceride from glycerol and a fatty acid.* * *Differentiate between a fat and an oil.* * *Write an equation to illustrate hydrogenation of a vegetable oil.* * *Define phospholipid.* * *Write the structure of a typical phosphoglyceride.* * *Describe how phospholipids help form the cell membrane.*

Steroids *Give the structures of some common steroids.* * *Describe some effects of steroids within the animal organism.* * *Differentiate between an androgen and an estrogen.*

Lipids are naturally occurring substances that are insoluble in water but soluble in relatively nonpolar organic solvents, such as diethyl ether. Carbohydrates, proteins, and many other natural products are insoluble in nonpolar solvents. Consequently, lipids can be extracted from crude cellular mixtures while carbohydrates and proteins remain behind.

A large variety of compounds are classified as lipids. In this chapter, we will discuss only the more important ones: the edible fats and oils (such as beef fat and corn oil), the phosphoglycerides (such as the lecithins), and the steroids (such as cholesterol and the sex hormones).

FATS AND OILS

Edible fats and oils (not to be confused with petroleum oils like mineral oil) are triesters of the triol glycerol and long-chain carboxylic acids called *fatty acids.* Fats and oils are referred to as *triglycerides.*

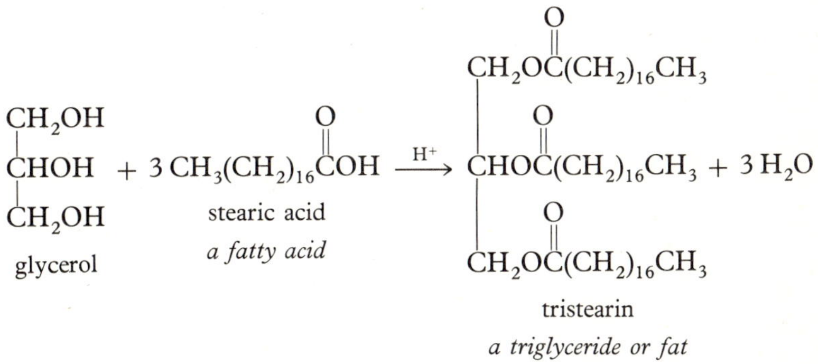

Variations in the structures (and thus in the properties) of fats and oils occur in the fatty acid portion of the ester. These long-chain acids may vary in the number of carbon atoms in the hydrocarbon chain and may, or may not, contain double bonds. (See Table 24-1 for some representative fatty acids.) In nature, many triglycerides are *mixed esters;* that is, the three fatty acid groups in the ester are not the same.

Sample Problem: What are the two possible structures for a triglyceride composed of two myristic acid residues and one palmitic acid residue?

Solution:

$$\begin{array}{l} CH_2O_2C(CH_2)_{12}CH_3 \\ | \\ CHO_2C(CH_2)_{12}CH_3 \\ | \\ CH_2O_2C(CH_2)_{14}CH_3 \end{array} \quad \text{and} \quad \begin{array}{l} CH_2O_2C(CH_2)_{12}CH_3 \\ | \\ CHO_2C(CH_2)_{14}CH_3 \\ | \\ CH_2O_2C(CH_2)_{12}CH_3 \end{array}$$

Table 24–1. Some Common Fatty Acids in Fats and Oils

Structure	Name of Acid	Number of Carbons	Number of Double Bonds
$CH_3(CH_2)_{12}CO_2H$	Myristic acid	14	0
$CH_3(CH_2)_{14}CO_2H$	Palmitic acid	16	0
$CH_3(CH_2)_{16}CO_2H$	Stearic acid	18	0
$CH_3(CH_2)_5CH{=}CH(CH_2)_7CO_2H$	Palmitoleic acid	16	1
$CH_3(CH_2)_7CH{=}CH(CH_2)_7CO_2H$	Oleic acid	18	1
$CH_3(CH_2)_4CH{=}CHCH_2CH{=}CH(CH_2)_7CO_2H$	Linoleic acid	18	2

By definition, a fat is a triglyceride that is solid or semisolid at room temperature; an oil is a triglyceride that is liquid at room temperature. Since room temperature varies from place to place and from season to season, the distinction between fats and oils is not sharp. Although coconut oil, for example, is a solid in temperate climates, it is a liquid in the tropics where it is prepared; thus, it is called an oil.

Saturated fatty acids form higher-melting triglycerides than do the unsaturated fatty acids. The saturated triglycerides tend to be solid fats, while the unsaturated triglycerides, particularly those that are polyunsaturated (with more than one double bond), tend to be oils.

In general, the triglycerides found in animals contain few double bonds and are solid; these are therefore referred to as *animal fats*. On the other hand, triglycerides found in plants tend to be polyunsaturated and liquid; hence we refer to these compounds as *vegetable oils*. The vegetable oils are found primarily in the fruit and seeds of the plants. This origin is apparent in the names of some of the common oils: olive oil, cottonseed oil, peanut oil, soybean oil, and so forth.

The composition of the fatty acid residues in fats and oils is exceedingly variable from species to species and within any given species itself. Dietary lipids are a major factor in the composition of the fat deposits in the body of an animal. Thus, the composition of fatty acids found in a particular type of animal is somewhat dependent on the season of the year and on the geographic region in which the animal is found. Table 24–2 lists the fatty acid compositions of some typical plant and animal tissues.

Table 24–2. Approximate Fatty Acid Composition in Plant and Animal Tissue

	Fatty Acid Composition (%)				
Source	Palmitic	Stearic	Palmitoleic	Oleic	Linoleic
Corn	10	5	1	45	38
Soybean	10	5	—	25	55
Lard	30	15	1	45	5
Whale	15	5	15	35	—
Human	25	8	6	46	10

The double bonds in an unsaturated triglyceride are easily hydrogenated, and, in this way, an oil may be converted to a fat. The hydrogenation, or hardening, of oils to fats, such as in the preparation of peanut butter or oleomargarine, is an important commercial process.

A hydrogenation reaction:

$$\begin{array}{l} CH_2O_2C(CH_2)_7CH{=}CH(CH_2)_7CH_3 \\ | \\ CHO_2C(CH_2)_7CH{=}CH(CH_2)_7CH_3 \\ | \\ CH_2O_2C(CH_2)_7CH{=}CH(CH_2)_7CH_3 \end{array} + 3\,H_2 \xrightarrow{\text{catalyst}} \begin{array}{l} CH_2O_2C(CH_2)_{16}CH_3 \\ | \\ CHO_2C(CH_2)_{16}CH_3 \\ | \\ CH_2O_2C(CH_2)_{16}CH_3 \end{array}$$

triolein — *unsaturated*; tristearin — *saturated*

In addition to being found in foodstuffs, fats and oils are used in the manufacture of soaps, such as $CH_3(CH_2)_{14}CO_2^-\ Na^+$, which are obtained by the saponification of triglycerides. (Because triglycerides can be saponified, they are often called *saponifiable lipids*. Steroids, by contrast, contain no ester groups and are called *nonsaponifiable lipids*.)

Sample Problem: Write the equation for the saponification of triolein, the triglyceride of oleic acid.

Solution:

$$\begin{array}{l} CH_2OC(=O)(CH_2)_7CH{=}CH(CH_2)_7CH_3 \\ | \\ CHOC(=O)(CH_2)_7CH{=}CH(CH_2)_7CH_3 \\ | \\ CH_2OC(=O)(CH_2)_7CH{=}CH(CH_2)_7CH_3 \end{array} + 3\,NaOH \longrightarrow$$

$$\begin{array}{l} CH_2OH \\ | \\ CHOH \\ | \\ CH_2OH \end{array} + 3\,CH_3(CH_2)_7CH{=}CH(CH_2)_7C(=O)ONa$$

Polyunsaturated vegetable oils, such as linseed oil, are also used as drying oils in paints and varnishes. When a drying oil is exposed to the oxygen in the air, a free-radical polymerization reaction takes place, and the molecules become crosslinked into a tough, hard film.

PHOSPHOLIPIDS. Phospholipids are a group of compounds that are phosphate esters of certain lipids. Phospholipids can be isolated from many sources; for example, the brain, nerve cells, egg yolks, yeast, wheat germ, and soybeans. One class of phospholipid, the *phosphoglycerides,* is closely related to the triglycerides. *Lecithins* and *cephalins* are the two most widespread phosphoglycerides. The structures of these two types of compound are similar—glycerol esterified with two long-chain fatty acids and an organophosphoric acid. The primary difference is that a lecithin has a quaternary nitrogen, while a cephalin has a primary nitrogen.

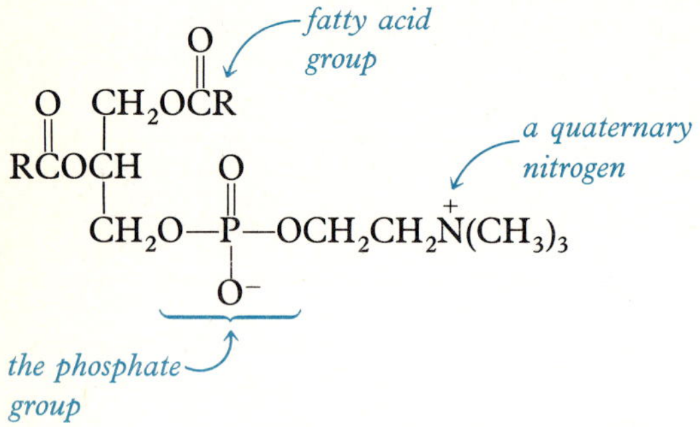

a lecithin

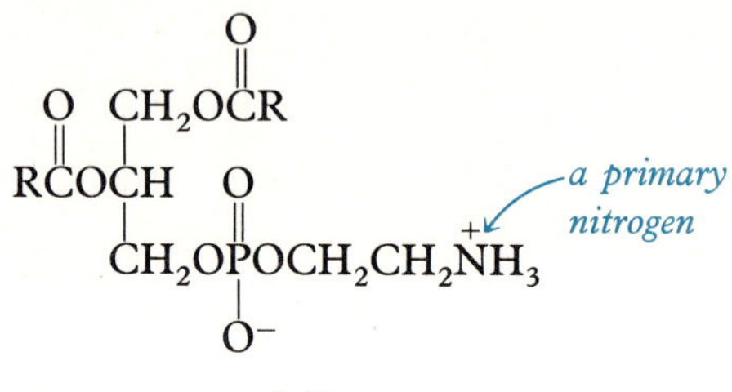

a cephalin

The fact that a phospholipid has an ionic end and a hydrocarbon end leads to the use of these compounds as emulsifying agents. In mayonnaise, it is the lecithins in the egg yolk that aid in emulsification of the oil in the vinegar.

The functions of phospholipids in animal organisms are only partially known. These compounds are an extremely important component of the cell plasma membrane, which is composed of approximately 40 percent phospholipid and 60 percent protein. The bilayer (double layer) of phospholipid molecules was shown in Figure 21-3.

STEROIDS

The steroids are a class of natural products that are found throughout the plant and animal kingdoms. In our bodies some steroids help regulate our sexuality and secondary sex characteristics. For example, a man who is given estrogen (a female hormone) may develop a high-pitched voice and enlarged breasts. Other steroids help regulate protein metabolism. Cortisone, for example, is a steroid that is found naturally in the body, but also may be administered to a patient to control inflammation.

Steroids are compounds that contain a distinctive ring system of three fused six-membered rings plus a fused five-membered ring. Branches and functional groups on these rings are what differentiate the steroids from each other. Although we will use flat rings to represent steroids, the fused rings are puckered, just as a simple ring such as cyclohexane would be.

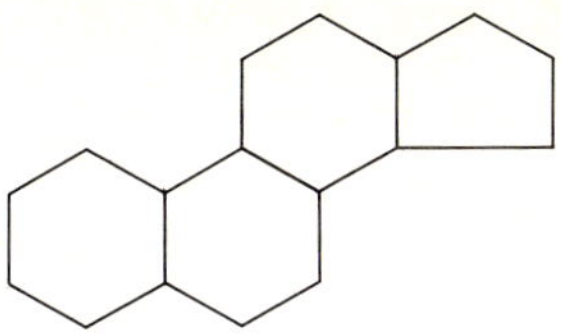

the steroid ring system, flat

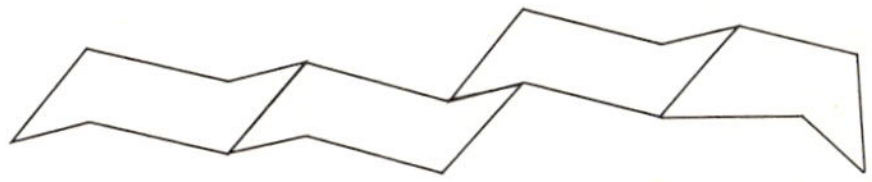

the steroid ring system, puckered

CHOLESTEROL. Cholesterol is a common steroid found in animals. It is present in the blood and is concentrated in the spinal cord and brain. Ten percent of the dry weight of the brain is from cholesterol. It is the principal constituent of gallstones, and its deposition in the arteries is a factor in hardening of the arteries.

Some nutritionists recommend that foods high in cholesterol (eggs and shellfish, for example) be eaten only in moderation. The theory is that less cholesterol in the system will lead to less cholesterol being deposited in the arteries. However, a low-cholesterol diet does not insure a low-cholesterol level in the blood because the body can synthesize its own cholesterol from other compounds such as carbohydrates or fats.

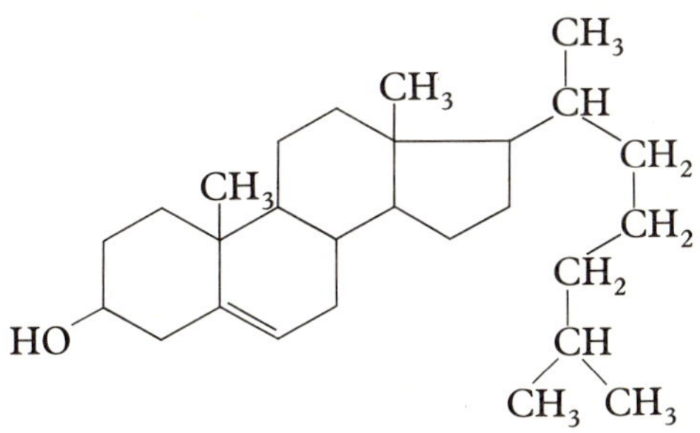

cholesterol

ADRENAL STEROIDS. One of the many hormones released into the bloodstream by the pituitary gland is the adrenocorticotropic hormone (ACTH), a protein. Among its other functions, ACTH stimulates the adrenal gland, located just above the kidney, to secrete a mixture of substances called *cortin*. At least thirty different steroids have been isolated from cortin. A lack of cortin in the body results in Addison's disease (a rare disorder characterized by a bronzing of the skin, weight loss, and lethargy). Excess cortin in children produces precocious sex development.

Two of the better-known steroids present in cortin are cortisone and cortisol (also called hydrocortisone). Both cortisol and cortisone alter protein metabolism. They are used extensively in treatment of inflammation and allergies and have been particularly useful in the symptomatic treatment of rheumatoid arthritis.

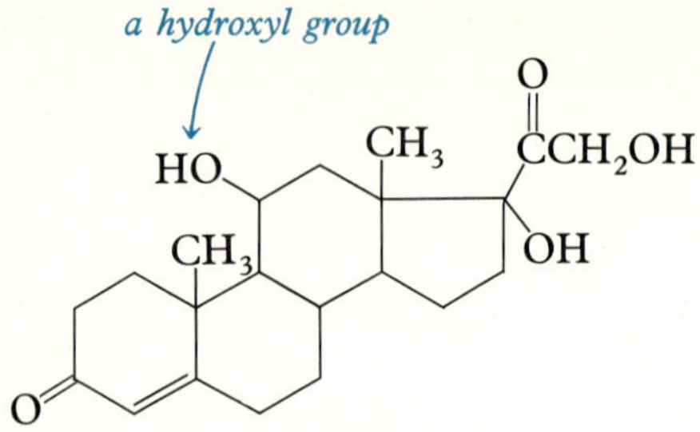

cortisol
(hydrocortisone)

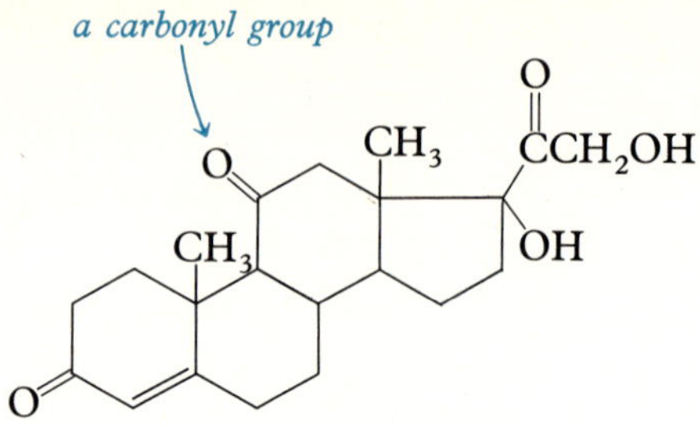

cortisone

SEX HORMONES. The sex hormones are produced primarily by the testes or ovaries of the male or female. Their formation is regulated by pituitary hormones, and they themselves regulate the sexual functions and impart the secondary sex characteristics. The male hormones are frequently referred to by the general term *androgens,* and the female hormones, *estrogens.*

As an example of the complex function of the sex hormones, we will use the estral cycle, which is under the control of the three principal female hormones, estrone, estradiol, and progesterone.

Some female hormones:

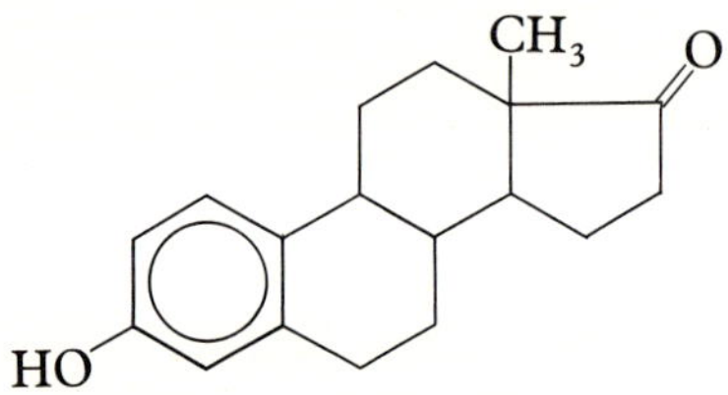

estrone

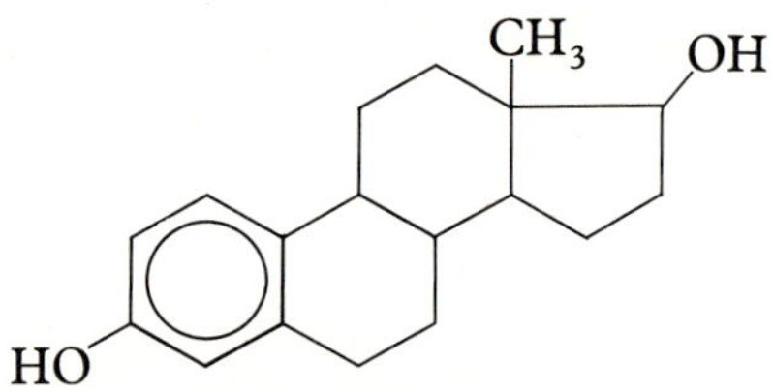

estradiol

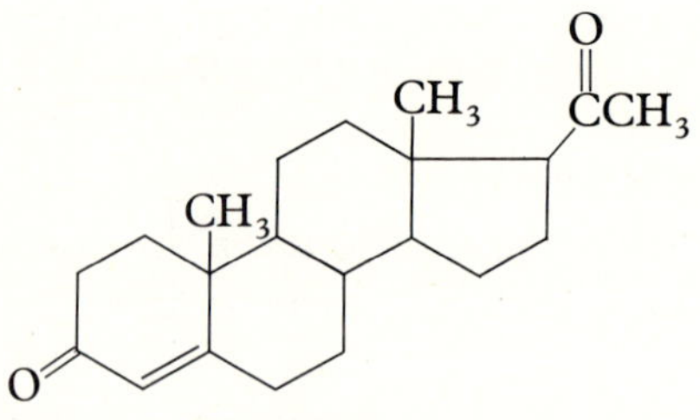

progesterone

The estral cycle is directed by the gonadotropin hormones, which are glycoproteins (sugar-protein complexes) released by the pituitary gland. One of these hormones stimulates the ovaries to produce the estrogens estrone and estradiol. These hormones, besides imparting feminine characteristics, cause proliferation of the uterine lining, the ripening of the ovum, and the development of the Graafian follicle, the covering of the ovum.

Prior to ovulation, the ovary is directed by another gonadotropin to secrete progesterone, which prevents other ova from ripening, prepares the uterus for possible pregnancy, and also further stimulates the pituitary gland to secrete a third gonadotropin that causes the ripened follicle to burst and release its ovum.

After the ovum is released, the follicle stays in the ovary and becomes the *corpus luteum* (yellow body), a temporary gland that also produces progesterone. If fertilization of the ovum does not occur, the corpus luteum dies in a few days. The lack of progesterone causes the uterine lining to disengage itself from the uterus and to be sloughed off in what is called the menstrual period.

If fertilization does occur, a gonadotropin hormone stimulates the corpus luteum to a longer lifetime of producing progesterone to maintain the pregnancy to full term.

The oral contraceptives are related to progesterone, and function by suppressing the gonadotropins that stimulate ovulation. One such oral contraceptive is norethynodrel (Enovid). It is given in conjunction with an estrogen.

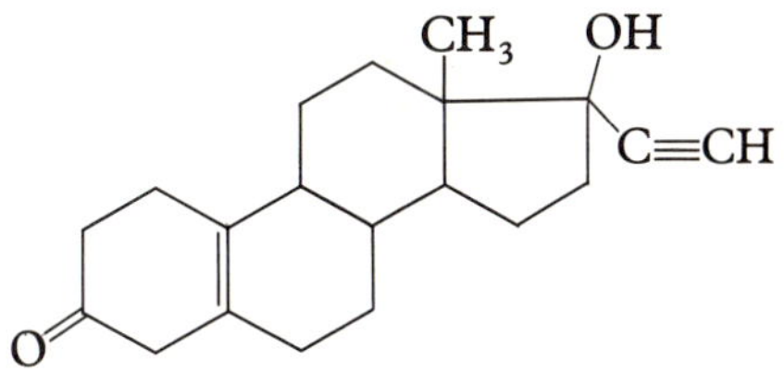

norethynodrel

There are at least five known male hormones, or *androgens*. Of these hormones, the most potent is testosterone, which is produced in the testes. In the adult male, the androgens induce sperm synthesis and cause the secondary masculine characteristics.

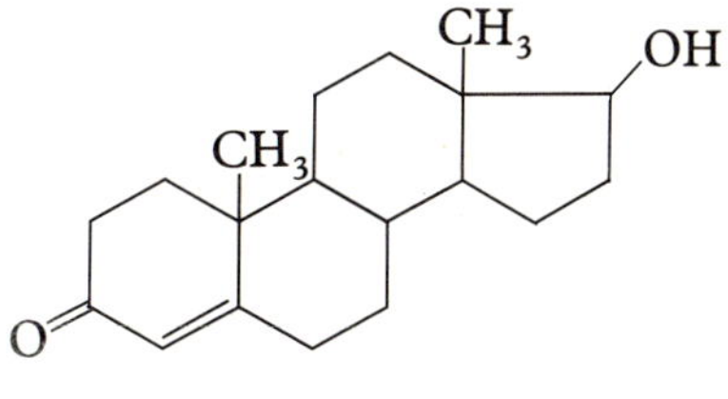

testosterone

an androgen, a male sex hormone

STUDY QUESTIONS

24-1. Match the following structures with their classes. (More than one match may be possible.)

(a)
$C_{17}H_{35}CO_2CH_2$
$C_{15}H_{31}CO_2CH$
$C_{17}H_{35}CO_2CH_2$

(b) $C_{17}H_{35}CO_2H$

(c) 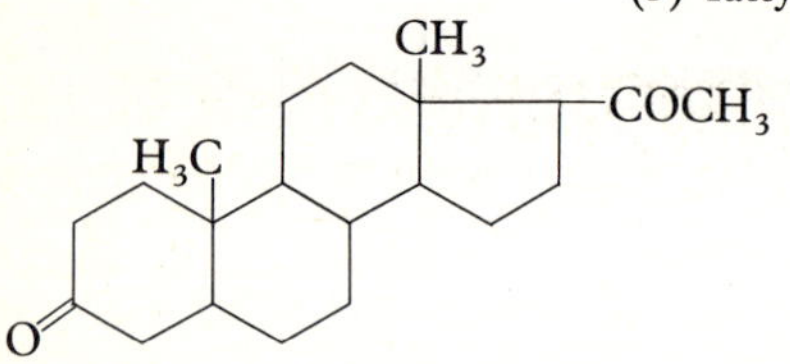

(1) lipid
(2) fat
(3) oil
(4) steroid
(5) fatty acid

24-2. Write the equation for the saponification of (a) ethyl acetate and (b) tripalmitin, the triglyceride of palmitic acid.

24-3. What is the difference between a petroleum oil and a vegetable oil?

24-4. Saturated triglycerides tend to be (solid/liquid) at room temperature while polyunsaturated triglycerides tend to be (solid/liquid).

24-5. A mixed triglyceride containing two units of stearic acid and one unit of palmitoleic acid (Table 24-1) is heated with a solution of NaOH. What are the products of the reaction?

24-6. Write an equation for the conversion of trilinolein (the triglyceride of linoleic acid) in corn oil to a semisolid fat that could be used for margarine.

24-7. Write the structures of the products of the acid hydrolysis of the lecithin on page 396.

24-8. Match the following:

(a) cortisone
(b) testosterone
(c) estrone

(1) a female hormone
(2) could cause facial hair growth
(3) increases during pregnancy
(4) could cause breast enlargement
(5) contains an aromatic ring
(6) contains a ketone group

24-9. Would you expect each of the following steroids to be an androgen or an estrogen?

(a) 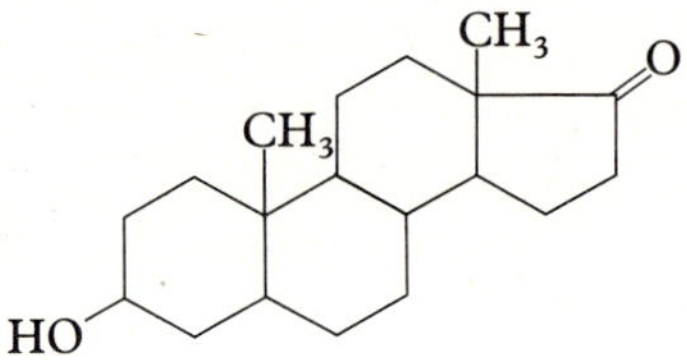

(b) 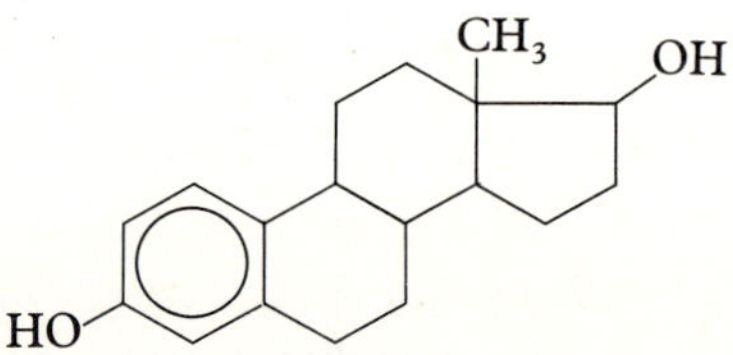

CHAPTER 25 BIOLOGICAL ENERGY

OBJECTIVES

Digestion and Metabolism *Define digestion, metabolism, catabolism, and anabolism.* * *Write general equations to illustrate the digestion and catabolism of a starch.*

Storage of Energy *Write the equation for the conversion of ATP to ADP, and tell why this reaction is important.*

The Respiration Cycle *Write the equation for the reduction of NAD^+. (Show the "working end" of the molecule.)* * *Write the equation for the reaction of $NADH_2$ with FAD.* * *Briefly outline the respiration cycle.*

A living organism is a biological machine that takes food and converts it to carbon dioxide, water, and energy. The energy is used for movement, for warmth, and for power to drive the chemical reactions that repair and maintain the health of the organism. Without biological energy, life cannot exist.

The food that we eat can be roughy classified into three groups: carbohydrates, fats, and proteins. All three types of food can be converted to energy by our bodies. Carbohydrates yield approximately 4.1 kilocalories of metabolic energy per gram; fats, 9.3 kilocalories per gram; and proteins, 4.4 kilocalories per gram.

DIGESTION AND METABOLISM

The process of food molecules being broken down into smaller molecules by enzymatic hydrolysis in the digestive tract (the mouth, stomach, and small intestine) is called *digestion.* For example, starch is broken down to glucose in the digestive tract. Digested food is absorbed into the bloodstream through the walls of the small intestine and then carried to the cells of the organism. The reactions that these nutrient molecules then undergo to provide cell growth, cell repair, or energy are called *metabolism reactions.*

Metabolism may be further subdivided into *catabolism,* the breaking down of large nutrient molecules into smaller molecules, usually with the production of energy, and *anabolism,* the building of large molecules (such as proteins) from smaller units, usually with the consumption of energy.

Glucose from ingested (and digested) starch can be metabolized in the cells to CO_2, H_2O, and energy. This conversion is an example of catabolism.

Digestion:

$$\text{starch} \xrightarrow[\text{saliva}]{H_2O} \text{maltose} \xrightarrow[\text{gastric enzymes}]{H_2O} \text{glucose}$$

Metabolism:

$$\text{glucose} + 6\,O_2 \xrightarrow{\text{many steps}} 6\,CO_2 + 6\,H_2O + 686\text{ kcal/mole}$$

As you probably realize, the oxidation of glucose is not a simple, straightforward reaction with oxygen. We do not burn glucose in our bodies in the same way we burn wood in the fireplace. Furthermore, since we do not eat continuously, the body must have a chemical method of *storing* energy. Before we look at the metabolic fate of glucose (Chapter 26), let us take a look at how the body stores energy to be used when we need it.

STORAGE OF ENERGY

Nutrients and oxygen are brought to the cells of the body by the blood. The oxidation of glucose and other nutrients occurs within the mitochondria of the cell. Each cell contains hundreds of these organelles. It is thought that a mitochondrion is rather like a factory, with the proper enzymes lined up on the inner membrane. (See Figure 25-1.) Nutrients pass through the enzyme line-up and are broken down in assembly-line fashion.

When an organic compound is oxidized in an organism, the released energy is used to synthesize high-energy compounds. These high-energy compounds are the storehouses of biological energy. The principal high-energy compounds in biological systems are anhydrides of phosphoric acid. When food is metabolized, the energy generated is used to synthesize an anhydride. When the body requires energy for movement, warmth, or chemical reactions, the anhydride is hydrolyzed. The hydrolysis reaction gives off energy which can then be used by the body.

$$\text{HO—P(=O)(OH)—O—P(=O)(OH)—OH} + H_2O \longrightarrow 2\ \text{HO—P(=O)(OH)—OH} + \text{Energy}$$

an anhydride of phosphoric acid *high energy* — phosphoric acid *low energy*

In biological systems, instead of phosphoric acid itself and its anhydrides, we find that the phosphoric anhydride is part of a larger molecule called adenosine triphosphate (abbreviated ATP). In the hydrolysis of ATP, usually only one of the phosphoric acid groups is cleaved. This inorganic cleavage product is referred to as inorganic phosphate, or P_i. The other cleavage product is adenosine diphosphate (ADP). Of course, phosphoric acid itself (as is shown in the equation) is not found in an organism; rather, phosphate *anions* are found. Adenosine triphosphate and other

Figure 25-1. Cutaway diagram of a mitochondrion.

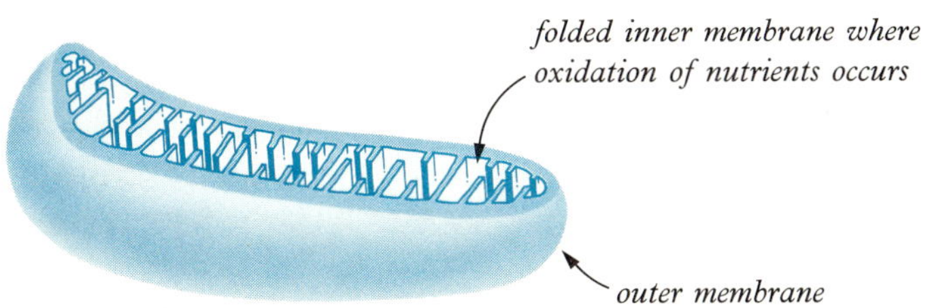

phosphates also are found in the anionic form, usually bonded to magnesium ions (Mg^{2+}). The number of hydrogen ions lost from the phosphate groups to the medium depends upon the pH of the medium.

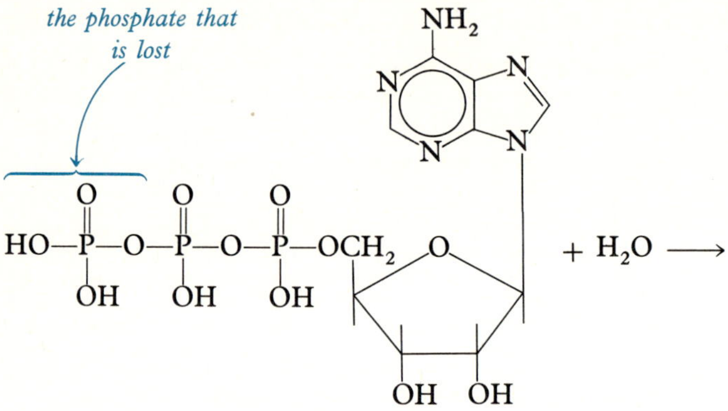

adenosine triphosphate (ATP)

high energy

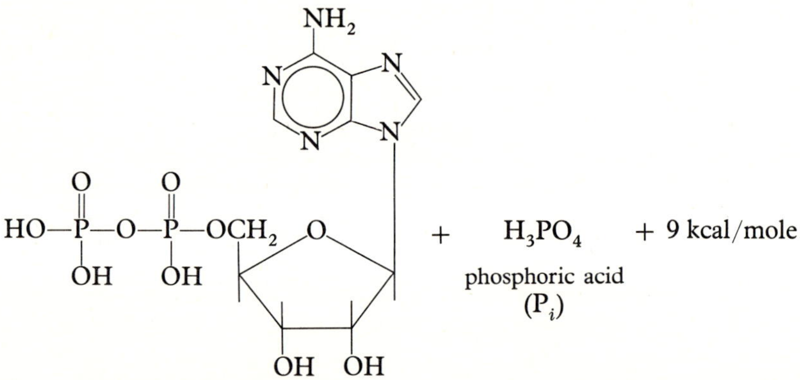

adenosine diphosphate (ADP)

low energy

or simply

$$ATP + H_2O \longrightarrow ADP + P_i + 9\ \text{kcal/mole}$$

Although the structures involved are complex, the principle of the hydrolysis reaction is simplicity itself. Energy is needed: ATP is hydrolyzed to ADP. Food is oxidized: ADP is converted back to ATP for later use.

Using energy:

$$\text{Relaxed muscle} + ATP \longrightarrow \text{Contracted muscle} + ADP + P_i$$

Storing energy:

$$\underset{\substack{\text{glucose}\\ \textit{from carbohydrates}}}{C_6H_{12}O_6} + 6\,O_2 + 38\text{ ADP} + 38\text{ P}_i \longrightarrow 6\,CO_2 + 6\,H_2O + 38\text{ ATP}$$

$$\underset{\substack{\text{palmitic acid}\\ \textit{from fats}}}{CH_3(CH_2)_{14}CO_2H} + 23\,O_2 + 130\text{ ADP} + 130\text{ P}_i \longrightarrow 16\,CO_2 + 16\,H_2O + 130\text{ ATP}$$

Note that the metabolism of one mole of glucose can theoretically yield 38 moles of high-energy ATP. Similarly, the metabolism of one mole of palmitic acid can theoretically yield 130 moles of ATP. (The principal difference between these two nutrients is that palmitic acid contains almost three times as many carbon atoms as glucose; therefore, more energy is produced when it is oxidized.) However, the number of ATP units produced by the oxidation of food is not actually as great as shown in the preceding two equations. Like all processes, biological oxidations are not 100 percent efficient. Of the potential energy that is available from the oxidation of food to carbon dioxide, the body retains only about half in the high-energy phosphate anhydride bond of ATP. The rest of the energy is given off as heat.

Sample Problem: In some biochemical reactions, ATP is hydrolyzed to adenosine monophosphate (AMP) and two phosphate groups. What is the structure of AMP?

Solution:

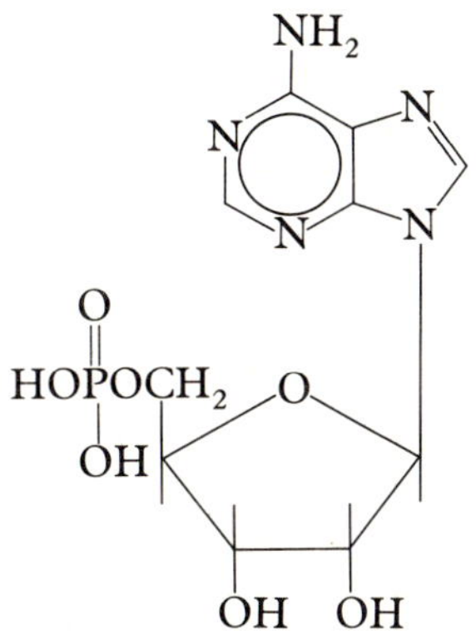

It is interesting that ATP itself is not present in any great quantity in muscle tissue. Instead, excess phosphate ions are held in reserve in a compound called *creatine phosphate*. Creatine phosphate and ADP react quickly to form ATP when it is needed by an organism. On the other hand, if the muscle cells already contain an abundance of ATP, then the reverse reaction occurs. In this way, the energy from ATP is readily stored, yet is also readily available.

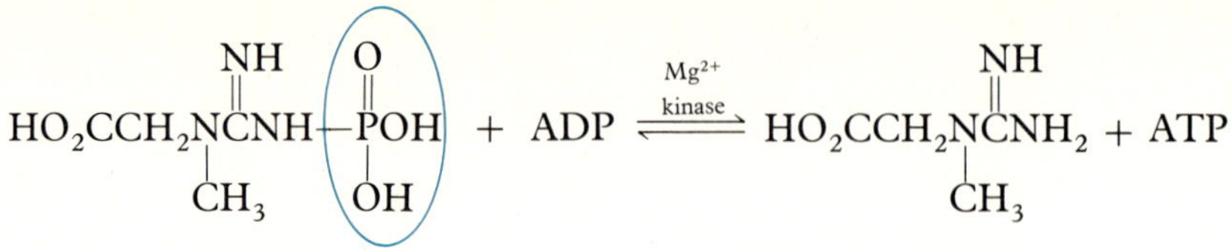

creatine phosphate creatine

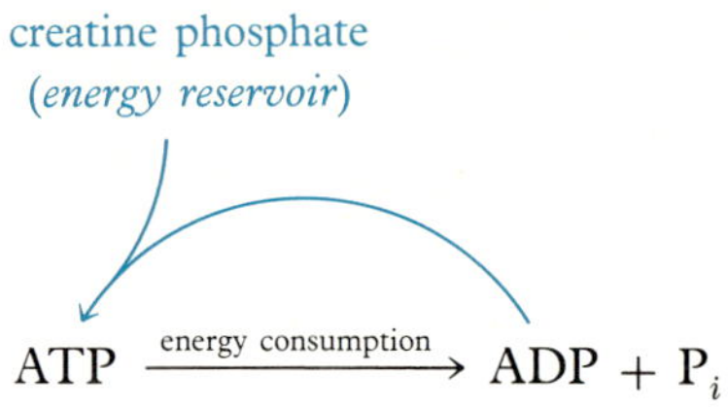

THE RESPIRATION CYCLE

Biological energy is obtained by the oxidation of nutrients in the cells, but in living systems glucose and other nutrients do not undergo reaction directly with oxygen gas. Instead, nutrients are broken down into fragments that are oxidized by other compounds, which are reduced in these reactions. These reduced compounds are reoxidized by yet other compounds, which, in turn, are reduced in the process. At the end of the line, we finally find a compound that reduces oxygen to water. In each oxidation and reduction reaction, the reduced product becomes in effect an electron carrier—a compound capable of being oxidized and giving electrons to the next compound in the sequence, much as a bucket brigade passes water buckets down the line to put out a fire. In the course of these reactions, energy that is released is trapped by the conversion of ADP to ATP. The whole system is referred to as the *cellular respiration cycle,* or the *electron-transport system* (ETS).

Conversion of a secondary alcohol to a ketone is a typical oxidation reaction that occurs in living systems as well as in the laboratory. Let us consider the respiration cycle in more detail and trace the results of the biological oxidation of an alcohol.

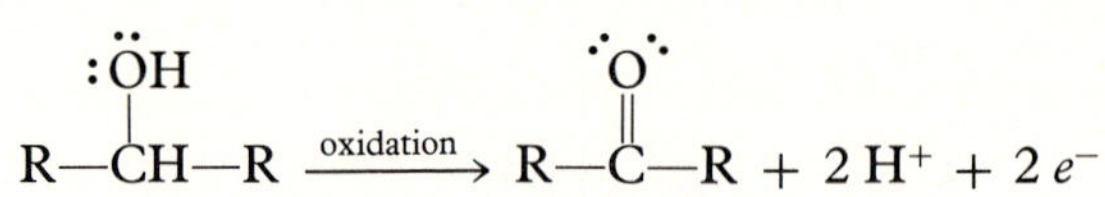

In this oxidation, two hydrogen ions and a pair of electrons are removed from the alcohol. These electrons cannot just leave, as the equation implies. Rather, as in all oxidation-reduction reactions, they must be accepted by another compound that is reduced. In a living system, compounds in the electron-transport system accept the electrons, and after a series of oxidation-reduction reactions the electrons are finally transferred to oxygen. The hydrogen ions in the above equation are also transferred to another compound, or in some cases they enter the surrounding cellular medium. At the end of the reaction sequence when oxygen is reduced to water, hydrogen ions are taken from the medium so that an overall ionic balance is maintained.

THE ROLES OF NAD⁺ AND FAD. The principal electron acceptors in the respiration cycle are two coenzymes: *nicotinamide adenine dinucleotide*, abbreviated NAD^+, and *flavin adenine dinucleotide*, abbreviated FAD.

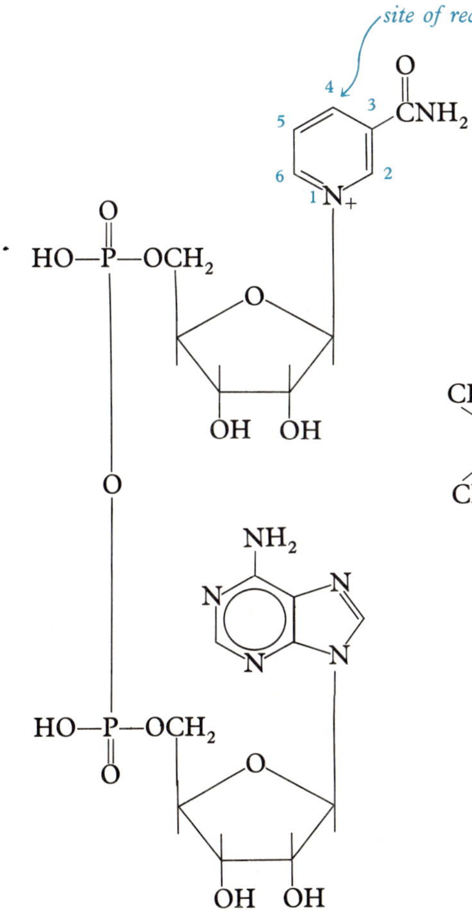

nicotinamide adenine dinucleotide
(NAD^+)

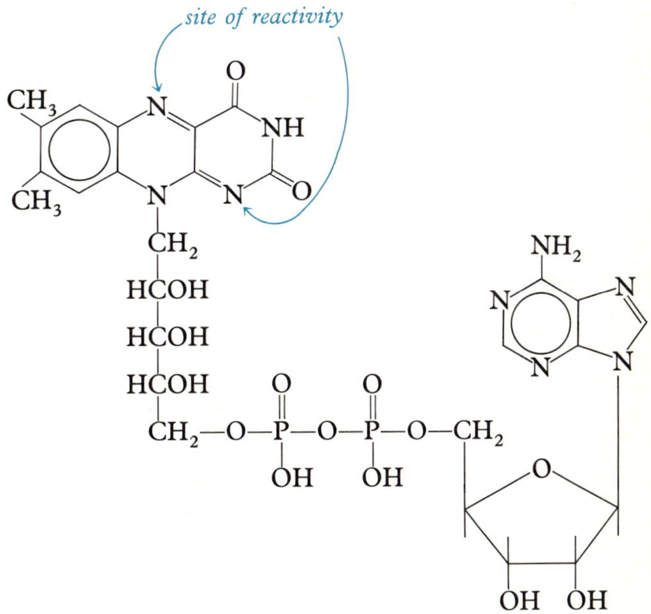

flavin adenine dinucleotide
(FAD)

Although NAD^+ and FAD look complex, the reactions they undergo are not. Each of these coenzymes has a working end, which is the only portion of the structure we need to consider. The working end of NAD^+ is nicotinamide while that of FAD is riboflavin. (Nicotinamide and riboflavin are both vitamins. Trace amounts of these two vitamins are necessary in the diet so that NAD^+ and FAD may be synthesized in the system and the processes of metabolism may be carried out.)

When a compound such as an alcohol is oxidized, NAD^+ is reduced to a compound called NADH. One proton and a pair of electrons from the alcohol go to position 4 of the pyridine ring (which already has one hydrogen). In NADH, the aromatic pi cloud of the pyridine ring is reduced to a diene, and the positive charge is lost.

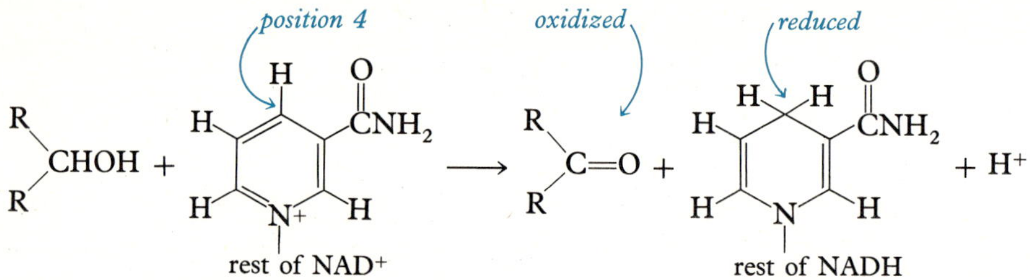

For simplicity in describing biochemical processes, these reactions can be written in an abbreviated form. The following symbols are taken to mean that when the alcohol is oxidized to a ketone, NAD^+ is reduced to NADH.

When an alcohol is oxidized to a ketone,
NAD^+ is reduced to NADH:

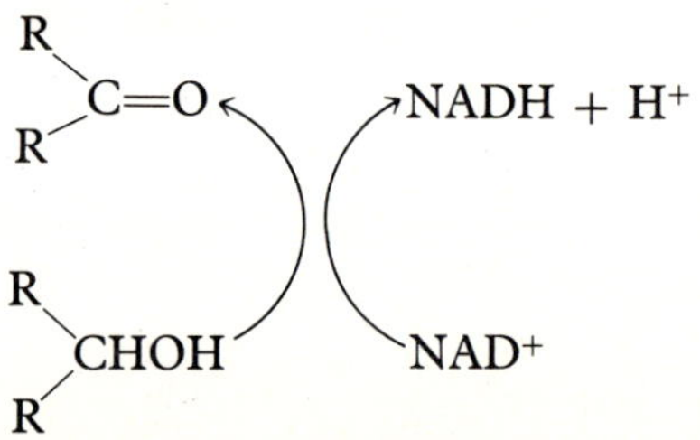

To keep the respiration cycle going, NADH must give up its hydrogen and electrons to some other compound. In other words, NADH must be oxidized back to NAD^+. The oxidizing agent in this case is FAD. When FAD reacts with NADH, the NADH is oxidized to NAD^+, and FAD is reduced to a compound called $FADH_2$. The net result is that electrons and hydrogen are passed from NADH to FAD.

$$\text{NADH} + \text{FAD} + \text{H}^+ \longrightarrow \underset{\textit{oxidized}}{\text{NAD}^+} + \underset{\textit{reduced}}{\text{FADH}_2}$$

In the reduction of FAD, H^+ (from the surrounding cellular medium) and H with a pair of electrons (from NADH) are transferred to the two double bonds circled. The two double bonds are reduced to one double bond.

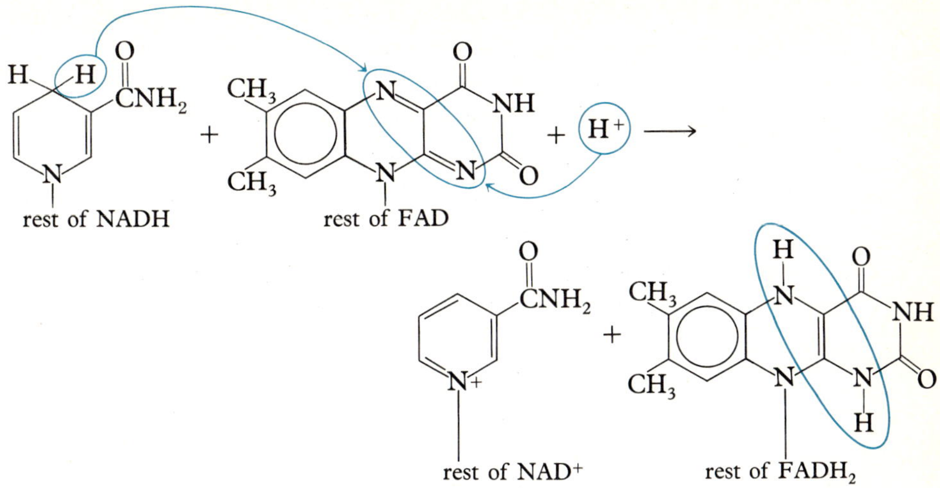

We can symbolize this reaction also in an abbreviated form and combine it with our previous abbreviated reaction:

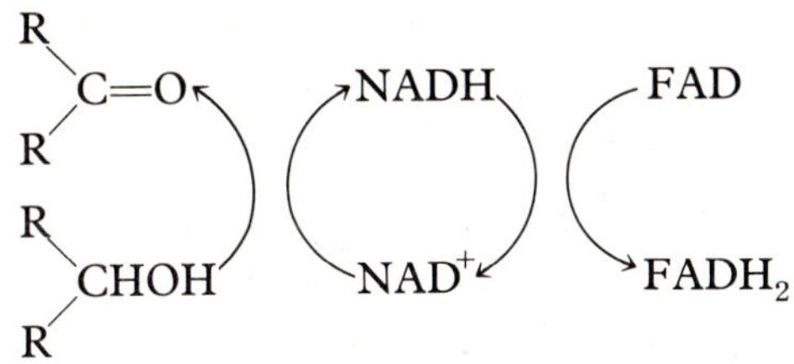

From $FADH_2$, the electrons and hydrogens pass to ubiquinone (coenzyme Q). From there, the electrons only are passed to a group of iron-containing structures, which are part of the *cytochrome system,* and finally to oxygen. The reactions of Fe^{2+} and ubiquinone were shown in Chapter 14. The entire sequence of reactions in the respiration cycle is shown in Figure 25-2.

Figure 25-2. The respiration cycle (also called the electron-transport system), showing also where ATP is synthesized from ADP.

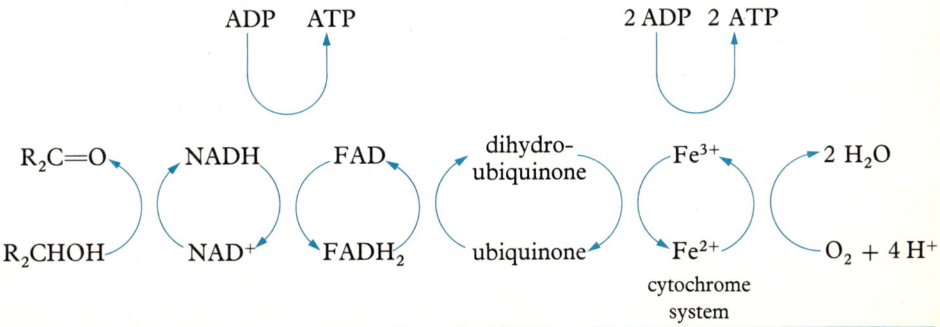

ATP SYNTHESIS IN THE RESPIRATION CYCLE. During the transport of the electrons to oxygen, energy is trapped at two steps—in the conversion of NADH to NAD^+ and in the cytochrome system. (See Figure 25-2.) For every pair of electrons transported (or for every oxygen atom reduced), three ATP units are generated from inorganic phosphate and ADP. We may write a net equation that shows three ATP units being produced for each NADH being oxidized. Of course, all the steps in the respiration cycle are not shown in this net equation.

Net reaction:

$$NADH + \tfrac{1}{2}O_2 + H^+ + 3\,ADP + 3\,P_i \longrightarrow NAD^+ + H_2O + 3\,ATP$$

Sample Problem: Which of the following equations represent oxidation reactions, and which reduction reactions?

(a) $Fe^{3+} \xrightarrow{\text{dihydroubiquinone}} Fe^{2+}$
(b) $ATP \longrightarrow ADP$
(c) $FADH_2 \longrightarrow FAD$
(d) $NAD^+ \longrightarrow NADH$

Solution: Reaction (c) is an oxidation because $FADH_2$ loses hydrogen. Reactions (a) and (d) are reductions. Reaction (b) is neither an oxidation nor a reduction; it is a hydrolysis reaction.

STUDY QUESTIONS

25-1. Label each of the following reactions as a digestion reaction or a metabolism reaction:

(a) phe-tyr-phe-ala $\xrightarrow{\text{carboxypeptidase}}$ phe-tyr-phe + ala

(b) phe-tyr-phe $\xrightarrow{\text{chymotrypsin}}$ 2 phe + tyr

(c) 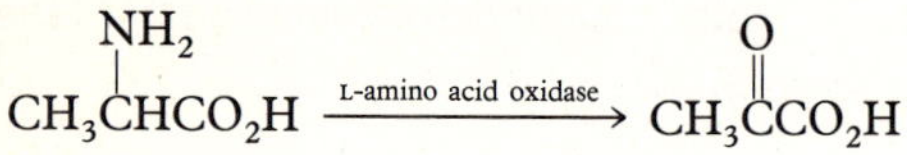

(d) sucrose $\xrightarrow{\text{invertase}}$ glucose + fructose

(e) glucose phosphate $\xrightarrow{\text{phosphoglucoisomerase}}$ fructose phosphate

(f) fructose phosphate $\xrightarrow{\text{aldolase}}$ $\xrightarrow{\text{triose phosphate isomerase}}$ 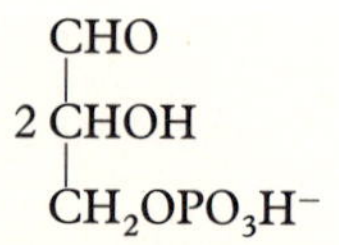

25-2. In which of the following reactions is energy liberated? In which is energy consumed?

(a) 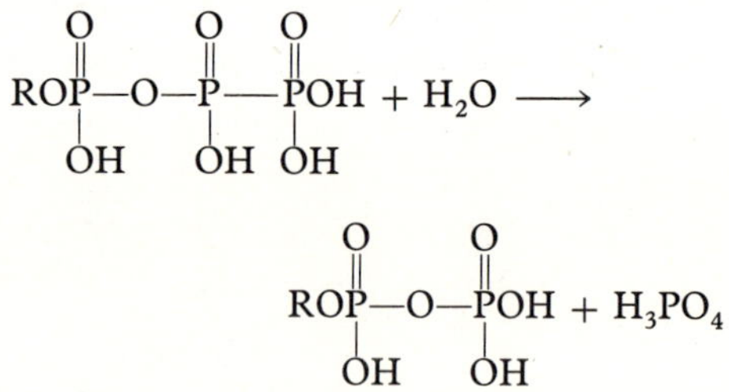

(b) 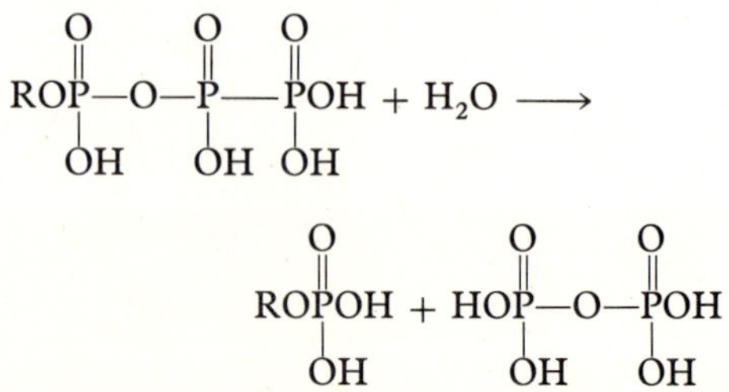

(c) 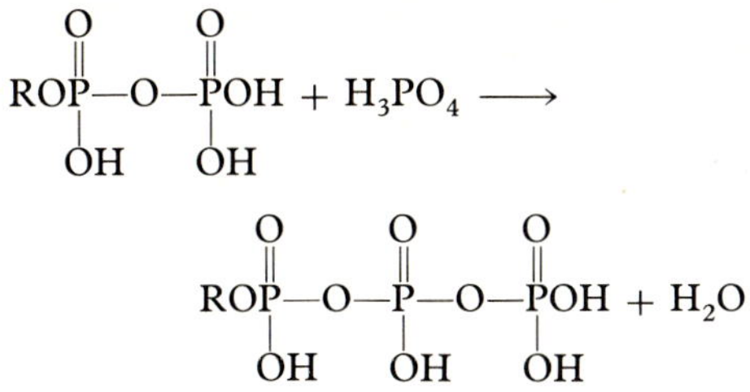

25-3. Which reaction occurs when a person is running?

(a) 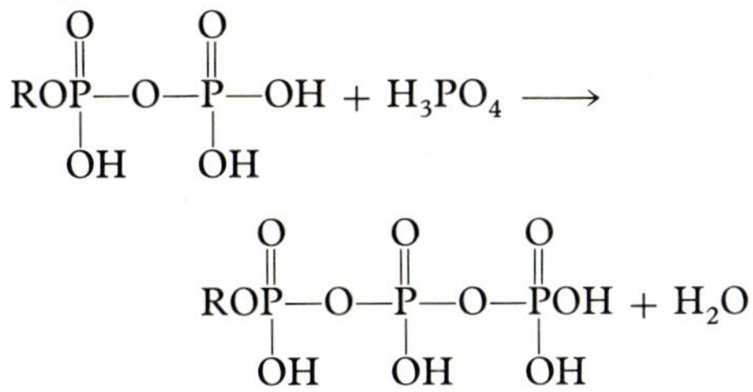

(b) 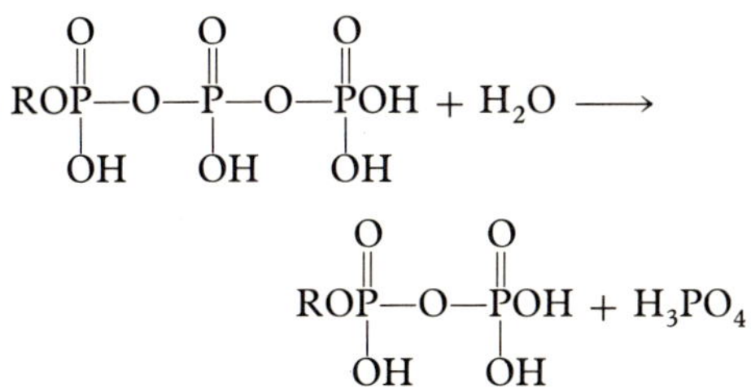

25-4. Which of the following phrases refer to ATP?

(a) a high-energy compound
(b) can be synthesized from ADP and creatine phosphate
(c) yields energy when it is hydrolyzed
(d) can be synthesized from ADP and P_i when extra energy is available

25-5. Which compound provides more energy per gram, tristearin or amylose?

25-6. Which compound provides more energy per mole, tristearin or glucose?

25-7. Classify each of the following reactions as oxidation or reduction:

(a) ubiquinone $\longrightarrow$ dihydroubiquinone

(b) FAD $\longrightarrow$ $FADH_2$

(c) 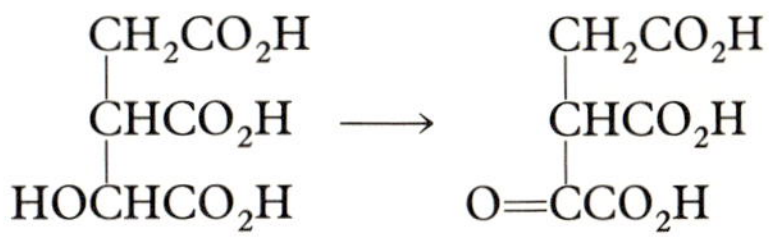

citric acid oxalosuccinic acid

(d) 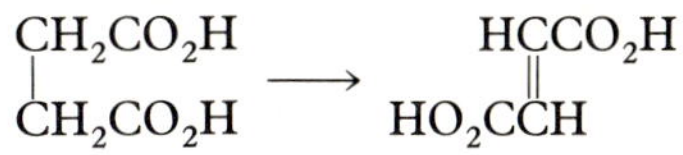

succinic acid fumaric acid

25-8. Complete the following equation:

$$NADH + FAD + H^+ \longrightarrow$$

25-9. Give the equation that shows the oxidation of NADH by FAD. (Show the structures of the "working ends" of these compounds.)

25-10. Complete the following equation:

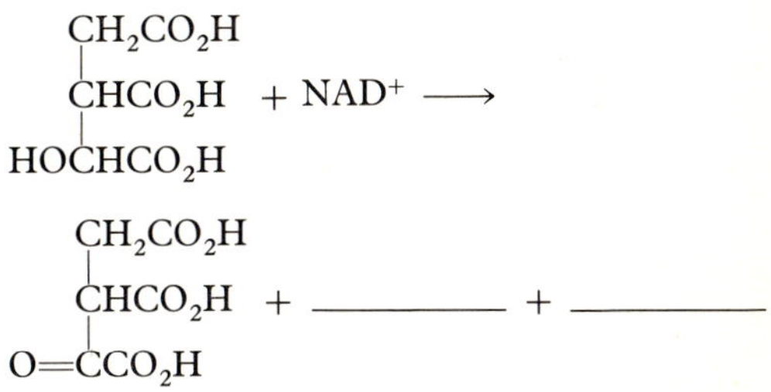

25-11. How many units of ADP are converted to ATP as a result of the reaction in the preceding problem?

CHAPTER 26 METABOLISM OF CARBOHYDRATES

OBJECTIVES

Digestion of Carbohydrates *List the steps in the digestion of starch. * Describe the structure and function of glycogen. * Write word equations illustrating glycogenesis and glycogenolysis.*

Blood Glucose Level *Tell how the level of glucose in the blood is controlled. * Describe the effects of ingestion of a large amount of carbohydrates by a normal person. * Define hypoglycemia and hyperglycemia. * Describe the effects of an inadequate insulin supply, as in diabetes mellitus.*

Galactosemia *Describe the cause and symptoms of galactosemia.*

Metabolism of Glucose *Write word equations showing the differences between the anaerobic and aerobic sequences of glucose metabolism. * Define glycolysis. * Describe the lactic acid cycle. * Describe how glucose is metabolized to pyruvic acid. * Write the equation for the conversion of pyruvic acid to the acetyl group in acetylcoenzyme A. * Describe the Krebs cycle.*

We have already considered the chemical structures of the major classes of food (carbohydrates, proteins, and fats). Now let us look at the chemical reactions these compounds undergo in living things. In this chapter, we will discuss the fate of carbohydrates after they have been ingested by an animal. In Chapter 27, we will consider the metabolism of fats and proteins.

DIGESTION OF CARBOHYDRATES

We obtain dietary carbohydrates principally from the starches and sugars found in grains, fruits, and vegetables and also from sucrose (table sugar) and lactose (in milk).

Starches are hydrolyzed initially to a mixture of small polymeric carbohydrates (called *limit dextrins*) and to maltose by the enzyme amylase (also called *ptyalin*) found in the saliva. Further digestion of starches occurs in the small intestine (not the stomach) and is catalyzed by enzymes originating in the pancreas and in the intestinal walls. (The intestines also secrete the enzymes sucrase and lactase, which catalyze the hydrolysis of sucrose and lactose.) The end products of the digestion of starch and sugars are monosaccharides (primarily glucose), which are absorbed into the bloodstream through the intestinal wall.

$$\text{starch} \xrightarrow[\text{(mouth)}]{\text{amylase}} \begin{matrix}\text{dextrins}\\ + \\ \text{maltose}\end{matrix} \xrightarrow[\text{(small intestine)}]{\text{maltase and dextrinase}} \text{glucose}$$

Once in the bloodstream, the glucose may be carried to cells and catabolized to CO_2, H_2O, and energy. How this occurs will be discussed shortly. An alternative pathway is that of storage; excess glucose can be polymerized into *glycogen,* a polysaccharide of 1500 to 20,000 units of 1,4-α-D-glucose, which is stored in the liver and muscles for future use. Glycogen resembles amylopectin in structure, but has even more cross-linking. The synthesis of glycogen from glucose or other compounds is called *glycogenesis* (from *glycogen* and *genesis,* meaning "the creation of glycogen").

When the body has need of glucose, enzymes break the glycogen back down to glucose. This process is called *glycogenolysis* (from *lysis,* meaning loosening or breaking).

If the blood contains adequate glucose and if the glycogen stores are also adequate, the excess glucose is converted to fat and other compounds such as steroids.

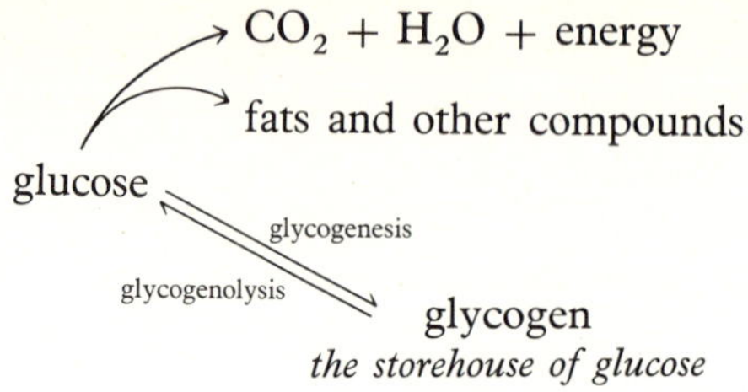

BLOOD GLUCOSE LEVEL

Many important tissues of the body, such as the brain, blood cells, and kidney, obtain virtually all of their energy from the oxidation of glucose; therefore, it is imperative that blood glucose levels be maintained. There are several hormones that regulate the various metabolism pathways available to glucose and thus the blood glucose level. We will mention just two here.

Epinephrine (also called *adrenaline,* see page 347) is an adrenal hormone that catalyzes hydrolysis of glycogen to glucose (that is, glycogenolyis) and increases the blood glucose level. *Insulin* is a pancreatic hormone that catalyzes the rapid utilization of glucose by cells. A large amount of insulin in the bloodstream causes a drop in the blood glucose level. In a normal person, the blood glucose level is kept in a state of balance at a fasting level of about 60–100 mg glucose/cc blood.

A *glucose tolerance test* can show if the body handles glucose in a normal way. After a normal person ingests 100 or more grams of carbohydrates, the blood glucose level rises. Then, as the pancreas supplies insulin to the bloodstream, the blood glucose level drops sharply. Finally, as the insulin is consumed by body processes, the blood glucose level returns to a normal fasting level.

Diabetes mellitus is a metabolic defect that is partially inherited. Approximately 3% of the population, principally the elderly, have some form of diabetes. A diabetic or a person with other pancreatic damage does not produce enough insulin to stimulate glucose metabolism. After ingestion of carbohydrates, the blood glucose level rises and stays high for a long period of time. (Other symptoms of diabetes will be discussed in Chapter 27.) Figure 26–1 shows graphs of normal and abnormal glucose tolerance.

An abnormally low level of blood glucose is called *hypoglycemia* and is characterized by dizziness and light-headedness. The symptoms arise from the brain cells being starved for glucose and thus not having enough energy to function properly. We may all suffer temporarily from these symptoms two or three hours after eating a high-carbohydrate snack, such as a candy bar. (Proteins and fats are only slowly

Figure 26–1. Blood glucose levels in glucose tolerance tests. (Shaded area indicates the normal fasting blood glucose range.)

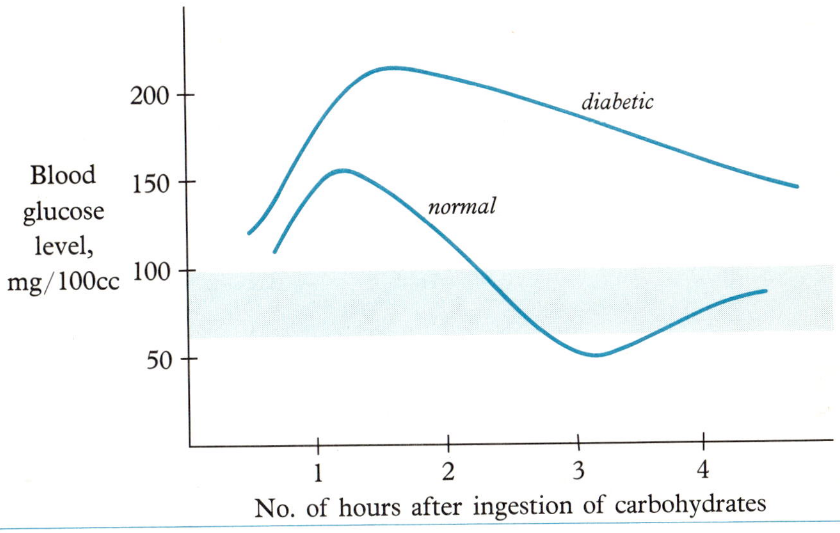

converted to blood glucose. These foods do not cause the rapid increase and decrease of the blood glucose level.)

An extreme case of hypoglycemia occurs when a diabetic has administered insulin, but has not eaten sufficient carbohydrates for the insulin to work on. In this case, blood glucose levels drop extremely low. The result is *insulin shock,* characterized by fainting, convulsions, shock symptoms, and even death.

A high blood glucose level, called *hyperglycemia*, arises normally after ingestion of a high-carbohydrate meal. A person with diabetes mellitus that is not controlled by diet, insulin, or other drugs has a chronic high glucose level in the blood. A blood glucose level of over 150 mg/cc is called the renal threshold of glucose. The term "renal threshold" refers to the upper limit of a substance present in the blood before the substance spills over into the urine. A person with a blood glucose level higher than the renal threshold exhibits *glucouria,* or glucose in the urine. Glucouria is, of course, one of the symptoms of diabetes.

Sample Problem: A *diabetic coma* can arise from extreme hyperglycemia. How would emergency treatment for a diabetic coma differ from that for insulin shock?

Solution: A person in a diabetic coma needs *insulin* to rid the blood of excess glucose. A person with insulin shock needs additional *glucose* in the blood.

GALACTOSEMIA

Lactose (milk sugar) is a disaccharide of glucose and galactose (see page 367). When lactose is digested, it is hydrolyzed enzymatically to glucose and galactose. In a normal metabolism, galactose is then converted to glucose in the cells and metabolized in the usual ways.

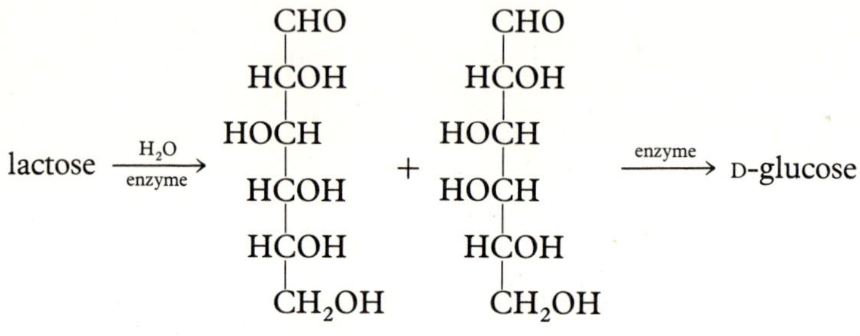

D-glucose D-galactose

Galactosemia is a hereditary disease observed in infants and is characterized by high levels of galactose in the blood and urine. The symptoms include vomiting when fed milk, failure to gain weight, and eventually dwarfism and mental retardation, sometimes leading to death. The cause of galactosemia is a deficiency of the enzyme that catalyzes the conversion of galactose to glucose. The treatment of this condition is the restriction of milk from the diet of affected infants. As these children grow, they may develop alternate metabolic paths such that they can metabolize galactose.

METABOLISM OF GLUCOSE

Now let us turn our attention to how glucose is catabolized. Two principal paths of glucose catabolism are observed. In the absence of oxygen, glucose undergoes catabolism by an *anaerobic path*. In the presence of oxygen, glucose is catabolized in a process called *aerobic oxidation*.

ANAEROBIC METABOLISM. The anaerobic path of glucose metabolism (also called the *Embden-Meyerhof path*) is used by the body in times of physical stress (such as running fast) when energy is needed, but when not enough oxygen is available for aerobic oxidation. The anaerobic path results in the glucose from the blood (and from stored glycogen) being converted to lactic acid. This conversion is called *glycolysis*. (Contrast this term to *glycogenesis*, the creation of glycogen, and *glycogenolysis*, the hydrolysis of glycogen.)

Figure 26–2. The biological oxidation of glucose.

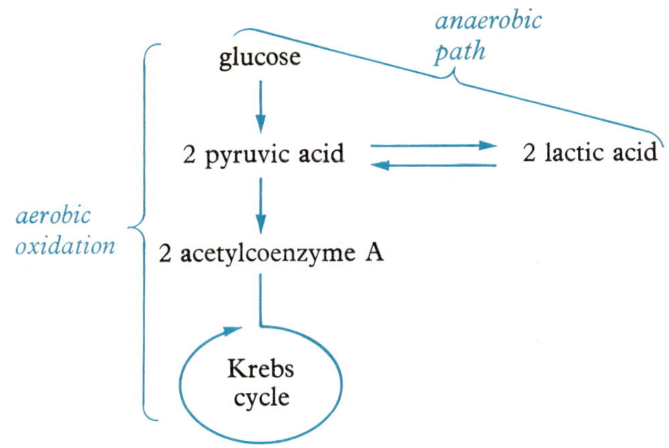

$$\text{glycogen} \underset{\text{glycogenesis}}{\overset{\text{glycogenolysis}}{\rightleftharpoons}} \text{glucose} \overset{\text{glycolysis}}{\rightleftharpoons} \underset{\text{pyruvic acid}}{CH_3\overset{\overset{\displaystyle O}{\|}}{C}CO_2H} \rightleftharpoons \underset{\text{lactic acid}}{CH_3\overset{\overset{\displaystyle OH}{|}}{C}HCO_2H}$$

The production of lactic acid is a dead-end as far as muscular energy is concerned. Glycolysis has a low energy yield compared to the aerobic oxidation pathway that we will discuss shortly. (The anaerobic sequence results in the net conversion of only two ADP units to ATP, compared to 38 ADP $\longrightarrow$ 38 ATP for the aerobic sequence!)

In the anaerobic sequence, lactic acid builds up in the muscle cells. This build-up is part of an *oxygen debt,* or insufficient oxygen. Heavy breathing and cessation of exercise restore the system to normal. The body is efficient and does not waste the lactic acid, which is eventually carried away by the blood to the liver. As soon as oxygen is again available, the lactic acid is metabolized. In the liver, part of the lactic acid is oxidized to CO_2, H_2O, and enough energy to convert the remaining lactic acid back to glucose and ultimately glycogen. This is one more example of glycogenesis. Figure 26–3 depicts this lactic acid cycle.

AEROBIC METABOLISM. As we have said, the anaerobic sequence of glucose metabolism has a low energy yield. Most of the energy obtained from dietary glucose comes from its aerobic oxidation, which results in CO_2, H_2O, and energy. The entire aerobic metabolism of one mole of glucose results in the conversion of 38 moles of ADP to ATP.

Figure 26-3. The lactic acid cycle.

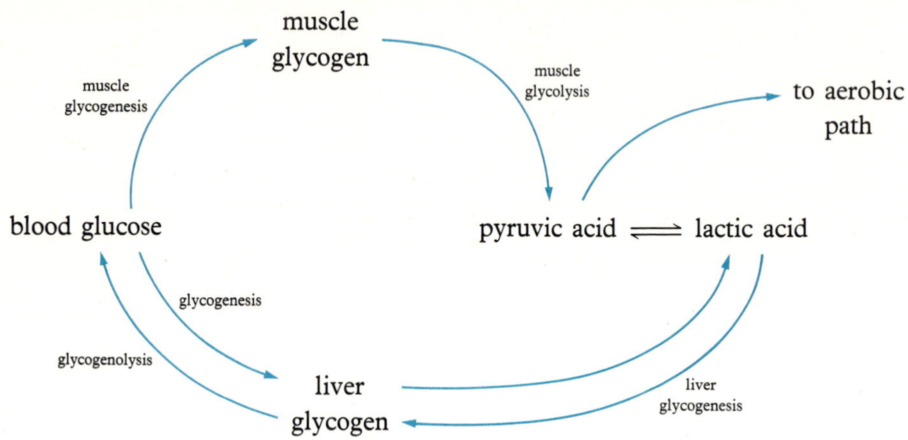

The first principal reaction sequence in either the aerobic path or anaerobic path is the conversion of glucose to pyruvic acid. The flow diagram in Figure 26-4 shows how this conversion is accomplished. Note that a net gain of eight ATP units is produced for each glucose unit in this sequence.

Sample Problem: Complete the following general equation, showing the structures of the intermediate and the product. (This type of reaction is found in the conversions of glucose 6-phosphate to fructose 6-phosphate and of dihydroxyacetone phosphate to glyceraldehyde 3-phosphate in Figure 26-4.)

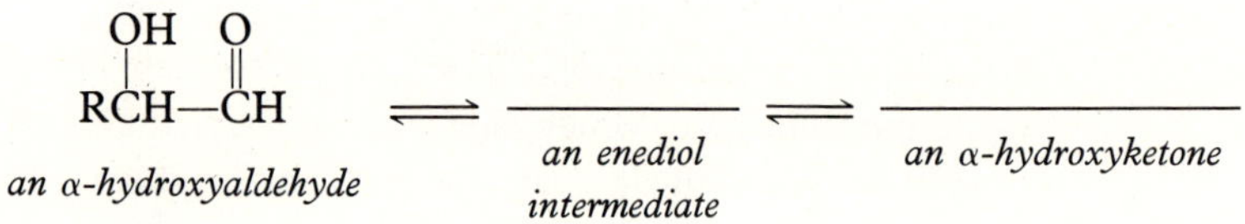

Solution:

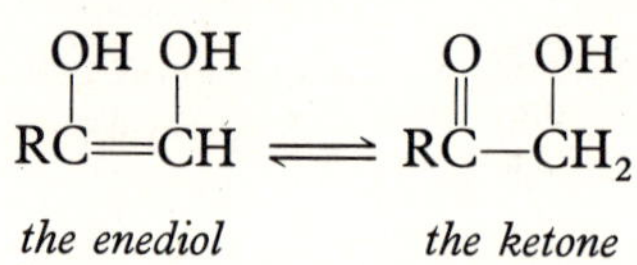

Let us summarize the reactions in Figure 26-4 in words:

$$\text{glucose} \xrightarrow{\text{ATP}} \text{glucose 6-phosphate} \longrightarrow \text{fructose 6-phosphate} \xrightarrow{\text{ATP}}$$

fructose 1,6-diphosphate ⟶ two glyceraldehyde 3-phosphate —(6 ATP)⟶

two glyceric acid 1,3-diphosphate —(2 ATP)⟶ two glyceric acid 2-phosphate —(2 ATP)⟶ two pyruvic acid

Conversion of glucose to pyruvic acid is only the first phase in aerobic oxidation. We are still a long way from carbon dioxide. The further oxidation of pyruvic acid will produce another thirty units of ATP.

Figure 26-4. Conversion of glucose to two molecules of pyruvic acid. Ⓟ in these formulas means a phosphate group, $—PO_3H_2$.

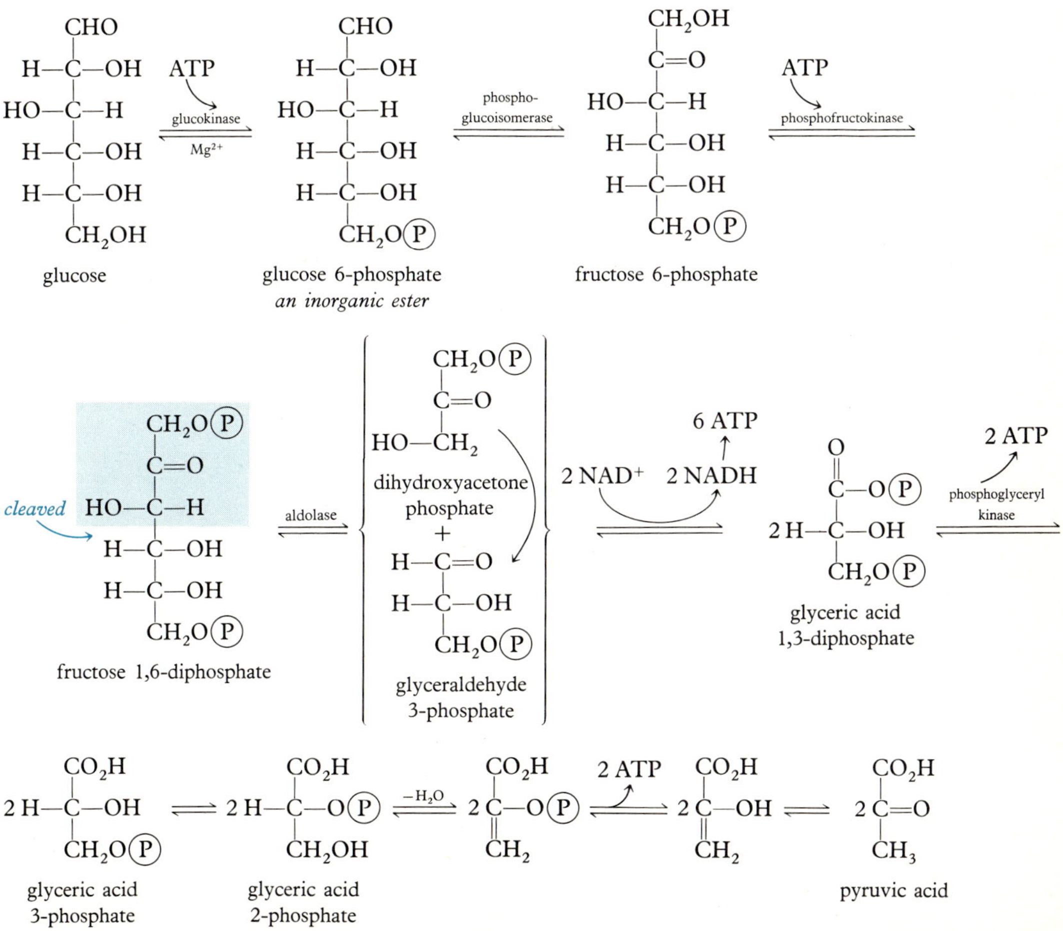

ACETYLCOENZYME A. The reaction of pyruvic acid with coenzyme A to yield acetylcoenzyme A is the second phase of aerobic oxidation. The structure of coenzyme A is shown on page 308. Recall that this compound is a thiol that can form a thioester.

Sample Problem: Write a general equation for the esterification of a carboxylic acid and a thiol.

Solution:

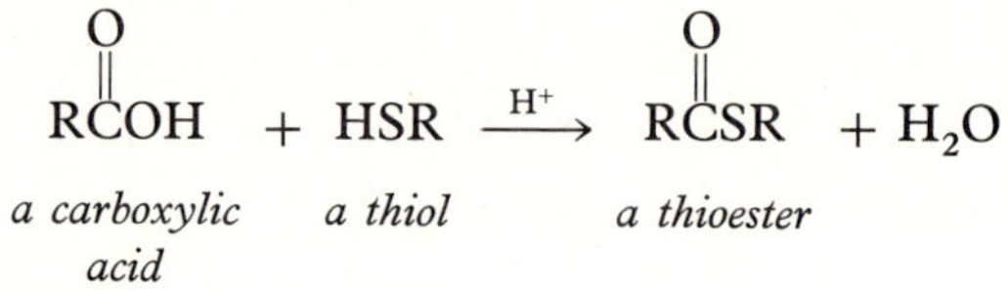

In the aerobic metabolism path, pyruvic acid undergoes reaction with coenzyme A to form acetylcoenzyme A and carbon dioxide. This reaction involves an oxidation of $-CO_2H$ to CO_2 by NAD^+.

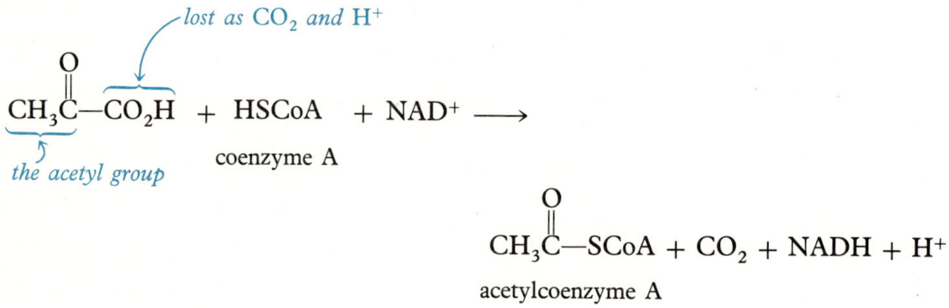

Each NADH molecule results in three ADP units going to ATP. Since *two* pyruvic acid molecules have arisen from one glucose molecule, a total of six ATP units is produced at this step.

THE KREBS CYCLE. The final phase in the oxidation of glucose is a series of enzyme-catalyzed reactions known as the *Krebs cycle,* the *citric acid cycle,* or the *tricarboxylic acid cycle.* This cycle is outlined in Figure 26-5. In this cycle, the remaining four carbons from the glucose (in the form of two acetyl groups in two molecules of acetylcoenzyme A) are oxidized to carbon dioxide.

The acetyl group from the acetylcoenzyme A is condensed with oxaloacetic acid to yield citric acid (step 1). Citric acid, in turn, is isomerized and oxidized to oxalosuccinic acid (steps 2 and 3). This β-keto acid undergoes decarboxylation (loss of CO_2) (step 4) and the product, α-ketoglutaric acid, undergoes oxidation and loss of carbon dioxide (step 5). In steps 4 and 5, two carbons are lost as carbon dioxide. Since each glucose molecule produced two acetylcoenzyme A units, a total of four carbon dioxide molecules have been lost in this cycle for each glucose. The actual carbon atoms lost may not be the same carbons as were in the original glucose, but these will

Figure 26-5. The Krebs cycle.

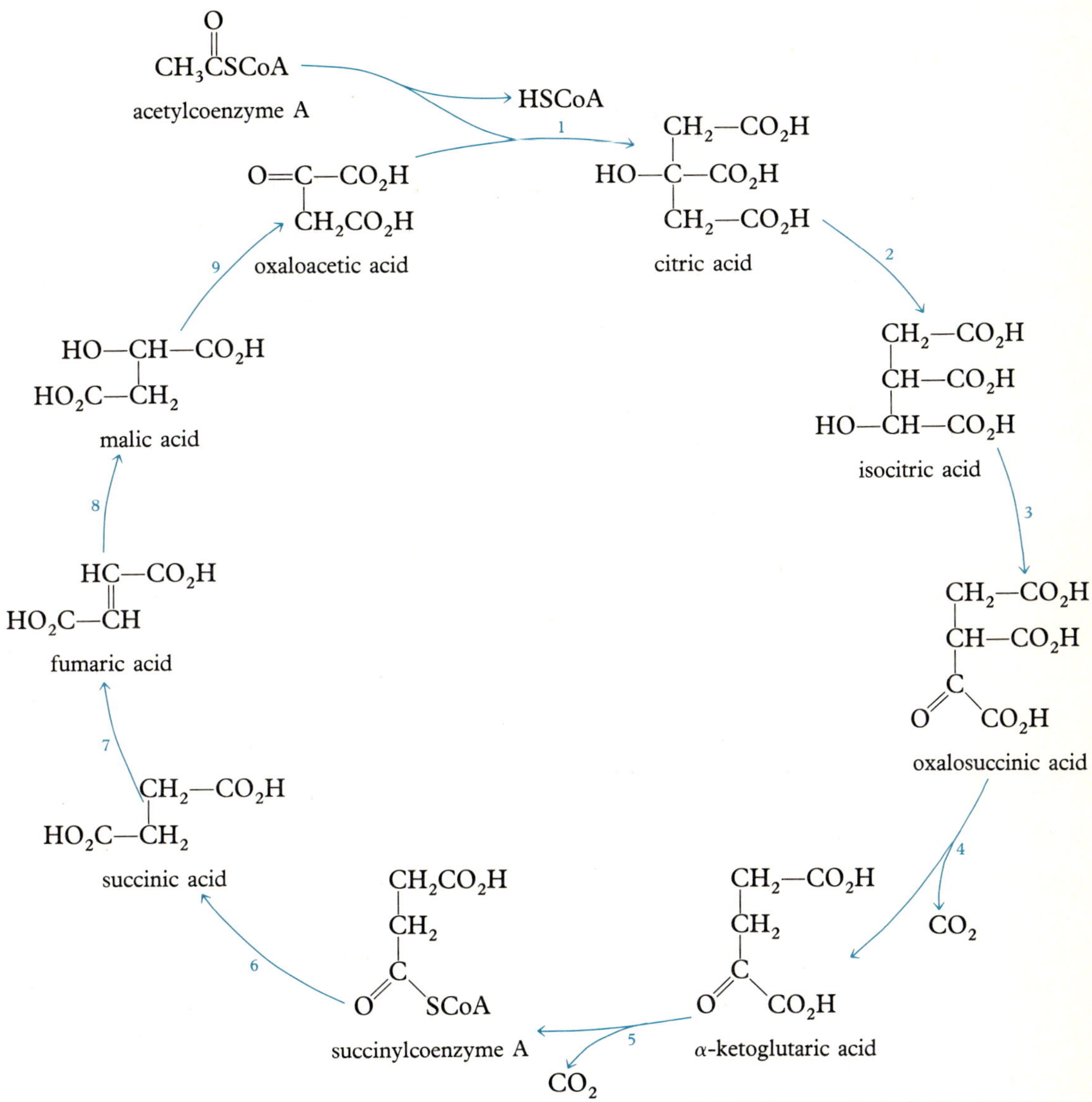

be lost in subsequent passes through the cycle. The rest of the cycle is involved in converting the succinyl group back to oxaloacetic acid so that another acetylcoenzyme A can be accepted and oxidized.

A total of 12 ATP units are produced in the Krebs cycle for each acetylcoenzyme A unit, or a total of 24 ATP units for the two acetyl groups that come from one unit of glucose. The total energy output from the aerobic oxidation of one molecule of glucose to carbon dioxide is sufficient to convert thirty-eight molecules of ADP to ATP, which can then be used by the organism at the time it requires energy for living. (See Table 26-1.)

Table 26-1. Total Energy Yield from the Aerobic Oxidation of Glucose

Reaction	ADP $\longrightarrow$ ATP
Glucose $\longrightarrow$ 2 Pyruvic acid	8
2 Pyruvic acid $\longrightarrow$ 2 Acetylcoenzyme A + 2 CO_2	6
2 Acetylcoenzyme A $\xrightarrow{\text{Krebs}}$ 2 Coenzyme A + 4 CO_2	24
(2 Oxaloacetic acid $\xrightarrow{\text{Krebs}}$ 2 Oxaloacetic acid)	
Total ATP	38

STUDY QUESTIONS

26-1. Describe in words how starch is digested and the various general metabolic paths available to the digested starch.

26-2. What is the effect on an animal organism if its system contains no insulin?

26-3. Under times of stress, epinephrine is secreted into the bloodstream by the adrenal gland. What is the effect of this secretion upon the blood glucose level?

26-4. Why should a person fast for several hours before a blood glucose test is carried out?

26-5. Match the following conditions with the definitions:

(a) hypoglycemia	(1) low blood glucose level
(b) hyperglycemia	(2) glucose in the urine
(c) glucouria	(3) spill-over of any substance into the urine
(d) renal threshold	(4) high blood glucose level

26-6. Which of the labelled steps in the following sequence is abnormal in galactosemia?

$$\text{lactose} \xrightarrow[\text{(a)}]{} \text{galactose} \xrightarrow[\text{(b)}]{\text{ATP galactokinase}}$$

$$\text{galactose 1-phosphate} \xrightarrow[\text{(many steps) (c)}]{\text{uridine triphosphate}}$$

$$\text{glucose 1-phosphate} \xrightarrow[\text{(d)}]{\text{(many steps)}} \text{pyruvic acid}$$

26-7. Define the *anaerobic sequence* and *aerobic sequence* of glucose metabolism.

26-8. In the conversion of glucose to pyruvic acid, which steps require the expenditure of ATP?

26-9. Write the equation, showing the intermediate, for the conversion that is brought about by the enzyme phosphoglucoisomerase.

26-10. What is the similarity between the isomerization of dihydroxyacetone phosphate and that of glucose 6-phosphate? (Write the formula of the intermediate.)

26-11. In the conversion of pyruvic acid to acetylcoenzyme A, is energy lost or gained by the system?

26-12. Circle the atoms that are lost as CO_2 in the formulas of oxalosuccinic acid and α-ketoglutaric acid. (See Figure 26-5.)

26-13. Refer to Figure 26-5 (but not to the textual description), and tell whether each of the following conversions is an oxidation, a reduction, an isomerization, an addition of H_2O, or a decarboxylation (loss of CO_2):

(a) citric acid $\longrightarrow$ isocitric acid
(b) isocitric acid $\longrightarrow$ oxalosuccinic acid
(c) succinic acid $\longrightarrow$ fumaric acid
(d) fumaric acid $\longrightarrow$ malic acid
(e) malic acid $\longrightarrow$ oxaloacetic acid

CHAPTER 27 METABOLISM OF FATS AND NITROGEN COMPOUNDS

OBJECTIVES

Metabolism of Fats *Write equations to show the hydrolysis of a triglyceride.* * *Describe the principal function of bile.* * *Describe the conversion of ingested triglycerides to an organism's tissue and depot lipids.* * *Diagram the fatty acid cycle.* * *Write equations that show the formation of ketone bodies from fatty acids and from acetylcoenzyme A.* * *Describe the causes and symptoms of ketosis and acidosis.* * *Write a flow diagram to show the biosynthesis of fatty acids.*

Metabolism of Nitrogen Compounds *List the ways that the amino acid pool is maintained.* * *Write an equation for the synthesis of new amino acids from old ones.* * *Diagram the urea cycle.* * *Show by an equation how an amino acid is oxidized to a keto acid and* NH_3. * *Exemplify by equations the catabolism of proteins.* * *Discuss the origin of uric acid in an animal system.* * *Describe the fate of replaced heme.*

METABOLISM OF FATS

Triglycerides are the most important class of lipid in our diet and are found in almost everything we eat—nuts, meat, milk, and so forth. A large part of excess carbohydrate in the diet also goes into the making of fat. So long as we do not accumulate too much fat, our fatty deposits serve useful purposes. They insulate our bodies against cold; they offer useful padding to bones and tissues; and they form an efficient energy reserve. Pound for pound, we can obtain more energy from our fat than we could from carbohydrate reserves.

DIGESTION OF FATS. The digestion of fats occurs in the small intestine, where the hydrolysis of triglycerides to diglycerides, monoglycerides, fatty acids, and glycerol is catalyzed by enzymes called *lipases*.

$$\begin{array}{l} CH_2O_2CR \\ | \\ CHO_2CR \\ | \\ CH_2O_2CR \end{array} + H_2O \xrightarrow{\text{lipase}} \begin{array}{l} CH_2O_2CR \\ | \\ CHO_2CR \\ | \\ CH_2OH \end{array} + RCO_2H$$

a triglyceride   *a diglyceride*   *a fatty acid*

Sample Problem: The preceding equation shows the hydrolysis of a triglyceride to a diglyceride. Write an equation for the hydrolysis of the diglyceride shown to a monoglyceride.

Solution:

$$\begin{array}{l} CH_2O_2CR \\ | \\ CHO_2CR \\ | \\ CH_2OH \end{array} + H_2O \xrightarrow[-RCO_2H]{\text{lipase}} \begin{array}{l} CH_2OH \\ | \\ CHO_2CR \\ | \\ CH_2OH \end{array} \quad \text{or} \quad \begin{array}{l} CH_2O_2CR \\ | \\ CHOH \\ | \\ CH_2OH \end{array}$$

monoglycerides

The digestion of triglycerides is aided by *bile*, which is formed in the liver, stored in the gall bladder, and released periodically into the small intestine. Bile is composed principally of salts of bile acids (which are steroids with acidic side chains), bile pigments, and cholesterol. The bile pigments, which are yellow to green, arise from the partial breakdown of hemoglobin in the liver.

In the intestines, bile serves to emulsify fats (break up large fat globules) so that hydrolysis proceeds smoothly. Bile also serves to remove cholesterol from the system. (Precipitation of the cholesterol in the gall bladder leads to one type of gallstone.)

Fats themselves cannot be absorbed through the intestinal wall, but the hydrolysis products are readily absorbed. As soon as they are absorbed, the fatty acids react with glycerol from the system's metabolic pool and re-form triglycerides. These new triglycerides become associated with proteins to form *lipoproteins.* They are then transported via the lymphatic system through the thoracic duct, into the bloodstream, and finally to the organs of the body. Within the organs, the triglycerides are again hydrolyzed and again resynthesized as they are deposited in the animal's own fatty tissue. Alternatively, the fatty acids and glycerol may be metabolized in the cells to carbon dioxide, water, and energy.

$$\underset{\textit{ingested}}{\text{Triglycerides}} \xrightarrow{+H_2O} \underset{\textit{in intestines}}{\left.\begin{matrix}\text{Fatty acids,}\\ \text{monoglycerides,}\\ \text{diglycerides}\end{matrix}\right\}} \xrightarrow{-H_2O} \underset{\textit{in lymph}}{\begin{matrix}\text{Re-formed}\\ \text{triglycerides}\end{matrix}} \longrightarrow$$

$$\text{Lipoproteins} \xrightarrow{+H_2O} \underset{\textit{in organs}}{\begin{matrix}\text{Fatty acids}\\ +\\ \text{glycerol}\end{matrix}} \begin{matrix}\nearrow \ CO_2 + H_2O + \text{energy}\\ \searrow \ \text{Triglycerides that form an animal's fat}\end{matrix}$$

An animal's lipids are of two types: *tissue lipids* (which are a functional component of the cells) and *depot lipids* (the stored fatty deposits used as padding and energy reserves). The depot lipids that are stored just beneath the skin are generally referred to as *adipose tissue.* The tissue lipids and the depot lipids are not stationary deposits; they are constantly being metabolized and redeposited. The half-life of the total lipid content in the human body is about 70 days.

THE FATTY ACID CYCLE. What happens to the fatty acids when the body converts them to energy? Like glucose, the fatty acids do not go directly to carbon dioxide and energy. Rather, a series of enzymatic reactions in the mitochondria, called the *fatty acid oxidation cycle,* result in the fatty acids being broken down two carbons at a time. The fatty acid is recycled through the pathway until all its carbon atoms are converted to acetylcoenzyme A.

Let us look at the reactions in the fatty acid cycle in more detail. The first reaction is the esterification of the fatty acid by coenzyme A. A unit of ATP is needed to provide the energy for the esterification. Instead of going to ADP, the ATP yields AMP (adenosine monophosphate) and pyrophosphate ions (composed of two phosphate units and designated PP_i).

$$RCH_2CH_2\overset{\overset{\large O}{\|}}{C}\text{—}\boxed{OH} + \boxed{H}SCoA + ATP \longrightarrow RCH_2CH_2\overset{\overset{\large O}{\|}}{C}SCoA + AMP + PP_i$$

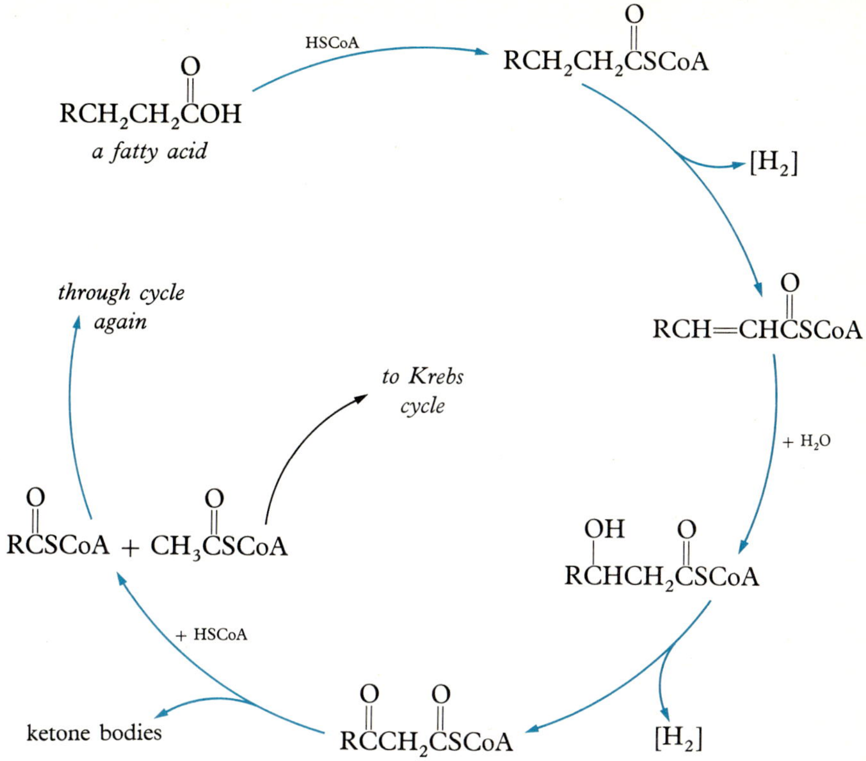

Sample Problem: Complete the following equation for the conversion of ATP and H_2O to AMP and PP_i:

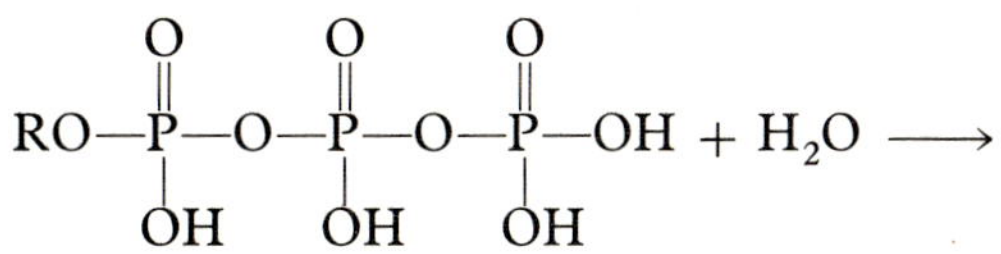

ATP

Solution:

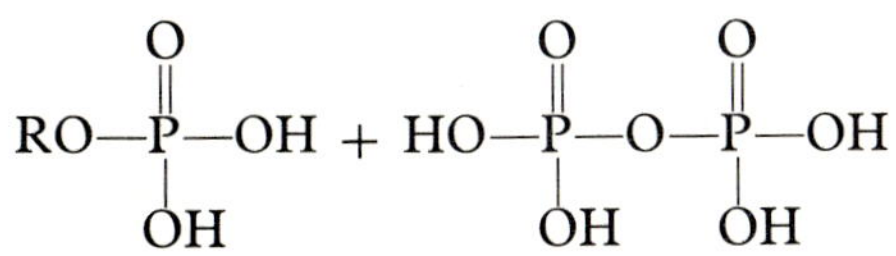

AMP PP$_i$

The second step in the fatty acid cycle involves a dehydrogenation (loss of hydrogen). This reaction is an oxidation reaction and requires FAD, which, in turn, is reduced to $FADH_2$.

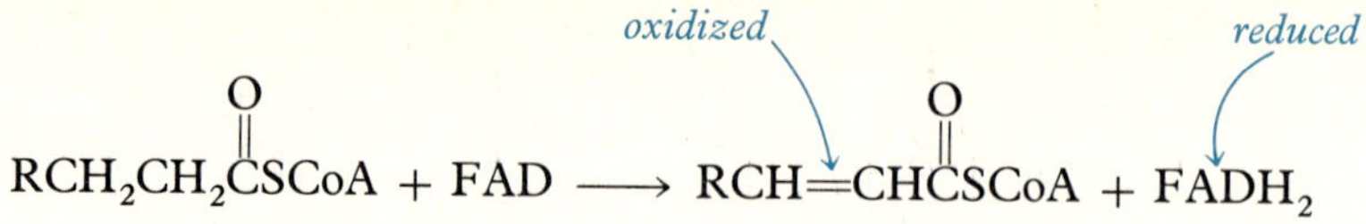

The next step is the addition of water across the double bond of the dehydrogenated fatty acid. This reaction is similar to the laboratory addition of water to an alkene. (See Chapter 13.)

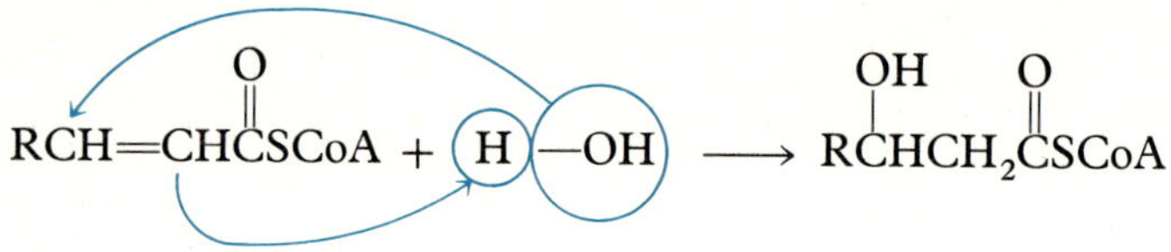

The hydroxyl group is then oxidized to a ketone group by NAD^+. This reaction leads to the formation of three ATP units when NADH enters the respiration cycle.

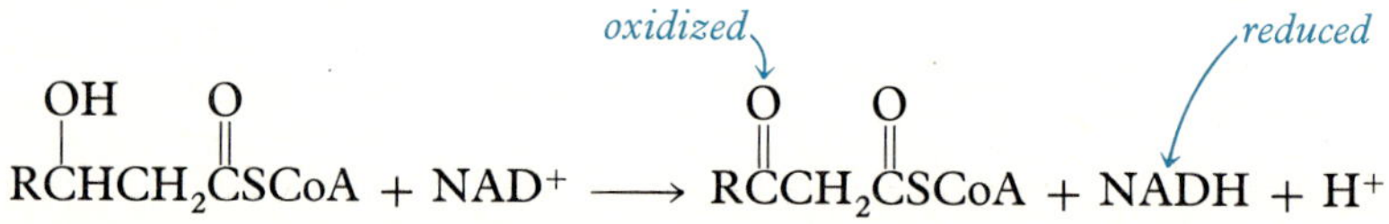

Another molecule of coenzyme A now enters the reaction sequence. The oxidized fatty acid molecule is cleaved, and the first two carbon atoms leave with the original coenzyme A as acetylcoenzyme A. Note that this reaction is an ester condensation in reverse. (See Chapter 17.) The new coenzyme A ester is formed from the rest of the fatty acid, which is now two carbons shorter.

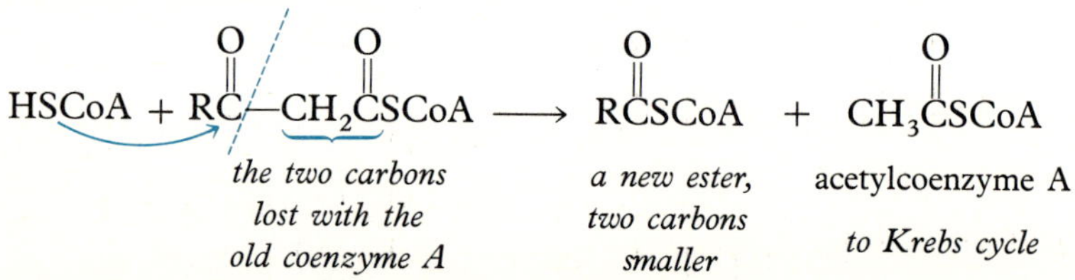

The new, shorter ester can enter the fatty acid cycle again and lose two more carbon atoms. Eventually, as it goes through the cycle over and over again, there will be nothing left of the fatty acid except for acetyl groups in acetylcoenzyme A.

The acetylcoenzyme A that is formed by each turn of the cycle can enter the Krebs cycle (Chapter 26), where the acetyl groups from the fatty acid will be oxidized to carbon dioxide. The oxidation of each acetyl group leads to 12 ATP units. In theory, a fatty acid such as stearic acid, $CH_3(CH_2)_{16}CO_2H$, would yield 147 ATP units. In reality, it is doubtful if this much ATP is actually gained, because the products of the fatty acid cycle can also be utilized in many other synthetic pathways.

$$CH_3(CH_2)_{16}CO_2H + 26\ O_2 + 147\ ADP + 147\ P_i \longrightarrow 18\ CO_2 + 18\ H_2O + 147\ ATP$$

Figure 27–1. The fatty acid cycle in relation to the Krebs cycle and the respiration cycle.

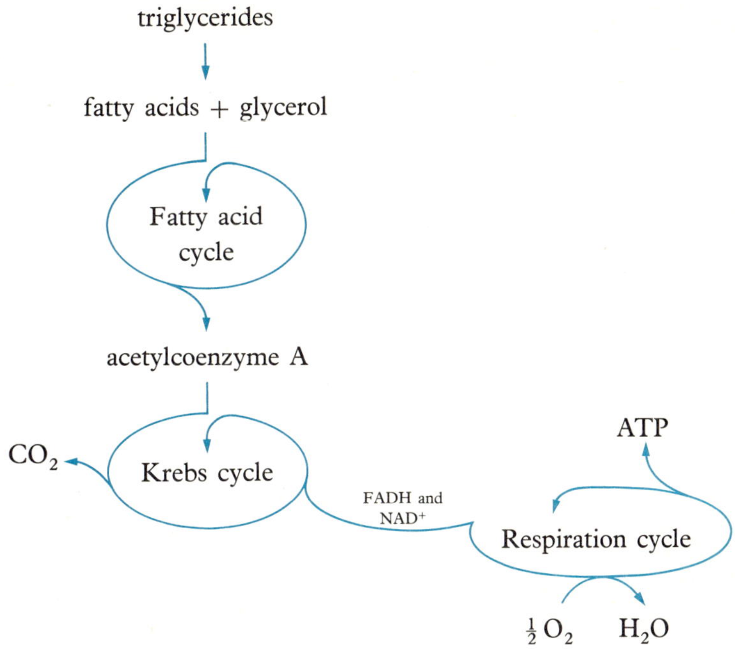

Figure 27–1 summarizes the overall metabolic fate of ingested triglycerides and shows the relationships between the fatty acid cycle, the Krebs cycle, and the respiration cycle.

KETOSIS AND ACIDOSIS. One of the products of the fatty acid oxidation cycle is a keto ester of coenzyme A called acetoacetylcoenzyme A. This keto ester is also synthesized in the liver by an ester condensation of acetylcoenzyme A. Cleavage of acetoacetylcoenzyme A results in three compounds that are collectively known as *ketone bodies:* acetone, acetoacetic acid, and β-hydroxybutyric acid (which does not actually contain a keto group).

Ketone bodies are a normal component of the bloodstream. The normal level for these compounds is about 1 mg/100 cc blood. A higher level of ketone bodies than normal is called *ketosis*. Ketonemia (ketone bodies in the blood), ketonuria (ketone bodies in the urine), and ketone breath (the smell of acetone on the breath) are all characteristic symptoms of ketosis.

Ketosis results from a deficiency of carbohydrates entering the Krebs cycle, which in turn may be due to starvation, fasting, a low-carbohydrate diet, or diabetes mellitus.

Formation of ketone bodies:

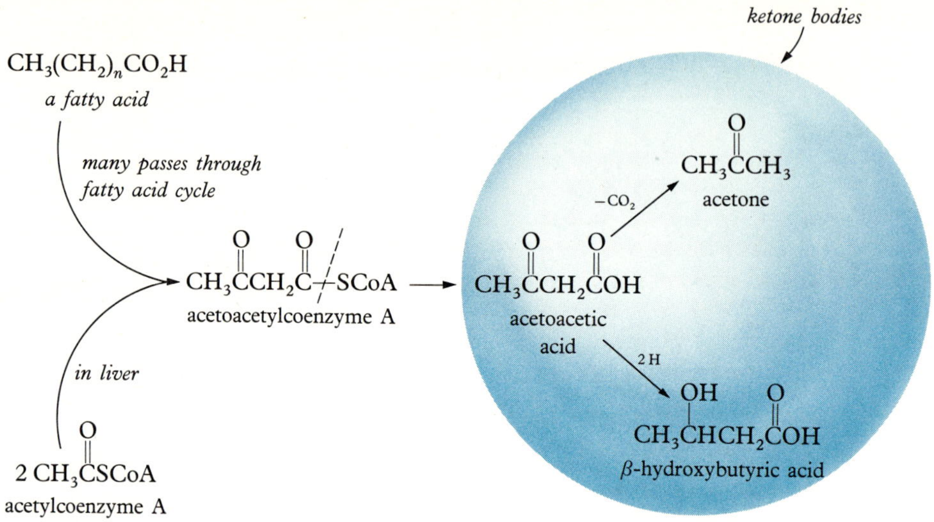

How does a lack of carbohydrates lead to ketosis? When the cells are not receiving sufficient energy from carbohydrate metabolism, the body calls upon its stores of fat to provide the needed energy. When large amounts of fatty acids are metabolized, there is a build-up of acetylcoenzyme A, which then leads to an excess of ketone bodies.

In a case of ketosis, the overabundance of acetoacetic acid and β-hydroxybutyric acid in the bloodstream may also cause a person to suffer from *acidosis,* a condition in which the pH of the blood approaches 7. (Chemically, pH 7 is considered neutral; however, a pH of 7 is on the acid side of the normal value of 7.4 for blood.) In cases of acidosis, the oxygen-carrying capacity of the blood decreases and the person may show symptoms of fatigue and can experience *dyspnea* (painful breathing). Dehydration may also occur because the kidneys use more water to excrete the excess ketone bodies. In serious cases of acidosis, coma and death can ensue.

BIOSYNTHESIS OF FATTY ACIDS AND TRIGLYCERIDES. The reversal of the fatty acid oxidation cycle can provide for the biosynthesis of fatty acids; however, most fatty acids are synthesized by an alternate route.

In the fatty acid synthesis cycle, acetylcoenzyme A is first converted to malonylcoenzyme A in a carboxylation reaction (addition of CO_2) that requires biotin (a vitamin), ATP, and CO_2.

$$\underset{\text{acetylcoenzyme A}}{CH_3\overset{O}{\overset{\|}{C}}SCoA} + CO_2 \xrightarrow[\text{acetylCoA carboxylase}]{\text{ATP, biotin}} \underset{\text{malonylcoenzyme A}}{HO\overset{O}{\overset{\|}{C}}CH_2\overset{O}{\overset{\|}{C}}SCoA}$$

Malonylcoenzyme A goes through a series of reactions with acetylcoenzyme A to yield fatty acids. (Because acetylcoenzyme A is the starting material in this sequence, fats can arise from the metabolism of carbohydrates.) A simplified scheme for the synthesis of a four-carbon carboxylic acid follows:

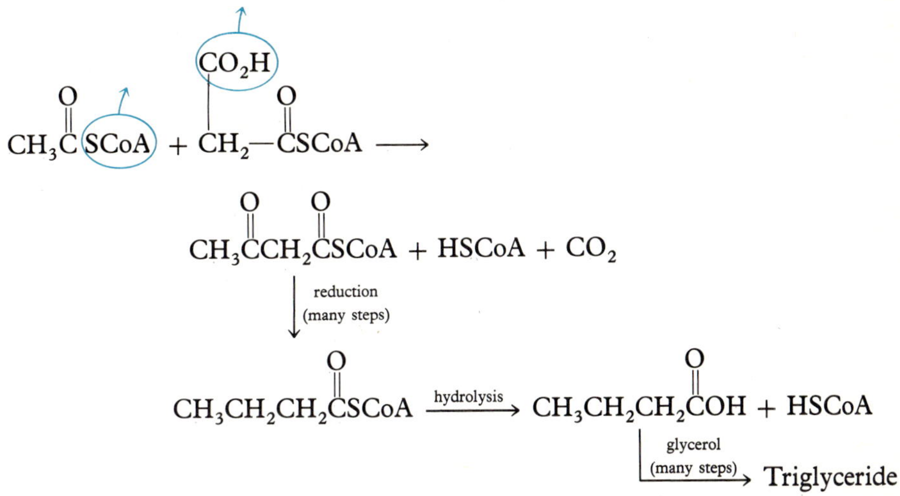

METABOLISM OF NITROGEN COMPOUNDS

Proteins form an important portion of our diet. We obtain proteins from meat, eggs, milk, and plant material, such as soybeans and nuts. In the body, proteins are broken down to amino acids, which are then used to build the system's own proteins and to synthesize other amino acids, enzymes, some hormones, nucleic acids, and other nitrogenous material. Like carbohydrates and fats, proteins can also be used for energy.

DIGESTION OF PROTEINS. Digestion of proteins begins in the stomach. The lining of the stomach is protected by a layer of *mucopolysaccharides,* which are polysaccharide-protein complexes, so that it too is not digested. Enzymes called *polypeptide depolymerases* break down large protein molecules into smaller peptides in the stomach and to individual amino acids in the small intestine. The amino acids are then absorbed through the intestinal wall. This complete breakdown of proteins all the way to amino acids is important. The body's natural defense mechanisms are based upon elimination of foreign protein material (bacteria, for example) from the system. If peptides or small proteins were absorbed into the system from the intestine, the organism would produce antibodies to ward off the foreign protein and severe allergic reactions would be the result.

AMINO ACID POOL. Once absorbed, the amino acids become a part of the *amino acid pool* of the system, which is not a storage site, but simply the available amino acids in the system. The amino acid pool is maintained by ingested proteins and by the natural breakdown and turnover of the body's proteins. Approximately half of the total body protein is broken down and replaced every 80 days. Individual proteins vary dramatically in their half-lives. For example, the half-life for insulin is 6 to 9 minutes, while that of muscle protein is about 180 days.

The amino acids in the amino acid pool are used in a number of ways. The most important use of the amino acids is, of course, the synthesis of new proteins for the organism. This topic will be discussed in Chapter 28. The amino acids are also used for the synthesis of nonprotein nitrogen compounds such as creatine, nucleic acids, and heme. Excess amino acids may be eliminated directly or as other nitrogen-containing excretion products. They may also be used to synthesize non-nitrogen compounds such as keto acids. The paths leading to and from the amino acid pool are summarized in Figure 27-2.

TRANSFORMATION OF AMINO ACIDS. Depending upon the needs of the system, some amino acids are incorporated into growing protein molecules; others may be converted to different amino acids in the liver. One important class of reactions for interconversion of amino acids is the group known as the *transamination reactions* (page 293). By these reactions, keto acids (from carbohydrates or fats) can be converted to new amino acids. The new keto acid formed in the reaction can enter the Krebs cycle and be oxidized to carbon dioxide and water, or it may be used to synthesize glucose or fat.

Figure 27-2. Paths leading to and from the amino acid pool.

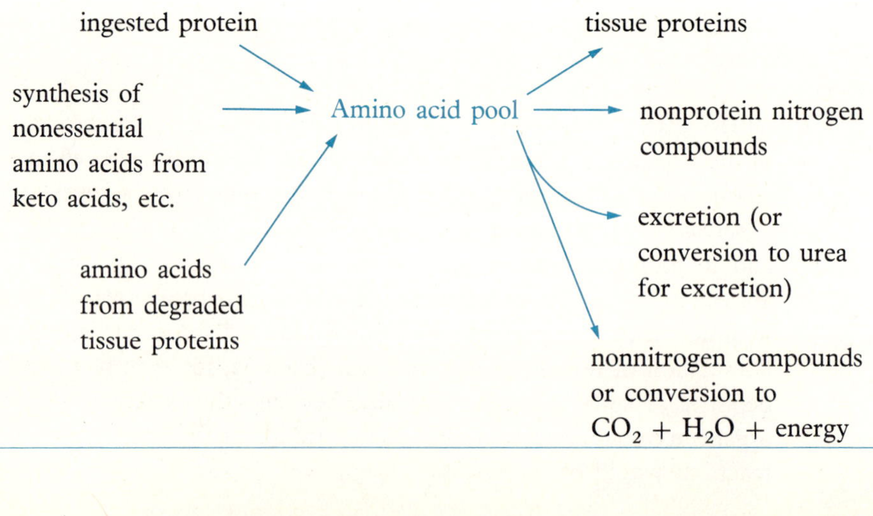

A transamination reaction:

$$\underset{\text{a keto acid}}{R\overset{O}{\overset{\|}{C}}CO_2H} + \underset{\text{an amino acid}}{R'\overset{NH_2}{\overset{|}{C}}HCO_2H} \xrightarrow{\text{transaminase}} \underset{\text{a new amino acid}}{R\overset{NH_2}{\overset{|}{C}}HCO_2H} + \underset{\text{a new keto acid}}{R'\overset{O}{\overset{\|}{C}}CO_2H}$$

Sample Problem: Write the net equation for the transamination reaction between glutamic acid (Table 23-1) and a keto acid to yield isoleucine and a new keto acid.

Solution:

$$\underset{\text{glu}}{HO_2CCH_2CH_2\overset{NH_2}{\overset{|}{C}}HCO_2H} + \underset{\text{a keto acid}}{CH_3CH_2\underset{CH_3}{\underset{|}{C}}H\overset{O}{\overset{\|}{C}}CO_2H} \longrightarrow$$

$$\underset{\text{a new keto acid}}{HO_2CCH_2CH_2\overset{O}{\overset{\|}{C}}CO_2H} + \underset{\text{ileu}}{CH_3CH_2\underset{CH_3}{\underset{|}{C}}H\overset{NH_2}{\overset{|}{C}}HCO_2H}$$

THE UREA CYCLE. Excess nitrogen from the ingestion of proteins must be excreted by a system. Fish and lower marine animals excrete unneeded nitrogen as ammonia. Birds, reptiles, and insects excrete nitrogen as uric acid. Higher animals, including human beings, excrete excess nitrogen principally in the form of urea. An adult male normally excretes about 30 grams of urea in a 24-hour period.

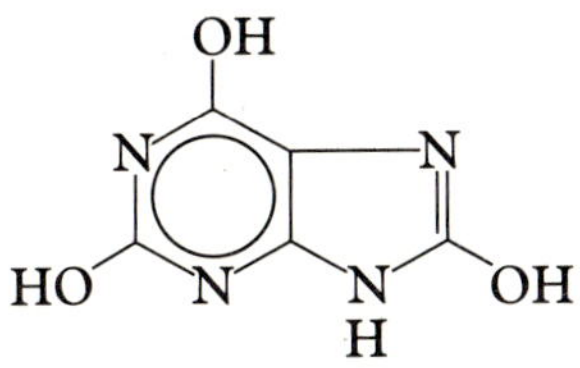

uric acid

$$H_2N-\overset{O}{\overset{\|}{C}}-NH_2$$

urea

The process of the conversion of amines and ammonia to urea, which occurs in the liver, is called the *urea cycle.* (See Figure 27-3.) The urea passes into the blood stream and is removed by the kidneys to be excreted in the urine.

Many amino acids undergo catabolism in the liver to yield keto acids (which can enter the Krebs cycle to yield energy) and ammonia, which in turn is converted to urea and excreted.

Figure 27-3. The urea cycle.

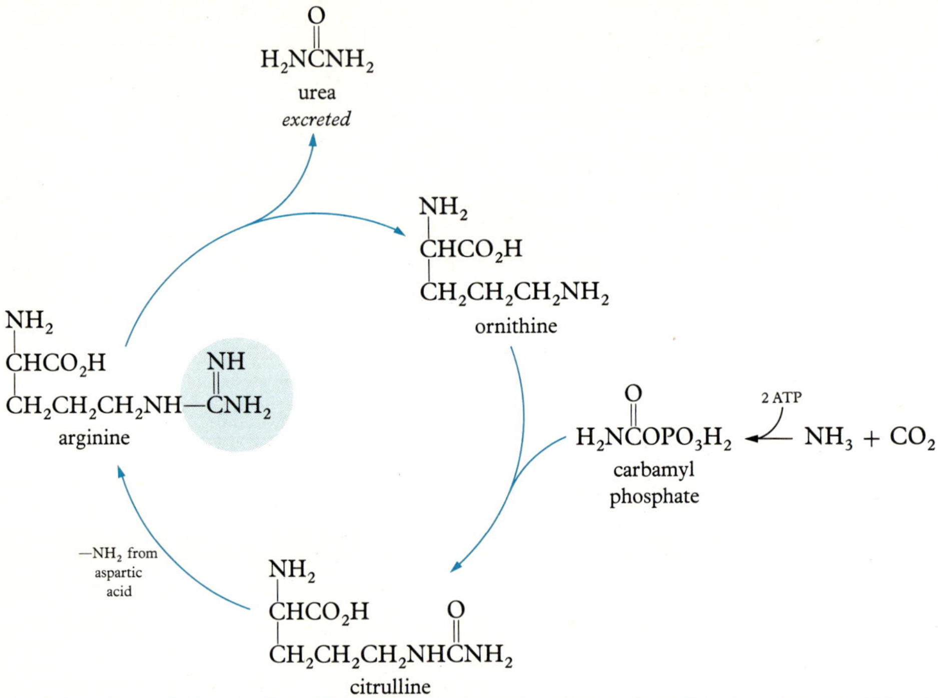

$$CH_3CH(NH_2)CO_2H + NAD^+ + H_2O \longrightarrow CH_3COCO_2H + NH_3 + NADH + H^+$$

alanine → pyruvic acid *to Krebs cycle*; NH_3 *to urea*; NADH *to respiration cycle*

METABOLISM OF PURINES AND PORPHYRINS. In addition to urea, some uric acid is also excreted by mammals. Uric acid is the end product of the metabolism of purine rings. Organ meats and legumes are rich in purines. In some people, sodium urate is deposited in the joints and causes great pain; this condition is known as *gout*.

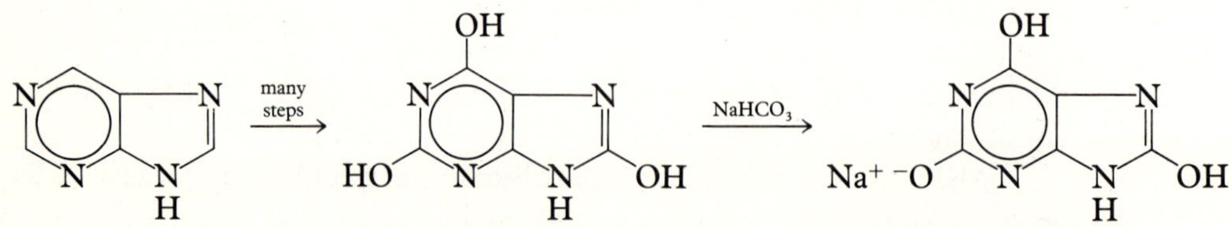

purine ring system — uric acid — a sodium urate

Heme in hemoglobin contains Fe^{2+} and the porphyrin ring system (page 187). When heme is metabolized, the Fe^{2+} ion combines with a protein to yield a complex called ferritin. In this manner, the iron is stored and reused at some later time.

In the metabolism of heme, the porphyrin ring system is converted in the liver to a reddish bile pigment called *bilirubin*. This compound is transferred by bile to the intestinal tract and is eliminated in the feces. The color of feces is due primarily to the presence of bilirubin. Some bilirubin is reabsorbed through the intestinal walls, then captured by the kidneys and eliminated in the urine. (The yellowish color of urine is also due primarily to bilirubin.)

If insufficient bile is produced by the liver to carry the bile pigments to the intestines, bilirubin builds up in the blood and gives a yellowish cast to the eyes and skin. This condition, called *jaundice*, may arise from liver disease, such as hepatitis.

STUDY QUESTIONS

27-1. (a) In what chemical form are fats absorbed through the intestinal wall? (b) In what chemical form are they transported within the system in the lymph? (c) In the blood?

27-2. At which steps in the fatty acid cycle does oxidation occur?

27-3. What compound is common to the fatty acid cycle and to the Krebs cycle?

27-4. Complete the following scheme:

$$C_{17}H_{35}CO_2H \xrightarrow{\text{HSCoA, ATP}} C_{17}H_{35}COSCoA \xrightarrow{\text{FAD}}$$

$$\underline{\text{(a)}} \xrightarrow{H_2O} \underline{\text{(b)}} \xrightarrow{\text{NAD}^+} \underline{\text{(c)}} \xrightarrow{\text{HSCoA}} \underline{\text{(d)}}$$

27-5. Branched chain fatty acids, such as the following acid, are not completely metabolized. Explain.

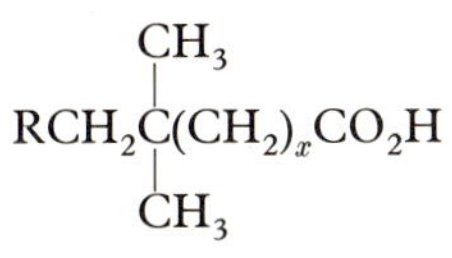

27-6. Describe (by equations) two ways that ketone bodies are produced. What is the underlying reason for the formation of an excess of these ketone bodies?

27-7. List the dietary causes of ketosis and acidosis.

27-8. Complete the following equation for hydrolysis:

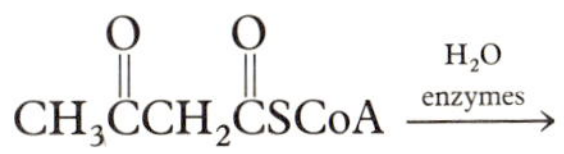

27-9. It has been found that most naturally occurring fatty acids have an even number of carbon atoms. Using the equation on page 431, show why this is true.

27-10. List the paths to and from the amino acid pool in an animal system.

27-11. Why is it important that proteins are completely broken down to amino acids before they are absorbed through the intestinal wall?

27-12. Using pyruvic acid and phenylalanine, write an equation that illustrates a transamination reaction.

27-13. Write equations for the biological conversion of arginine to citrulline.

27-14. What type of compound is metabolized and excreted as uric acid in an animal system? What are some dietary sources of these compounds?

27-15. Describe the fate of heme when it is broken down in the system.

27-16. Although Fe^{2+} ions are important in hemoglobin and in the cytochrome system, the dietary requirement of Fe^{2+} for an adult male is low unless he has suffered from internal or external bleeding. Explain.

CHAPTER 28 THE CHEMISTRY OF HEREDITY

OBJECTIVES

The Structure of DNA *Give the names and structures of the sugar and the bases in DNA. * Define nucleoside and nucleotide. * Describe with a word diagram how the component units are attached to each other to form a DNA chain. * Describe the hydrogen bonding between DNA chains.*

Replication of DNA *Diagram the replication of DNA.*

Structure of RNA *Diagram the structure of RNA.*

The Biosynthesis of Proteins *Diagram the synthesis of mRNA from DNA. * Define codon and anticodon. * Describe how amino acids are incorporated into growing protein molecules.*

Genetic Recombination *Describe how genes undergo recombination.*

Genetic Disorders *Describe the origin of genetic disorders, and list some examples.*

Viruses *Tell how a virus is reproduced. * Define the lysogenic state of viral DNA.*

Why is a dog a dog and not a cat? Why do some people have blue eyes and not brown? We need only look in the mirror to see that the synthesis of proteins is not a random process. From a chemical standpoint, how does the body know what particular types of protein to synthesize? How is this information transmitted from one generation to the next? The study of the chemistry of heredity is one of the most fascinating fields of research today. In recent years, great strides have been made in understanding the processes of heredity, growth, and life itself, but much of the workings of living systems still eludes our understanding.

It was recognized in the 19th century that the nucleus of a living cell contains stringy particles, which scientists called *chromosomes*. Human beings have 46 chromosomes in each cell. It was proposed as early as 1884 that the chromosomes are the transmitters of hereditary traits. It was later proposed that each chromosome is composed of a large number of *genes,* each of which directs the synthesis of a particular protein in the body.

Only in more recent years has it been discovered that chromosomes are composed of *nucleic acids* (so called because they come from the nucleus of the cell and are acidic), which we call specifically *deoxyribonucleic acids,* or DNA. The term DNA can refer to the class of this type of nucleic acid or to an individual molecule. Like proteins, DNA molecules are polymers that are different from, but also very similar to, each other. The differences among DNA molecules, like those among protein molecules, arise from variations in the sequence of the repeating units.

Different types of DNA have molecular weights from about 2 million up to 3 billion and even higher. Each molecule of DNA contains thousands of genes, which carry the codes that direct the syntheses of specific proteins within a particular organism. A DNA molecule is capable of self-duplication so that the same information can be carried by a new cell or by an offspring.

For protein synthesis, the information contained in the DNA is transmitted in a process called *transcription* to another nucleic acid called *messenger ribonucleic acid,* or mRNA. The mRNA acts as a template (pattern) for the incorporation of amino acids in the proper sequence in a growing protein chain. The amino acids are brought to the mRNA in the cell by yet another nucleic acid called *transfer ribonucleic acid,* or tRNA. The synthesis of proteins by RNA is called *translation.*

$$\text{DNA} \xrightarrow{\text{transcription}} \text{mRNA} \xrightarrow{\text{translation}} \text{proteins}$$

THE STRUCTURE OF DNA

Like most natural macromolecules, the nucleic acids are polymers composed of repeating units. Partial hydrolysis of a nucleic acid breaks down the polymer into *nucleotides*. The nucleotides can be further hydrolyzed into *nucleosides* and phosphate ions. The nucleosides in turn can be broken down into a mixture of aromatic heterocyclic nitrogen compounds, which are called *bases,* and one of two sugars: ribose from RNA or deoxyribose from DNA.

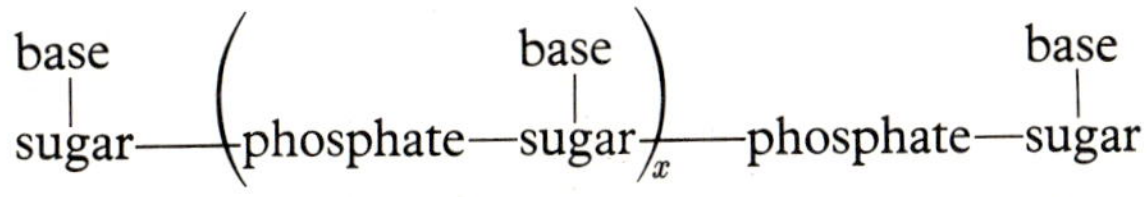

a nucleic acid

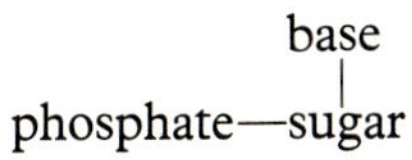

a nucleotide

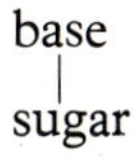

a nucleoside

THE SUGAR. Starting from the bottom of the sequence and working up, we will examine the structure of DNA in more detail. Only one sugar, 2-deoxy-β-D-ribose, is found in DNA. The name of the nucleic acid, deoxyribonucleic acid, is derived from the name of this sugar.

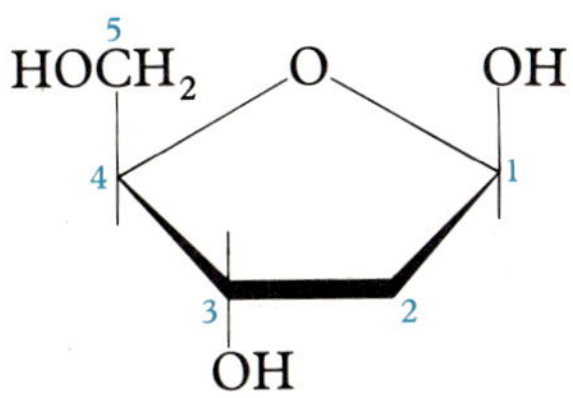

2-deoxy-β-D-ribose
the sugar found in DNA

THE BASES. Only four principal bases are found in DNA. Each is a derivative of either purine or pyrimidine.

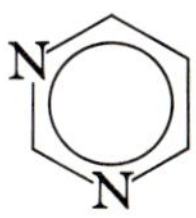

pyrimidine

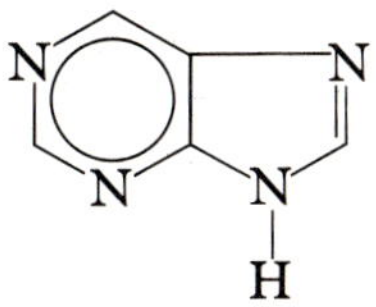

purine

The four bases found in DNA:

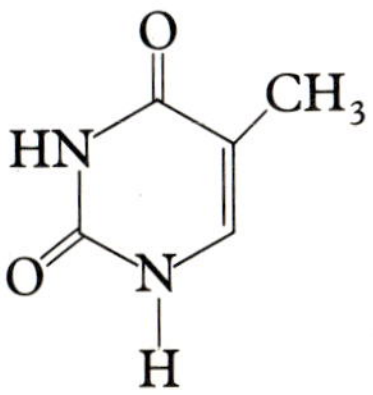

thymine

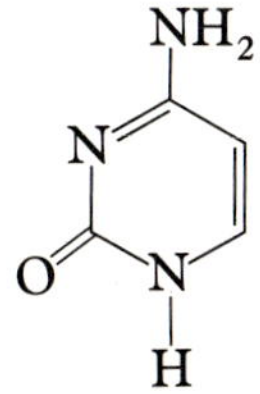

cytosine

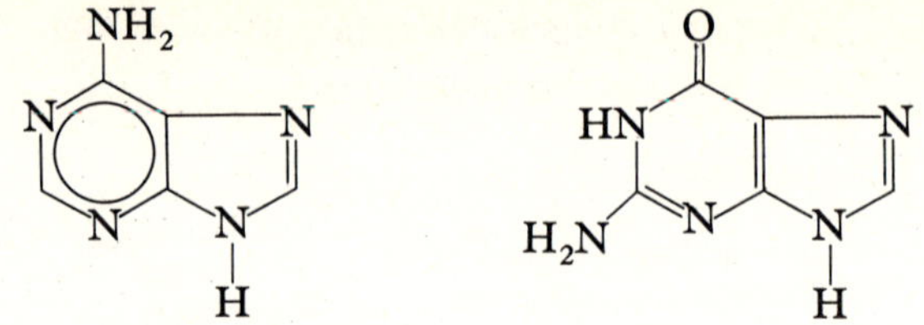

adenine guanine

NUCLEOSIDES. The nucleosides of DNA are composed of the sugar deoxyribose, bonded by a β link from the sugar to a nitrogen of the base.

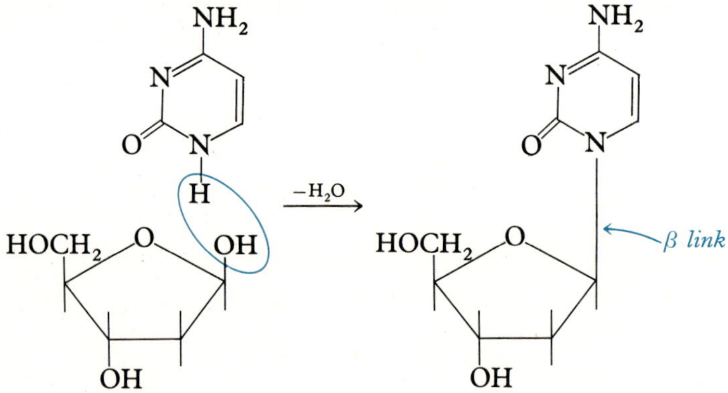

cytosine
+
2-deoxy-β-D-ribose

deoxycytidine
a nucleoside found in DNA

Because only four principal bases are found in DNA, the polymeric structure is derived primarily from repeating units of four nucleosides: deoxycytidine, deoxyadenosine, deoxyguanosine, and deoxythymidine.

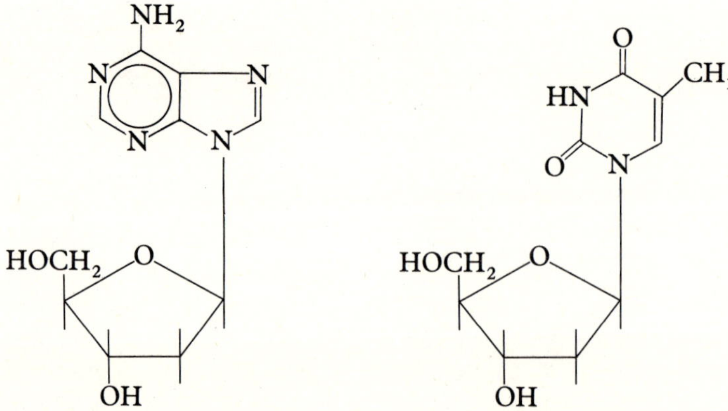

deoxyadenosine deoxythymidine

Sample Problem: We have shown the structures of three of the four nucleosides in DNA. Draw the structure of the fourth nucleoside, deoxyguanosine.

Solution:

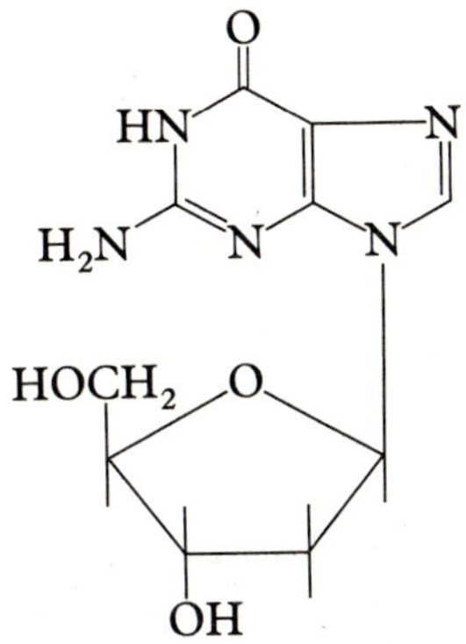

NUCLEOTIDES. A nucleotide is simply the phosphate derivative of a nucleoside. It is the acidic properties of the phosphate group, shown in the completely protonated form, that give the nucleic acids their acidic properties.

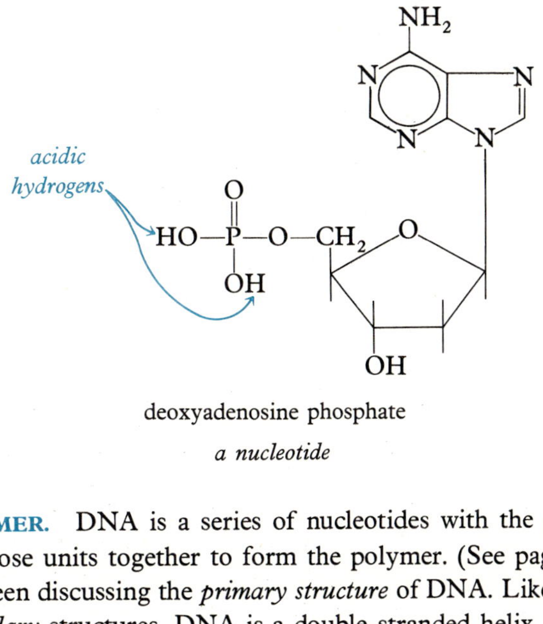

deoxyadenosine phosphate

a nucleotide

THE DNA POLYMER. DNA is a series of nucleotides with the phosphate groups joining deoxyribose units together to form the polymer. (See page 442.)

We have been discussing the *primary structure* of DNA. Like proteins, nucleic acids have *secondary* structures. DNA is a double-stranded helix. Two nucleic acid chains are wound about each other and held together by hydrogen bonds between pairs of bases and by the energy associated with the "stacking" of the helix. The overall secondary structure resembles a flexible ladder—the base pairs and the hydrogen bonds represented by the rungs of the ladder. (See Figure 28-1.)

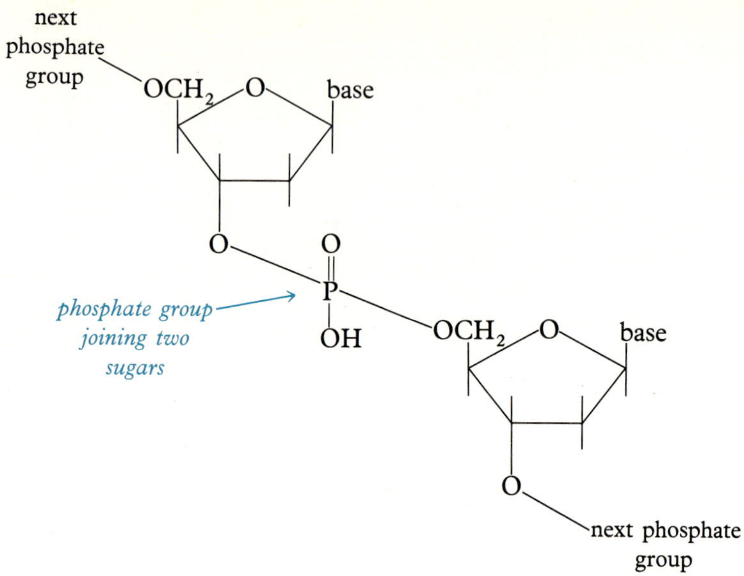

The bases in one strand of DNA do not form random hydrogen bonds with bases in the other strand in the helix. *The hydrogen bonds are specific between pairs of appropriate bases.* Guanine and cytosine hydrogen bond with each other, as do adenine and thymine. Adenine and guanine, however, cannot form such favorable hydrogen-bond interactions. In DNA, adenine and guanine do not hydrogen bond with each other; neither do adenine and cytosine. Only the two pairs of bases form the hydrogen bonds that hold the two-stranded helix together.

Figure 28–1. DNA is a double-stranded helix held together by hydrogen bonding.

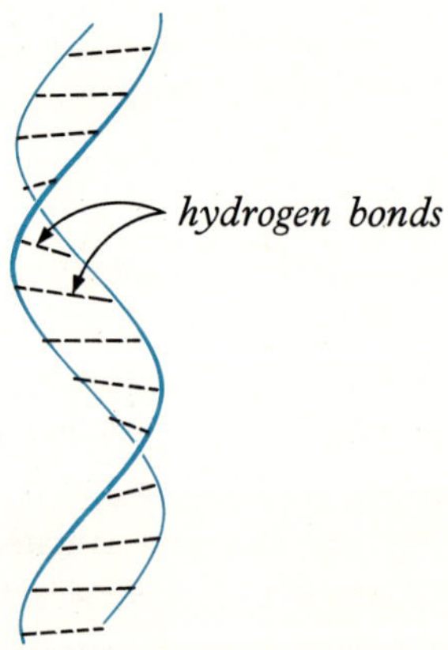

Hydrogen bonds between guanine and cytosine:

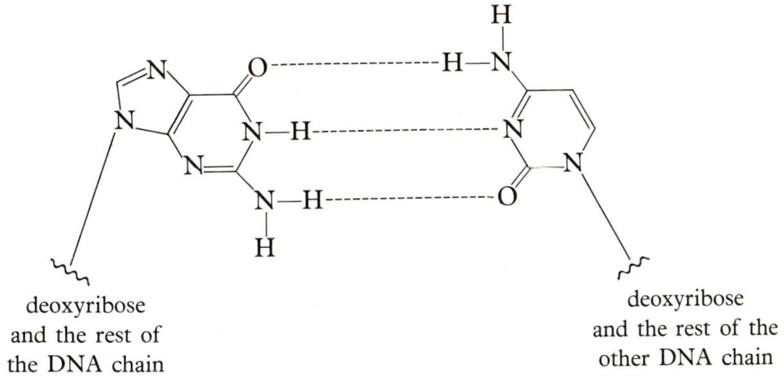

Hydrogen bonds between adenine and thymine:

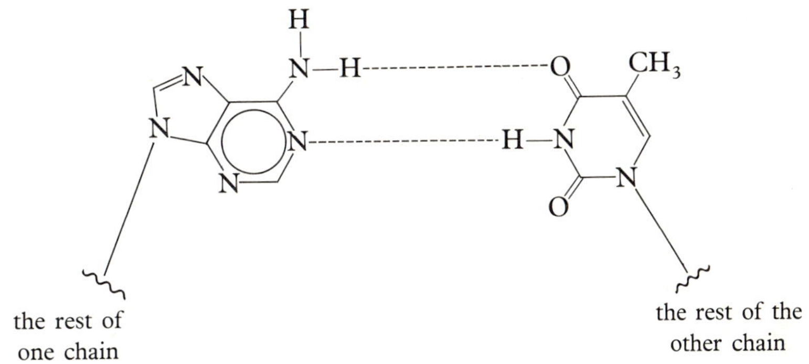

Guanine and thymine are unsuitable for strong hydrogen bonding between the two:

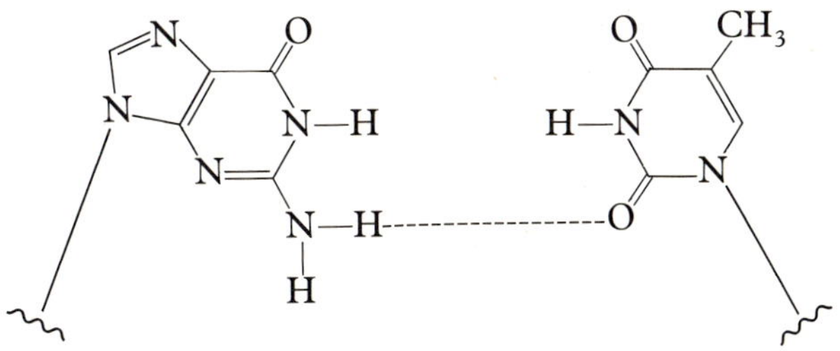

Adenine can hydrogen bond only with thymine; therefore, when an adenine base is present on one strand, a thymine must be opposite it in the other strand. The same holds true for guanine and cytosine. The sequence of bases in one strand dictates the sequence of bases in the other strand. The strands are completely complementary. This complementary nature of the strands is the chemical basis of heredity.

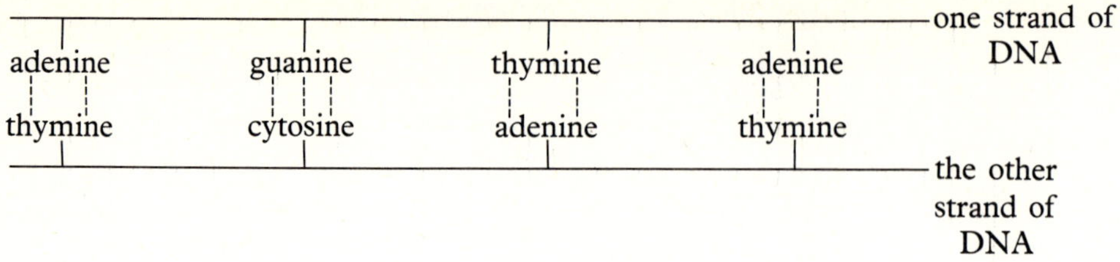

REPLICATION OF DNA

In ordinary cell division (*mitosis*), cells divide to form new cells with identical genes and therefore identical DNA helices. The information contained in the DNA polymer (that is, the sequence of bases) can be transferred to new cells because DNA is capable of duplicating itself in a process called *replication.*

At the start of replication, the DNA helix begins to unwind. As it is unwinding, new nucleotides containing triphosphate groups line up along each single strand of the original DNA. The enzyme DNA polymerase catalyzes the successive polymerization of the nucleotides into a new DNA chain.

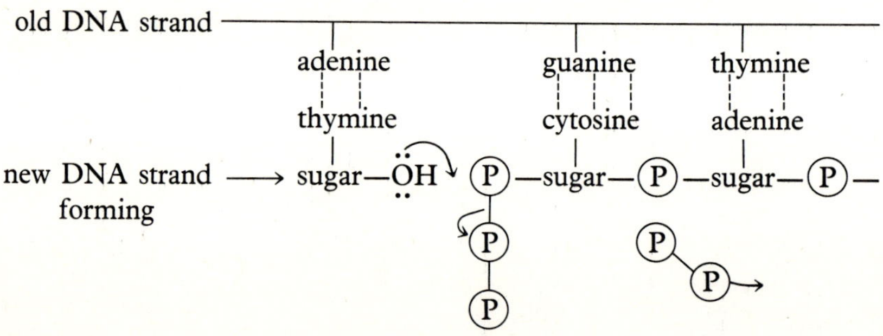

Because of the favorable hydrogen bonding between base pairs, the nucleotides in the newly forming DNA line up along the original strand in a complementary fashion—thymine to adenine and cytosine to guanine. The new DNA chain is the direct complement of the strand that synthesized it; it is identical to the second strand in the original DNA helix. As the helix unwinds, two identical DNA helices form. Each helix contains one old strand plus one new strand. When replication is complete, two new strands have been formed and two identical helices produced where only one existed before. One DNA helix has produced its replica, which will be part of the genetic code for another cell. (See Figure 28-2.)

Figure 28–2. DNA replicates by unwinding and bringing in complimentary nucleotides, which form a new strand.

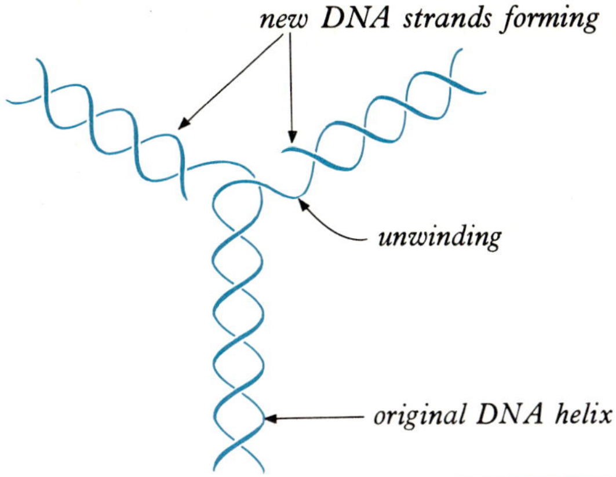

STRUCTURE OF RNA

Ribonucleic acid (RNA) has a structure very similar to that of DNA. RNA, like DNA, is a series of sugar units joined together by phosphate groups. A base is attached to each sugar. The sugar forming the backbone of RNA is β-D-ribose instead of 2-deoxy-β-D-ribose. Ribose has a hydroxyl group at carbon 2, where 2-deoxyribose has none.

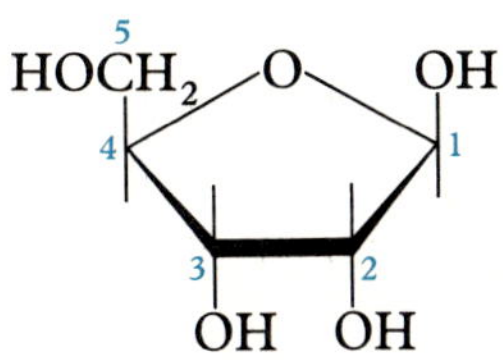

β-D-ribose

The four bases in RNA are two purine derivatives, adenine and guanine, and two pyrimidine derivatives, cytosine and uracil. The first three of these bases are found in DNA. The fourth, uracil, replaces the thymine found in DNA.

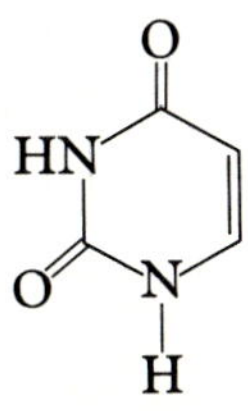

uracil
found in RNA

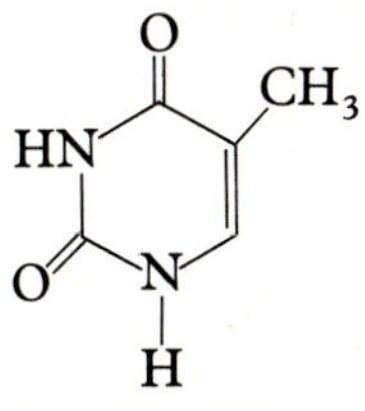

thymine
found in DNA

The nucleosides of RNA are connected in a directly analogous fashion to those of DNA. RNA is a polymer with phosphate bridges from one ribose unit to the next.

THE BIOSYNTHESIS OF PROTEINS

DNA carries the genetic code, but it never leaves the nucleus of the cell to direct the synthesis of proteins. DNA transfers the code by the synthesis of mRNA, which does leave the nucleus.

DNA undergoes replication by producing a pair of entirely new but completely complementary strands. When DNA synthesizes mRNA, only a portion of the helix unwinds, and only a portion of one single strand is used as a template. That portion constitutes a particular gene or series of genes. As in replication, nucleotides (ribonucleotides containing ribose as the sugar in this case) line up along the unwound strand in a complementary fashion. Uracil has the same favorable hydrogen bonding characteristics toward adenine as thymine. Instead of thymine pairing with adenine as in the replication of DNA, uracil pairs with adenine in the synthesis of RNA.

In RNA, uracil replaces thymine as the base that forms hydrogen bonds with adenine:

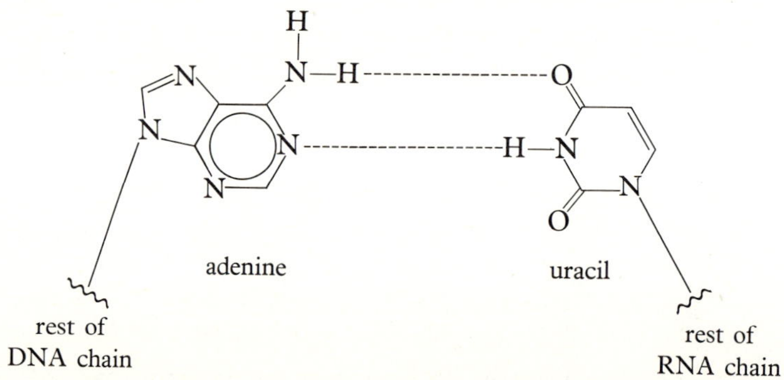

Because of the hydrogen bonding of the base pairs, the strand of DNA produces a complementary mRNA. Except that ribose replaces deoxyribose and uracil replaces thymine, the strand of mRNA that is formed is the complement of a portion of the DNA strand.

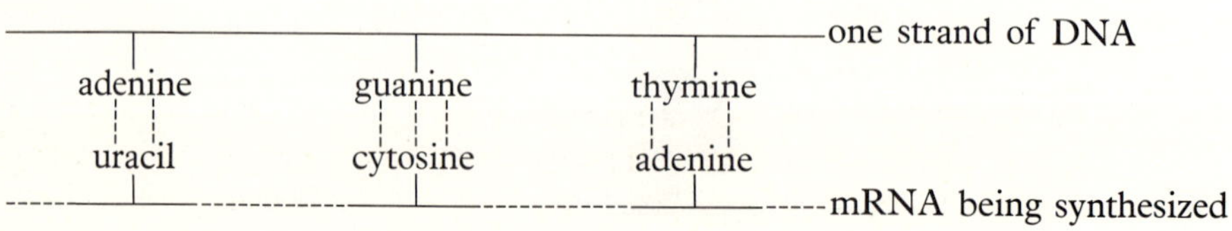

Figure 28–3. Ribosomes become attached to mRNA to form a polysome.

When the strand of mRNA is complete, it is released from the DNA and moves out of the nucleus into the endoplasmic reticulum, a series of small interconnecting channels that permeate most of the cytoplasm of a cell. These channels are studded with granules called *ribosomes,* which become attached to the mRNA molecule.

A ribosome is composed of about 60 percent *ribosomal* RNA (rRNA) and about 40 percent complex proteins. It is shaped rather like an acorn, a sphere with a smaller sphere attached. The smaller sphere is the part that becomes attached to the mRNA. A cluster of ribosomes connected to a strand of mRNA is referred to as a *polysome.* Although the function of ribosomes is not completely understood, it is believed that they aid in reading and transmitting the information contained in the mRNA and help hold the molecules in the proper positions as the protein is being synthesized. More will be said about this topic shortly.

mRNA is the template for the growing protein chain. The sequence of bases in the mRNA determines the sequence in which amino acids will be incorporated into the protein molecule. The bases are read three at a time. A specific sequence of three bases in a row specifies one particular amino acid. Such a sequence of three bases is called a *codon.* The codons for the various amino acids have been determined. For example, a codon for the amino acid phenylalanine is uracil-uracil-uracil. The three uracil bases in a row in mRNA signal that phenylalanine is to be placed in a protein at that spot. A codon for alanine is guanine-cytosine-cytosine. This sequence signals the placement of alanine in the protein. Most amino acids have more than one codon. The sequence uracil-uracil-cytosine is a second codon for phenylalanine, and guanine-cytosine-uracil is a second codon for alanine.

Repeating sequences of three bases in a row are thus the fundamental genetic units in living systems. For example, a series of particular sequences determines that our hair will be hair and not stomach lining.

The proper amino acid is brought to the mRNA in the cell by other nucleic acids called *transfer* RNA, or tRNA. With the help of activating enzymes and ATP, the amino acid forms an ester with the hydroxyl group of the ribose in tRNA. The activating enzymes are specific—they determine that the correct amino acid becomes attached to the correct tRNA.

$$\text{amino acid} + \text{ATP} \xrightarrow{-\text{PP}_i} \text{amino acid—AMP complex} \xrightarrow[-\text{AMP}]{\text{tRNA}} \text{amino acid—tRNA ester}$$

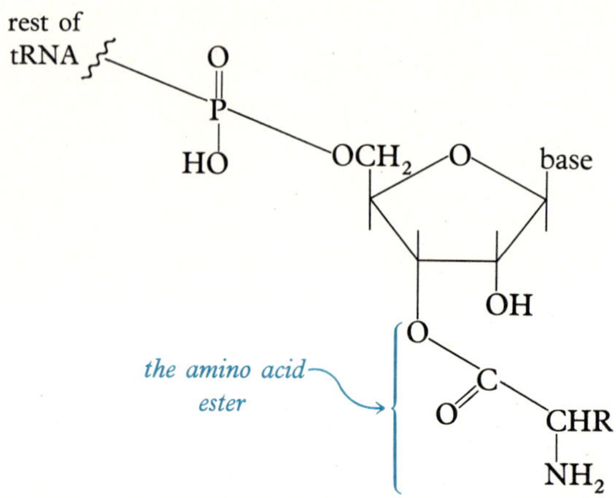

The series of bases in tRNA are complementary to a portion of the mRNA. A sequence of three bases in tRNA is called an *anticodon,* since it is the complement of the codon in mRNA.

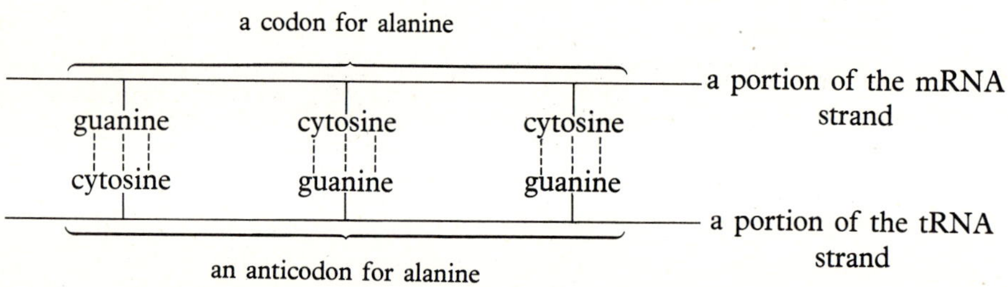

Sample Problem: Two codons for lysine are adenine-adenine-adenine and adenine-uracil-guanine. What are two anticodons for lysine?

Solution: The anticodons are the complements of the codons; therefore, the anticodons for lysine are uracil-uracil-uracil and uracil-adenine-cytosine.

Because of the complementary anticodon, the tRNA becomes bonded to the proper sequence (the codon) in mRNA. With the help of the ribosome, the new amino acid is correctly positioned at a location called the *amino acid binding site,* or the *A* site, on the ribosome. The ribosome also holds the growing end of the peptide at the *polypeptide binding site,* or *P* site. The end of the polypeptide chain becomes bonded to the new amino acid, and thus is shifted to the *A* site. (See Figure 28-4.) The ribosome

Figure 28–4. The ribosome moves along the mRNA to the right, adding amino acids from tRNA molecules to the growing peptide chain.

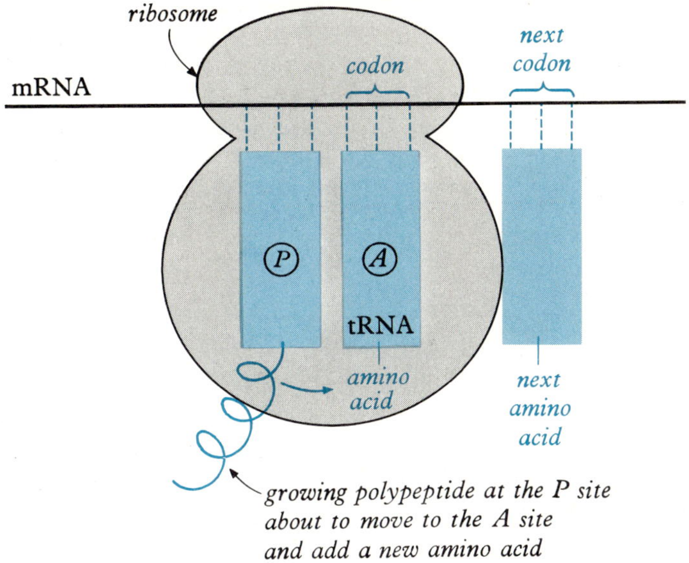

with the attached polypeptide moves on to the next codon, and the polypeptide reassumes its position at the *P* site, ready for the next amino acid.

As one protein chain is being synthesized, other molecules of the same protein may be synthesized in assembly line fashion as other ribosomes and growing proteins follow the first. When a protein chain is complete, the ribosome drops off the end of the mRNA, and the completed protein molecule is released. (See Figure 28–5.)

It is evident that there must be a system of controlling the biosynthesis of proteins. All genes do not work full time frantically producing proteins. Rather, proteins are produced as they are needed. By the synthesis (or nonsynthesis) of particular enzymes, the DNA can (and must) regulate many of the body's processes. When energy is needed, for example, DNA starts the process of synthesizing the activating enzymes that catalyze the hydrolysis of ATP. Or, if certain amino acids are needed and are not present, transamination enzymes are formed or activated so that the needed amino acids may be synthesized.

It is thought that the normal state of genes is that of *repression,* that is, turned off. Possibly a repressor molecule is associated with an *operator gene* (a gene that controls other genes) and effectively keeps the switch locked in the "off" position. When the signal comes (by means of a steroid hormone released by the endocrine system, for example) that the operator gene is to be turned to "on," an *inducer* molecule pushes the repressor molecule out of the way and unlocks the off switch. In this way, the synthesis of a particular protein can be strictly regulated.

Figure 28–5. A series of ribosomes move along the mRNA chain. Amino acids are added as the ribosomes move along. When the ribosome reaches the end of the mRNA molecule, it drops off and releases the completed protein. Adapted from A. White, P. Handler, and E. Smith, *Principles of Biochemistry,* 4th edition, McGraw-Hill Book Co.

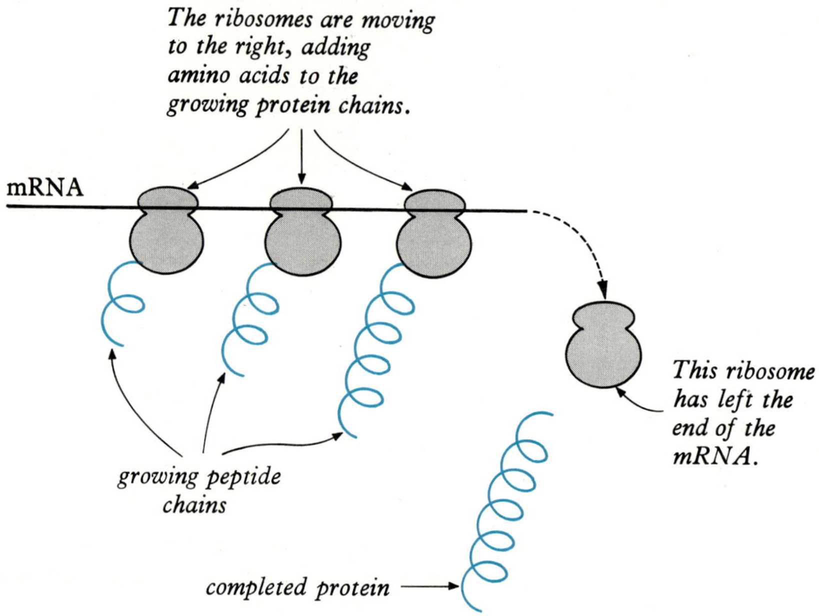

The process of protein synthesis may seem complex, but consider the result! A living organism can be formed, it can grow, it can be maintained, its needs can be met—a truly amazing result of organized molecules at work.

GENETIC RECOMBINATION

In some circumstances, two different chromosomes can exchange some of their genes. On the molecular level, we would say that two different DNA molecules split and then rejoin with each other. This process, called *genetic recombination,* can occur in sexual reproduction.

Meiosis (as opposed to mitosis) is the division of cells to yield *germ cells,* which are used to form sperm cells and ova (egg cells). These cells each have half the usual number of chromosomes for that organism. Humans have 46 chromosomes in the cells of tissues; therefore, human sperm cells and ova contain only 23 chromosomes.

In sexual reproduction, a sperm cell and an ovum unite to form a new cell, called a *zygote,* which has the normal number of chromosomes. The zygote, of course, multiplies by mitosis and in time becomes an offspring.

In the formation of the zygote, the chromosomes of the sperm and ovum undergo recombination. The new chromosomes, called *recombinants,* contain some genetic code from each of the original chromosomes. Genetic recombination thus assures a variety of combinations of genes in the genetic pool of a species.

Chemists are also experimenting with genetic recombination in the laboratory—for example, by inserting new genes into *E. coli* chromosomes. This type of research leads to some interesting speculations. Already chemists are forcing *E. coli* to produce human insulin. Could hereditary disorders be cured by manipulation of human genes? Could highly virulent microorganisms with no natural enemies escape from the laboratory and contaminate the earth? The moral and social implications of artificial genetic recombinations are certain to create political controversy.

GENETIC DISORDERS

The processes of the formation, replication, transcription, and translation of the genetic code of an organism are quite complex. Errors in the sequence of bases in DNA molecules can lead to defective genes. These defective genes, in turn, give rise to abnormalities that can be transmitted to offspring. Such inherited disorders are called *genetic disorders.*

A genetic disorder is initially caused by a *mutation,* which results from a change in the base sequence of DNA. Mutations can be spontaneous or they may be due to damage caused to the DNA in germ cells by radiation or drugs. A natural mutation is a very rare event; a gene may be duplicated a million times before a mutation can be detected. Infrequent as they are, however, mutations are the molecular foundation of evolution. Most mutations are undesirable and are sometimes fatal for the offspring.

Sickle cell anemia, diabetes, phenylketonuria (PKU), and galactosemia are examples of genetic disorders that have been discussed elsewhere in this book.

VIRUSES

Our discussion so far has been limited to the theories of how nucleic acids in typical animal cells usually behave. The nucleic acids in viruses can behave somewhat differently. A virus is a small particle composed of either a DNA or a RNA molecule associated with many polypeptide molecules, some of which form a coat around the virus body. All organisms, plant and animal, are subject to viral infections. For humans, these infections range from exasperating conditions (such as colds, flu, and cold sores) to those that are fatal (such as rabies).

A virus cannot reproduce itself, nor can it synthesize its protein molecules. Therefore, a virus cannot truly be called a living thing. Some scientists say that a virus is "on the threshold of life."

If a virus cannot reproduce itself, how can there be such events as flu epidemics? The reason is that a virus forces another cell, called a *host cell,* to reproduce viruses. In most cases, the virus forces the host cell to stop all its normal activities, such as replication or synthesis of the organism's own proteins, and become a slave to the virus. The infected cell then devotes all its energy toward producing viral nucleic acids and viral proteins. This form of slavery usually results in the eventual death of the host cell.

The viruses that have been the most extensively studied are *bacteriophages,* viruses that infect bacteria. In one well-studied case of bacterial viral infection, the bacteriophage injects viral DNA through the bacterial cell wall into the cytoplasm of the bacterium. Approximately 20 minutes later, the bacterial cell bursts and 100 or more new viruses are released. These new viruses can then infect other bacteria.

Some viruses carry genes for their own enzymes. This type of virus thus directs the synthesis of its enzymes by a host cell. Other types of virus simply use the enzymes of the host cell to catalyze their reproduction.

Some viruses also contain genes for enzymes that catalyze the hydrolysis of the host DNA and thus the destruction of the host cell. (On the other hand, some host cells contain enzymes that can destroy the viral nucleic acids in a similar fashion.)

The number of genes involved in viral reproduction is rather small, from about five up to about 200. By way of comparison, the one chromosome of *E. coli* contains approximately 4000 genes, and the chromosomes of a mammalian cell contain around 10^6 to 10^7 genes.

How the virus accomplishes its own replication at the expense of the host cell is the subject of current research. In some cases, viral RNA acts as a template for viral DNA that can be inserted into the DNA of the host cell. When this host cell divides, the new cell also contains abnormal DNA. It is postulated that some cancer cells are formed by the incorporation of a viral DNA into the host DNA.

Instances are known where the combination of viral and host DNA does not affect the metabolism of the host cell appreciably. In this case, the viral DNA is repressed, or in a *lysogenic state*. Even though the viral DNA is replicated right along with host DNA, no harm arises unless the repression of the viral DNA is ended. It is known, for example, that ultraviolet light can terminate the lysogenic state of some bacteriophages and result in harmful viral infection.

STUDY QUESTIONS

28-1. What are the general functions within an organism of the following nucleic acids?
(a) DNA (b) mRNA (c) tRNA

28-2. Which of the following could arise from the complete or partial hydrolysis of DNA?
(a) D-ribose (b) 2-deoxy-D-ribose
(c) phosphate ions (d) nucleotides

28-3. Give the name and the structure of the sugar unit found in DNA.

28-4. Which of the following structures are nucleosides?

(a) 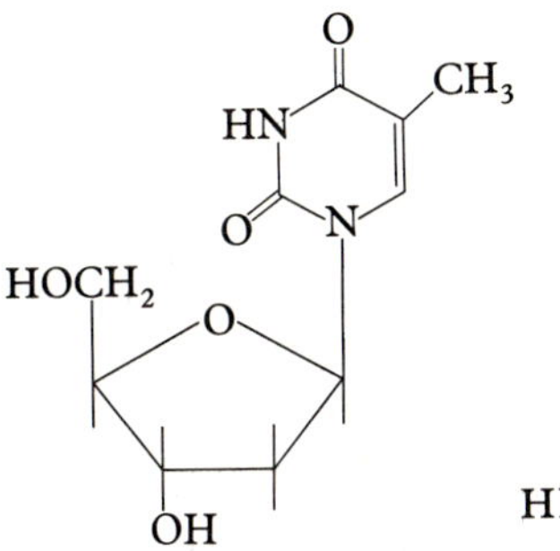

(b) 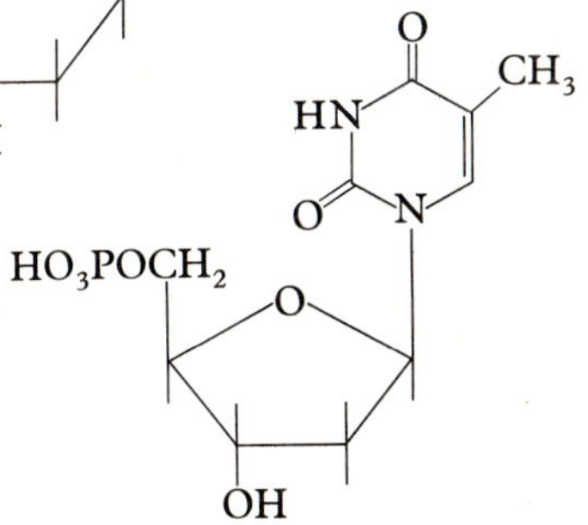

(c) 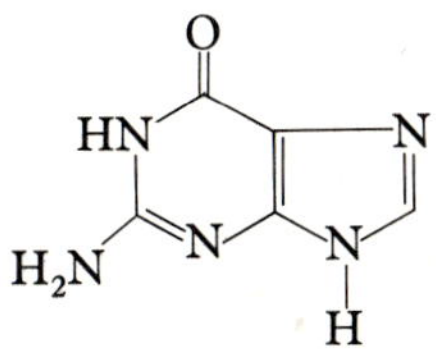

(d) 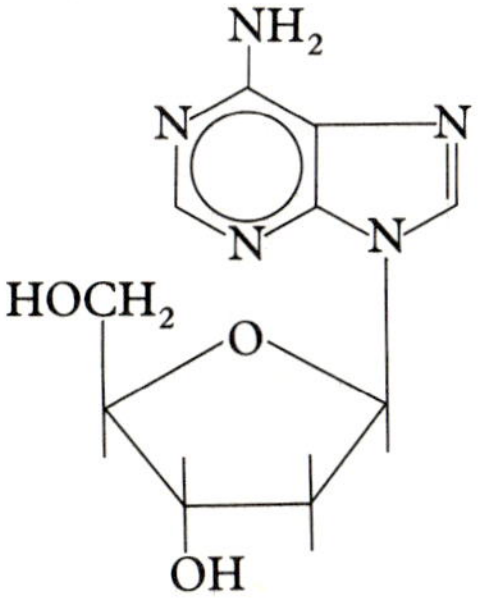

28-5. Draw the structure of α-deoxycytidine.

28-6. Concerning the nucleosides formed from the hydrolysis of DNA, which of the following statements are true?
(a) There are an unlimited number of nucleosides that can be isolated after the hydrolysis.
(b) There are principally four nucleosides that can be isolated after the hydrolysis.
(c) A nucleoside of DNA is composed of 2-deoxy-D-ribose connected by a β link to one of four purine or pyrimidine derivatives.
(d) If the nucleoside deoxycytidine is hydrolyzed further, the products are 2-deoxy-D-ribose and adenine.

28-7. Which positions in the following structures are used in the formation of nucleoside links?

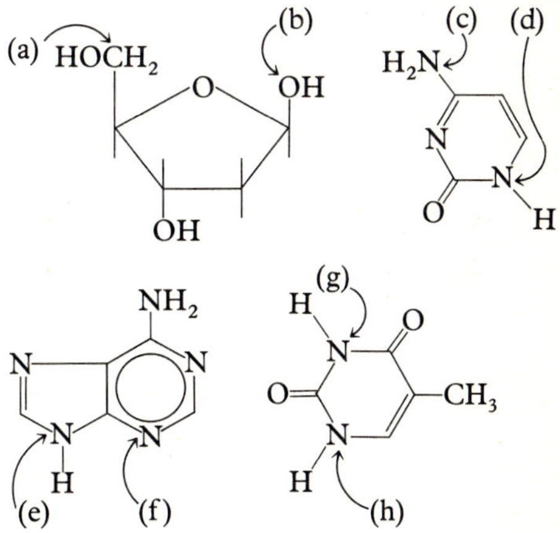

28-8. What structural feature is responsible for the acidic properties of DNA?

28-9. Which pair of bases would you expect to be held the most firmly together by hydrogen bonding?
(a) guanine and cytosine
(b) guanine and thymine
(c) adenine and thymine

28-10. Concerning the representation of DNA below, which of the following statements are true?

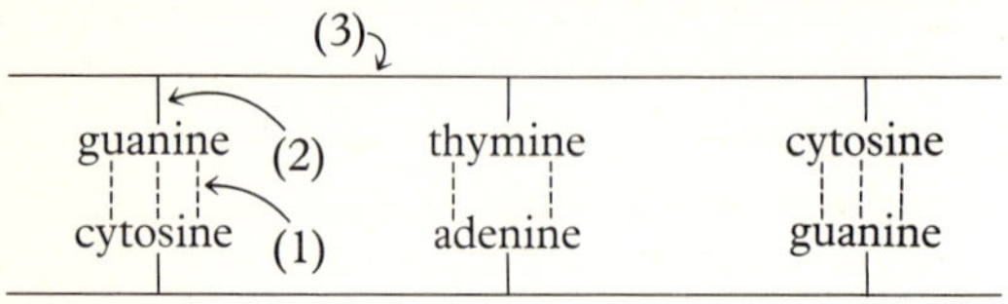

(a) The drawing shows a portion of two separate molecules held together by hydrogen bonds.
(b) The hydrogen bonds are shown by the solid line at (2).
(c) Guanine is bonded to a unit of 2-deoxy-D-ribose at (2).
(d) Guanine is bonded to thymine at (3).
(e) Phosphate connecting links are found at the dotted line at (1).
(f) The solid line at (3) represents units of deoxyribose joined together by phosphate links.

28-11. A double helix of DNA can replicate itself exactly because:

(a) hydrogen bonding occurs between specific base pairs and only those pairs.
(b) certain nucleotides are complementary to each other.
(c) either single strand of DNA can dictate the specific order of nucleotides incorporated into a new strand.
(d) an enzyme named *DNA polymerase* guides the incoming bases to their correct location on the newly forming chain.

28-12. Below we have represented a portion of a double strand of DNA. Which of the following statements are true?

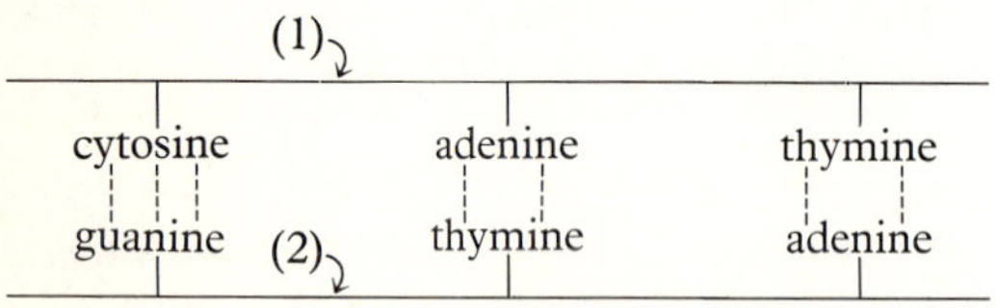

(a) When the strand unwinds for replication, hydrogen bonds are broken.
(b) In the replication, (1) will be used to form a strand identical to itself, that is, a new (1).
(c) In the replication, (2) will dictate that new nucleotides containing cytosine, adenine, and thymine will be lined up in that order.
(d) When the replication is complete, the original (1) and original (2) will be part of one helix and the new (1) and new (2) will be part of another identical helix.

28-13. Draw the structure of the sugar found in RNA, and give its name.

28-14. In RNA, the base uracil undergoes hydrogen bonding with one of the other three bases. Draw the structures of uracil and this other base, showing the hydrogen bonding.

28-15. Can ATP (Chapter 25) be obtained by the hydrolysis of RNA?

28-16. Tell whether each of the following phrases refers to DNA, to mRNA, to tRNA, or to ribosomes:

(a) carries the codon that signals the incorporation of an amino acid into a protein molecule.
(b) is a complement to a portion of a DNA strand.
(c) is a complement to a portion of a mRNA strand.
(d) remains in the cell nucleus.
(e) forms an ester with a specific amino acid.
(f) is the carrier of the genetic code from an old cell to a newly formed cell.

28-17.

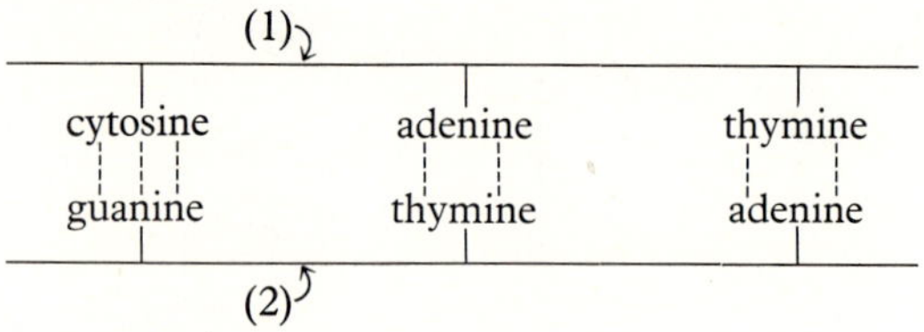

The hydrogen-bonding characteristics of a portion of mRNA that was synthesized from strand (1) of DNA (above) would be similar to the hydrogen-bonding characteristics of

(a) strand (1). (b) strand (2).
(c) neither of these. (d) both of these.

28-18. A codon
(a) is a sequence of four bases in a row.
(b) signals that three particular amino acids be incorporated into a growing peptide.
(c) helps position the new amino acids correctly by hydrogen bonding with an anticodon of tRNA.

28-19. An anticodon
(a) is a sequence of three bases in a row.
(b) is found in mRNA.
(c) is the complement of a codon.
(d) helps position amino acids by hydrogen bonding with a portion of mRNA.

28-20. Match the following terms with their definitions:

(a) zygote (c) mitosis
(b) meiosis (d) recombinant

(1) duplication of a cell by division
(2) formation of a germ cell with half the chromosomes of the parent cell
(3) the cell arising from the combination of two germ cells
(4) a chromosome containing genes from two different original DNA molecules

28-21. Which of the following sequences can arise from a viral infection?
(a) viral RNA $\xrightarrow{\text{host cell}}$ viral DNA
(b) viral DNA $\xrightarrow{\text{viral enzymes}}$ viral RNA
(c) host DNA $\xrightarrow{\text{viral enzymes}}$ nucleotides
(d) viral DNA + host DNA $\xrightarrow{\text{viral enzymes}}$ recombinant DNA
(e) viral protein $\xrightarrow{\text{host cell}}$ viral DNA

CHAPTER 29 BLOOD AND URINE

OBJECTIVES

The Body Fluids *List examples of circulating and noncirculating body fluids.*

Blood *Write equations that show how the pH of blood is maintained.* * *Write an equation that shows the dependency of O_2 transport on pH.* * *List the principal components and functions of blood.*

Blood Cells *List the three types of blood cells.* * *Define hematocrit value and sedimentation rate.* * *Describe the differences between the blood types.*

Blood Plasma *List the types of protein in blood plasma.* * *Describe the function of albumin in regulating osmotic pressure.* * *Write an equation to represent the biuret test.*

Blood Clotting Factors *Diagram the blood clotting mechanism, and describe the action of some anticoagulants.*

Kidney Function and Urine *Describe the functions of the kidneys, and how they are regulated.* * *List the properties and constituents of the urine.* * *Describe some conditions that give rise to abnormal urine.*

Spectrophotometry *Define electromagnetic radiation and wavelength.* ∗ *Describe the absorption of electromagnetic radiation by organic compounds and how it gives rise to color.*

THE BODY FLUIDS

Approximately 75 percent of a mammalian organism is composed of water. This water is found in the blood and the interstitial fluids (fluids such as lymph that are found between cells). These are *circulating* fluids that are responsible for supplying nutrients to the cells and for removing waste products.

Some body fluids are *noncirculating*. For example, our digestive system depends on digestive "juices," noncirculating solutions and colloids containing digestive enzymes. Another example of a noncirculating fluid is urine, which carries away many of a mammal's water-soluble waste products.

In this chapter, we will limit our discussion to the blood and the urine. We will also consider a few of the common clinical tests performed on these two fluids. At the end of the chapter, we will briefly discuss the spectrophotometer, an instrument used in clinical testing.

BLOOD

An adult human contains approximately 5.7 liters (six quarts) of blood. The blood serves a variety of functions in a mammal. It carries nutrients and oxygen to the cells and carries wastes and CO_2 away. It transports hormones, which regulate the activity of the cells, and antibodies, which ward off infection. The blood also helps regulate the organism's temperature and pH.

ACID-BASE BALANCE OF BLOOD. The pH of normal blood is 7.3 to 7.5 (slightly alkaline). This slight alkalinity is absolutely necessary if the blood is to perform its function of delivering O_2 to cells.

Let us consider the oxygen-carbon dioxide exchange of hemoglobin. The air in the lungs is rich in O_2 and deficient in CO_2. Because the solubility of a gas is proportional to the pressure of that gas (Chapter 6), O_2 in the lungs tends to dissolve in blood, while CO_2 is released to be exhaled. However, blood carries far more O_2 than can be explained by simple gas solubility. The binding of O_2 to hemoglobin explains why the blood can carry such a large amount of O_2.

When hemoglobin (Hb) binds with oxygen in the lungs, a hydrogen ion is lost to the surrounding medium.

$$\underset{\text{hemoglobin with } H^+}{HHb^+} + O_2 \rightleftharpoons \underset{\text{oxyhemoglobin}}{HbO_2} + H^+$$

The oxygenation of hemoglobin is reversible. An alkaline medium removes H^+; consequently, HbO_2 is favored. On the other hand, in an acidic medium (as in acidosis), the equilibrium is shifted to the left—that is, to HHb^+. Therefore, blood that is acidic cannot assimilate as much oxygen.

Now let us consider what happens when oxygen-rich arterial blood reaches the capillaries. Cells use O_2 and give off CO_2 as a waste product; therefore, the environment around a cell is low in O_2 but rich in CO_2. These relative concentrations help blood release O_2 to the cells and take in CO_2. Again, this process alone is not efficient enough to accomplish the gas exchange needed.

The pH of fluids around the cells also helps hemoglobin release O_2. The CO_2 formed in the cells dissolves in water to form H_2CO_3, which is acidic ($H_2CO_3 \rightleftharpoons H^+ + HCO_3^-$). This increased acidity at the cell site is desirable because it shifts the $HHb^+ \rightleftharpoons HbO_2$ equilibrium to the left, releasing O_2 to the cells. Thus, we can see why the pH of the blood must be rigidly regulated; the pH determines both the uptake and the release of O_2 by hemoglobin.

The pH of the blood is maintained primarily by the buffering action of the bicarbonate ion generated in part by CO_2 released from the cells. Most foodstuffs are metabolized to acidic products. Excess acid in an organism reacts with HCO_3^- to yield H_2CO_3, which is in equilibrium with H_2O and CO_2. Excess CO_2 can then be eliminated by way of the lungs.

$$H^+ + \underset{\text{bicarbonate ion}}{HCO_3^-} \xrightleftharpoons{\text{blood}} \underset{\text{carbonic acid}}{H_2CO_3} \xrightleftharpoons{\text{lungs}} H_2O + CO_2$$

The kidneys also play a role in maintaining the bicarbonate-carbonic acid equilibrium. The kidneys remove hydrogen ions from H_2CO_3 in the blood. These hydrogen ions acidify the urine. (This is a topic we will discuss later in this chapter.) The kidneys then return HCO_3^- to the blood to continue their buffering action.

$$H_2CO_3 \xrightleftharpoons{\text{kidneys}} \underset{\textit{to urine}}{H^+} + \underset{\textit{to blood}}{HCO_3^-}$$

In addition to the bicarbonate buffer, some proteins in blood, including hemoglobin, act as buffers.

$$\text{protein-}\ddot{\text{N}}\text{H}_2 + \text{H}^+ \rightleftharpoons \text{protein-}\overset{+}{\text{N}}\text{H}_3$$

an amino group in a protein

To a small extent, phosphate ions also help keep blood at the proper pH (see page 166).

COMPOSITION OF BLOOD. Whole blood is composed of *blood cells* suspended in *plasma*. Among other substituents, plasma contains proteins called *clotting factors*. Blood with the blood cells and clotting factors removed is called *serum*. (Serum can be observed as the straw-colored fluid oozing from a scab.) Most clinical blood tests are performed on either blood plasma or the serum.

whole blood
- plasma
  - serum
  - clotting factors
- blood cells

Let us briefly look at some of the components found in each of these parts of whole blood.

BLOOD CELLS

Blood cells, sometimes called *formed bodies,* are of three types: red blood cells (*erythrocytes*), which we will discuss here; white blood cells (*leukocytes*), which function to fight infectious bacteria; and platelets (*thrombocytes*), which are necessary for blood clotting.

Erythrocytes are the most numerous blood cells. A cubic centimeter of blood contains 4–5 million of these cells (about 15 g/100 cc blood). The erythrocytes contain hemoglobin and thus are responsible for transporting O_2 to the tissues and removing CO_2.

Normal erythrocytes have about a four-month life span. They are synthesized in the bone marrow and destroyed in the liver and spleen at a balanced rate. When erythrocytes are destroyed, the organic portion of heme (the porphyrin ring) is metabolized in the liver and eliminated (see page 435). Clinical tests for porphyrin pigments provide information on liver function.

Too few erythrocytes (*anemia*) may be caused by a number of factors. Loss of blood by the system is one cause. A low rate of erythrocyte synthesis, as in leukemia, is another.

The concentration of erythrocytes in blood can be estimated by simple tests. One such test is the *hematocrit value,* which measures the relative volumes of cells and plasma in blood. The test consists of the centrifugation of a sample of whole blood (pre-treated with an anticoagulant to prevent clotting) and measurement of the height of the sediment (the cells) relative to the height of the entire sample.

Another clinical test performed on whole blood (again pre-treated with an anticoagulant) is the measuring of the *sedimentation rate,* the speed at which erythrocytes settle to the bottom of a tube. Normal blood has a very low sedimentation rate. If a person has a condition in which cell destruction occurs at a faster rate than normal, the sedimentation rate increases. This test is useful in diagnosing rheumatic fever, arthritis, and myocardial infarction (a heart attack in which a clot blocks the blood flow to a portion of the heart muscle).

BLOOD TYPE. It was observed early in the history of medicine that recipients of whole-blood transfusions sometimes lived, but sometimes died. The reason for many of these deaths was the incompatibility of the donor's blood and the recipient's blood because of a difference in *blood type.*

There are four major blood types: A, B, AB, and O. (There are also some other rare blood types and other factors, such as the Rh factor, which we will not discuss here.) The serum of these four different types of blood contains antibodies, called *agglutinins,* that can cause precipitation of the erythrocytes of other types of blood. Type A blood, for example, has antibodies that cause the erythrocytes of Type B or Type AB blood to precipitate. Type B blood, in turn, contains antibodies that cause Type A or Type AB erythrocytes to precipitate.

A person with Type O blood is called a "universal donor" because none of the other blood types contain antibodies that cause it to precipitate. A person with Type O blood can safely receive only Type O blood, however, because this type of blood will cause Type A, B, or AB to precipitate.

Type AB blood, by contrast, is "universal-recipient" blood because it does not have antibodies that precipitate A, B, or O blood.

Blood types are determined by adding a sample of whole blood to standard serum samples of the various blood types in test tubes and noting the presence or absence of a precipitate. (See Figure 29-1.)

Sample Problem: A blood sample is added to Type A, B, AB, and O serums. The serums of Types A and O show a precipitate, while those of Types AB and B do not show a precipitate. What is the blood type of the sample?

Solution: The sample must be Type B. (It is not A or O because it is precipitated by these serums. It is not Type AB because AB blood would be precipitated by the antibodies in the Type B serum.)

Figure 29–1. Determination of blood types.

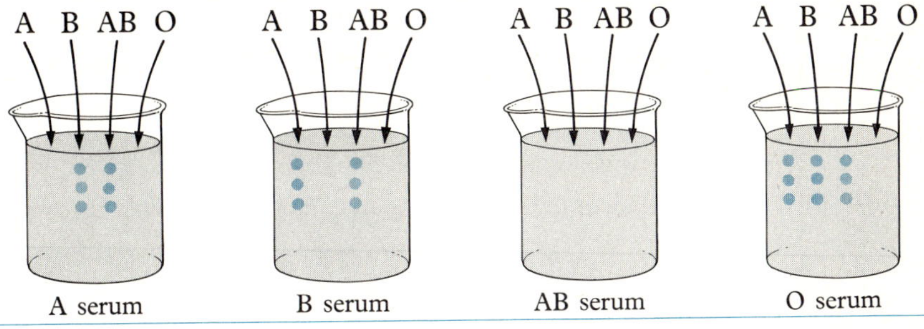

Why are some of the blood combinations incompatible? The body has a natural defense system, called the immunological system, by which a foreign compound is recognized by antibodies and then destroyed. A foreign compound eliciting this response is called an *antigen* and is most commonly a protein. Antibodies are also proteins, and their synthesis is induced by the presence of the antigen. Erythrocytes can act as antigens and therefore can elicit antibody formation in a recipient's blood stream. (Rejection of organ transplants is also due to antigen activity.)

Antigen activity results from the molecular shape of the surface of a substance. If the surface is recognized as strange, or foreign, then antibodies are formed to rid the organism of the foreign substance.

The surface differences of erythrocytes arise from the different shapes of the polysaccharide coatings of these cells. For example, the polysaccharide chain of a Type A erythrocyte has an end group of galactose with an N-acetyl attachment. Type B erythrocytes have plain galactose as the end group. Type AB blood contains both these end groups, and Type O erythrocytes are lacking the galactose end group entirely.

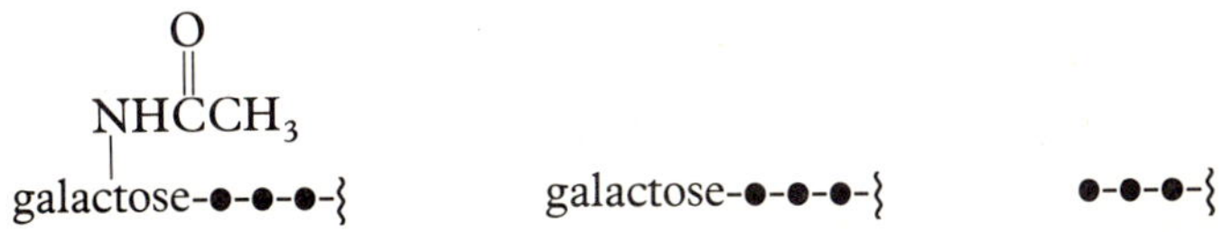

Type A Type B Type O

BLOOD PLASMA

Blood plasma is a water solution of a variety of organic compounds, including proteins, carbohydrates, enzymes, vitamins, and inorganic ions. Of course, the principal constituent of plasma is water (about 90%). Proteins are the next most abundant constituent and account for about 8 percent by weight of plasma. The proteins in plasma are of three principal types:

> *Albumins,* about 4 g/100 cc blood. (These regulate osmotic pressure and transport metal ions and fatty substances.)
>
> *Globulins,* about 3 g/100 cc blood. (α and β globulins transport other compounds and ions, while γ globulins contain antibodies).
>
> *Fibrinogen,* about 0.3 g/100 cc blood. (This is responsible for clotting of blood.)

Albumins are soluble proteins. Blood is a fairly concentrated solution of these compounds, while interstitial fluids do not contain a high concentration of proteins. Recall from Chapter 6 that osmosis is the natural tendency of water to migrate across a membrane from a solution of low concentration to a solution of high concentration. Therefore, by osmosis, interstitial fluids flow into the capillaries in an attempt to dilute the concentrated albumins within the blood. Cell wastes are carried into the blood stream along with the interstitial fluid (by dialysis) and are disposed of by the kidneys, lungs, etc., as the blood circulates through these organs. This osmosis occurs at the venous end of the capillaries where the blood pressure (caused by the pumping of the heart) is low.

This process of fluid entering capillaries is balanced by an equal amount of solution being pushed out of the arterial ends of the capillaries by the blood pressure, which is greater than the osmotic pressure at these locations. The fluid leaving the capillaries and being pushed toward the cells contains nutrients and oxygen for the cells.

$$\underset{\textit{high albumin}}{\text{blood}} \quad \underset{\substack{\text{osmosis and dialysis return} \\ \text{fluids and wastes}}}{\overset{\substack{\text{blood pressure} \\ \text{pushes out fluid and nutrients}}}{\rightleftharpoons}} \quad \underset{\textit{low albumin}}{\text{cell}}$$

If the blood is low in albumin proteins, the process of osmosis is not as efficient; therefore, water collects in the tissues instead of being returned to the blood. The result is *edema,* a swelling of the feet, legs, hands, and possibly the rest of the body.

What can cause low-protein blood? Too little protein in the diet is one cause. For example, *kwashiorkor* (an African name) is a protein-deficiency disease observed in children in a deprived environment. Kwashiorkor appears after a child is weaned and is switched to a starchy diet low in protein. The symptoms are a generalized edema along with a very swollen abdomen and discolored skin.

Another cause of edema is damaged kidneys, which allow albumins to pass out of the blood and into the urine. Urine specimens are commonly tested for the presence of proteins as a check on kidney function.

The *biuret test* is a typical clincial assay for proteins in blood plasma or urine. In this test, the sample is treated with dilute solutions of base and a copper(II) salt. Compounds that contain at least two peptide bonds react to form a violet-colored complex.

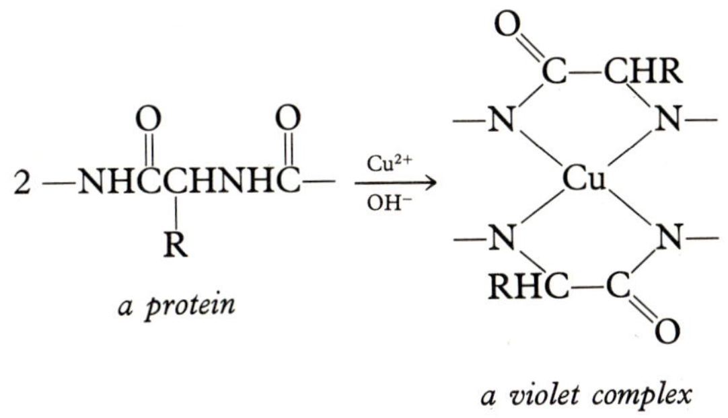

BLOOD CLOTTING FACTORS

Fibrinogen is a soluble protein found in blood. When a blood vessel is injured, the fibrinogen changes to an insoluble, bushy mass of threads called *fibrin*. The fibrin traps blood cells and solidifies into a clot, then finally into a scab on an open wound. The chemistry of blood clotting is complex. We will only briefly summarize what occurs.

Blood clotting is initiated by alteration of the platelets (thrombocytes), small cytoplasmic fragments in the blood. The alteration may be a rough spot caused at the site of injury. In the presence of calcium ions, this platelet alteration causes the enzyme prothrombokinase to be converted to thrombokinase. This enzyme, in turn, catalyzes the conversion of prothrombin to thrombin. (Vitamin K is known to be instrumental in the synthesis of prothrombin.) Finally, thrombin acts as the catalyst for the conversion of fibrinogen to fibrin.

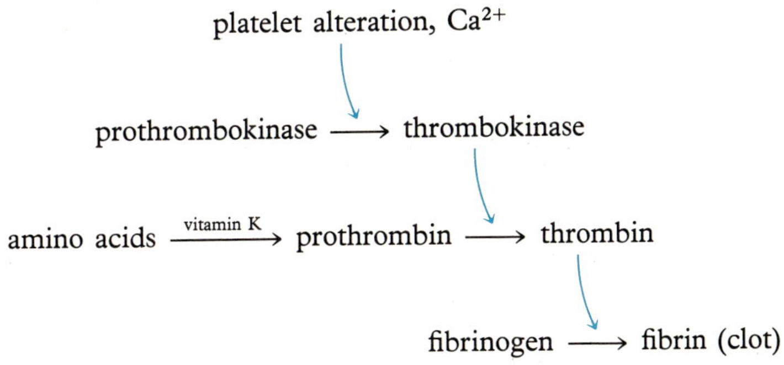

Anticoagulants are agents that prevent blood clots, either in a patient or in a test tube. Anticoagulants work to block the clotting mechanism in a number of different ways.

Heparin is a compound related to the sugars that blocks the catalysis of thrombin and thus interferes with clotting. Salts that precipitate calcium ions prevent the formation of thrombokinase and thus also interrupt the clotting sequence. Sodium oxalate is a typical anticoagulant that causes the precipitation of Ca^{2+} as calcium oxalate.

$$Ca^{2+} + {}^{-}O\overset{\overset{\large O}{\|}}{C}-\overset{\overset{\large O}{\|}}{C}O^{-} \longrightarrow CaC_2O_4$$

oxalate ion — calcium oxalate, *insoluble*

The rat poison *warfarin* is a substance that interferes with the action of vitamin K. Rats that consume warfarin die from internal hemorrhages. Antagonists of vitamin K are also used medically as anticoagulants.

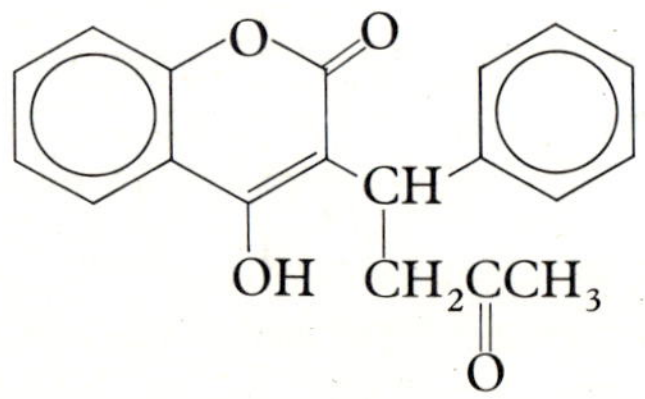

warfarin

The clotting ability of blood can be measured in a number of ways. For example, it may be measured directly on the patient by determining the length of time for a small capillary puncture to stop bleeding. This length of time is called the *bleeding time*. The time required for whole blood to coagulate in a test tube is called the *coagulation time*.

KIDNEY FUNCTION AND URINE

The kidneys serve an indispensable role in mammalian metabolism. It is widely known that the kidneys rid the blood of toxic waste products by transferring them to the urine so that they may be excreted; however, this is only one of the functions of the kidneys. They also help control the electrolyte balance (the relative concentrations of ions) and the water balance of the blood.

The blood passes through the kidneys at a rate of about 1700 liters per day. Therefore, each drop of blood, on the average, is filtered by the kidneys more than once every hour. About 1–2 liters of fluid are retained each day by the kidneys to be passed as urine and rid the body of urea (from metabolized protein), uric acid (from metabolized purines), excess sodium chloride and other salts, and other waste materials. Table 29-1 lists some normal constituents of urine.

Table 29–1. Some Normal Constituents of Urine

Inorganic cations:
Na^+ K^+ Ca^{2+} Mg^{2+} NH_4^+ traces of others
Inorganic anions:
Cl^- $H_2PO_4^-$ SO_4^{2-} traces of others
Organic compounds:
urea, uric acid, and other N-containing compounds
sulfur-containing compounds
bile pigments
traces of glucose, cholesterol, etc.

Kidney function is regulated by a variety of hormones. If, for example, the blood is too concentrated (*hypertonic*), the pituitary hormone vasopressin (page 380) is secreted to signal the kidneys to excrete less water in the urine.

On the other hand, if the blood is too dilute (*hypotonic*), vasopressin secretion is diminished, and an increased secretion of adrenal hormones stimulates *diuresis* (the production of urine). More water is thus excreted. At the same time, a greater percent of ions is resorbed into the blood from the kidneys to maintain the electrolyte balance.

The composition of the urine reflects the health of the kidneys and of the body in general. Urine is easily collected from a patient and consequently is the subject of many clinical tests.

Normal urine has a pH of 4.8–7.5 (usually about 6.0). The slight acidity helps keep the salts in solution. Urine that is more acidic than normal can indicate excessive acid in the system (acidosis), which can result from a high-protein diet, diabetes, etc. (see page 429). After a meal, the urine is usually alkaline.

The specific gravity of urine is used to estimate the quantity of dissolved solid material. Normal urine has a specific gravity of 1.008–1.03. Drinking large amounts of liquid increases the volume of the urine. As urine becomes more dilute, the specific gravity approaches 1.000 (the specific gravity of pure water). By contrast, high fever or other conditions of dehydration lead to more-concentrated urine and a specific gravity greater than normal.

In diabetes mellitus, the volume of urine is large, but the specific gravity is quite high because of dissolved glucose. (A normal person passes only a trace of carbohydrates, although a high-carbohydrate diet can increase the normal quantity.) The presence of glucose in the urine can be detected by enzyme test paper.

Normal urine is clear, but becomes cloudy on standing. If the turbidity is due to the normal precipitation of metal salts, it is cleared up by acidification. If the cloudiness persists after acidification, we would suspect the presence of pus or protein, neither of which is normal.

Pus is an accumulation of white blood cells (leukocytes); its presence in urine indicates an infection in the kidneys or urinary tract. The presence of leukocytes, erythrocytes, and casts (fibrous material) is determined by microscopic examination of the urine sediment after centrifugation.

Blood albumin can pass into the urine from the kidneys or bladder when they are diseased. Other causes for the appearance of albumin in the urine are violent exercise or heart disease, which can lead to interference in the normal circulation to the kidney. A routine urinalysis includes a qualitative test for albumin, such as by the biuret method (page 462).

Although not a clinical test, even the smell of urine can indicate something about its composition. Certain foods, such as asparagus, can lend an unpleasant odor, which arises from thiols. Microorganisms can convert urea in the urine to NH_3. Therefore, an ammoniacal odor can mean a bacterial infection, or it can indicate that the urine sample is not fresh. Preservatives or chilling retard bacterial action on collected urine specimens.

A sweet, fruity smell is the result of ketone bodies (ketonuria) and is indicative of diabetes mellitus, starvation, or a low-carbohydrate diet (page 429). A sweet smell can also arise from diethyl ether used for anesthesia or from other conditions.

SPECTROPHOTOMETRY

One of the most useful tools in clinical analysis is the *spectrophotometer*—an instrument that measures the absorption of visible or ultraviolet light by a sample.

On page 462, we mentioned that proteins react with Cu^{2+} ions to yield a violet complex (the biuret test). The intensity of the violet color is directly proportional to the number of peptide bonds in the sample—that is, if more protein material is present, the color is more intense. The quantity of protein present may therefore be estimated by simply comparing the color with that of a known sample. Spectrophotometry allows us to compare the quantities of the known and the unknown with much greater accuracy than we can obtain by visual comparisons.

THE ELECTROMAGNETIC SPECTRUM. Energy can be transmitted in the form of *electromagnetic radiation.* The electromagnetic spectrum includes radio waves, infrared radiation, visible light, ultraviolet light, *x*-rays, gamma rays, and cosmic rays. The light we see and the colors that brighten our world are only a tiny portion of this spectrum.

All electromagnetic radiation travels in waves and travels at the same speed, the speed of light (3×10^{10} cm/sec). The difference between one type of radiation and another is due to a difference in the *wavelengths* of the radiation. A wavelength is the measure of the distance from the crest of one wave to the crest of the next. Cosmic rays are transmitted by very short waves of about 10^{-10}, or 0.0000000001, cm. Standard-broadcast radio waves, at the other end of the spectrum, are transmitted by extremely long waves almost a thousand meters long. (A mile is 1609 m.)

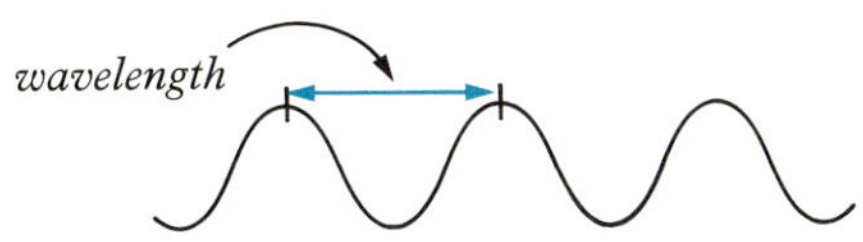

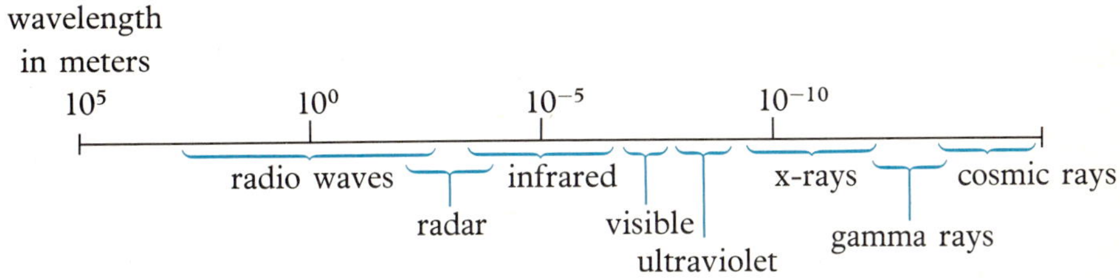

The wavelength of radiation determines its energy content. The shorter the wavelength, the higher is the energy. Looking at the above chart, we can see that ultraviolet radiation is of higher energy than visible light. Both are of much higher energy than radio waves.

An organic molecule can absorb electromagnetic radiation. The absorbed energy may cause electrons to be promoted into higher-energy orbitals or may cause greater vibrations of the bonds in the molecule. The important feature is that, because of structural differences, *different organic molecules absorb radiation of different wavelengths*. The reason for this is that different structures require different amounts of energy (therefore, different wavelengths) to undergo vibrational or electronic changes. Because of this different absorption, chemists can analyze compounds for certain structural features. Another result of energy absorption is that some compounds are colored.

Sample Problem: Promotion of electrons to higher-energy orbitals requires more energy than that required for increased vibration of bonds. Would infrared or ultraviolet radiation be more likely to cause electronic transitions?

Solution: Ultraviolet light contains more energy than infrared radiation and causes electronic transitions in some molecules. Infrared radiation causes vibrational changes.

COLOR. The wavelengths of electromagnetic radiation that allow us to see colors are in the range of 400–700 nanometers (nm), where 1.0 nm = 1.0×10^{-9} m.

400 nm = violet

500 nm = green

700 nm = red

What we perceive as white light is a mixture of the wavelengths of visible light. This mixture of wavelengths can be separated into the different colors by a prism or by raindrops. Then we see the spectrum of visible light as a rainbow.

When we see a colored object, such as a tomato, the color is rarely due to the object giving off a specific wavelength of colored light. Rather, the object absorbs one or more wavelengths of light and reflects the rest. We see the reflected light, not the light absorbed by the object. The color of the object is therefore due to the reflected light, which is white light minus the wavelengths that are absorbed. (See Figure 29-2.)

infrared ← 700 nm — 600 nm — 500 nm — 400 nm → ultraviolet

color absorbed:	red	orange	yellow	green	blue	violet
color observed:	green	blue	violet	red	orange	yellow

The protein-copper complex from the biuret test absorbs light at 546 nm. This wavelength of light is in the yellow portion of the spectrum. Because the yellow portion of the visible spectrum is removed by the complex, we observe a violet color.

Sample Problem: What range of the visible spectrum is absorbed by a lemon?

Solution: According to the chart above, a yellow object absorbs the violet wavelengths (about 400 nm) from the visible spectrum.

Figure 29-2. Color is usually caused by an object absorbing some wavelengths and reflecting the rest.

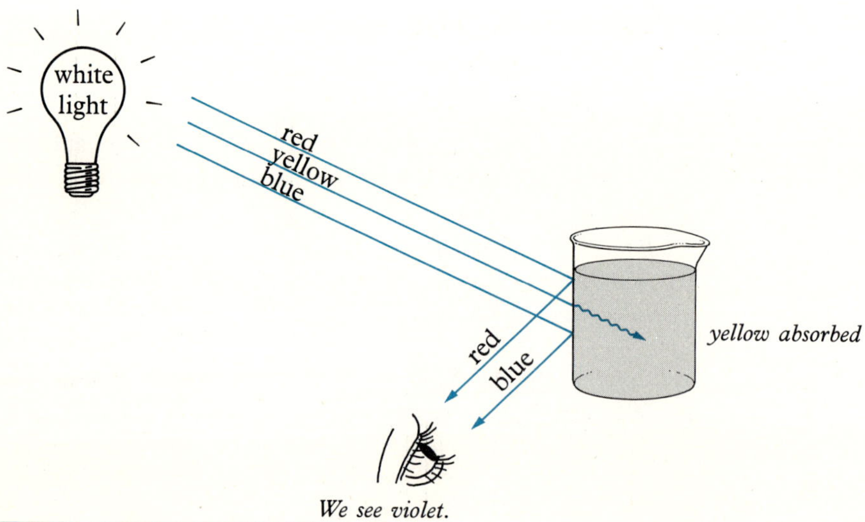

THE SPECTROPHOTOMETER. Consider the following experiment. Suppose that we build an apparatus that generates one wavelength of light (for example, 546 nm). Suppose that we pass the beam of light through a sample of protein-copper complex. Will the light be as intense after passing through the sample as it was when it was first generated? The answer is "No." The sample of protein-copper complex absorbs light with a wavelength of 546 nm; therefore, the intensity of light diminishes as it passes through the sample.

How much the beam diminishes in intensity depends upon the quantity of protein-copper complex in the sample. If there is a large amount of complex in the sample, then the sample removes a large percentage of the light. In this case, we see a large decrease in intensity as the beam passes through the sample. Conversely, if the quantity of the protein-copper complex is small, then only a small decrease in intensity will be noted. If no complex is present, then the intensity of the beam will not be changed as it passes through the sample.

A spectrophotometer is an instrument designed to generate a beam of light, pass it through a sample and measure the amount of absorption of light (called *absorbance*) at a specific wavelength. Most spectrophotometers measure the absorption by a sample over a range of wavelengths. The absorption is graphed by a recorder to give us a *spectrum* of the sample. An example of this type of spectrum is shown in Figure 29-3.

Alternatively, the operator can choose a single wavelength and compare the absorbance of an unknown sample with that of a standard sample of known concentration. Spectrophotometric techniques are very sensitive and can be used to measure quantitatively small amounts of compound present in a sample.

Figure 29–3. Ultraviolet spectrum of tryptophan, showing absorbance (which is a measure of light absorption) at a variety of wavelengths.

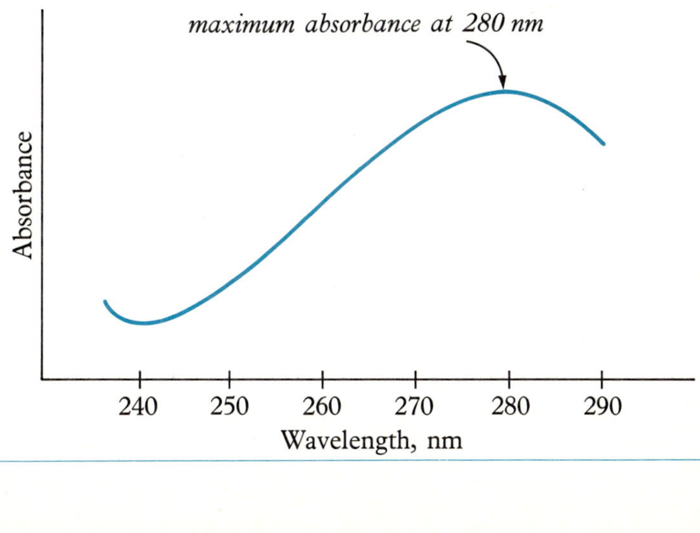

As we have said, different organic compounds absorb light at different wavelengths. A large number of biologically important compounds absorb light in the ultraviolet and visible range. Spectrophotometry is widely used to test for different compounds (sometimes in conjunction with chemical reactions, as in the biuret method). For example, tryptophan, tyrosine, and some other amino acids absorb ultraviolet light at about 280 nm. This absorption can be used to analyze directly for protein. Sugars, uric acid, barbiturates, and porphyrins are just a few other compounds that can be analyzed spectrophotometrically.

STUDY QUESTIONS

29-1. Label each of the following body fluids as circulating or noncirculating:
(a) blood (b) lymph (c) urine

29-2. What reactions occur in the blood to counteract an excess of acid?

29-3. Match the following terms with their definitions:

(a) blood serum	(1) a clotting factor
(b) blood plasma	(2) red blood cells
(c) fibrinogen	(3) blood cells instrumental in clotting
(d) erythrocytes	(4) white blood cells
(e) leukocytes	(5) whole blood with the cells and clotting factors removed
(f) platelets	(6) whole blood with only the cells removed

29-4. Tell whether each of the following test samples is Type A, B, AB, or O blood:
(a) precipitates with Types B and O serum, but not with A or AB serum.
(b) does not precipitate with any of the test serums.
(c) precipitates with A, B, or O serum, but not with AB serum.

29-5. Write the equation for the reaction of ala-ala with copper(II) chloride in alkaline solution. Describe in words what you would observe.

29-6. Which of the following circumstances might interfere with normal blood clotting?
(a) low platelet level
(b) vitamin K deficiency
(c) calcium deficiency
(d) a dose of heparin

29-7. Suggest one technique for preventing the clotting of a blood sample in a test tube.

29-8. Without referring to the text, list the functions of the kidneys.

29-9. Give the effects on the urine of (a) drinking a large amount of fluid, (b) drinking very little fluid, (c) diabetes mellitus, (d) acidosis, (e) eating a pound of liver, (f) ketosis.

29-10. Name some causes of cloudy urine.

29-11. β-Carotene, which is found in carrots, is a compound that can be converted to vitamin A (page 476) by animal organisms. β-Carotene absorbs visible light at 451 nm. What color is β-carotene? Explain your answer.

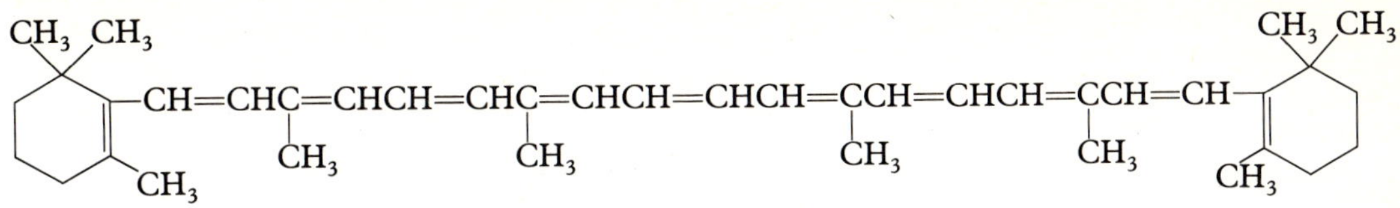

β-carotene

29-12. Ultraviolet light with a wavelength of 250 nm is passed through the following samples. Tell what would be observed in each case.

(a) a dilute solution of a compound that absorbs at 250 nm.

(b) a concentrated solution of a compound that absorbs at 250 nm. [Compare with (a).]

(c) a concentrated solution of a compound that does not absorb at 250 nm.

CHAPTER 30 MINERALS, VITAMINS, AND DRUGS

OBJECTIVES

Minerals *Name and list the functions and sources of some of the minerals that are necessary to a balanced diet.* * *Describe flame photometry and atomic absorption photometry.*

Vitamins *Define vitamin, and describe the general function of vitamins.* * *Describe the functions and sources of (1) the water-soluble vitamins and (2) the fat-soluble vitamins.*

Drugs *Describe the action of different types of drugs, such as insulin, thyroid, and penicillin.* * *Describe some of the characteristics of drug response (for example, species differences, tolerance, and so forth).* * *Describe how the shape of a molecule can cause drug activity.*

Carbohydrates, proteins, and lipids make up the bulk of animal foodstuffs. In addition to these, animals must have small amounts of certain minerals and vitamins in their diets. In this chapter, we will consider the roles of minerals and vitamins in nutrition. We will also take a brief look at drugs, compounds that are used to keep the body functioning as we would like.

MINERALS

In nutrition, a *mineral* is defined as an inorganic substance necessary for the proper functioning of an organism. The mineral that most adults ingest in the largest quantities is sodium chloride, which is necessary for the proper electrolytic balance of the body fluids, a balance necessary for the regulation of osmotic pressures. Potassium salts, abundant in fruits and vegetables, are also necessary for this balance. (See Table 30–1.)

Calcium ions are found principally in the extracellular fluids. A deficiency of this ion can readily be detected by analysis of either blood serum or plasma. Transmission of nerve impulses, muscle contractions, and coagulation of blood all involve calcium ions. In addition, a balanced ratio of calcium and phosphate ions along with vitamin D is required for the growth and maintenance of bone structure. Milk and cheese are sources of these two elements. Certain vegetables are also rich in calcium, while meat contains an abundant amount of phosphorus. A glass or so of milk per day supplies the average adult with all the necessary calcium and phosphorus. Of course, growing children and lactating mothers require more calcium salts than the average adult.

Magnesium ions are found as part of the structure of chlorophyll; thus, green, leafy vegetables supply this nutrient. Drinking water, which almost always contains magnesium ions, is another source. Magnesium is used in the blood cells, bones, and body fluids and also aids in enzyme action.

Table 30–1. Minerals Required by the Body

Mineral[a]	Approximate Daily Adult Requirement	Where Needed	Good Source
Na^+	5 g	Body fluids	Table salt
K^+	3 g	Body cells	Plants
Ca^{2+}	0.8 g	Bones	Milk
P[b]	0.7 g	Bones	Milk
Mg^{2+}	0.3 g	Body fluids	Plants
Fe^{2+}	0.01 g	Blood	Liver

[a] In lesser quantities, the following are also essential: Cu, Cl, I, Co, Mn, Zn, and Mo.
[b] As HPO_4^{2-} and $H_2PO_4^-$.

Iron is important in red blood cells, where it serves to help transport oxygen and carbon dioxide. Iron is also a necessary part of the cytochrome system (Chapter 25). Our dietary iron requirement is not unusually high because the body is capable of recycling iron that is used in the body processes. A deficiency of iron results in *anemia*, or "iron-poor blood." This condition is easily remedied by the addition of an iron salt supplement or by the inclusion of iron-rich foods (such as liver) in the diet.

The other minerals required by human beings are needed in only trace amounts. With the possible exception of iodide ion and zinc ion, these minerals are present in adequate amounts in an ordinary diet. The need for iodide ion was discussed in Chapter 9.

Zinc plays an essential cofactor role in many enzyme systems. The average person obtains about 10 mg of zinc ions per day from meat and vegetables, which is an adequate amount. In some areas, the zinc salts in the soil have been depleted. Vegetables grown in these areas may be deficient in zinc. As a result, some nutritional authorities recommend that vegetarians take zinc salt supplements.

In clinical laboratories, analysis for ions is often accomplished with a *flame photometer*. The sample to be analyzed is aspirated into a hot flame. The electrons of the ions present in the sample are promoted to higher-energy orbitals by the heat of the flame. When these electrons return to the ground state (lowest-energy state), they release their captured energy in the form of light. Instead of measuring the energy absorbed as we do in absorption spectrometry, we measure the intensity of the light emitted at certain wavelengths.

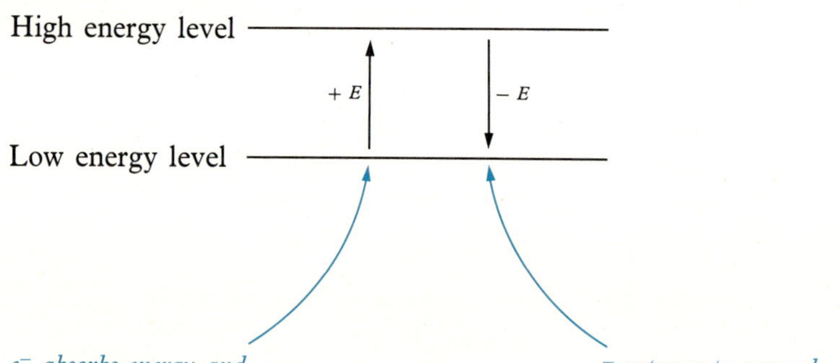

The key feature of flame photometry is that different ions emit light at different and characteristic wavelengths. We can thus monitor specific wavelengths and analyze a sample for a variety of inorganic ions.

Another common method for the analysis of ions is the use of *atomic absorption photometry* (called AA). As in flame photometry, the sample to be analyzed is aspirated into a hot flame. Atomic absorption differs from flame photometry in that a light beam is passed through the flame. The decrease in intensity of the light beam is measured at specific wavelengths. In this respect, atomic absorption is similar to visible and absorption spectrophotometric techniques.

VITAMINS

A *vitamin* is an organic compound (besides a carbohydrate, protein, or lipid) that is normally found in foodstuffs and is necessary in small quantities for the functioning of an organism. Vitamins serve as coenzymes in the metabolic and regulatory pathways within the system.

Vitamins are classified as *fat-soluble* or as *water-soluble*. (See Table 30–2). The water-soluble vitamins, such as vitamin C, are not stored to any appreciable extent in the body. These vitamins should be supplied daily in the diet. The fat-soluble vitamins, such as vitamins A and D, can be stored in the tissues of the body to be drawn upon when needed. Ingestion of too great a quantity of a fat-soluble vitamin can result in *vitaminosis,* the overaccumulation of the vitamin in the tissues. The symptoms of vitaminosis vary; they depend upon which vitamin has been taken to excess. For example, greatly excessive amounts of vitamin D can lead to dissolution of the bones.

Vitamin A is necessary for growth, healthy skin, and proper vision. It is found in green, orange, and yellow fruits and vegetables and in cod liver oil. You may have heard that the eating of carrots, rich in carotene (page 471), which produces vitamin A,

Table 30–2. Vitamins

Vitamin	Function	Good Source
Fat-soluble:		
A_1 and A_2	Necessary for vision	Green and yellow vegetables
D_2 and D_3	Ca and P metabolism	Sunlight, enriched milk
E	Cofactor	Green, leafy vegetables
K	Blood clotting	Vegetables
Water-soluble:		
B_1 (Thiamine)	Nerve activity; carbohydrate metabolism	Cereals, liver
B_2 (Riboflavin)	Coenzyme	Milk, vegetables
B_6	Coenzyme	Cereals, liver
PP (Nicotinamide)	Metabolism of ATP	Meat
B_{12}	Growth factor	Eggs, meat
C (Ascorbic acid)	Coenzyme	Citrus fruits
Biotin	Growth factor; coenzyme	Liver, dried peas
Folic acids[a]	Cofactors	Liver, chicken
Pantothenic acid	Growth factor; in coenzyme A	Liver, eggs
Inositol[a]	Growth factor	Plants

[a] Role in human nutrition not fully established.

prevents and cures night blindness—and so it will, if the night blindness was caused by a vitamin A deficiency. In Chapter 13, we discussed the reason for this.

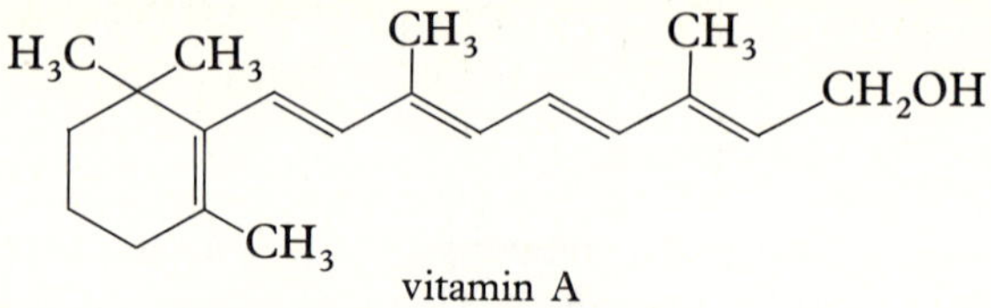

vitamin A

Vitamin B_1, also called *thiamine*, functions as a coenzyme that, for example, is instrumental in converting pyruvic acid to acetylcoenzyme A in the metabolism of glucose. Lack of vitamin B_1 can result in beriberi and nervous system abnormalities. Vitamin B_1 is found in fruits, vegetables, whole grains, and enriched bread. Although it is quite widely distributed, this vitamin can be destroyed by overcooking.

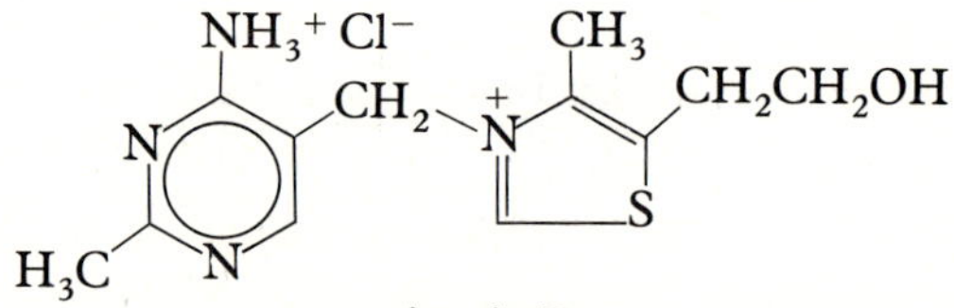

vitamin B_1

Vitamin B_2, or *riboflavin*, forms a portion of the coenzyme FAD. (See Chapter 25.) This vitamin is contained in milk, liver, eggs, and green, leafy vegetables.

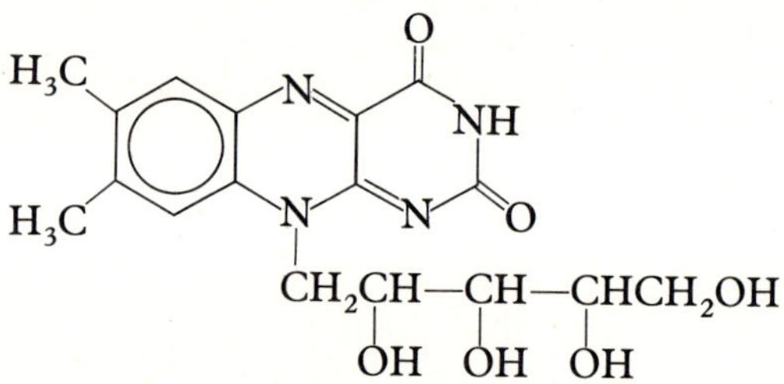

vitamin B_2

Nicotinamide, also called *niacinamide*, forms the working end of the coenzyme NAD^+. (See Chapter 25.) Lack of this vitamin results in the condition known as pellagra. Nicotinamide is found in many meats and vegetables. The body is also able to synthesize this compound from nicotinic acid or from the amino acid tryptophan.

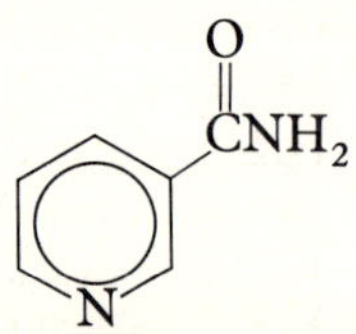

nicotinamide
(niacinamide)

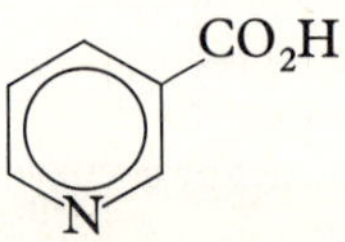

nicotinic acid
(niacin)

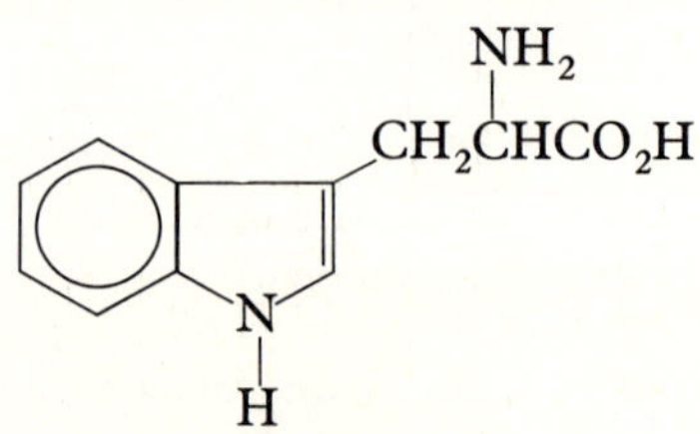

tryptophan

Vitamin B_6, or *pyridoxal,* forms a coenzyme that catalyzes transamination reactions. It is not known whether or not the body can synthesize this vitamin. However, vitamin B_6 occurs in whole grains, meat, vegetables, and milk. Because it is so widely distributed, a deficiency in this vitamin would be unlikely.

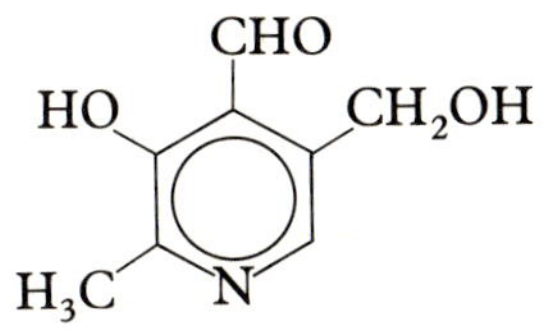

vitamin B_6

Vitamin B_{12}, or *cobalamin,* contains a chelated cobalt ion (hence, the name cobalamin). Lack of vitamin B_{12} leads to pernicious anemia, a disease that was often fatal before the cause was known. This vitamin is found in many animal products; therefore, a deficiency is unlikely in a person who has normal metabolism.

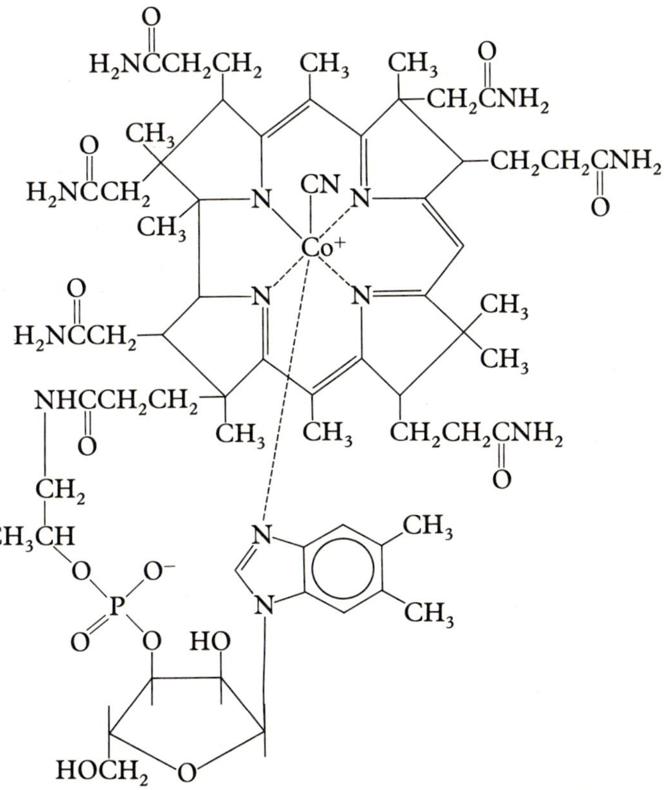

vitamin B_{12}

Vitamin C (ascorbic acid), one of the best known ‚itamins, is a cyclic ester (a *lactone*) and also contains an enediol grouping.

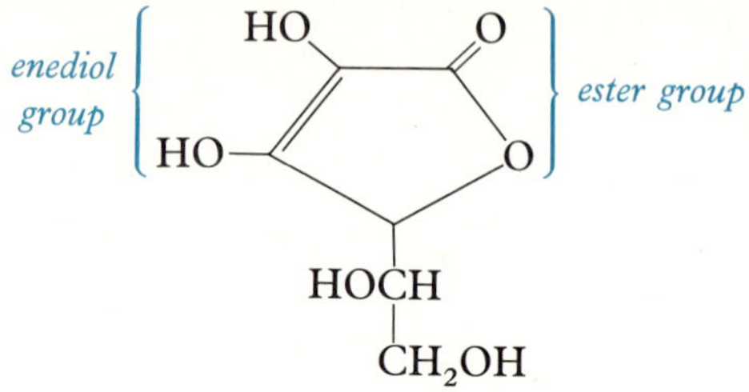

vitamin C

Most animals can synthesize vitamin C from carbohydrates in their diets. Dogs, horses, cattle, and other animals have this ability. Humans, other primates, and certain rodents (for example, guinea pigs) cannot synthesize vitamin C; consequently, it must be supplied in their diets. Because this vitamin is not stored in the body's tissues, an individual's daily diet should include some source of vitamin C. Orange juice is one traditional source of vitamin C, but certain other vegetable products (such as tomatoes, potatoes, and cabbage) are also fairly rich in this vitamin. Synthetic vitamin C is sometimes added to fruits as a preservative.

Vitamin C is readily air-oxidized to a diketone at the enediol portion of the molecule, especially in alkaline solution. The oxidized product does not possess vitamin activity. Exposure to air and cooking causes oxidation of the vitamin C in many foods, thus diminishing their vitamin C activity. The vitamin C in tomatoes and citrus fruits (both acidic) is more stabilized; these foods retain much of their vitamin C even when heated.

Sample Problem: Suggest a structure for the product of the oxidation of vitamin C.

Solution:

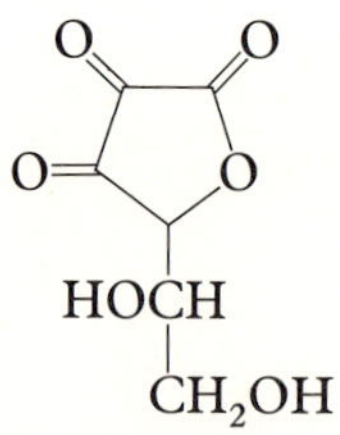

Vitamin C is instrumental in the synthesis of collagen, which forms the connective tissue in the body. Connective tissue is in a constant state of being broken down and replaced. Lack of vitamin C leads to scurvy, a condition in which there is a reduction in the amount of connective tissue because of its lack of replacement. The symptoms of scurvy include bleeding of the gums, gastrointestinal tract, skin, and muscles.

In the "age of exploration" (the 16th through the 18th centuries), death from scurvy was common among sailors, who sometimes went for months without fresh fruits or vegetables. The British discovered that lemons and limes prevented and cured scurvy. The practice of carrying these fruits on voyages led to the name "limey" to describe a British sailor.

The use of "megadoses" of vitamin C to prevent colds or to alleviate the symptoms of a cold has been widely publicized. Whether or not large doses of vitamin C will perform these miracles (without adverse side effects) is still a controversial topic.

Vitamin D is another vitamin that is well known to most of us. Lack of vitamin D in children and young animals leads to rickets, a condition in which the bones do not grow and harden normally. Sunlight is one source of vitamin D; ultraviolet light catalyzes the formation of this vitamin from steroids in the skin. (Subscript numbers in a vitamin name—for example, vitamin D_3—usually refer to different forms of a vitamin with slight variations in structure.)

A synthetic vitamin D, vitamin D_2 (irradiated ergosterol), which is usually added to milk, is another source of this vitamin.

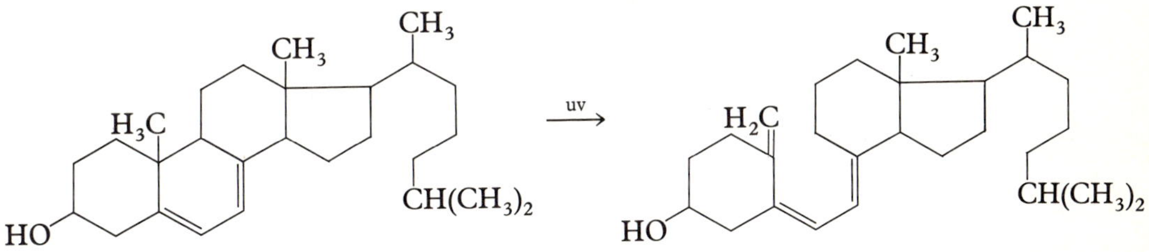

7-dehydrocholesterol
in the skin

vitamin D_3

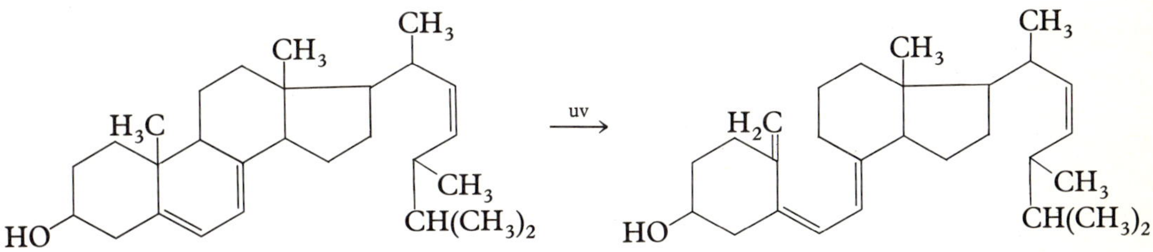

ergosterol
found in yeast, soybeans, and the fungus ergot

vitamin D_2
irradiated ergosterol

Vitamin K is necessary for proper clotting of the blood. This compound is found in many foods and is synthesized by intestinal bacteria. Deficiencies are rarely encountered.

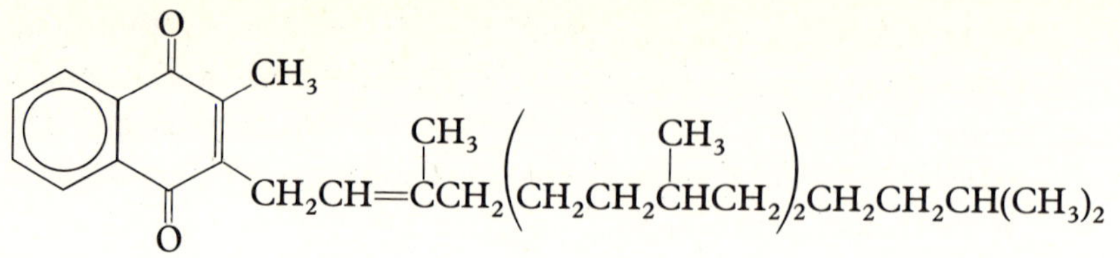

vitamin K_1

Biotin serves as a coenzyme in the synthesis of fatty acids in the body. This vitamin is widely distributed in foods, and deficiencies are rare. However, biotin in the system is effectively destroyed by the protein in raw egg whites. The practice of ingesting large amounts of raw eggs (as in homemade eggnog) or of routinely feeding raw eggs to dogs as a coat conditioner is to be discouraged. (Cooked egg whites, which are denatured, do not tie up biotin and are safe to eat.)

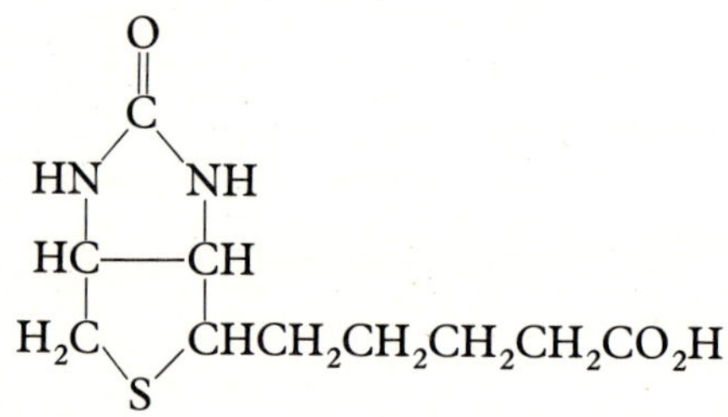

biotin

DRUGS

A *drug* is a compound that is given to a person or an animal to prevent or cure a disease or to elicit some unusual or specific physiological effect. Because of the diversity of their functions, drugs vary considerably in their structures, properties, and effects.

The drugs that have the most obvious effects on the body are the *replacement drugs,* that is, substances used to replace or supplement a normal hormone or other compound that is in short supply in a particular patient's system. *Insulin* is an example of a replacement drug. A diabetic produces an insufficient amount of the pancreatic hormone insulin, which promotes the metabolism of glucose. Severe cases of diabetes can be controlled by injections of insulin. Mild cases are often treated with oral products that stimulate the production of insulin in the patient's body. Careful monitoring of carbohydrate intake is also important.

Another common replacement drug is *thyroid,* which is obtained from dried and powdered thyroid glands from domestic animals killed for meat. If the thyroid gland does not produce enough of the regulatory hormone thyroxine, an individual will suffer from hypothyroidism. The entire metabolism will be depressed and the individual may be lethargic and gain weight. Treatment with thyroid (which contains thyroxine) or synthetic thyroxine itself results in remarkable improvement in the well-being of these patients and allows them to lead normal lives.

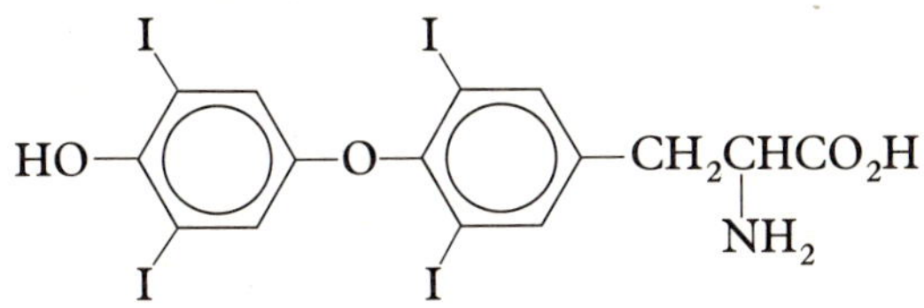

thyroxine

The body has a number of defenses against invading organisms such as bacteria or viruses. The skin and mucous membranes are the first lines of defense. These tissues protect an organism by keeping invaders out mechanically—they form the moat around the castle. If an invader gets past this first line of defense, *phagocytes,* a type of white blood cell, may attack and antibodies may be produced by the system.

Vaccination with killed viruses or with modified live viruses (viruses that have been purposely weakened before injection) stimulates a system to generate its own antibodies against a particular disease without actually infecting the system with the disease itself.

Antibodies occasionally do harm to their own systems. We have already mentioned the role of antibodies in the rejection of blood of the wrong type (Chapter 29). Allergic reactions to penicillin or to bee stings are also the result of antibodies being formed to reject foreign protein material. Severe allergic reactions can, of course, be fatal. In transplantation of organs (such as kidney transplants), antibody formation must be suppressed by drugs or by radiation; otherwise, the recipient of the organ will reject the transplant as foreign protein.

In addition to taking advantage of a system's natural defense mechanisms, scientists have also developed artificial defenses that can be administered to a system to cure or prevent disease. One of the triumphs of modern medical science was the discovery that synthetic, nonnaturally occurring organic compounds can be used to suppress or to destroy invaders of the human body.

Examples of compounds that are more toxic to an invader than the host are the *penicillins,* which act by blocking the synthesis of the bacterial cell wall. (See Chapter 21.) The penicillin compounds selectively inhibit the growth of bacteria, and the natural mechanisms of the body can then get rid of them.

CHARACTERISTICS OF DRUG ACTION. The specialized cells of all organisms do not always behave in the same manner when exposed to a drug. *Species differences* are common. Most drug testing is carried out using rats and mice, not human beings.

However, a compound may not elicit the same response in a mouse as it does in a human. For example, LSD (lysergic acid diethylamide) appears to have no unique activity in mice, yet it is a potent hallucinogen in humans. LSD has been given to all types of animals including clams, snails, spiders, flies, Siamese fighting fish, guppies, monkeys, and an elephant. As far as we can tell, none react in just the same way that humans do. The hallucinogenic properties of this drug could not have been predicted by animal testing.

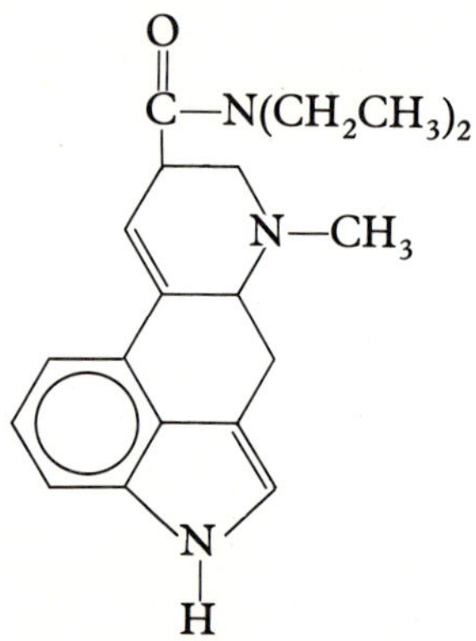

lysergic acid diethylamide (LSD)

Generally, a drug will elicit the same responses from members of the same species. However, an occasional individual may exhibit an atypical response, called an *idiosyncrasy*. Morphine acts as a central nervous system depressant in almost all humans, but an occasional person becomes excited when given morphine. *Hypersensitivity,* or drug allergy, is another atypical response to a drug. The allergic reaction of some people to penicillin is an example of hypersensitivity.

Many drugs have undesirable, but predictable, *side effects*. These side effects are not atypical and are to be watched for in any individual receiving that drug.

In addition to known side effects of drugs, there are three other aspects to be considered in the use of drugs: tolerance, habituation, and addiction. *Tolerance* is defined as unusual resistance to the ordinary dose of a drug and is frequently acquired from constant use of the drug. The cells affected by a drug may become acclimated to the presence of the drug or, in some cases, the body learns to detoxify and eliminate the drug at a more rapid than usual rate. In either case, higher than normal doses are required to elicit an expected response. Tolerance to nicotine, morphine, and barbiturates are not at all unusual.

When some drugs are taken by a person for a period of time, a psychological dependence on the drug develops. This dependence is called *habituation*. The dependence of many individuals on coffee, which contains the stimulant caffeine, or dependence on cigarettes, which contain nicotine, are examples of habituation.

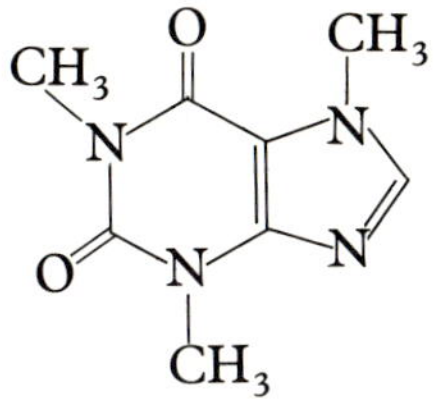

caffeine

a stimulant to the central nervous system

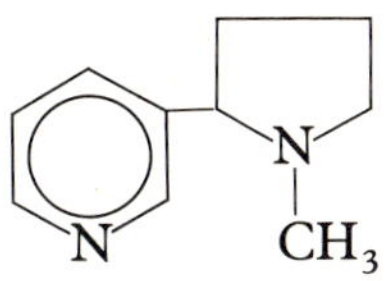

nicotine

a drug that first stimulates, then depresses, the central nervous system

Addiction is the physiological, as well as psychological, dependence of an individual on a drug such as morphine or heroin. The body develops a tolerance for the drug and so completely acclimates itself to the presence of the drug that the drug is actually needed to keep the body in some state of equilibrium. Complete and sudden withdrawal of the drug results in the so-called withdrawal symptoms, which are physical as well as mental in nature. The reasons for this physical dependence on some drugs are not entirely understood. Fortunately, few drugs are truly addictive.

There is not a clean dividing line between habituation and addiction. For example, for some individuals the nicotine in tobacco is only habitual; for other individuals nicotine is thought to be addictive.

STRUCTURAL RELATIONSHIPS OF DRUGS. It is interesting to note the similarities in structure among the addictive drugs morphine, codeine, and heroin, called the *opiates*. Morphine and codeine occur in the opium poppy. Heroin does not occur naturally, but is synthesized from morphine.

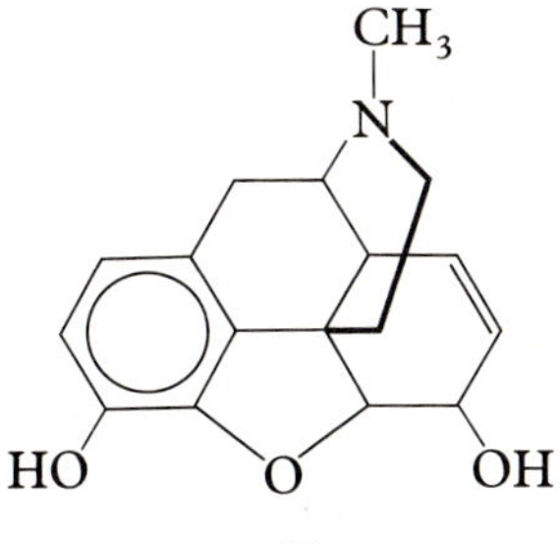

morphine

an addictive pain reliever

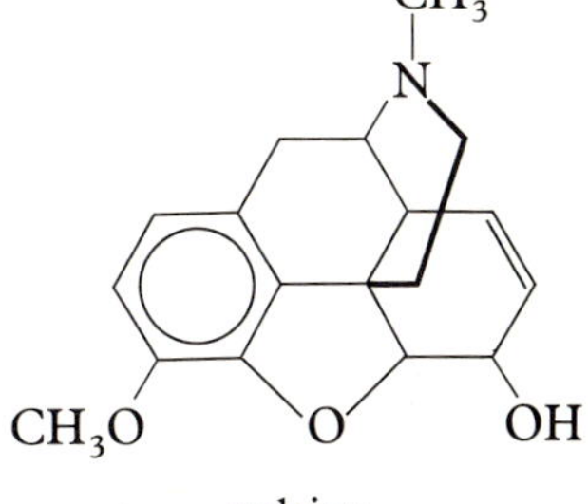

codeine

an addictive pain reliever and cough suppressant

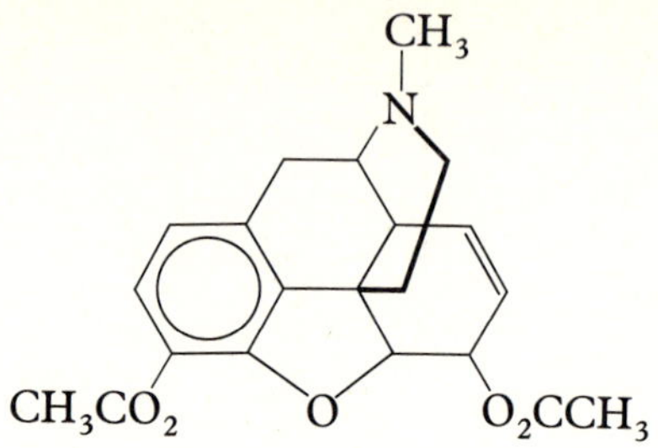

heroin
(diacetylmorphine)
a highly addictive drug

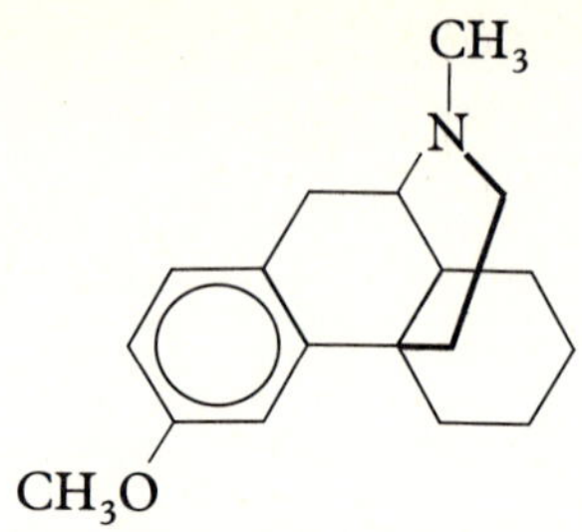

dextromethorphan
a nonaddictive cough suppressant

The action of opiates is only partially understood. It is thought that these molecules become attached at sites in the brain called *opiate receptors.* It has been discovered that certain normally occurring peptides (called *enkephalins*) can compete with opiates at the opiate receptors and also relieve pain. Possibly, the opiates actually mimic the normal action of these brain peptides and modulate nerve impulses to pain-sensing portions of the brain.

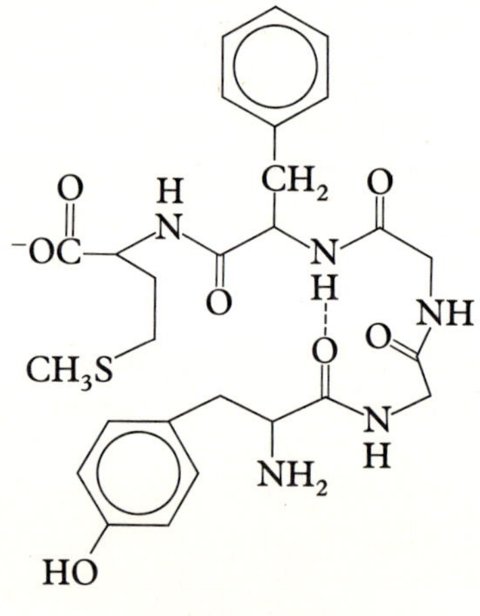

tyr-gly-gly-phe-met
an enkephalin, a natural peptide with opiate activity

Other drugs also show structure-activity relationships. Norepinephrine is one of the compounds used to transmit nerve impulses to the organs of our bodies by way of the *sympathetic nervous system*—the system that causes the pupils to dilate, the veins to constrict, the heart to beat faster, and otherwise prepare the body for "fight or flight" in time of danger. A related compound is epinephrine (adrenaline), which is secreted by the adrenal gland and also stimulates the sympathetic nervous system.

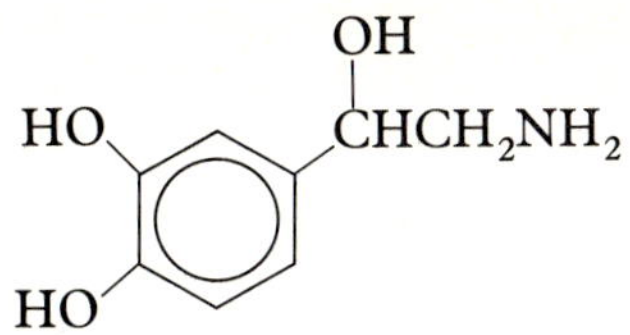

norepinephrine

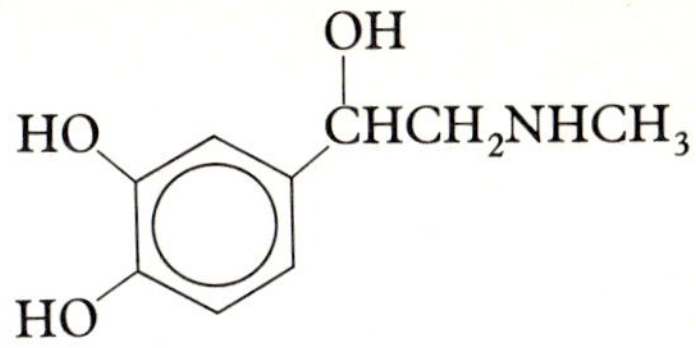

epinephrine
(adrenaline)

Both these naturally occurring compounds are derivatives of β-phenylethylamine.

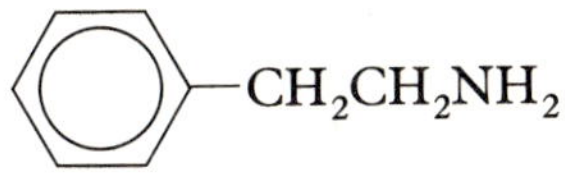

β-phenylethylamine

Not only these two natural compounds, but also many synthetic derivatives of β-phenylethylamine, have physiological activity. Often, these synthetic compounds mimic the activity of norepinephrine or epinephrine on the sympathetic nervous system, probably by fitting themselves to the protein receptors that are designed for norepinephrine. These synthetics are referred to as *sympathomimetic amines*. Amphetamine and dexedrine are two common drugs of this class. Amphetamine is a racemic mixture; dexedrine is the dextrorotatory isomer. Both compounds increase blood pressure, dilate pupils, decrease salivation, and otherwise mimic preparation for fight or flight. Ephedrine, a decongestant used in nose drops, is another example of a sympathomimetic amine. Ephedrine causes constriction of the nasal capillaries, shrinking of swollen tissues, dilation of the bronchioles, and inhibition of nasal secretions.

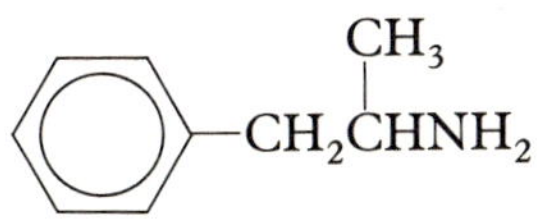

amphetamine
(Benzedrine)

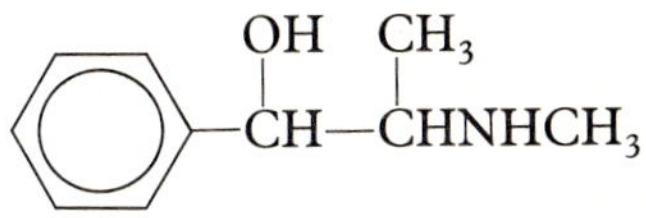

ephedrine

STUDY QUESTIONS

30-1. What are the principal roles played by the following minerals in body processes?
(a) magnesium (b) iron
(c) calcium (d) phosphorus

30-2. Define (a) *vitamin* and (b) *vitaminosis*.

30-3. Which of the following vitamins are classified as fat-soluble vitamins?
(a) nicotinamide (b) thiamine
(c) ascorbic acid (d) panthothenic acid

30-4. Match the following vitamin-deficiency conditions with their respective vitamins.

(a) nicotinamide	(1) pernicious anemia
(b) vitamin C	(2) scurvy
(c) vitamin B_1	(3) pellagra
(d) cobalamin	(4) beriberi

30-5. Some scientists feel that vitamin D should not be classed as a vitamin. Suggest an alternative classification.

30-6. Which of the following substances can be considered as a drug against infectious diseases?
(a) phagocytes (b) thyroid
(c) sulfathiazole (d) benzylpenicillin

30-7. Define each of the following terms as related to drug action:
(a) species difference (b) idiosyncrasy
(c) hypersensitivity (d) side effect
(e) tolerance (f) habituation
(g) addiction

30-8. Which of the following compounds are β-phenylethylamines?

(a) $C_6H_5CHCH_2NH(CH_3)_2$
$|$
CH_3

NH_2
$|$
(b) $C_6H_5CHCH_3$

(c) $C_6H_5CHCH_2CH_3$
$|$
$CH_2NHCH_2CH_3$

(d) $(C_6H_5)_2CHCH_2NH_2$
(e) dextromethorphan
(f) codeine

30-9. Acetylcholine is one of the compounds responsible for transmitting nerve impulses. Part of the action of acetylcholine is to increase sweating and salivation and to increase intestinal activity (sometimes resulting in diarrhea).

$$CH_3\overset{\overset{O}{\|}}{C}OCH_2CH_2\overset{+}{N}(CH_3)_3\ Cl^-$$

acetylcholine chloride

It has been found that $(CH_3)_4N^+\ Cl^-$ mimics the action of acetylcholine to some extent. Which of the following compounds would be even more likely to mimic the action of acetylcholine?

(a) CH_3CH_2OH

(b) $CH_3CH_2CH_2\overset{+}{N}(CH_3)_3\ Cl^-$

(c) 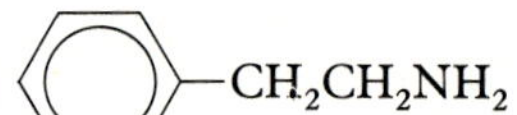

(d) 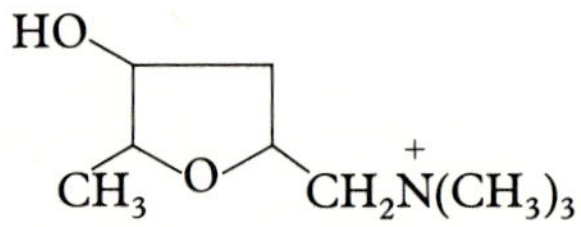

muscarine

found in the poisonous mushroom Amanita muscaria

30-10. Mescaline is a hallucinogen found in the peyote cactus. Suggest a reason for its physiological activity.

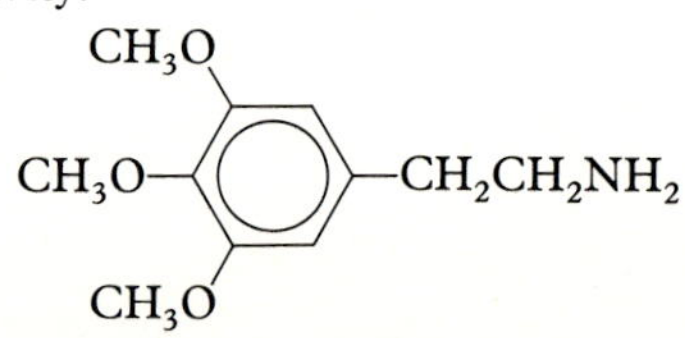

mescaline

APPENDIX A REVIEW OF SOME ALGEBRAIC MANIPULATIONS

The ability to use simple algebra is essential to the study of chemistry. This review is intended for those students who may feel insecure about their mathematical background. We would also encourage these students to refer to a good, beginning algebra text.

MULTIPLICATION

The process of multiplication may be represented in a number of ways. (Note the difference between the algebraic x, which represents an unknown value, and the operational $\times$, which indicates multiplication.)

$$5 \times 10 \qquad 5 \cdot 10 \qquad 5(10) \qquad (5)(10)$$

All mean 5 times 10.

$$xy \qquad x \cdot y \qquad x(y) \qquad (x)(y)$$

All mean x times y.

If a number is used to multiply a series of terms, such as in the expression $10(a + b)$, then *each* term must be multiplied by the number.

$$10(a + b) \text{ means } 10a + 10b$$

FRACTIONS

A whole number is a number (such as 1, 2, 25, or 132) that does not contain a decimal or a fraction. Thus, 1.25 and $2\frac{1}{2}$ are not whole numbers. In chemistry, fractions and decimals are encountered frequently.

A fraction is merely an expression of a division, that is, division of the numerator (the top) by the denominator (the bottom).

$$\tfrac{10}{5} \text{ means 10 divided by 5}$$

When another term is to be multiplied by a fraction, the term is multiplied by the numerator and divided by the denominator.

$$\frac{5}{9}(32) \text{ means } \frac{5 \times 32}{9}, \qquad \text{or} \qquad \frac{160}{9}$$

Just as in any multiplication, when a fraction is used to multiply a series of terms, each term must be multiplied by the fraction.

$$\tfrac{5}{9}(^\circ\text{F} - 32^\circ) \text{ means } \tfrac{5}{9}(^\circ\text{F}) - \tfrac{5}{9}(32^\circ) \qquad \text{or}$$

$$\frac{(5)(^\circ\text{F})}{9} - \frac{(5)(32)}{9}$$

Note that $\dfrac{x+y}{x} = \dfrac{x}{x} + \dfrac{y}{x}$, or $1 + \dfrac{y}{x}$

$\dfrac{x+y}{x}$ does NOT equal y.

By the same rules, $\dfrac{5 \times 6}{6} = 5$

but $\dfrac{56}{6} = 9.33$, and not 5.

A *ratio* is nothing more than a fraction. Let us say that we have 10 oranges in one basket and 5 in the other. The ratio of oranges in the two baskets is 10 to 5, or 10:5, or 10/5. The terms in a ratio, as those in a fraction, may be multiplied or divided by any number, so long as the numerator (top) or denominator (bottom) are multiplied by the same number.

$$\frac{10}{5} = \frac{10(100)}{5(100)} = \frac{2}{1} = \frac{8}{4}$$

$$\frac{x}{y} = \frac{10x}{10y} = \frac{38.4x}{38.4y}$$

$$10{:}5 = 100{:}50 = 2{:}1 = 8{:}4$$

When we speak of the ratio of atomic masses of hydrogen to oxygen in water, the ratio may be expressed as 1:8 or as 2:16.

$$\text{Hydrogen:Oxygen} = 1{:}8 = 2{:}16$$

$$\frac{\text{Hydrogen}}{\text{Oxygen}} = \frac{1}{8} = \frac{2}{16}$$

DECIMALS

A decimal is simply a fraction of 10 or a fraction of a multiple of 10.

$$0.1 = \tfrac{1}{10} \qquad 0.01 = \tfrac{1}{100} \qquad 0.02 = \tfrac{2}{100} \qquad 5.3 = 5\tfrac{3}{10}$$

In solving problems involving either whole numbers or decimals, note should be taken of the number of significant figures. (This is particularly true when a pocket calculator is used, because these calculators show more digits than are significant or meaningful.) The definition of significant figures is best shown by examples.

0.02 — *one significant figure*

0.023 — *two significant figures* (Initial zeros are not significant.)

1.023 — *four significant figures*

As may be seen in the examples, the placement of the decimal point and the number of *initial* zeros is immaterial in determining the number of significant figures. However, the number of intermediate or terminal zeros *are* counted as significant figures.

1.00003 — *six significant figures*

1.0300 — *five significant figures*

1030 — *three significant figures (but maybe four)*

The answer to a mathematical problem should contain no more significant figures than are found in any term of the initial data.

$$\underbrace{6.19}_{\text{three}} \times \underbrace{3.000}_{\text{four}} = \underbrace{18.6}_{\text{three}} \text{ and not } 18.57$$

MULTIPLICATION AND DIVISION WITH DECIMALS

A floating-point pocket calculator positions a decimal automatically. In slide-rule calculations or hand calculations, however, the placing of a decimal in the answer must be determined.

When a number is to be multiplied by a decimal fraction, first carry out the multiplication and *then* place the decimal in the answer. To place the decimal, *add* the number of decimal places in each of the multiplied terms and use this sum as shown in the examples.

Example: 0.05×25

Step 1: Multiply. $5 \times 25 = 125$

Step 2: Sum the decimal places in the original terms.

0.05 25.

2 places + 0 places = 2 places

Step 3: Insert the decimal in the answer.

1.25 or 1.25

2 places

Example: 0.002×0.15

Step 1: Multiply. $2 \times 15 = 30$

Step 2: Sum the decimal places in the original terms.

0.002 0.15

3 places + 2 places = 5 places

Step 3: Insert decimal in the answer.

0.00030 or 0.00030

5 places

To divide terms with decimals, first divide. Then *subtract* the number of decimal places in the denominator from the number of decimal places in the numerator.

Example: 5.04/0.20

Step 1: Divide. $504 \div 2 = 252$

Step 2: Subtract the decimal places in the original terms.

5.04 0.2

2 places − 1 place = 1 place

Step 3: Insert decimal.

252 or 25.2

1 place

Example: 0.0054/2.0

Step 1: Divide. $54/2 = 27$

Step 2: Subtract.

0.0054 2.0

4 places − 0 places = 4 places

Step 3: Insert decimal.

00027 or 0.0027

4 places

PERCENTAGES

A percent is a fraction multiplied by 100. To calculate what percent by weight of a mixture of nuts is peanuts, you would need the following information: (1) the weight of the peanuts in the mixture and (2) the total weight of all of the nuts together (peanuts included).

$$\%(\text{w/w})\ \text{Peanuts} = \frac{\text{Wt. of peanuts}}{\text{Total wt. of nuts}} \times 100$$

Similarly, to calculate the approximate percent of oxygen in water, divide the atomic mass of oxygen by the molecular weight of water, and multiply by 100.

$$\begin{aligned}\%\ \text{O} &= \frac{\text{Atomic mass of O}}{\text{Molecular wt. of } H_2O} \times 100 \\ &= \frac{16}{18} \times 100 \\ &= 89\%\end{aligned}$$

In solution problems, the same technique is used. If 200 ml of solution contains 6.5 ml of ethyl alcohol, what is the percent by volume of ethyl alcohol?

$$\begin{aligned}\%(\text{v/v})\ \text{Ethyl alcohol} &= \frac{\text{Volume of ethyl alcohol}}{\text{Total volume}} \times 100 \\ &= \frac{6.5}{200} \times 100 \\ &= 3.2\%\end{aligned}$$

ALGEBRAIC EQUATIONS

An algebraic equation lends itself to almost any mathematical manipulation. The equality between the two sides of the equation will still be present—*as long as the same operation is carried out on each side of the equation.*

If $x = y$, then $23.7x = 23.7y$

$$\text{or} \quad \frac{x}{8.2} = \frac{y}{8.2}$$

$$\text{or} \quad x + 693.1 = y + 693.1$$

$$\text{or} \quad (x + 5)^2 = (y + 5)^2$$

An equation may be rearranged by, for example, the addition or subtraction of a term from both sides of the equation:

Original equation: $P_{\text{total}} = P_A + P_B$

Subtract P_B from both sides of the equation: $P_{\text{total}} - P_B = P_A + P_B - P_B$

Rearranged equation: $P_{\text{total}} - P_B = P_A$

There is no need to go through the above steps to rearrange the original equation if it is kept in mind that, when a term is shifted from one side of the equation to the other, the sign of the term is changed.

$$P_{\text{total}} = P_A + P_B$$

$$P_{\text{total}} - P_B = P_A \qquad (\textit{Note change in sign of } P_B)$$

A term that is a multiplier or a divider may be shifted to the other side of the equation by multiplying or dividing both sides of the equation by that term:

Original equation: $P_1V_1 = P_2V_2$

Divide both sides by V_1: $\dfrac{P_1V_1}{V_1} = \dfrac{P_2V_2}{V_1}$

Since $V_1/V_1 = 1$, the left side of the equation becomes P_1, and the rearranged equation becomes:

$$P_1 = \frac{P_2V_2}{V_1}$$

To shift a divider, multiply both sides of the equation by that term:

Original equation: $\dfrac{V_1}{T_1} = \dfrac{V_2}{T_2}$

$$\textit{Multiply both sides by } T_1\text{:} \quad \frac{V_1 T_1}{T_1} = \frac{V_2 T_1}{T_2}$$

$$\textit{Rearranged equation:} \quad V_1 = \frac{V_2 T_1}{T_2}$$

As in addition or subtraction, the middle step may be eliminated. A multiplier may be simply shifted to the other side of the equation as a divider. Or a divider may be similarly shifted; it becomes a multiplier.

$$\text{(1) } \textit{Equation:} \quad P_1 V_1 = P_2 T_2$$

$$\textit{Shift:} \quad P_1 = \frac{P_2 T_2}{V_1}$$

$$\text{(2) } \textit{Equation:} \quad \frac{V_1}{T_1} = \frac{V_2}{T_2}$$

$$\textit{Shift:} \quad V_1 = \frac{V_2 T_1}{T_2}$$

CANCELLING

Identical terms in a fraction may be cancelled. In the following two examples, 1.8 may be cancelled where indicated because 1.8/1.8 = 1:

$$\text{(1) } K = \frac{\cancel{(1.8)}(10^{-5})(1.8)(10^{-5})}{\cancel{(1.8)}(10^{-4})} \quad \text{or} \quad \frac{(10^{-5})(1.8)(10^{-5})}{10^{-4}}$$

$$\text{(2) } K = \frac{(\cancel{1.8} \times 10^{-5})(1.8 \times 10^{-5})}{(\cancel{1.8} \times 10^{-4})} \quad \text{or} \quad \frac{(10^{-5})(1.8)(10^{-5})}{(10^{-4})}$$

Similarly, a multiplier or divider may be cancelled if it appears on each side of an equation:

$$\text{(1) } (\cancel{1.8} \times 10^{-4})\,K = (\cancel{1.8} \times 10^{-5})(1.8 \times 10^{-5})$$

$$\text{(2) } \frac{(1.0 \text{ atm})(5.0 \text{ liters})}{\cancel{273°\text{K}}} = \frac{P(2.0 \text{ liters})}{\cancel{273°\text{K}}}$$

An added or subtracted term may also be cancelled if the same term appears on each side of an equation. (This is true with some types of chemical equations as well as with algebraic equations.)

$$x + y - \cancel{z} = m + n - \cancel{z}$$

$$\cancel{Na^+} + Cl^- + Ag^+ + \cancel{NO_3^-} \longrightarrow AgCl\downarrow + \cancel{Na^+} + \cancel{NO_3^-}$$

UNITS IN AN EQUATION

The units of numbers (g, atm, etc.) should be carried along in an algebraic equation if possible.

$$\frac{P_{old}V_{old}}{T_{old}} = \frac{P_{new}V_{new}}{T_{new}}$$

where $P_{old} = 1.5$ atm, $V_{old} = 2.0$ liters, $T_{old} = 300°K$, $P_{new} = 2.0$ atm, $T_{new} = 400°K$, $V_{new} = ?$

note units included

Substitute: $$\frac{(1.5 \text{ atm})(2.0 \text{ liters})}{300°\text{K}} = \frac{(2.0 \text{ atm})\ V_{new}}{400°\text{K}}$$

Rearrange: $$\frac{(1.5 \text{ atm})(2.0 \text{ liters})(400°\text{K})}{(300°\text{K})(2.0 \text{ atm})} = V_{new}$$

A *unit* that appears on the top and bottom, as well as a *number* that appears on the top and bottom, may be cancelled.

Cancel units: $$\frac{(1.5 \text{ atm})(2.0 \text{ liters})(400°\text{K})}{(300°\text{K})(2.0 \text{ atm})} = V_{new}$$

Solve: $$V_{new} = \frac{(1.5)(2.0 \text{ liters})(400)}{300(2.0)}$$

$$= \frac{(1.5)(400) \text{ liters}}{300}$$

$$= 2.0 \text{ liters}$$

Carrying the units along is a check on the answer. If the new volume had turned out to be in °K, we would know an error had been made!

SETTING UP A PROPORTION

Proportions are useful in chemical calculations. A proportion is an extension of a ratio. For example, two moles of hydrogen react with one mole of oxygen. The ratio is 2 : 1. How many moles of hydrogen react with five moles of oxygen?

We say "2 is to 1 as x is to 5," and we write:

$$\frac{2 \text{ moles } H_2}{1 \text{ mole } O_2} = \frac{x \text{ moles } H_2}{5 \text{ moles } O_2}$$

Given a balanced chemical equation, we may set up proportions directly from the prefix numbers in the equation:

$$2\,NaOH + H_2SO_4 \longrightarrow Na_2SO_4 + H_2O$$

2 moles is to 1 mole
as
x moles is to y moles

$$\frac{2 \text{ moles NaOH}}{1 \text{ mole } H_2SO_4} = \frac{x \text{ moles NaOH}}{y \text{ moles } H_2SO_4}$$

Proportions may also be used to calculate the number of moles in a given weight of a substance:

$$\text{Molecular weight } H_2SO_4 = 2 + 32 + 64 = 98$$

98 g is to 1.0 mole H_2SO_4
as
x g is to y moles H_2SO_4

$$\frac{98 \text{ g}}{1.0 \text{ mole}} = \frac{x \text{ g}}{y \text{ moles}}$$

In any of the preceding examples, the proportional equations may be solved according to standard algebraic techniques.

Care must be taken that a stated proportion is translated to an algebraic equation in the proper order. When the statement is made, "10 is to 5 as x is to 28," it means "$10/5 = x/28$" and *not* "$10/5 = 28/x$." The possibilities for error are minimized if units are included in the algebraic expression. (Note that, in a proportion, the units should be able to be cancelled.)

Example: How many moles Na_3PO_4 could be produced from 10 moles NaOH?

$$\underset{10 \text{ moles}}{3\,NaOH} + H_3PO_4 \longrightarrow \underset{x \text{ moles}}{Na_3PO_4} + 3\,H_2O$$

$$\frac{3 \text{ moles NaOH}}{1 \text{ mole } Na_3PO_4} = \frac{10 \text{ moles NaOH}}{x \text{ moles } Na_3PO_4}$$

EXPONENTS OF 10

Exponents to the base 10 are often used when extremely large or small numbers are encountered. Positive exponents are used to denote large numbers:

$$
\begin{aligned}
10^1 &= 10 \text{ (1 with one zero)}\\
10^2 &= 100 \text{ (1 with two zeros)}\\
10^3 &= 1000 \text{ (1 with three zeros)}\\
5 \times 10^2 &= 500 \text{ (5 with two zeros)}\\
5 \times 10^6 &= 5{,}000{,}000 \text{ (5 with six zeros)}\\
6.023 \times 10^{23} &= 602{,}300{,}000{,}000{,}000{,}000{,}000{,}000
\end{aligned}
$$

Negative exponents are used to represent small numbers:

$$
\begin{aligned}
10^{-1} &= 0.1 \text{ (decimal moved back one place from 1)}\\
10^{-2} &= 0.01 \text{ (decimal moved back two places from 1)}\\
10^{-3} &= 0.001 \text{ (decimal moved back three places from 1)}\\
5 \times 10^{-3} &= 0.005 \text{ (decimal moved back three places from 5)}
\end{aligned}
$$

For use in problem solving, exponents may be changed by proper manipulation of the multiplier.

$$
\begin{aligned}
1 \times 10^{-5} &= 10 \times 10^{-6}\\
1 \times 10^{3} &= 0.1 \times 10^{4}\\
5.61 \times 10^{6} &= 56.1 \times 10^{5}
\end{aligned}
$$

MULTIPLYING WITH EXPONENTS

When two numbers with exponents are to be multiplied, the exponents are *added:*

$$
\begin{aligned}
(10^5)(10^3) &= 10^8 \quad \curvearrowleft\ 5+3\\
(10^{-3})(10^2) &= 10^{-1} \quad \curvearrowleft\ -3+2\\
(2 \times 10^{-2})(5 \times 10^{-3}) &= 10 \times 10^{-5} \quad \curvearrowleft\ -2+(-3)
\end{aligned}
$$

DIVIDING WITH EXPONENTS

When numbers with exponents are divided, the exponents are *subtracted* (denominator from numerator):

$$\frac{10^3}{10^2} = 10^1 \quad \curvearrowleft\ 3-2$$

$$\frac{10^{-3}}{10^2} = 10^{-5} \quad \curvearrowleft\ -3-2$$

$$\frac{10^6}{10^{-2}} = 10^8 \quad \curvearrowleft\ 6-(-2)$$

$$\frac{4 \times 10^6}{8 \times 10^{-2}} = \frac{4}{8} \times 10^8 = 0.5 \times 10^8$$

SQUARE ROOTS

Square roots are often encountered in equilibrium problems. A square root is a fractional exponent: $\sqrt{10} = 10^{\frac{1}{2}}$. Thus, the square root of a term with an exponent may be calculated by dividing the exponent by 2.

$$\sqrt{10^{-8}} = 10^{-4}$$
$$\sqrt{10^{-4}} = 10^{-2}$$
$$\sqrt{3 \times 10^{-8}} = \sqrt{3} \times 10^{-4}$$

If a number contains an odd exponential number, rather than an even number, the exponent may be changed to an even number as shown in the following examples:

$$\sqrt{10^{-5}} = \sqrt{10 \times 10^{-6}} = \sqrt{10} \times 10^{-3}$$
$$\sqrt{30 \times 10^{-9}} = \sqrt{3.0 \times 10^{-8}} = \sqrt{3.0} \times 10^{-4}$$

The square roots of the nonexponential expressions may be found in a table of square roots, on most slide rules, or by the use of some pocket calculators.

LOGARITHMS

A logarithm is an exponent. A logarithm to the base 10 ($\log_{10}$, or simply log) is an exponent of 10.

$$10^5$$

the exponent (5)

the base (10)

The log of 10^5 is 5, or $\log 10^5 = 5$

$$10^{-3}$$

the exponent (−3)

the base (10)

$$\log 10^{-3} = -3$$

The term pH is defined as the *negative* log of hydrogen ion concentration (in molarity, M).

the exponent is -3

$$[H^+] = 10^{-3}\,M$$

$$pH = -\log 10^{-3} = -(-3) = +3$$

Numbers with exponents are multiplied by addition of the exponents ($10^2 \times 10^3 = 10^5$). Therefore, the log of a number such as 2.0×10^{-3} is calculated by *addition* of the logs of the two terms. (The value for log 2.0 taken from a log table, slide rule, or calculator.)

$$\begin{aligned} \log(2.0 \times 10^{-3}) &= \log 2.0 + \log 10^{-3} \\ &= 0.30 - 3 \\ &= -2.7 \end{aligned}$$

If $[H^+]$ is expressed as $2.0 \times 10^{-3}\ M$, then the pH may be calculated as follows:

$$\begin{aligned} \text{pH} &= -\log[H^+] \\ &= -\log(2.0 \times 10^{-3}) \\ &= +2.7 \end{aligned}$$

GRAPHS

Graphs are used to show pictorially a mathematical expression or any other expression that depends on two variables. An airplane takes off; as time proceeds, the plane gains altitude.

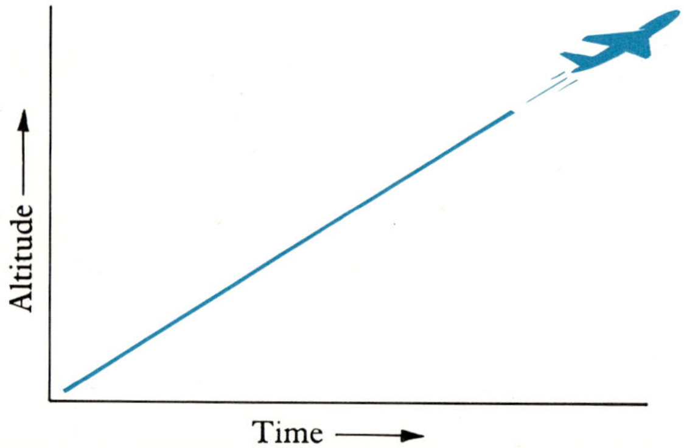

The altitude of the airplane may be plotted against time to give a picture of the two variables: altitude versus time.

Related to the airplane graph is the energy diagram, in which the average energy of molecules is plotted against the progress of the reaction:

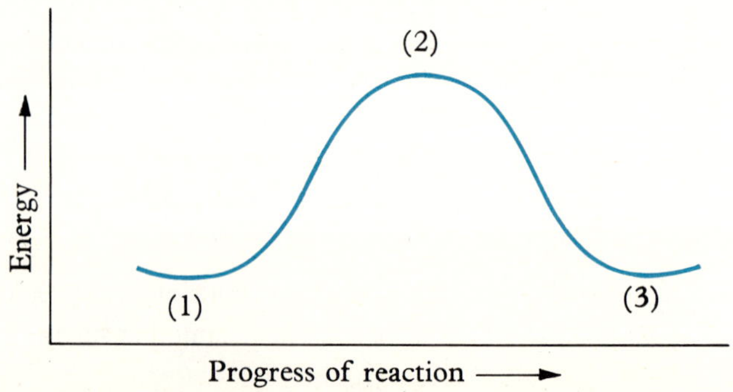

1. At the start of the reaction, the average energy of the reactant molecules is at (1).
2. As the molecules change from reactants to products, the average energy of the molecules is at its highest point, (2).
3. At the end of the reaction, the average energy of the product molecules is at (3).

ANSWERS TO ODD-NUMBERED STUDY QUESTIONS

CHAPTER 1

1-1. (a) law (b) hypothesis

1-3. (a) potential (b) kinetic

1-5. (a) 5.0 mi ÷ 0.62 mi/km = 8.1 km
(b) 1.0 ft × 1.0 cm/0.394 in. × 12 in./1.0 ft = 30.5 cm
(c) $1.0 \times 10^6\ \mu$ (d) $1.94 \times 10^{-4}\ \mu$
(e) 7.6×10^{-9} in. (f) 2.5×10^8 Å

1-7. (a) 0.95 liter (b) 3.8 liters
(c) 5.92 gal (d) 5×10^{-3} liter

1-9. 37°C

1-11. 2200 cal, 2.2 kcal

1-13. (a) 2.15 kg (b) 1.66 kg
(c) 2.1 kg (d) 28.6 kg

1-15. No, its density is 11.3 g/cc, not that of silver.

CHAPTER 2

2-1. (a) diatomic element
(b) compound
(c) compound
(d) diatomic element

2-3. (a) (4) (b) (1) (c) (2) (d) (2)

2-5. (a) does not (b) does
(c) does not (d) does

2-7. (a) 7800 cal (b) 445 cal
(c) 9670 cal

2-9. 1000 cal

2-11. 5.7 kcal

2-13. (a) evaporation (b) condensation
(c) an equilibrium

2-15. (a) (b) (e) (f) (g)

2-17. 2.85 liters

2-19. 1 atm, or 760 mm

2-21. 37 liters

2-23. 1360 liters

CHAPTER 3

3-1. (a) formula (b) formula
(c) atomic symbol (d) formula

3-3. (a) both (b) neither (an electron)
(c) neutron (d) both

3-5. (b) (c) (d)

3-7. (a) and (b): reductions
(c) and (d): oxidations

3-9. (a) (b) (d)

3-11. (b) (d)

3-13. (a) zero (b) three
(c) two (d) two

3-15. (a) Te, Se, S, O (b) Na, Mg, Al, P
(c) I, Br, Cl, F (d) K, Ca, Fe, Cu

3-17. 8

3-19. (a) (d)

3-21. Cl

CHAPTER 4

4-1. (a) metallic (b) ionic
(c) covalent (d) Na^+ OH^- (*ionic* ↗ ↖ *covalent*)

4-3. (a) calcium oxide
(b) copper(II) fluoride
(c) iron(III) sulfide
or ferric sulfide

4-5. O

4-7. (a) 86.8 (b) 78.0

4-9. Zn, +2; Br, −1; Fe, +2; S, −2; Al, +3; O, −2; Hg, +1; Cl, −1

4-11. (a) 2 (b) 4 (c) 2

4-13. (a) 1 (b) 7 (c) 5 (d) 3

4-15. four

4-17. sp^3, s

4-19. Three sp^2 orbitals and one p orbital.

4-21. One half of the pi bond is above the plane of the sigma bonds; the other half of the pi bond is below the plane.

4-23. (a) sigma (b) one sigma and two pi

4-25. They lie in a plane.

4-27. six

4-29. two in each case

4-31. (a) $\overset{\delta+}{H}—\overset{\delta-}{Br}$ (b) not polar

(c) 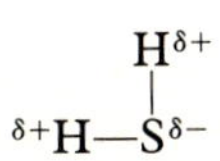 (d) 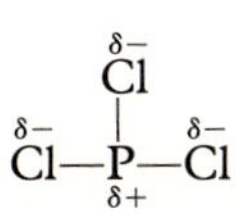

CHAPTER 5

5-1. (a) Two atoms of sodium metal undergo reaction with two molecules of water to yield two molecules of sodium hydroxide and one molecule of hydrogen gas.
(b) Two molecules of ammonium nitrate under pressure undergo a fast reaction to yield two molecules of nitrogen gas, one molecule of oxygen gas, and four molecules of water.

5-3.

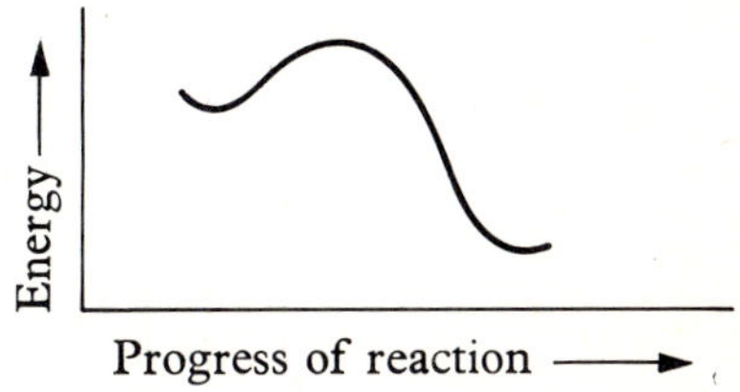

5-5. (c) (d)

5-7. (a) CH≡CH, C_2H_2, CH
(b) $(HO)_3PO$, H_3PO_4, H_3PO_4
(c) $CH_3CH_2CH_3$, C_3H_8, C_3H_8
(d) $CH_3CH{=}CH_2$, C_3H_6, CH_2
(e) $CH_3\underset{\displaystyle OH}{\underset{|}{C}}HCH_3$, C_3H_8O, C_3H_8O

5-9. (a) 74.8% (b) 80.0% (c) 52.1%
(d) 92.3% (e) 26.7%

5-11. (a) $2\,K + 2\,H_2O \longrightarrow 2\,KOH + H_2$
(b) $H_2 + Br_2 \longrightarrow 2\,HBr$
(c) $2\,CH_3CH_3 + 7\,O_2 \longrightarrow 4\,CO_2 + 6\,H_2O$
(d) $Zn + 2\,HCl \longrightarrow ZnCl_2 + H_2$

5-13. (a) 2.00 g (b) 22.1 g (c) 38.8 g
(d) 14.9 g

5-15. 11.9 g pure HCl

5-17. $2\,NaCl \longrightarrow 2\,Na + Cl_2$

$$\frac{142\ g}{70.9\ g/mole} = 2.00\ mole\ Cl_2$$

4.00 moles NaCl = 234 g NaCl

CHAPTER 6

6-1. The solvent is water; the solutes are sugar and the soluble compounds in the coffee. Sugar is soluble; the coffee grounds are partially soluble.

6-3. (a) and (b) saturated,
(c) unsaturated

6-5. The carbon dioxide would be approximately one-third as soluble; the excess would bubble out of solution.

6-7. $\frac{15 \text{ ml}}{70 \text{ ml}} \times 100 = 21\%$

6-9. (a) Dilute 70 ml ethyl alcohol to 200 ml with water.
(b) Dilute 0.5 g KOH (dissolved) to 50 ml.
(c) Dissolve 0.5 g NaOH in 9.5 g water.

6-11. 17.5 g

6-13. 1000 gal/sec × 3.78 liters/gal × 604,800 sec/wk = 2.29×10^9 liters/wk; 2.29×10^9 liters/wk × 3.0 mg/liter × 2.205 lb/10^6 mg = 15,000 lb

6-15. 2.0 g

6-17. 2.0 *M*

6-19. 0.167 mole

6-21. 56.1 × 2.00 *M* × 100/1000 = 11.2 g

6-23. (a) 0.05 (b) 0.075 (c) 0.18

6-25. (a) 16 (b) 2.0 (c) 0.8 (d) 3.0

6-27. (a) 63 g (b) 120 g
(c) 26 g (d) 60 g

6-29. 4 *N*

6-31. 37.5 ml

6-33. (a) 75 ml (b) 25 ml
(c) 5667 ml (d) 0.033 ml

6-35. No. The final volume would be 5.0 ml—in other words, *less* than the starting volume.

6-37. (a) −1.86°C (b) −1.86°C
(c) −13.4°C (d) −22.3°C

6-39. (a) −3.7°C, 101°C
(b) −1.9°C, 100.5°C
(c) −3.7°C, 101°C

6-41. (a) from the bag (b) contract
(c) decrease

6-43. Brownian motion and electrostatic repulsion between particles.

6-45. (a) back and forth
(b) from side 2 to side 1
(c) from side 2 to side 1
(d) stays on side 1

CHAPTER 7

7-1. (a) (d)

7-3. (a) salt (b) acid salt
(c) acid salt (d) base

7-5. (a) soluble (b) soluble
(c) insoluble (d) soluble
(e) soluble

7-7. $NH_3 + H_2O \rightleftharpoons NH_4^+ + OH^-$

7-9. $Cu^{2+} + S^{2-} \longrightarrow CuS\downarrow$

7-11. (a) $K = \frac{[H^+][CN^-]}{[HCN]}$

(b) $K = \frac{[Ag^+][Cl^-]}{[AgCl]}$ or $[Ag^+][Cl^-]$

since [AgCl] is constant

(c) $K = \frac{[CH_3CH_2CO_2^-][H^+]}{[CH_3CH_2CO_2H]}$

(d) $K = \frac{[CH_3NH_3^+][OH^-]}{[CH_3NH_2]}$

7-13. 0.18 *M*

7-15. (a) $-\log(3.5 \times 10^{-5}) = 5 - \log 3.5 = 4.5$
(b) 1.2 (c) 12 (d) 1.3
(e) 3.2 (f) 9.3 (g) 0 (h) −0.7

7-17. (a) acidic (b) acidic
(c) basic (d) basic

7-19. (a) 10^{-12} *M* (b) 10^{-1} *M*
(c) 3.2×10^{-10} *M* (d) 1.6×10^{-7} *M*

7-21. (a) (c)

7-23. (a) 25.00 ml (b) 25.00 ml
(c) 50.00 ml (d) 3.00 ml

7-25. $CH_3\overset{O}{\overset{\|}{C}}CH_2\overset{O}{\overset{\|}{C}}OH + HCO_3^- \rightleftharpoons$

$CH_3\overset{O}{\overset{\|}{C}}CH_2\overset{O}{\overset{\|}{C}}O^- + H_2CO_3 \rightleftharpoons H_2O + CO_2$

CHAPTER 8

8-1. (a) +3 (b) 0 (c) 0 (d) +1

8-3. (a) Cu red., Fe ox. (b) Cl red., Br ox.
(c) Na ox., H red. (d) C ox., O red.

8-5. (a) $CuCl_2$ ox. agent, Fe red. agent
(b) Cl_2 ox. agent, Br^- red. agent
(c) Na red. agent, H_2O ox. agent
(d) CH_4 red. agent, O_2 ox. agent

8-7. (a) $Br_2 + SO_2 + 2\,H_2O \longrightarrow 4\,H^+ + 2\,Br^- + SO_4^{2-}$
(b) $H_2S + H_2O_2 \longrightarrow S + 2\,H_2O$
(c) $2\,HBrO \longrightarrow 2\,H^+ + 2\,Br^- + O_2$
(d) $H_2SO_4 + 2\,HBr \longrightarrow SO_2 + Br_2 + 2\,H_2O$

8-9. $2\,SO_2 + O_2 \longrightarrow 2\,SO_3$

8-11. $4\,Ag + 2\,H_2S + O_2 \longrightarrow 2\,Ag_2S + 2\,H_2O$

8-13. (a) and (c) oxidation
(d) reduction
(b) and (e) neither

CHAPTER 9

9-1. Conductors of heat and electricity, malleable, ductile, lustrous, dense.

9-3. The oxides of these metals form basic, or alkaline, solutions.

9-5. (a) $CsOH + H_2$ (b) Rb_2O
(c) $CH_3OLi + H_2$ (d) $NaCl + H_2$

9-7. Hard water contains Ca^{2+}, Mg^{2+}, or some other cation that will form a precipitate with soap; soft water does not contain such ions.

9-9. $Mg(OH)_2 + 2\,HCl \longrightarrow MgCl_2 + H_2O$

9-11. Those metals with an incomplete inner electronic shell.

9-13. In NH_3 solution, the soluble silver-ammonia complex $Ag(NH_3)_2^+$ is formed.

9-15. In each case, +5.

9-17. $N_2 + O_2 \longrightarrow 2\,NO$

or

$N_2 + 2\,O_2 \longrightarrow 2\,NO_2$

9-19. (a) 0 (b) +4 (c) +6
(d) −2 (e) +6 (f) +4

9-21. $2\,Br^- + Cl_2 \longrightarrow 2\,Cl^- + Br_2$

CHAPTER 10

10-1. (a) (2) (4) (b) (3) (5)
(c) (1) (3)

10-3. (a) ${}_{-1}^{0}e$ (b) ${}_{86}^{222}Rn$

10-5. ${}_{3}^{6}Li + {}_{0}^{1}n \longrightarrow {}_{1}^{3}H + {}_{2}^{4}He$

10-7. 1.25 g

10-9. 22,280 yr

10-11. (a) ${}_{1}^{2}H$ (b) ${}_{92}^{235}U$ (c) ${}_{11}^{24}Na{}_{1}^{2}H$
(d) ${}_{92}^{238}U$

CHAPTER 11

11-1. An ionic bond is formed from electron transfer:

$$Na\cdot + \cdot\ddot{\underset{..}{Cl}}: \longrightarrow Na^+ \; :\ddot{\underset{..}{Cl}}:^-$$

A covalent bond is formed from electron sharing:

$$H\cdot + \cdot H \longrightarrow H:H$$

11-3. (a) sp^3: four equivalent orbitals (at 109° angles) formed from one $2s$ orbital and three $2p$ orbitals; CH_4.

(b) sp^2: three equivalent orbitals (at 120° angles) formed from one 2*s* orbital and two 2*p* orbitals; one *p* orbital, at right angles to the plane of the sp^2 orbitals, is available for the pi-bond formation; $CH_2{=}CH_2$.

(c) *sp*: two equivalent orbitals (at 180° angle) formed from one 2*s* orbital and one 2*p* orbital; two *p* orbitals are available for pi-bond formation; $CH{\equiv}CH$.

11-5. (a) C_3H_8O, 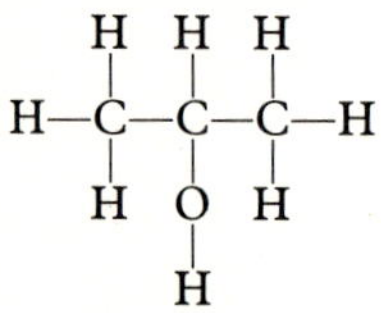

(b) C_3H_8O, $CH_3OCH_2CH_3$

(c) C_3H_8O, 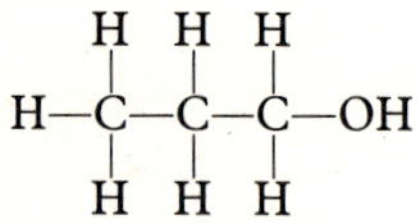

(d) Any of the above.

11-7. (a) (c)

11-9. (d)

11-11. In each case, one of the carbon atoms has five bonds.

11-13. $CH_3CH_2CH{=}CH_2$, $CH_3CH_2C{\equiv}CH$, $CH_3CH_2CH_2CH_2Cl$, $CH_3CH_2CH_2CH_2OH$, $CH_3CH_2CH_2CH_2NH_2$, $CH_3CH_2CH_2CO_2H$ (*Note:* There are other correct answers.)

CHAPTER 12

12-1. (a) alkane (b) aromatic (c) unsaturated (d) cycloalkane

12-3. No, it is a mixture of saturated hydrocarbons.

12-5. See text (Table 12-1).

12-7. (a) 6 (b) 7 (c) 8 (d) 8

12-9. (a) isopropyl (b) *t*-butyl (c) *sec*-butyl (d) isobutyl

12-11. (a) 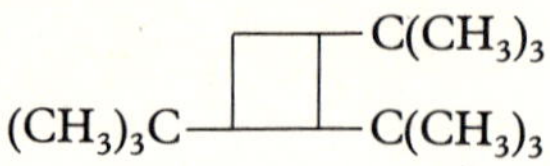

(b) $(CH_3)_3CC(CH_3)_3$

(c) 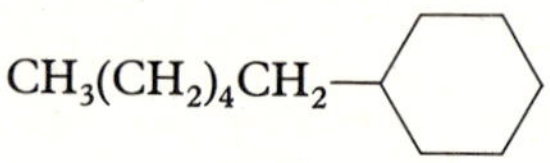

(d) 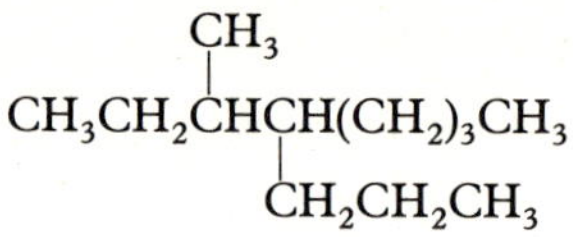

(e) 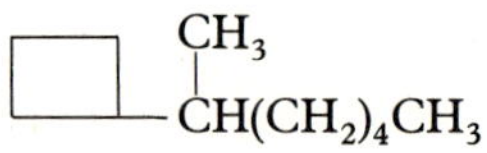

CHAPTER 13

13-1. (a) 3-methyl-1-butene (b) 5-cyclohexyl-4-isopropyl-2-pentene (c) 3-methyl-1-butyne (d) 2-methyl-1,3-heptadiene (e) 3-*t*-butyl-1,4-hexadiyne

13-3. (a) *cis:* 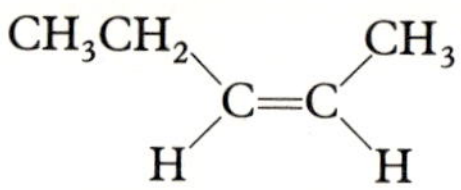

trans: 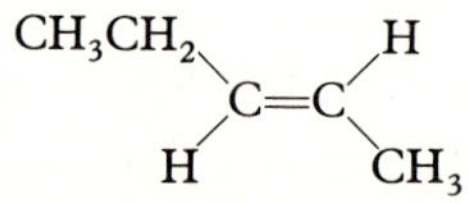

(b) *cis:* 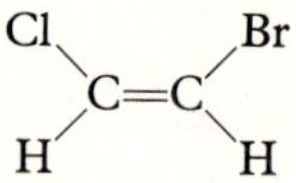

trans: 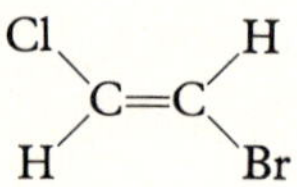

(c) *cis:* 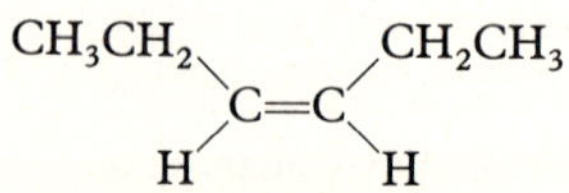

trans: 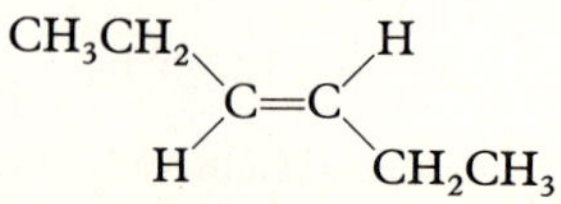

(d) *cis:*

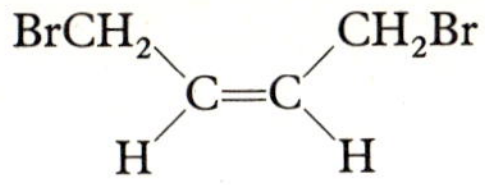

trans:

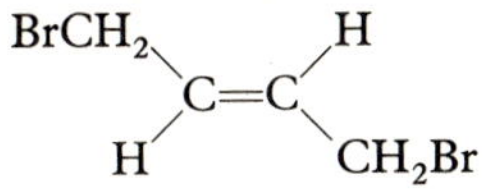

13-5. (c) (d)

13-7. (a) *cis; trans:*

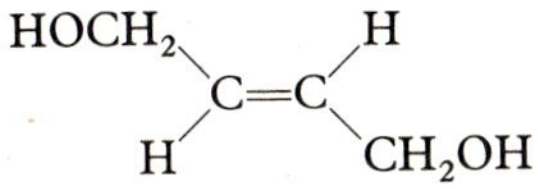

(b) *cis; trans:*

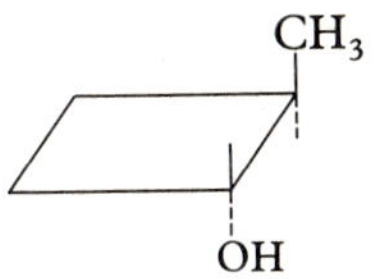

(c) *trans; cis:*

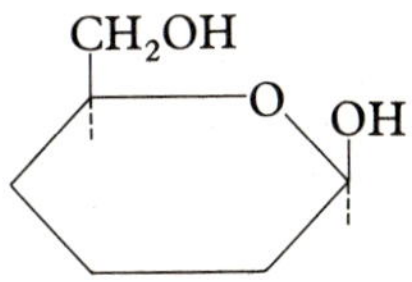

(d) *cis; trans:*

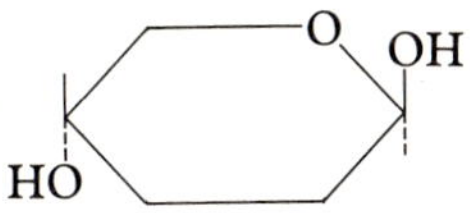

13-9. (a)  (b) $CH_3CH_2\underset{CH_3}{\overset{CH_3}{C}}CH{=}CH_2$

(c) $(CH_3)_2C{=}CH_2$ (d) 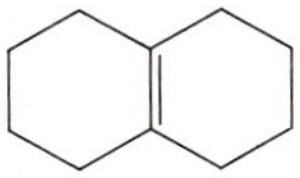

13-11. (a) and (c) symmetrical
(b) and (d) unsymmetrical

13-13. $CH_3CH{=}CH(CH_2)_4CH_3 + 12\ O_2 \longrightarrow 8\ CO_2 + 8\ H_2O$

13-15. (a) $\left(\underset{CH_2CH_3}{CH}{-}CH_2\right)_x$

(b) $(CCl_2CHCl)_x$

(c) $(CH_2CBr{=}CHCH_2)_x$

13-17. (a) $CH_2{=}CHCH_3$

(b) $CH_2{=}CHO\overset{O}{\overset{\|}{C}}CH_3$

CHAPTER 14

14-1. (a) aryl halide
(b) and (c) alkyl halides

14-3. (a) $(CH_3)_3CF$ (b) $(CH_3)_2CHBr$

(c) 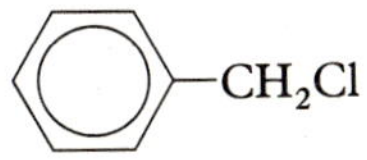

(d) $ClCH_2\underset{CH_3}{\overset{CH_3}{C}}CH_2CH_2CH_3$

(e) 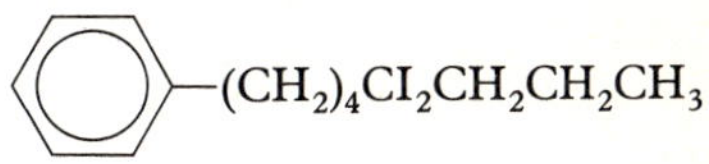

14-5. (a) $Cl_2 \xrightarrow{uv} 2\ Cl\cdot$

(b) $Cl\cdot + Cl_2CHCH_2Cl \longrightarrow Cl_2\dot{C}CH_2Cl + HCl$

$Cl_2\dot{C}CH_2Cl + Cl_2 \longrightarrow Cl_3CCH_2Cl + Cl\cdot$

$Cl\cdot + Cl_3CCH_2Cl \longrightarrow Cl_3C\dot{C}HCl + HCl$

$Cl_3C\dot{C}HCl + Cl_2 \longrightarrow Cl_3CCHCl_2 + Cl\cdot$

$Cl\cdot + Cl_3CCHCl_2 \longrightarrow Cl_3C\dot{C}Cl_2 + HCl$

$Cl_3C\dot{C}Cl_2 + Cl_2 \longrightarrow Cl_3CCCl_3 + Cl\cdot$

14-7. (a) Substitution reaction: a reaction in which one group is substituted for another.

$$CH_3CH_2Br + OH^- \longrightarrow CH_3CH_2OH + Br^-$$

(b) Elimination reaction: a reaction in which a small molecule is lost from a larger molecule.

$$(CH_3)_3CBr + OH^- \longrightarrow CH_3C(=CH_2)CH_3 + H_2O$$

14-9. (a) $C_6H_5OCH_3 + I^-$
(b) $CH_3CH_2SCH_3 + Br^-$
(c) $C_6H_{11}CN + Br^-$
(d) $CH_3CH_2CH(CH_3)OH + Cl^-$
(e) $C_6H_5CH_2OCH_2C_6H_5 + Br^-$

14-11. Alkene products:

(a) $(CH_3CH_2)_2C{=}CHCH_3$

(b) 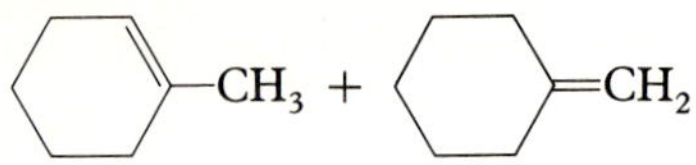

(c) 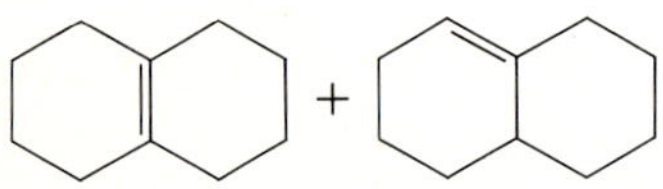

CHAPTER 15

15-1. (a) alcohol (b) ether
(c) phenol (d) alcohol

15-3. (a) 2-propanol, isopropyl alcohol
(b) 1-butanol, butyl alcohol
(c) cyclohexanol, cyclohexyl alcohol
(d) 2-methyl-2-propanol, *t*-butyl alcohol

15-5. (b) (c)

15-7. Oxidation-reduction

$$CH_3CH_2OH + Na \longrightarrow CH_3CH_2O^-\ Na^+ + \tfrac{1}{2}H_2$$

(H of OH: *reduced*; Na: *oxidized*)

15-9. (a) KOH (b) Na metal

15-11. (a) 1 (b) 2 (c) 1 (d) 2

15-13. $C_6H_5CO_2H + HOCH(CH_3)CH_2CH_3 \overset{H^+}{\rightleftharpoons} C_6H_5C(=O)OCH(CH_3)CH_2CH_3 + H_2O$

15-15. (a) $CH_3CH_2CH{=}CH_2 + CH_3CH{=}CHCH_3$
(b) $CH_3CH_2CO_2C_6H_{11} + H_2O$
(c) $C_4H_9OSO_3H + H_2O$
(d) $CH_3OH + H_2SO_4$
(e) $CH_3OCH_2CH_3 + H_2SO_4$
(f) $CH_3CH{=}CH_2 + H_2SO_4$
(g) $C_6H_{11}ONO_2 + H_2O$

CHAPTER 16

16-1. (c) (d)

16-3. (a) 2-bromoethanal
(b) 3-pentanone
(c) 1-bromo-3-methyl-2-butanone
(d) 4-methyl-1-cyclohexanone

16-5. (a) $(CH_3)_2CHC(=O)CH(CH_3)_2$

(b) $(CH_3)_2CHC(=O)CH(CH_3)_2$

(c) 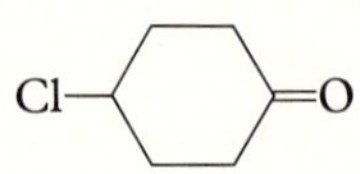

(d) $CH_3CH_2CH_2C(CH_3)(CH_2CH_3)CH_2CHO$

16-7. (a) (b) (c) (d)

16-9. No, because the carbon in question has three alkyl groups attached and cannot accommodate two more bonds.

16-11. $CH_3CH_2\overset{\overset{O}{\|}}{C}H$

16-13. (a) CH_3CH_2 (CH_2) CHO

(b) $CH_3CH_2O_2C$ (CH_2) $CO_2CH_2CH_3$

(c) (CH_3) $\overset{\overset{O}{\|}}{C}$ (CH_3)

(d) C_6H_5 (CH_2) CHO

16-15. (a) $CH_3CH{=}CHOH$

(b) 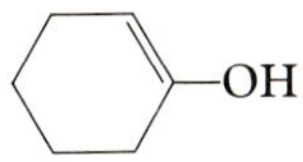

CHAPTER 17

17-1. (a) $C_6H_5CO_2^-\ Li^+ + H_2O$

(b) $Na^+\ {}^-O_2CCO_2^-\ Na^+ + 2\ H_2O$

(c) $CH_3CH_2CH_2CO_2^-\ K^+ + H_2O$

(d) 

17-3. (a) 3-bromo-2-methylbutanoic acid
(b) propanoic acid
(c) methanoic acid

17-5. Primary alcohols and aldehydes.

17-7. (a) 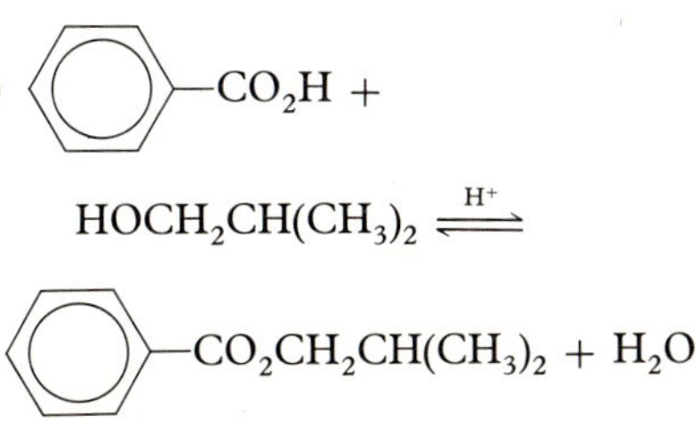

(b) $\overset{H^+}{\rightleftharpoons}$ $C_6H_5CO_2CH(CH_3)_2 + H_2O$

(c) 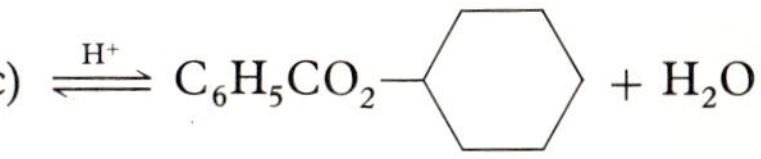

(d) $\overset{H^+}{\rightleftharpoons}$ $C_6H_5CO_2CH_2C_6H_5 + H_2O$

17-9. (a) methyl pentanoate
(b) propyl 3-methylbutanoate
(c) ethyl 3,3,5-trimethylheptanoate

17-11. Heat the ester with aqueous NaOH; then acidify.

17-13. (a) propanamide or propionamide
(b) benzamide
(c) formamide or methanamide
(d) 2-methylbutanamide

17-15. 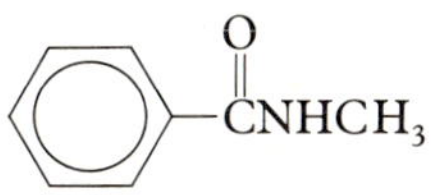

17-17. 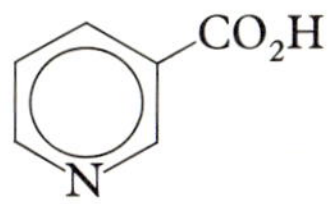

CHAPTER 18

18-1. 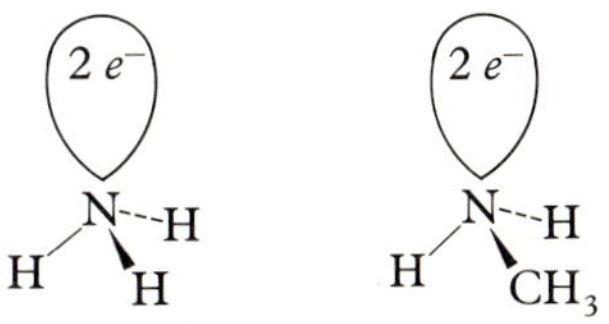

18-3. (a) ethanol or ethyl alcohol
(b) ethylamine
(c) isopropylamine or 2-propylamine
(d) diethylamine
(e) *t*-butylamine or 2-methyl-2-propylamine
(f) *t*-butyl alcohol or 2-methyl-2-propanol
(g) N-methyl-*t*-butylamine
(h) N,N-dimethyl-*t*-butylamine
(i) piperidine

18-5. 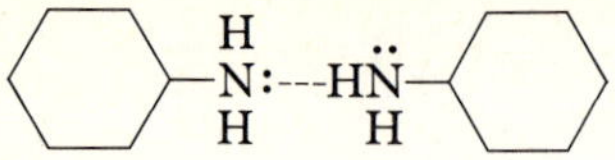

18-7. (a) cannot undergo hydrogen bonding because it has no partially positive hydrogen. (b) can undergo hydrogen bonding with water.

$$(CH_3CH_2)_3N\text{:---}HOH$$

18-9. $K = \dfrac{[R_3NH^+][OH^-]}{[R_3N]};$

$$\frac{[OH^-]^2}{0.100\,M} = 6 \times 10^{-5};$$

$[OH^-] = 2.5 \times 10^{-3};$

$$[H^+][OH^-] = 10^{-14};$$

$[H^+] = 4.0 \times 10^{-12};$

$pH = -\log(4.0 \times 10^{-12}) = 11.4$

CHAPTER 19

19-1. (b) (c)

19-3. ten

19-5. (a) 3 (b) 5 (c) 7 (d) 7

19-7. (a) $C_6H_5CO_2H$

(b) 

(c) 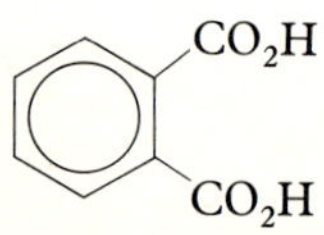

19-9. (a) *o*-bromonitrobenzene
(b) *p*-dimethylbenzene or *p*-xylene
(c) *m*-nitrobenzoic acid
(d) 1-bromo-2-methylcyclohexane

19-11. (a) $C_6H_5Cl + HCl$
(b) $C_6H_5NO_2 + H_2O$

19-13. (a) 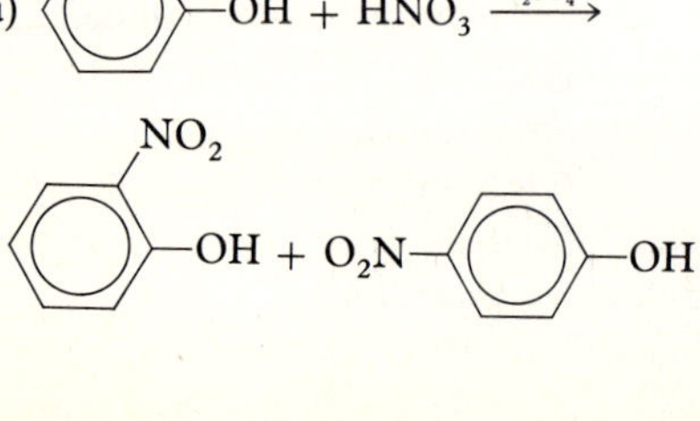

(b) 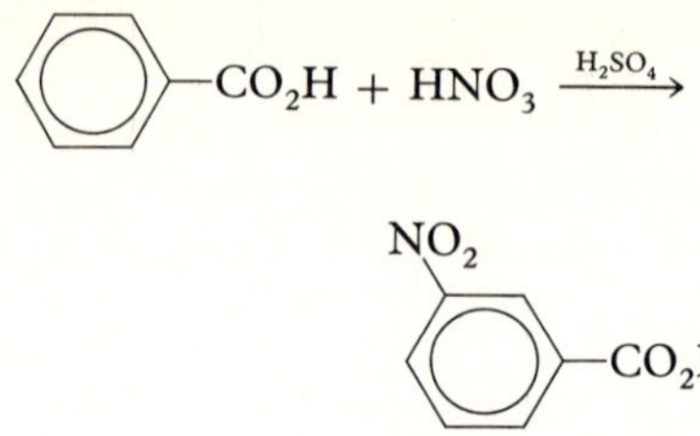

(c) 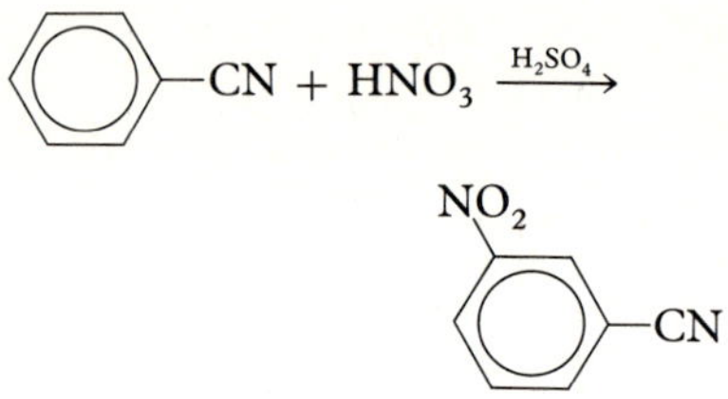

19-15. (b) (c)

CHAPTER 20

20-1. A term used to describe (a) two or more compounds (isomers) with the same molecular formula, but different structures:

$$CH_3CH_2OH \text{ and } CH_3OCH_3$$

(b) isomers in which the order of attachment of atoms within the molecules varies. For example, see (a). (c) two isomers in which groups may be on the same side (*cis*) of a pi bond or a ring or on the opposite sides (*trans*):

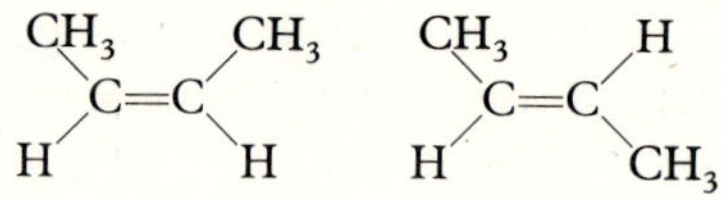

cis *trans*

(d) isomers that lack a plane of symmetry and thus can vary in the right- or left-handed arrangement of their atoms:

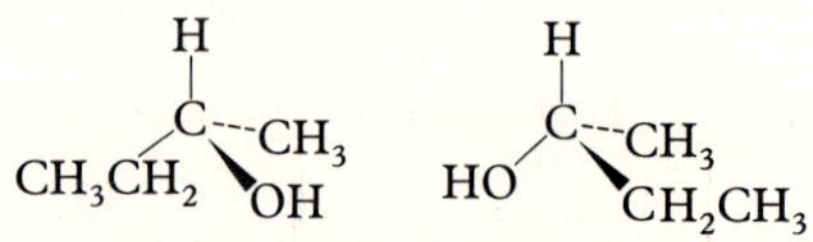

20-3. (a) an infinite number (b) 3 (c) 1

20-5. (a) none (b) 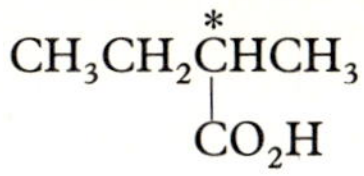

(c) 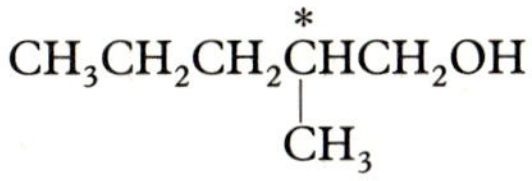 (d) none

20-7. (a) (3) (b) (2) (c) (1) (d) (3)

20-9. (1) (3)

20-11. (a) 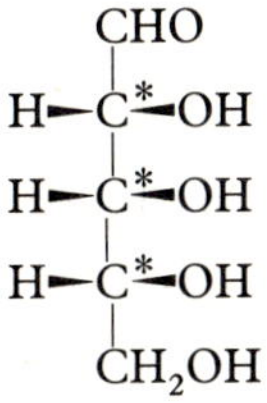 (b) 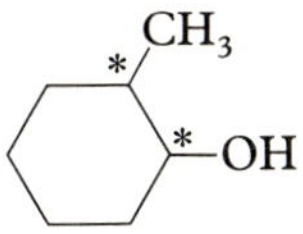

(c) 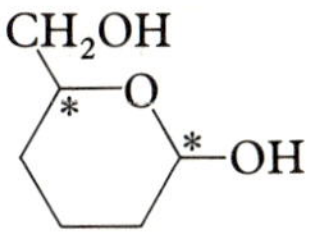 (d) none

20-13. (b)

CHAPTER 21

21-1. 16; C, H, O, N

21-3. $\left(\text{NHCHR}\overset{\overset{\text{O}}{\|}}{\text{C}}\right)_x$, amide

21-5. A biological catalyst; a type of protein.

21-7. (a) (4) (b) (3) (c) (2) (d) (1)

CHAPTER 22

22-1. (a) (1) (2) (3) (b) (1) (4)
(c) (1) (3) (d) (1) (2) (3)

22-3. Refer to text.

22-5. (a) aldose (b) ketose
(c) aldose (d) ketose

22-7. 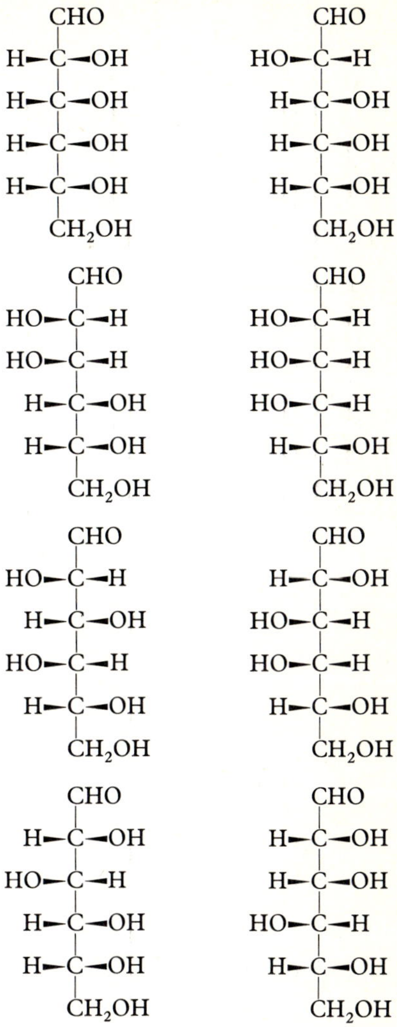

22-9. (a) 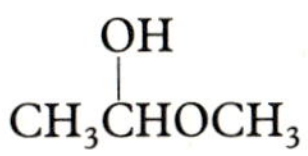 (b) 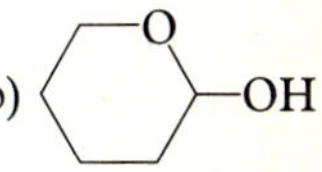

(c) 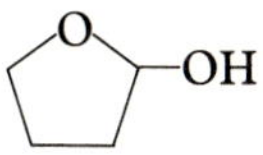 (d) 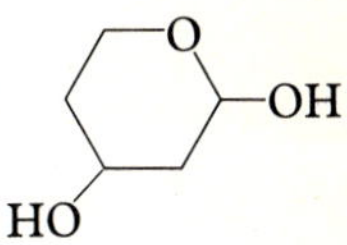

22-11. No, they are not mirror images; the configuration differs only at carbon 1.

22-13. 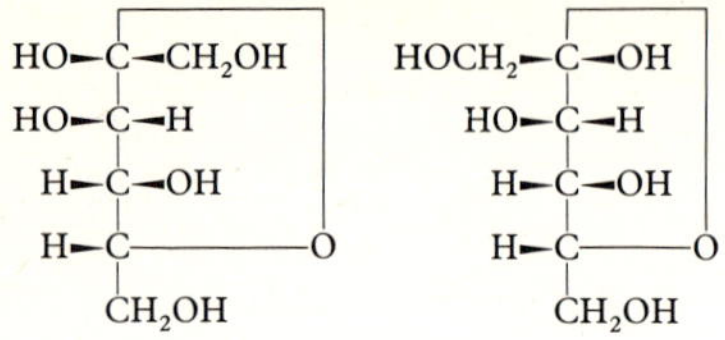

22-15. (a) (c)

22-17. 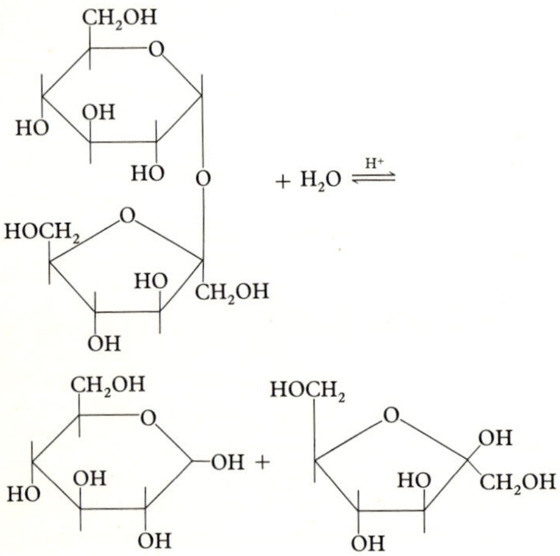

22-19. 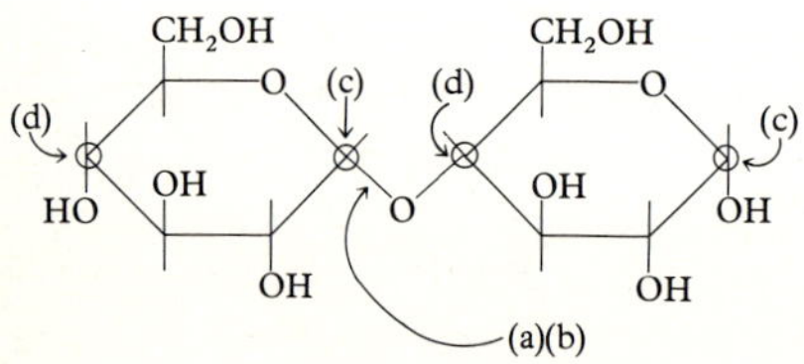

22-21. 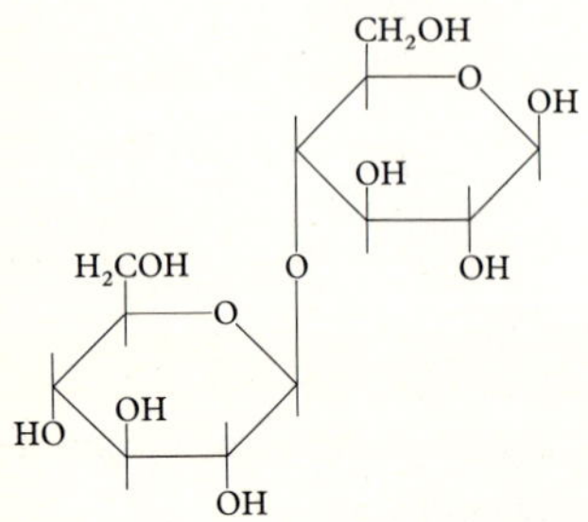

CHAPTER 23

23-1. (a) $CH_3C(=O)N(CH_3)_2 + H_2O + H^+ \longrightarrow CH_3CO_2H + H_2N^+(CH_3)_2$

(b) 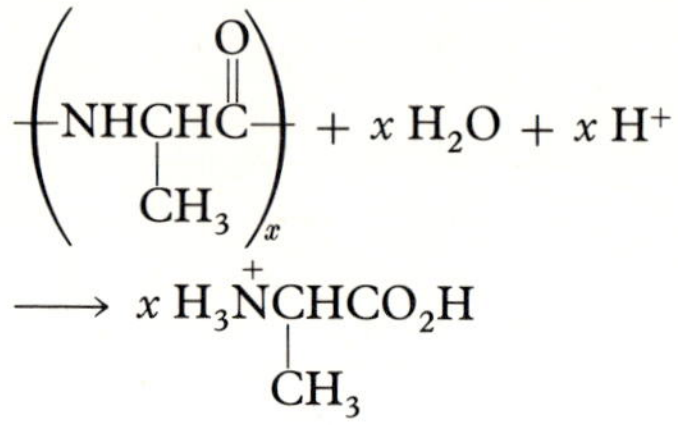

23-3. See text for structures;
(a) alanine (b) phenylalanine
(c) tyrosine (d) serine

23-5. (a) $CH_3CH(NH_3^+)CO_2H$

(b) $CH_3CH(NH_2)CO_2^- + H_2O$

23-7. 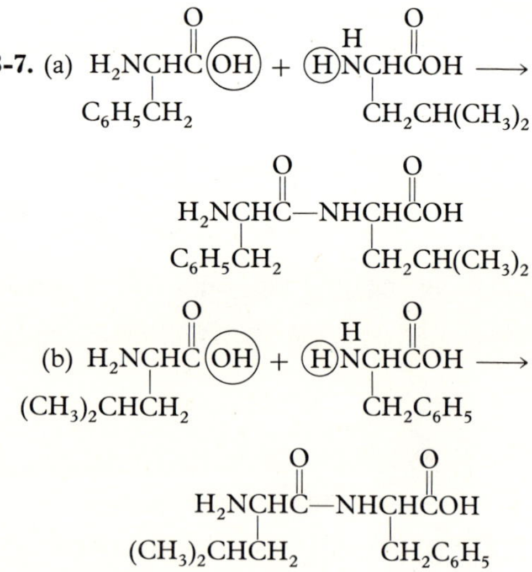

23-9. ala-ser-val-glu

23-11. (a) amino-acid sequence (b) shape assumed by the protein chain (c) interactions between portions of the protein chain that cause folding or bending

23-13. (a) (3) (b) (4) (c) (1) (d) (2)

23-15. (a) $H_2N\underset{\substack{|\\ CH_2C_6H_5}}{C}HCO_2H + H_2N\underset{\substack{|\\ CH_3}}{C}HCO_2H$

(b) $H_2N\underset{\substack{|\\ CH_2C_6H_5}}{C}HCO_2H + H_2NCH_2CO_2H$

(c) $H_2N\underset{\substack{|\\ CH_2C_6H_5}}{C}HCO_2H + H_2N\underset{\substack{|\\ CH(CH_3)_2}}{C}HCO_2H$

23-17. (a) (b) (c) (e)

CHAPTER 24

24-1. (a) (1) (2) (b) (5) (c) (1) (4)

24-3. A petroleum oil is a mixture of alkanes. A vegetable oil is a triester of glycerol, or a triglyceride.

24-5. $2\ CH_3(CH_2)_{16}CO_2^-\ Na^+$

$+ HOCH_2\overset{\substack{OH\\ |}}{C}HCH_2OH$

$+ CH_3(CH_2)_5CH{=}CH(CH_2)_7CO_2^-\ Na^+$

24-7. $\begin{array}{l} CH_2OH \\ | \\ CHOH \\ | \\ CH_2OH \end{array} + 2\ RCO_2H + H_3PO_4$

$+ HOCH_2CH_2\overset{+}{N}(CH_3)_3$

24-9. (a) androgen (b) estrogen

CHAPTER 25

25-1. (a), (b), and (d) are digestion reactions, while (c), (e), and (f) are metabolism reactions.

25-3. (b)

25-5. tristearin

25-7. (a) and (b) reduction
(c) and (d) oxidation

25-9.

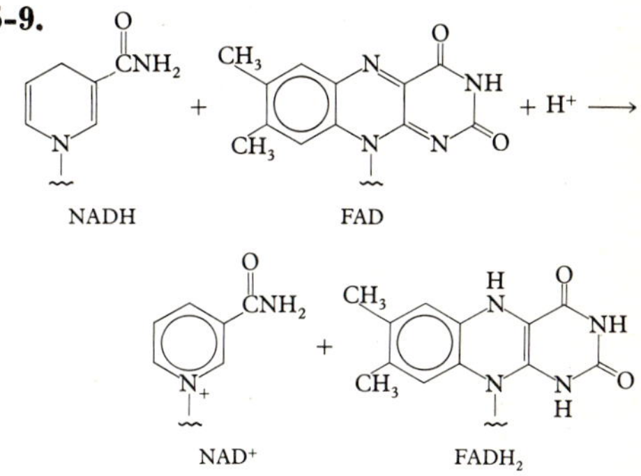

25-11. three

CHAPTER 26

26-1. The digestion of starch is a series of enzyme-catalyzed hydrolysis reactions leading to dextrins, maltose, and finally glucose. Glucose is absorbed into the bloodstream where it may remain as blood sugar or may be converted to glycogen, primarily in the liver. Blood glucose may also be catabolized in the cells to CO_2, H_2O, and energy or it may be converted to acetyl-coenzyme A and then to keto acids, fats, steroids, proteins, or other compounds.

26-3. It increases the blood glucose level by promoting glycogenolysis.

26-5. (a) (1) (b) (4) (c) (2) (d) (3)

26-7. The *anaerobic sequence* is the conversion of glucose to lactic acid. It occurs in the absence of oxygen and has a low energy yield. The *aerobic sequence* is the oxidation of glucose to CO_2 and H_2O. This sequence has a high energy yield.

26-9.

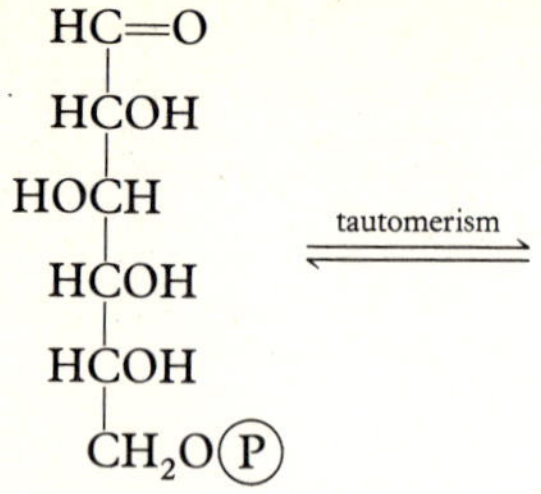

glucose
6-phosphate

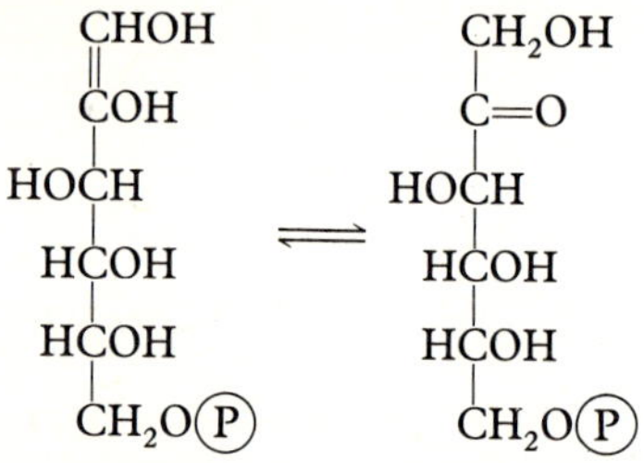

enediol intermediate

fructose 6-phosphate

26-11. Energy is gained.

26-13. (a) isomerization
(b) oxidation
(c) oxidation
(d) addition of water
(e) oxidation

CHAPTER 27

27-1. (a) as fatty acids, monoglycerides, diglycerides, and glycerol. (b) and (c) as lipoproteins.

27-3. acetylcoenzyme A

27-5. A carbon atom with three or four alkyl groups attached cannot be oxidized to a carbonyl group or to a double bond.

27-7. *Ketosis* is caused by a deficiency in glucose metabolism from a lack of dietary carbohydrates or from a lack of insulin (diabetes). *Acidosis* results from the same causes; its symptoms result from an excess of acidic ketone bodies in the bloodstream.

27-9.

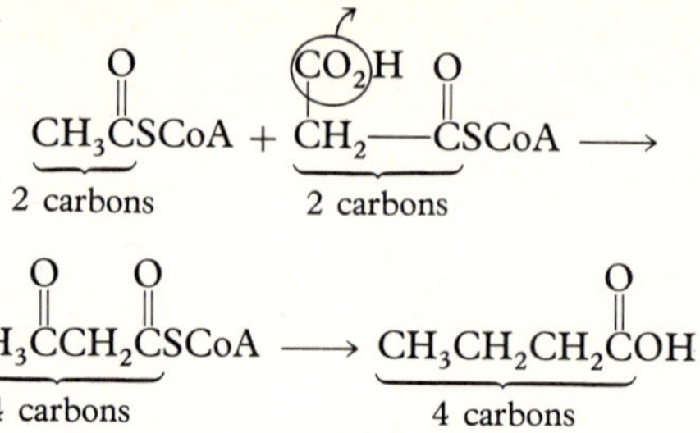

In each pass, two carbon atoms are added to the growing fatty acid chain, resulting in an even number of carbons.

27-11. Because peptides elicit a foreign-protein response: antibody production, allergic symptoms, and even shock or death.

27-13.

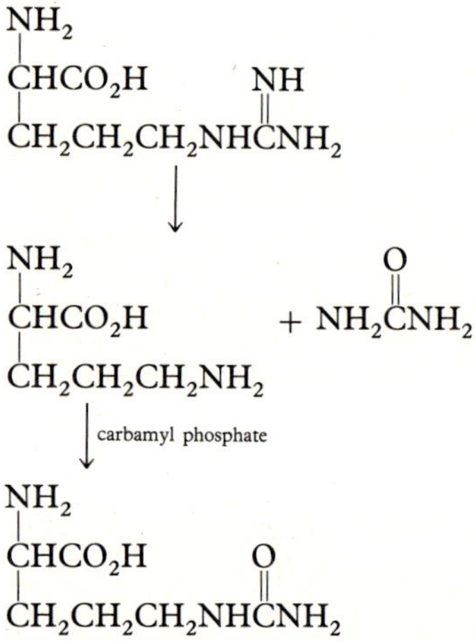

27-15. The Fe^{2+} ions are stored as an iron-protein complex (ferritin). The porphyrin ring system is converted to bilirubin, which is eliminated in the feces and urine.

CHAPTER 28

28-1. (a) to store the genetic code
(b) to act as a template in protein biosynthesis
(c) to carry amino acids to the site of protein biosynthesis

28-3. 2-deoxy-β-D-ribose,

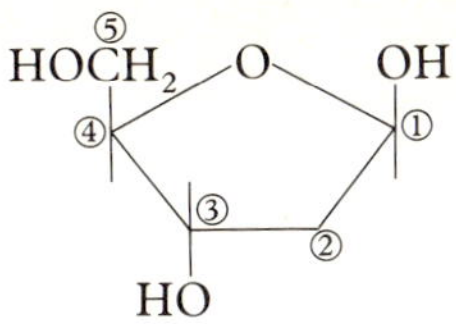

28-5.

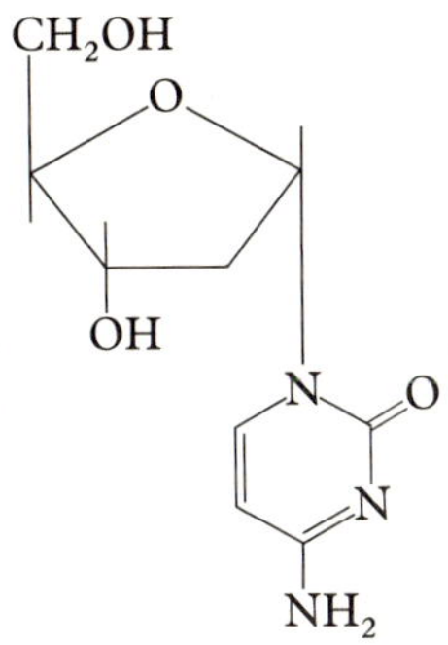

28-7. (b) (d) (e) (h)

28-9. (a)

28-11. (a) (b) (c) (d)

28-13. β-D-ribose,

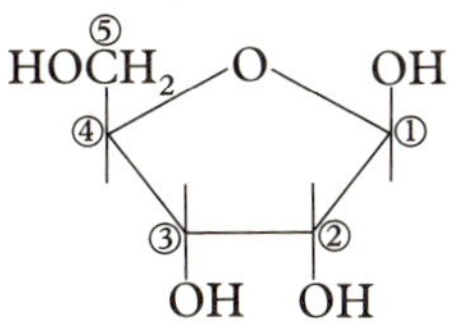

28-15. No, but adenosine monophosphate (AMP) could be obtained.

28-17. (b)

28-19. (a) (c) (d)

28-21. (a) (b) (c) (d)

CHAPTER 29

29-1. (a) circulating
(b) circulating
(c) noncirculating

29-3. (a) (5) (b) (6) (c) (1)
(d) (2) (e) (4) (f) (3)

29-5.

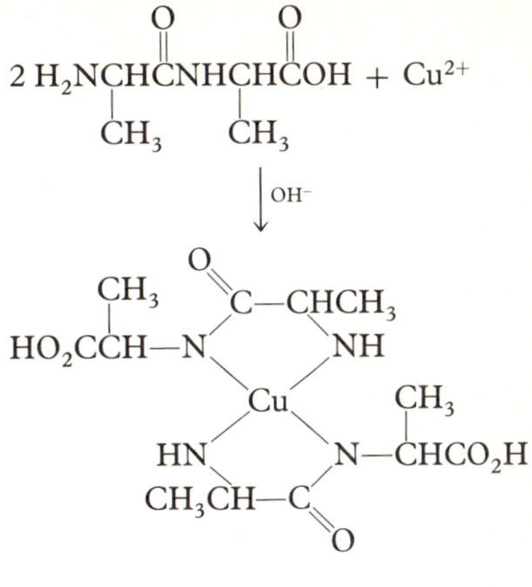

(*Note:* A violet color would be observed.)

29-7. Add sodium oxalate to remove Ca^{2+} ions.

29-9. (a) Volume is increased, and specific gravity is decreased.
(b) Volume is decreased, and specific gravity is increased.
(c) Volume is increased, and specific gravity is also high (because of dissolved glucose).
(d) The pH is lower than normal.
(e) The amount of uric acid is increased.
(f) The amount of ketone bodies is greater than normal.

29-11. β-Carotene absorbs the wavelength of 451 nm (blue light) from the visible spectrum and so appears orange.

CHAPTER 30

30-1. (a) used in blood cells, bones, and body fluids, and aids in enzyme actions
(b) used in hemoglobin and in the cytochrome system
(c) in bones and teeth, and also for nerve impulse transmission, muscle contraction, and blood clotting
(d) in bones and teeth, in ATP, and in nucleic acids

30-3. none

30-5. a steroid

30-7. (a) different responses by different species
(b) an atypical response
(c) an unusual sensitivity
(d) an undesirable, but typical, reaction
(e) an unusual resistance to the action of a drug
(f) a psychological dependence
(g) a physiological dependence

30-9. (b) (d)

INDEX

Some common units of length in SI

UNIT	SI EQUIVALENT	ENGLISH-SYSTEM EQUIVALENTS
meter (m)	. . .	39.37 inches
centimeter (cm)	0.01 m	0.3937 inch
millimeter (mm)	0.001 m	0.03937 inch
kilometer (km)	1000 m	0.6214 mile
micron (μ, or μm)	10^{-6} m	. . .
Angstrom (Å)	10^{-8} m	. . .
nanometer (nm)	10^{-9} m	. . .

Some comparisons between the Fahrenheit and Celsius scales

	°F	°C
Absolute zero	–459.67	–273.15
Freezing water	32	0
Room temperature	72	22
Body temperature	98	37
Boiling water	212	100

Some common units of mass and volume in SI

UNIT	SI EQUIVALENT	ENGLISH-SYSTEM EQUIVALENT
Mass:		
gram (g)	. . .	0.03527 ounce
milligram (mg)	0.001 g	0.01543 grain
kilogram (kg)	1000 g	2.205 pounds
Volume:		
liter (l)	1000 cm^3, or 1000 ml	1.057 quarts
milliliter (ml)	1.0 cm^3, or 0.001 l	0.0338 fluid ounce